About Island Press

Island Press is the only nonprofit organization in the United States whose principal purpose is the publication of books on environmental issues and natural resource management. We provide solutions-oriented information to professionals, public officials, business and community leaders, and concerned citizens who are shaping responses to environmental problems.

In 2002, Island Press celebrates its eighteenth anniversary as the leading provider of timely and practical books that take a multidisciplinary approach to critical environmental concerns. Our growing list of titles reflects our commitment to bringing the best of an expanding body of literature to the environmental community throughout North America and the world.

Support for Island Press is provided by The Nathan Cummings Foundation, Geraldine R. Dodge Foundation, Doris Duke Charitable Foundation, Educational Foundation of America, The Charles Engelhard Foundation, The Ford Foundation, The George Gund Foundation, The Vira I. Heinz Endowment, The William and Flora Hewlett Foundation, Henry Luce Foundation, The John D. and Catherine T. MacArthur Foundation, The Andrew W. Mellon Foundation, The Moriah Fund, The Curtis and Edith Munson Foundation, National Fish and Wildlife Foundation, The New-Land Foundation, Oak Foundation, The Overbrook Foundation, The David and Lucile Packard Foundation, The Pew Charitable Trusts, The Rockefeller Foundation, The Winslow Foundation, and other generous donors.

The opinions expressed in this book are those of the author(s) and do not necessarily reflect the views of these foundations.

CLIMATE CHANGE POLICY

CLIMATE CHANGE POLICY

A SURVEY

Edited by
Stephen H. Schneider
Armin Rosencranz
John O. Niles

ISLAND PRESS
Washington • Covelo • London

Library of Congress Cataloging-in-Publication Data
Climate change policy : a survey / edited by Stephen H. Schneider, Armin Rosencranz, John O. Niles.
p. cm.
Includes bibliographical references and index.
ISBN 1-55963-880-X (hardcover : alk. paper) — ISBN 1-55963-881-8 (pbk. : alk. paper)
1. Climatic changes—Government policy. I. Schneider, Stephen Henry. II. Rosencranz, Armin. III. Niles, John O.
QC981.8.C5 C511416 2002
363.738'745--dc21 2002005344

British Cataloguing-in-Publication Data available.

Printed on recycled, acid-free paper

Manufactured in the United States of America
09 08 07 06 05 04 03 02 8 7 6 5 4 3 2 1

Contents

Acknowledgments

The editors gratefully acknowledge the contributions of numerous people who made this book possible, particularly our peer reviewers, Christian Azar, Paul Baer, John Barton, Sandra Brown, Elsa Cleland, Philip B. Duffy, Riley Dunlap, Anita Engels, Marc Feldman, Geoffrey Heal, Daniel Kammen, Orie Loucks, Aaron McCright, Evan Mills, Ronald Mitchell, Richard Moss, Norman Myers, Brian O'Neill, Vern Ruttan, Bob van der Zwan, and Erika Zavaleta.

For their assistance, we are grateful to the students who helped reorganize particular chapters: Hunt Allcott, David Halsing, Sean Joy, Jonathan Neril, Sierra Peterson, and Anne-Lise Quach.

Our thanks also go to the students who submitted articles but whose work could not be included, primarily for lack of space: Aimée Christensen, Sahra Girshick, Robb Kapla, Chris Lee, Mike Mastrandrea, Lydia Olander, and Maya Trotz.

We acknowledge the many contributions to this survey by the Intergovernmental Panel on Climate Change (IPCC), whose Third Assessment Report (TAR) was the prime source for much of the data in our chapters and was a highly accessible and invaluable resource for our students.

Finally, we thank Françoise Fleuriau-Halcomb for her invaluable help, coordination, and advice. We couldn't have produced this book without her steadfast support.

Introduction

Climate change is a worldwide environmental, social, and economic challenge. It touches on aspects of air pollution, land use, toxic waste, transportation, industry, energy, government policies, development strategies, and individual freedoms and responsibilities. Human use of the atmosphere as an unpriced dumping space has led to the buildup of gases and particles that can alter the radiant energy exchange between the earth's surface and space. Carbon dioxide (CO_2), methane (CH_4), and water vapor are the principal heat-trapping greenhouse gases. Carbon, the main element in the top two human-enhanced greenhouse gases (CO_2 and CH_4), is the underpinning of most fuels used in transportation and power production. Carbon also makes up about half of the dry weight of most vegetation. Thus, the carbon that cycles through life, air, waters and soils is both an essential nutrient and a potential problem. Human modification of the carbon cycle has far-reaching implications for human welfare and the health of the biosphere. Given the short-term planning horizon of many political and economic institutions, such a pervasive and technically complex issue as climate change presents major policy challenges. This volume is designed to clarify the primary issues embedded in those challenges.

The topics raised in this volume often transcend temporary political regimes. The first such topic is that global climate change is indeed global. Greenhouse gases arise out of every nation, from every city and village, and are quickly dispersed. Fossil fuel burning and tropical deforestation, the leading causes of greenhouse gas emissions, are geographically widespread. Other human activities that bear on our climate (such as alteration of greenhouse gas sinks) also occur all around the world. Given that the atmosphere and oceans rapidly transform and transport greenhouse gases, this planetary condition necessitates a planetary response. Thus, climate change policy is a far broader matter than any environmental or economic issue tackled at a national or subnational level.

Despite the scope and complexity of the climate change issue at Bonn, Germany, in July 2001, most nations agreed to do something to address climate change. That brings us to a second issue that we think will influence climate change policy for a long time: Who is going to do what? Even if the global character of climate change implies that every nation needs to *play* in the process of remedy, the different economic status across nations suggests that not every nation can be expected to *pay* the same amount to clean things up. This idea has been recognized in international law since the UN Conference on Environment and Development of 1992 (commonly known as the Rio Earth Summit) as the principle of common but differentiated responsibility. Although we all live on the same planet and share the same changing atmosphere, that does not mean we all have equal sacrifices to make. The responsibility each nation should take for mitigating a problem rests on the relative contribution it made toward that problem and the relative ability of states to solve problems. For instance, some nations have dumped much more greenhouse gas into the atmosphere than the rest of the world. If the polluter-pays principle is invoked, then the larger polluters should bear a greater share of the cleanup responsibility.

The case for common but differentiated responsibility becomes even more obvious when one considers that the relative contribution to global warming is correlated with the ability to pay for mitigation. We live in a highly unequal world where some countries enjoy very different standards of living. The average Nigerian typically is much poorer than a very poor American. Environmental conservation, no matter how important to the preservation of nature, often seems to be a luxury to those who cannot afford a few extra dollars at the gas pump to get to work. Moreover, what made the wealthy countries wealthy was the very dumping of greenhouses gases into the atmosphere from polluting industries and the cutting down of their forests. If the rich used the Victorian industrial revolution—with its sweatshops, land clearing, dirty coal burning, and internal combustion engines—to get fabulously wealthy, then why should developing countries be prevented from copying the same pattern? One obvious worry is that 10 billion people acting like the rich world of the twentieth century will more than triple CO_2 in the atmosphere in the next 150 years.

One promising solution to the environment versus development dilemma is leapfrogging. This is a strategy in which developing countries don't copy historical inefficiencies and environmental threats but rather jump over the old ways to newer, cleaner economies. This might entail using cell phones instead of stringing copper wires, or powering nascent auto industries with fuel cells rather than internal combustion engines. Leapfrogging could also mean saving a diverse carbon-rich tropical forest from loggers' saws by promoting ecotourism or certified and sustainable forestry, reemploying the former cutters as forest stewards. How to provide incentives for leapfrogging and who should pay for this is a big part of the climate policy debate, and it will have a major impact on the future development of nations. Similarly, the impacts of climate change will also be unevenly distributed among nations and sectors of the economy, making equity—fairness—an issue comparable to efficiency in dealing with climate policy.

So international climate change policy increasingly will have a central tension revolving around these global equity concerns. Every country has an obligation and a responsibility to be involved in building solutions, but the level of obligation won't necessarily be equal for everyone. This will be complicated by at least two other factors. First, wealthy countries are in a position of competitive negotiating advantage. At the bargaining table, it is a whole lot easier for the European contingent to crunch numbers and lobby for favorable positions than it is for the probably underfunded and less cyberconnected delegation from Zambia. The second complication is that in coming decades, wealthier economies probably will accelerate their decarbonization while emerging and modernizing economies increase their carbon emissions.

This brings us to the third climate policy consideration that we think will persist through transient political regimes: cost. People generally want more goods and services than there are resources to make or distribute them. Even global climate change policy can't ignore this fundamental economic force, although some environmental groups may act as if they believed otherwise. Any climate policy that is to stand a chance politically must embrace some cost-effectiveness measures. Commonly called flexibility measures, these are policies that can help minimize the cost of whatever level of climate change insurance (i.e., emission reductions or sink enhancements) the nations of the world agree to purchase. Policy tools such as tradable permits, the inclusion of multiple gases, land use measures, and international collaboration all lower costs. That is, rather than having a rich country cut all of its own emissions domestically only in industrial sectors where it might be expensive, flexible policies could allow this country to pay another amenable country to reduce equal amounts of emissions. For example, the United States could build an efficient gas power plant in China that emits less CO_2 than a coal-burning plant that China would have built for itself. And it is usually cheaper to build a new clean plant than it is to upgrade existing capital stock. The price of things matters, and the pursuit of least-cost mitigation measures will continue. However, these flexibility measures often have additional concerns and risks.

For instance, although it may be cheaper to cut emissions from a power plant in China than it is to cut emissions from a power plant in Indiana, to some this seems to be a buyout of responsibility. Moreover, direct costs and benefits are not the only things that matter. In the example of a clean plant in China, Chinese not only get a newer plant but probably will enjoy the ancillary benefits of less air pollution and better health around a cleaner plant. A possible secondary cost (to China) would occur if developing countries took on emission reduction commitments. In such a case, China might trade away some of its cheaper options for lowering greenhouse gases. Also, just because one country builds an efficient power plant or protects a patch of carbon-storing primary forest in another country is not a guarantee that other patches will not be deforested or that inefficient plants supposedly displaced by the clean ones won't continue to operate. Therefore, international compliance monitoring must also be part of the strategy. Each party or nation must carefully weigh the pros and cons of proposed flexibility policies and projects.

Another persistent issue that governments eventually must face is that all the flexibility in the world won't prevent climate change unless the world's biggest polluters reduce their own emissions. However, to get the big countries to agree to participate, these policies must be politically acceptable. These two realities will result in calls for the unlimited use of flexibility tools being tempered by limits on flexibility and rules to ensure that flexible programs are carefully screened, monitored, and enforced.

These rules on the use of flexibility probably will also account for non–greenhouse gas considerations such as local employment, biodiversity, regional sovereignty, and access to information. And throughout this back-and-forth negotiating, it is almost certain that some industries and countries will seek loopholes or try to evade the rules. An effective policy also must account for relative differences in enforcement or monitoring systems in various countries. Certification bodies and flexibility measures have already been some of the key battlegrounds in the last 10 years of international climate change negotiations. Effective international monitoring and compliance measures will be essential to make any global climate change mechanism enforceable. This is no easy task because nations have different views about the amount of controls that should be ceded by sovereign states to international authorities.

In addition to all the technical and economic complexities, climate policy is embedded in deeply divided ideological conceptions of world development, ranging from strong social welfare models to philosophies that focus on the individual. One of the tasks of this volume is to help the reader determine the extent to which disputes arise from analytic imprecision or from value differences. One example of the importance of this distinction is obvious in the forest and land use debate. This issue is extremely complex, and it was arguably the main reason that key international talks (COP-6) broke down at The Hague in November 2000. They were also a main component of the eventual agreement among all major industrialized countries (excluding the United States) at Bonn in July 2001. Although some of the disputes arose from differences in various models, definitions, and calculations, it was also apparent that negotiating blocs had vastly different moral opinions on the role of forests and agriculture in solving climate change.

We are pleased to present this collection of chapters that surveys the main issues surrounding global climate change policy. We recognize that this survey covers a large array of topics and viewpoints. Yet it still doesn't cover every aspect of international climate policy debates. To attempt to cover the entire range of opinions and subjects on a matter as complex as global climate change would be intractable and confusing. Contributors to this book address what we believe are the core issues and ideas in the evolution of global climate change policy. Although each contribution was peer reviewed, there are varying degrees of complexity in each presentation and some variations in writing style and viewpoint. Despite the unevenness this may occasionally entail, we believe it is better to cover more core topics at various levels of depth and span broader views than it is to rely on a few experts. Moreover, we are especially pleased that a few of these chapters were written by students who participated in climate change policy seminars we taught and that their work passed the peer review required for this collection.

We hope you enjoy learning about climate change policy as much as we have enjoyed putting this book together. Achieving a more equitable and sustainable development path that minimizes dangerous human interference with the climate system is a challenge that is worth the effort to comprehend, then to negotiate, and finally to implement corrective action. To fathom this debate is to involve oneself in a critical component of global governance in the twenty-first century.

Stephen H. Schneider
Armin Rosencranz
John O. Niles
Stanford, California

PART I

SCIENCE AND IMPACTS

CHAPTER 1

Understanding Climate Science

Richard Wolfson and Stephen H. Schneider

This book is a survey of climate change policy. But developing, advocating, and implementing viable policies is impossible without some understanding of the science that underlies the climate change debate. This chapter provides just such an understanding. Whether or not you become involved formally in teaching about climate change, you will gain a sufficiently high level of expertise to help others grasp the subject at the level needed by an informed citizenry. This chapter has two explicit goals. The first is to educate you in the science of climate change. The second is to equip you as a citizen for a role in educating the broader public—including government officials and others charged with making policy—so that their decisions may be firmly grounded in the most current scientific knowledge of climate change.

Implicit here is a third, broader goal: to provide a concrete example of the policymaking context for a complex sociotechnological problem marked by conflicting claims of experts and the use of science to justify very different political ends. Whether the issue is genetically engineered food, missile defense, energy policy, or climate change, the burden on you, the informed citizen, is the same. You need to be literate enough about the nature of the debate and the underlying science to have your views counted in the political process. It is through the political process that society decides whether to take a given risk and determines who will be most exposed to the potential dangers. If the decision is to avert risk, then society decides how to do so and who should pay. Although each issue has its own particular scientific aspects, the associated policy processes have many common elements. This book will help you become more environmentally and scientifically literate not only on issues of climate change but also on a host of issues whose understanding is essential to full citizenship in the democratic process of the twenty-first century.

Is Earth s Climate Changing?

The Global Temperature Record

Modern temperature records, derived from thermometers sufficiently accurate and geographically dispersed to permit computation of a global average temperature, date back to the mid-nineteenth century. Extracting a global average from the data is complicated by many factors ranging from the growth of cities, with their "heat island" warming of formerly rural temperature measuring stations, to such mundane effects as changes in the types of buckets used to sample seawater temperature from ships. Early data suffer from a dearth of measurements and a bias toward the more developed regions of the planet. But climatologists understand how to account for these complications, and essentially all agree that Earth's average temperature increased by approximately 0.6°C since the mid-nineteenth century (we'll use Celsius temperatures throughout this book; 1°C is 1.8°F, so a rise of 0.6°C is about 1°F). Figure 1.1 shows the global temperature record as a plot of the yearly deviations from the 1961–1990 average temperature.[1]

A glance at Fig. 1.1 shows that Earth's temperature is highly variable, with

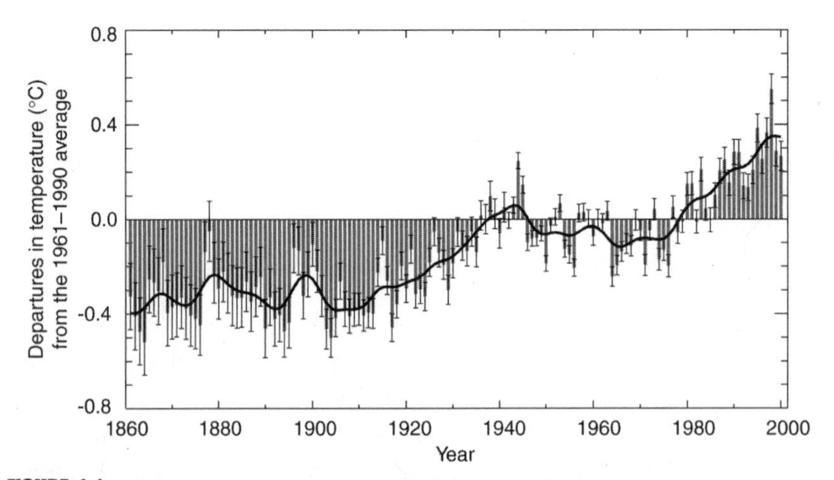

FIGURE 1.1. Variation in Earth's average global temperature from 1860 to 1999. Data are taken from global networks of thermometers, corrected for a variety of effects, and combined to produce a global average for each year. Wider, solid bars represent temperature deviations for each year, relative to the 1961–1990 average temperature, and narrow gray bars show uncertainties in the yearly temperatures. Black curve is a best fit to the data. (Adapted from IPCC, 2001a.)

year-to-year changes often masking the overall rise of approximately 0.6°C. Nevertheless, the long-term upward trend is obvious. Especially noticeable is the rapid rise at the end of the twentieth century. Indeed, all but 3 of the 10 warmest years on record occurred in the 1990s, with 1998 marking the all-time record high through 2000. There is good reason to believe that the 1990s would have been even hotter had not the eruption of Mt. Pinatubo in the Philippines put enough dust into the atmosphere to cause global cooling of a few tenths of a degree for several years. Looking beyond the top 10 years, Fig. 1.1 shows that the 20 warmest years include the entire decade of the 1990s and all but 3 years from the 1980s as well. Clearly the recent past has seen substantial surface warming.

A Natural Climate Variation?

Could the warming shown in Fig. 1.1, especially of the past few decades, be a natural occurrence? Might Earth's climate undergo natural fluctuations that could result in the temperature record of Fig. 1.1? Increasingly, we are finding that the answer to that question is "no." We would be in a better position to determine whether the temperature rise of the past century is natural if we could extend the record further back in time. Unfortunately, direct temperature measurements of sufficient accuracy or geographic coverage simply don't exist before the mid-1800s. But by carefully considering other quantities that do depend on temperature, climatologists can reconstruct approximate temperature records that stretch back hundreds, thousands, and even millions of years.

Figure 1.2 shows the results of a remarkable study, completed in 1999, that attempts to push the Northern Hemisphere temperature record back a full thousand years.[2] In this work, climatologist Michael Mann and colleagues performed a complex statistical analysis involving 112 separate indicators related to temperature.[3] These included such diverse factors as tree rings, the extent of mountain glaciers, changes in coral reefs, sunspot activity, volcanism, and many others. The resulting temperature record of Fig. 1.2 is a "reconstruction" of what one might expect had thermometer-based measurements been available. Although there is considerable uncertainty in the millennial temperature reconstruction, as shown by the error band in Fig. 1.2, the overall trend is most consistent with a gradual temperature decrease over the first 900 years, followed by a sharp upturn in the twentieth century. That upturn is a compressed representation of the thermometer-based temperature record shown in Fig. 1.1. Among other things, Fig. 1.2 suggests that the 1990s was the warmest decade not only of the twentieth century but of the entire millennium. Taken in the context of Fig. 1.2, the temperature rise of last century clearly is an unusual occurrence.

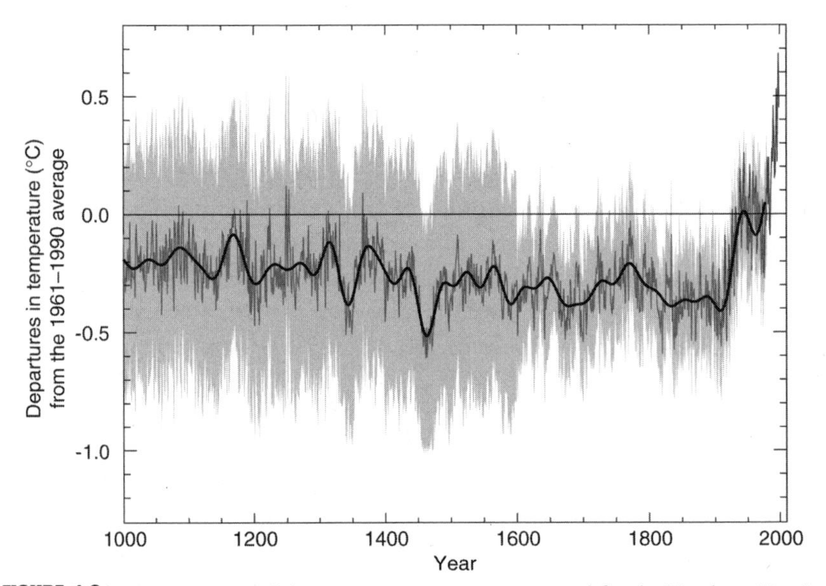

FIGURE 1.2. Reconstruction of the 1,000-year temperature record for the Northern Hemisphere. Black curve is a best fit to the millennial temperature record; gray is the 95% confidence interval, meaning that there is a 95% chance that the actual temperature falls within this band. Date from the mid-nineteenth century on are from the thermometer-based temperature record of Fig. 1.1. (Adapted from Mann et al. as shown in IPCC, 2001a.)

But is it unnatural? Mann et al. approached that question by correlating their temperature reconstruction with several factors known to influence climate, including solar activity, volcanism, and humankind's release of heat-trapping gases (greenhouse gases; more on this later in the chapter). They found that solar variability and volcanism were the dominant influences in the first 900 years of the millennium but that much of the twentieth-century variation could be attributed to human activity. Given the indirect, statistical nature of the study, this result can hardly be taken as conclusive evidence that humans are to blame for twentieth-century global warming. But the Mann et al. result does provide independent corroboration of computer climate models that also suggest a human influence on climate.

Climate Science: Keeping a Planet Warm

How can human activity affect Earth's climate? What ultimately determines climate, and specifically Earth's temperature? That question is at the heart of climate science and of the issues surrounding human-induced climate change and policies to prevent, ameliorate, or mitigate its effects.

Energy Balance

What keeps a house warm in the winter? After all, heat is continually flowing out through the walls and roof, through the windows and doors. So why doesn't the house get colder and colder? Because some source—a gas furnace, a heat pump, a woodstove, sunlight, an oil burner, electric heaters—supplies heat at just the right rate to replace what's being lost. In other words, the house is in energy balance: Energy enters the house at the same rate at which it's being lost. Only under that condition of energy balance will the house temperature remain constant.

The same idea holds for Earth and other planets. Energy, essentially all of it in the form of sunlight, arrives at Earth. In turn, Earth loses energy to the cold vacuum of space. When there's a balance between the incoming sunlight and the energy lost to space, then Earth's temperature remains constant (Fig. 1.3).

Why should there be a balance? Because the rate at which Earth loses energy depends on its temperature. That loss rate is given by a well-known and fundamental law of physics stating that all objects lose energy to their surroundings in the form of radiation. The higher the temperature, the greater the loss rate. Suppose Earth were to be so hot that it loses energy at a greater rate than the incoming sunlight supplies it. Then there is a net loss of energy, so the planet cools. As

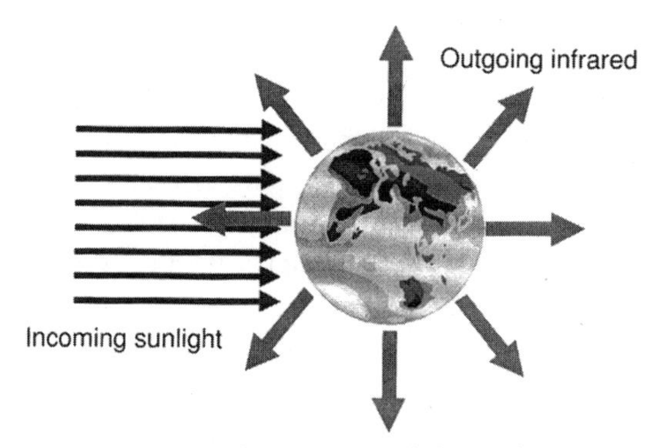

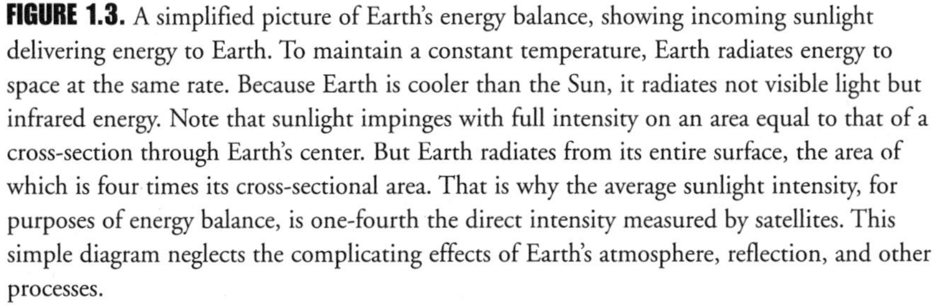

FIGURE 1.3. A simplified picture of Earth's energy balance, showing incoming sunlight delivering energy to Earth. To maintain a constant temperature, Earth radiates energy to space at the same rate. Because Earth is cooler than the Sun, it radiates not visible light but infrared energy. Note that sunlight impinges with full intensity on an area equal to that of a cross-section through Earth's center. But Earth radiates from its entire surface, the area of which is four times its cross-sectional area. That is why the average sunlight intensity, for purposes of energy balance, is one-fourth the direct intensity measured by satellites. This simple diagram neglects the complicating effects of Earth's atmosphere, reflection, and other processes.

it cools, the energy loss rate drops. Eventually the loss becomes equal to the energy supplied in the incoming sunlight, and at that point Earth is in energy balance at a fixed, lower temperature. If the planet is too cool, so it loses energy at a lower rate than the incoming sunlight supplies it, then Earth experiences a net energy gain and heats up. As it heats, the loss rate goes up until it just balances the incoming sunlight. Again, Earth achieves energy balance at a fixed, higher temperature.

What is that fixed temperature? Knowing the rate at which solar energy reaches Earth and knowing the mathematical form of the law for the energy loss, it's a simple matter to equate the two and solve for the temperature. The result, for Earth, is a predicted global average temperature of about –18°C, or about 0°F. That may sound quite cold, and it is, for reasons we'll explore shortly.

Our estimate of a –18°C global average temperature is based on the simplest possible climate model. The model assumes that Earth is a single point, characterized by a single temperature. Ignored are variations with latitude, longitude, and altitude. Also ignored are the tilt of Earth's axis and the resulting seasonal climate variations. So are the existences of separate land and ocean areas, and of the atmosphere, and of air and water currents that transport heat across the planet. Despite these simplifications, the model nevertheless provides a reasonable estimate of Earth's global average temperature as would be seen by a space traveler passing by the planet.

The Greenhouse Effect

Our simple energy balance model predicts a temperature that, though not absurd, seems cold. Too much of Earth's surface is well above freezing for a global average of –18°C or 0°F to be right. In fact, Earth's average surface temperature is about 15°C (59°F), some 33°C higher than the simplest model predicts. Why the discrepancy?

The answer lies in the atmosphere, and to understand it one needs to know more about how objects lose energy. Not only is the energy loss rate dependent on temperature, but so is the specific form of the energy being lost. Any object surrounded by a vacuum loses energy by radiation—more precisely, electromagnetic radiation. Electromagnetic radiation includes visible light, the radio and microwaves used in communication, the invisible infrared and ultraviolet (UV) "light" that lie just outside the visible range, and the penetrating X rays and gamma rays. All these forms of radiation are essentially the same; they differ only in the frequency of the electromagnetic vibrations or, equivalently, in their wavelength (distance between wave crests). Radio waves have the lowest frequency and longest wavelength, followed by microwaves, infrared, visible light, ultraviolet, X rays, and gamma rays.

Here's the climatologically important point: The hotter an object, the higher the frequency and shorter the wavelength of the dominant radiation it emits. The Sun, at 6,000°C, emits primarily visible light. Some bizarre astrophysical objects are so hot they emit primarily X rays. A hot stove burner glows a dull red and emits a mix of infrared and visible light. Your own body emits primarily infrared radiation, which sensitive instruments can detect for use in medical diagnosis. Similarly, infrared cameras image buildings to determine where heat loss occurs. And Earth itself, a cooler object, emits primarily infrared radiation, as shown in Fig. 1.3. For energy balance, the rate at which the planet loses energy in the form of infrared radiation must equal the rate at which it receives solar energy in the form of sunlight.

The gases that make up Earth's atmosphere are largely transparent to visible light. That's obvious because we can see the Sun, Moon, and stars from the ground. Therefore, much of the incident sunlight penetrates the atmosphere to reach the surface (we'll get more specific about this shortly). Once absorbed, this solar energy warms the atmosphere, and particularly the surface, which then re-emits the energy as infrared radiation. But the atmosphere is not so transparent to infrared. Certain naturally occurring gases absorb infrared radiation and limit its ability to escape from Earth. These gases—and cloud particles also—re-emit some of the infrared downward. As a result, Earth's surface warms further, emitting infrared radiation at a still greater rate, until the emitted radiation is again in balance with the incident sunlight. But because of the atmosphere with its infrared-absorbing and re-emitting gases, the resulting surface temperature is higher than it would be otherwise. That is what accounts for the 33°C difference between our simple prediction and Earth's actual surface temperature.

Because the atmosphere functions roughly like the heat-trapping glass of a greenhouse, this excess heating has earned the name *greenhouse effect*, and the gases responsible are called greenhouse gases. The most important natural greenhouse gas is water vapor, followed by carbon dioxide and, to a lesser extent, methane. (The greenhouse analogy is not such a good one; a greenhouse traps heat primarily by preventing the wholesale escape of heated air, with the blockage of infrared playing only a minor role.) We'll explore the role of the greenhouse effect in Earth's energy balance in more detail shortly.

The 33°C warming caused by natural greenhouse gases and particles in the atmosphere is the natural greenhouse effect, and it makes our planet much more habitable than it would be otherwise. What we're concerned about now is the *anthropogenic* greenhouse effect arising from additional greenhouse gases emitted by human activities. Such emissions add to the blanket of heat-trapping gases, further increasing Earth's temperature. Before we turn to the details, though, it's important to recognize that the basic greenhouse phenomenon is well understood and solidly grounded in basic science.

A Tale of Three Planets

We can't carry out controlled experiments with Earth's greenhouse effect because we have only one Earth and because such experiments would take decades or longer for definitive results. (Of course, we are in the midst of an *uncontrolled* experiment with Earth's climate as we pour greenhouse gases into our atmosphere.) But our two neighbor planets, Mars and Venus, conveniently provide us with natural greenhouse "experiments." Mars, somewhat farther from the Sun than Earth, should be correspondingly cooler. A simple energy balance calculation neglecting Mars's atmosphere suggests a surface temperature around –60°C. In fact, Mars's temperature is only a little warmer, at about –50°C. That's because Mars's atmosphere is so thin that it provides very little greenhouse warming. Venus, on the other hand, is closer to the Sun, and the simple calculation suggests a surface temperature around 50°C. But Venus's surface temperature is a much hotter: 500°C. Why? Because Venus's atmosphere is very thick and is composed primarily of the greenhouse gas carbon dioxide (CO_2). Consequently, Venus has a "runaway" greenhouse effect that greatly increases its temperature. Earth lies, physically and climatologically, between Venus and Mars. Our atmosphere is 100 times denser than Mars's, but the dominant gases (nitrogen and oxygen) do not absorb significant amounts of infrared radiation. In Earth's atmosphere the greenhouse gases occur in trace amounts, less than 0.1 percent for CO_2 and up to a few percent (varying with humidity) for water vapor. Thus we have a modest greenhouse warming of about 33°C, compared with Mars's 10°C and Venus's dramatic 450°C. This comparison with our neighbor planets confirms our basic scientific understanding of the greenhouse effect and increases confidence in our ability to calculate quantitatively the warming caused by changes in atmospheric greenhouse gases.

Incidentally, Earth's atmosphere is unique in another important way. Unlike the atmospheres of Mars and Venus, which result from geophysical processes, Earth's present atmosphere is strongly biologically controlled. More than 3 billion years ago, the first photosynthetic organisms began emitting oxygen, at that time just a byproduct, and to them a toxic one at that. Later organisms evolved to use the new atmospheric oxygen in a higher-energy metabolic process that ultimately made possible the rapid mobility of animal species. Today's atmospheric composition—about 80 percent nitrogen, 20 percent oxygen, and traces of other gases including CO_2—is significantly regulated by biogeochemical cycles that include plant photosynthesis and respiration by both plants and animals. Without life, atmospheric oxygen would disappear in the geologically short time of a few million years.

Earth's Energy Balance

The simplest way to understand the greenhouse effect is to consider greenhouse gases as a moderately insulating blanket that traps heat. Just as a blanket covers your body and keeps you warm, so the greenhouse gases blanket Earth and keep it warmer than it would be without those gases. Adding more greenhouse gases—as humans have been doing since the industrial revolution—is like making the blanket thicker. For the general public, that explanation is sufficient to capture the essence of the phenomenon and to show why anthropogenic greenhouse gas emissions should lead to climate change. Even for elementary school students, the greenhouse effect at this level is eminently comprehensible. We emphasize again that this picture of the greenhouse effect is solidly grounded in basic physics and confirmed by observations of Venus, Earth, and Mars.

However, the level of this book calls for a more sophisticated understanding of the greenhouse effect, including a detailed look at Earth's energy balance. On average, the rate at which solar energy arrives at the top of Earth's atmosphere is nearly 1,368 watts on every square meter oriented at right angles to the incident sunlight. (For several decades this figure has been accurately monitored by satellites; it varies by about 0.1 percent over the 22-year solar activity cycle and has been speculated to vary by up to 0.5 percent over century-long timescales.) Accounting for Earth's spherical shape and the fact that only the daytime half the planet faces the Sun results in an average solar energy incident on the planet of 342 watts per square meter (W/m^2). For energy balance, Earth must return energy to space at exactly this rate. Figure 1.4 shows the details of how this happens.[4] (Numbers given in Fig. 1.4 and in the text discussion are approximate, and some are uncertain by as much as 10 percent.) Of the incident sunlight energy, some 31 percent is reflected back into space, most of it by clouds but some by ice, snow, deserts, and other light-colored surfaces. This reflected energy is never converted to heat, so it plays essentially no role in climate. That leaves some 235 W/m^2 that is absorbed by the Earth–atmosphere system and must be returned to space. Incidentally, a change in the 31 percent reflectance figure—resulting, for example, from ice melting in response to global warming—could have significant climatic effects.

Another 20 percent or so of the incident solar energy is absorbed in the atmosphere, directly heating it. The remainder—nearly 50 percent—reaches and warms the surface. The warm surface warms the atmosphere, which, in turn, cools by emitting infrared radiation. This helps to explain why air tem-

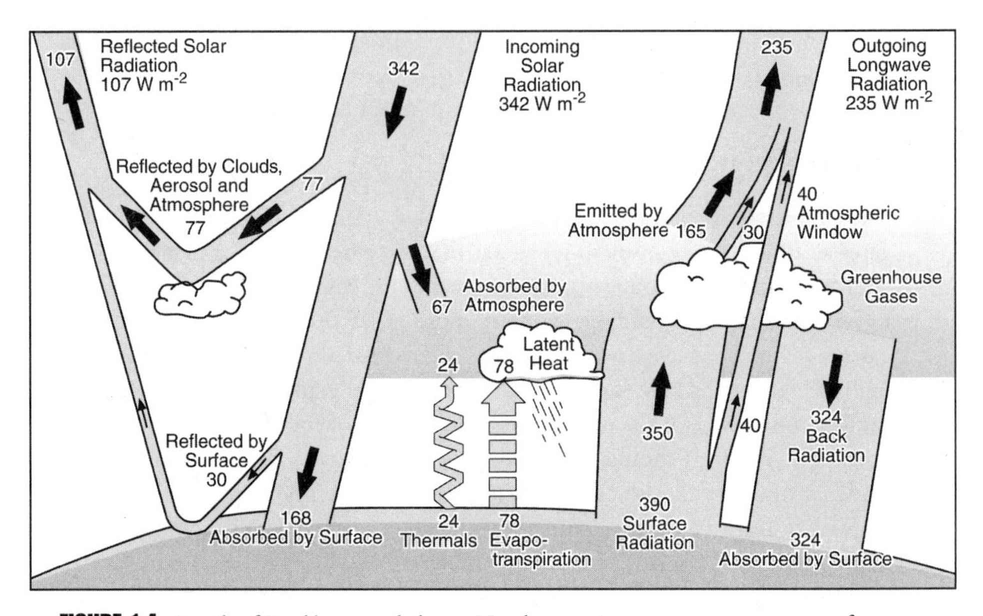

FIGURE 1.4. Details of Earth's energy balance. Numbers are in watts per square meter of Earth's surface, and some may be uncertain by as much as 10%. The greenhouse effect is associated with the absorption and reradiation of energy by atmospheric greenhouse gases, resulting in a higher downward flux of infrared radiation from the atmosphere to the surface and therefore in a higher surface temperature. Note that the total rate at which energy leaves Earth (107 W/m^2 of reflected sunlight plus 235 W/m^2 of infrared [long-wave] radiation) is equal to the 342 W/m^2 of incident sunlight. Thus Earth is in energy balance. (From Kiehl & Trenberth, 1997.)

perature usually decreases with altitude. Figure 1.4 shows that some heat is transported into the atmosphere by bulk air motions, which physically raise warm air from the surface and, more importantly, carry evaporated water and the latent energy it contains. When this water recondenses to form clouds, energy is released to the air. This energy transport process is what powers hurricanes, for example. The atmosphere, warmed by direct heating and by heating from the surface, in turn radiates energy to space to help maintain energy balance. In the absence of greenhouse gases, the surface would also radiate a significant amount of infrared energy directly to space. But clouds and greenhouse gases block much of this outgoing infrared, instead absorbing the energy and thus heating the atmosphere. The atmosphere, in turn, radiates the absorbed energy in all directions, again in the form of infrared radiation. Some escapes to space, but some heads downward, further warming the surface. The result, in the steady state depicted in Fig. 1.4, is a warmer surface that produces a larger flow of infrared radiation upward, not quite balanced by the smaller but still substantial flow downward from the atmosphere overhead. The difference between the upward and downward energy flows, in the steady state, is just the right amount to maintain energy balance between absorbed solar radiation, evaporation, thermal energy lost via rising plumes of heated air, and the net infrared radiation balance. So Earth is in nearly perfect energy balance but with a surface temperature significantly higher than it would be in the absence of greenhouse gases. This scientific theory is firmly established.

Past Climates

Just how much will increasing greenhouse gas concentrations affect climate? We can get clues by looking at past climates. The last 140 years, as shown in Fig. 1.1, have been a period of significant warming. Also, as Fig. 1.5 shows, atmospheric carbon dioxide has increased by more than 30 percent during the same period.[5] The reality of this CO_2 increase is unquestioned, and virtually all climatologists agree that the cause is human activity, predominantly the burning of fossil fuels and to a lesser extent deforestation and other land use changes, along with industrial activities such as cement production. (Although water vapor is the predominant greenhouse gas, its concentration is affected only indirectly by human-induced warming. Carbon dioxide, therefore, is the most important anthropogenic greenhouse gas that results directly in global warming, although we'll later take a look at some other significant heat-trapping gases.)

Note the units and numbers in Fig. 1.5. The unit of atmospheric CO_2

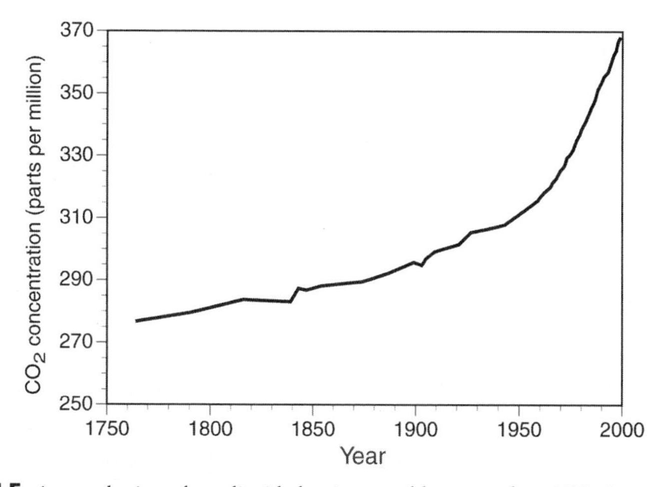

FIGURE 1.5. Atmospheric carbon dioxide has increased by more than 30% since preindustrial times. (Data are from Neftel et al., 1994, and Keeling & Whorf, 2000.)

concentration is the part per million (ppm). This describes the number of volume units of CO_2 in a million units of air. For example, the CO_2 concentration of some 370 ppm at the start of the twenty-first century means that out of every million liters of air, 370 of them are carbon dioxide. This level of 370 ppm is up from about 280 ppm at the beginning of the industrial era.

Figures 1.1 and 1.5 taken together show contemporaneous increases in global temperature and carbon dioxide concentration, both occurring during an era of rapid industrialization. So are anthropogenic CO_2 emissions a direct cause of recent warming? As the study summarized in Fig. 1.2 suggested, it looks increasingly like the answer is "yes." But the connection between the past 140 years' warming and the coincident rise in CO_2 is not so obvious. For example, global temperature actually declined in the period after World War II, a time of rapid industrialization when CO_2 concentrations began an especially rapid increase. On the other hand, temperature rises should lag CO_2 increases, so we shouldn't expect to find that recent temperature and CO_2 are instantaneously correlated. Moreover, there are other factors that can influence climate fluctuations or trends, and all of these are confounded in the data shown in Figs. 1.1 and 1.5. Separating the anthropogenic "signal" of climate change from the "noise" of natural fluctuations can be a tricky process.

We can get a better understanding of the temperature–CO_2 relationship by looking much further back in time. Ice cores bored from the Greenland and Antarctic ice sheets provide estimates of both quantities going back hundreds of thousands of years. Variations in ice density associated with seasonal snowfall patterns provide a year-to-year calibration of the time associated with a given point in the ice core. CO_2 measurement is easy: Analysis of air bubbles trapped in ancient ice gives an indication of CO_2 concentration. Temperature inference is a bit more subtle and usually is accomplished by comparing the ratio of two different forms (isotopes) of oxygen whose uptake in evaporation, and therefore concentration in precipitation and thus in the ice itself, is sensitive to temperature. The result of an ice core analysis, shown in Fig. 1.6, gives dramatic evidence that temperature and carbon dioxide concentration are correlated over the long term.[6]

Are the CO_2 variations in Fig. 1.6 the cause of the temperature changes? That's not clear from the graph alone. Sometimes a CO_2 increase precedes a warming, but sometimes not. In fact, climatologists suspect a feedback process whereby a slight increase in temperature, probably caused by subtle changes in Earth's orbit, results in an increase in atmospheric CO_2 through a variety of mechanisms such as the release of CO_2 dissolved in the oceans. The increased atmospheric CO_2, in turn, leads to greenhouse warming, amplifying the initial temperature increase. The result is a nearly simultaneous and substantial increase

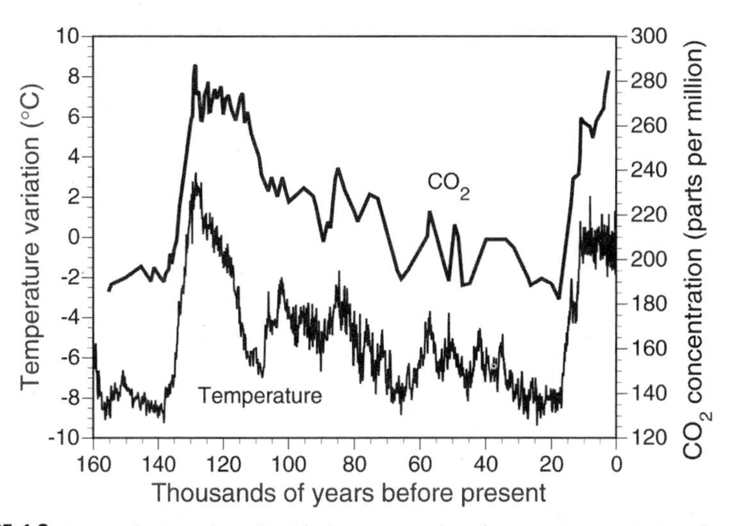

FIGURE 1.6. Atmospheric carbon dioxide (upper curve) and temperature variation (lower curve) over the past 160,000 years, from ice cores taken at Vostok, Antarctica. The record shows long stretches of low temperature (ice ages) separated by brief, warm interglacial periods. The correlation between CO_2 and temperature is quite obvious. Note also the small change, averaging perhaps 6°C, between the present warm climate and the recent ice age. Data do not extend to the present, but stop well before the industrial era. (CO_2 data are from Petit et al., 2000; temperature data from Jouzel et al., 1987, as reproduced in the Carbon Dioxide Information Analysis Center.)

in both CO_2 and temperature. Eventually orbital changes trigger a modest temperature decrease, and again feedback mechanisms amplify the decrease, driving down both CO_2 and temperature. Some paleoclimatologists believe that an initial cooling causes a drying of the continents, which therefore produce more windblown dust. This dust contains minerals needed by phytoplankton in the oceans. As dust settles on the ocean surface, it fertilizes these tiny oceanic organisms. The phytoplankton, in turn, increase their productivity by drawing down atmospheric CO_2, thus making the move toward an ice age even more rapid and deep. Such biotic feedback mechanisms illustrate how complex the actual climate system is and help us to understand why in the policy debates to be presented later, many claims will be made by advocates incompletely selecting bits of this complex story to suit certain value positions. (More on that in later chapters.) But despite the complexity, there is still much regularity in the record. The pattern of varying temperature and carbon dioxide concentration shown in Fig. 1.6 is believed to repeat on a timescale of roughly every 100,000 years over most of the past million years, at least in part as a result of periodic changes in Earth's orbit and inclination of its polar axis.

Note that Fig. 1.6 shows brief periods of warmth punctuated by much longer, cooler ice ages. They are characterized by dramatically different climatic conditions, with ice sheets 2 kilometers thick covering what is now Canada, the northeastern United States, and northwestern Europe and engulfing high mountain plateaus all around the world. Today we enjoy the warmth of an interglacial period, but not long ago, geologically speaking, conditions were very different.

What sort of global temperature change characterizes the contrast between an ice age and our present interval of warmth? A look at Fig. 1.6 shows that change to be on the order of 6°C (11°F). You can quibble by a few degrees, but it's certainly no more than 10°C and, on average, quite a bit less. This point is crucial because climate models driven by standard assumptions about population, land use, and energy consumption project a warming over the next century of 1.5°C to 6°C. The difference between the higher and lower ends of this range has substantial implications for sea level rise, extreme weather, redistribution of species ranges, and other impacts. Policymakers and the general public often ask how a few degrees can matter all that much. Figure 1.6 provides one startling answer: Downward changes on the same order as the largest projected warming are enough to make the difference between our current climate and an ice age. A few degrees, sustained in time and taken over the entire globe, can make a big difference.

A second important point follows from comparing Figs. 1.6 and 1.5. Note in Fig. 1.6 that the maximum CO_2 concentration in the ice core record of the past 160,000 years (and probably for at least millions of years) is under 300 ppm. This does not include the very recent past, but only the preindustrial period. Now look at Fig. 1.5, with its present-day concentration of 370 ppm—far above anything Earth has seen, probably, for millions of years. Figure 1.7

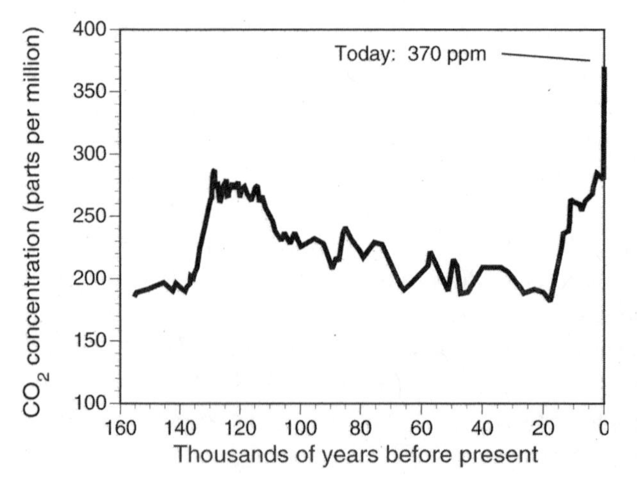

FIGURE 1.7
The CO_2 record of Fig. 1.6, with data to 1999 included. The CO_2 rise of Fig. 1.5 shown here as a dramatic jump to levels not seen on Earth for hundreds of thousands (and probably millions) of years.

shows the effect of adding the recent rise in CO_2 to the ice core data. Clearly the anthropogenic increase in CO_2 concentration is unprecedented in both its size and its rapidity. We have made truly dramatic changes in Earth's atmosphere over the past century or so, and we can almost certainly expect significant climate change to result.

Projecting Future Climate: Greenhouse Gases and Feedbacks

We know that human activities have increased the concentration of atmospheric carbon dioxide. Given the many decades of inertia built into social and industrial systems, they will almost certainly continue to do so for at least decades to come. We know that much of the extra CO_2 remains in the atmosphere for centuries. We also understand the molecular properties of CO_2 and can therefore predict how much infrared radiation over how long a period a given injection of CO_2 should absorb. If that were the whole story, it would be a simple matter to predict Earth's future climate.

However, anthropogenic carbon dioxide is not the whole story. Although CO_2 is the most significant anthropogenic greenhouse gas, accounting for some 60 percent of the enhanced infrared blockage, a host of other greenhouse gases also result from human activities. Another major complication in predicting future climate is feedback effects, whereby human-induced greenhouse warming may cause other processes that either exacerbate or dampen the warming. Finally, other human activities—most notably the emission of particulate pollution from cars and fossil-fueled power plants—can result in regional cooling that may mask or reduce the effects of greenhouse warming. To project future climate confidently, we must take these and many other effects into account. Unfortunately, not all uncertainties can now be, or soon will be, resolved, adding further to confusion in the public policy debate (see the discussion in Chapter 2).

Greenhouse Gases and Radiative Forcing

Although carbon dioxide is the most important of the anthropogenic greenhouse gases in terms of its direct effect on climate, other gases play a significant role, too. On a molecule-to-molecule basis, most other greenhouse gases (except water vapor) are far more potent absorbers of infrared radiation than is carbon dioxide, but they are released in much lesser quantities, so their overall effect on climate is smaller. Climatologists characterize the effect of a given atmospheric constituent by its radiative forcing, the rate at which it alters absorbed solar or outgoing infrared energy. Currently anthropogenic CO_2 produces a radiative forcing estimated at about 1.5 watts for every square meter of Earth's surface (all forcings cited in this section are from the Intergovernmental Panel on Climate Change [IPCC] Third Assessment Report).[7] Relative to the 235 W/m^2 of solar energy that is absorbed by Earth and its atmosphere, the CO_2 forcing is a modest perturbation of the overall energy balance. Very crudely, one can think of that 1.5 W/m^2 of CO_2 forcing as having roughly the same effect as would an increase in the incoming sunlight energy by an average of 1.5 W on every square meter. The global warming resulting from a specified amount of radiative forcing, after the climate has settled into a new equilibrium state, is called climate sensitivity. If we knew the climate sensitivity and the concentration of all atmospheric constituents that affect radiative forcing, then we could more credibly predict future global warming.

Another anthropogenic greenhouse gas is methane (CH_4), produced naturally and anthropogenically when organic matter decays anaerobically (that is, in the absence of oxygen). Such anaerobic decay occurs in swamps, landfills, rice paddies, land submerged by hydroelectric dams, the guts of termites, and the stomachs of ruminants such as cattle. Methane is also released by oil and gas drilling, coal mining, volcanic eruptions, and the warming of methane-containing compounds on the ocean floor. One methane molecule is roughly 30 times more effective at blocking infrared than is one CO_2 molecule, although this comparison varies with the timescale involved and the presence of other pollutants. Whereas CO_2 concentration increases tend to persist in the atmosphere for centuries or longer, the more chemically active methane typically disappears in decades, making its warming potential relative to that of CO_2 lower on longer timescales. Currently methane accounts for about 0.5 W/m^2 of anthropogenic radiative forcing, about one-third that of CO_2.

Other anthropogenic greenhouse gases include nitrous oxide, produced from agricultural fertilizer and industrial processes, and the halocarbons used in refrigeration. (A particular class of halocarbons—the chlorofluorocarbons—is also the leading cause of stratospheric ozone depletion. Newer halocarbons do not cause severe ozone depletion but are still potent greenhouse gases.) Together, nitrous oxide and halocarbons account for roughly another 0.5 W/m^2 of radiative forcing. A number of other trace gases contribute roughly 0.05 W/m^2 of additional forcing. All the gases mentioned so far are well mixed, meaning that they last long enough to be distributed in roughly even concentrations throughout the lowest 10 km of so of the atmosphere.

Another greenhouse gas is ozone (O_3), familiar because of its depletion by anthropogenic chlorofluorocarbons. Ozone occurring naturally in the stratosphere (some 10–50 km above the surface) absorbs incoming ultraviolet radiation and protects life from UV-induced cancer and genetic mutations, hence the concern about ozone depletion and in particular the polar "ozone holes." Unfortunately, ozone depletion and global warming have become confused in the public mind, even among political leaders and some environmental policymakers. But the two are very distinct problems. The ozone depletion problem is not the same as the global warming problem! Ozone depletion eventually will come under control because of the 1987 Montreal Protocol, an international agreement that bans the production of the chlorinated fluorocarbons that destroy stratospheric ozone. Whether similar agreements can be forged for climate-disturbing substances is what the current debate—and this book—are about.

Because ozone is a greenhouse gas, there are some direct links between greenhouse warming and anthropogenic changes in atmospheric ozone. Ozone in the lower atmosphere—the troposphere—is a potent component of photochemical smog, resulting largely from motor vehicle emissions. Tropospheric ozone contributes roughly another 0.35 W/m^2 of radiative forcing, although unlike the well-mixed gases, tropospheric ozone tends to be localized where industrialized society is concentrated. In the stratosphere, the situation is reversed. Here the anthropogenic effect has been ozone depletion, resulting in a negative forcing of approximately –0.15 W/m^2. Thus stratospheric ozone depletion, on its own, would cause a slight global cooling. Taken in the context of the more substantial positive forcings of other gases, though, the effect of stratospheric ozone depletion is a slight reduction of the potential for global warming, an effect that will diminish as the ozone layer gradually recovers under the Montreal Protocol's ban on chlorofluorocarbons. The net effect of all anthropogenic ozone (both tropospheric and stratospheric) probably amounts to a slight positive forcing. The net forcing to date from all anthropogenic gases probably is about 3 W/m^2 and is expected to become much larger if business-as-usual development scenarios are followed in the twenty-first century.

Aerosols

Fuel combustion, and to a lesser extent agricultural and industrial processes, produce not only gases but also particulate matter. Coal-fired power plants burning high-sulfur coal, in particular, emit gases that become sulfate aerosols that reflect incoming solar radiation and thus results in a cooling trend. Natural aerosols from volcanic eruptions and the evaporation of seawater also produce a cooling effect. However, diesel engines and some biomass burning produce black aerosols such as soot, which can warm the climate. Recent controversial estimates suggest that these could offset much of the cooling from sulfate aerosols, especially in polluted parts of the subtropics.[8] The IPCC estimates the total radiative forcing resulting directly from all anthropogenic aerosols very roughly at about –1 W/m^2. However, this figure is much less certain than the radiative forcings caused by the greenhouse gases. Furthermore, aerosol particles also exert an indirect effect in that they act as "seeds" for the condensation of water droplets to form clouds. Thus the presence of aerosols affects the size and number of cloud droplets. An increase in sunlight reflected by these aerosol-altered clouds may result in a cooling due to the associated –2 W/m^2 of radiative forcing. Similarly, soot particles mixed into clouds can make the droplets absorb more sunlight, producing some warming. Taken together, aerosols add an element of uncertainty into anthropogenic radiative forcing of about 1 W/m^2 and complicate attempts to discern an anthropogenic signal of climatic change from the noise of natural climatic fluctuations.

Solar Variability

Variation in the Sun's energy output affects Earth's climate. Variations caused by the 22-year solar activity cycle amount to only about 0.1 percent and are too small and occur too rapidly to have a significant climatic effect. Long-term solar variations, either from variability at the Sun itself or from changes in Earth's orbit and inclination, have substantially affected Earth's climate over geologic time. Although accurate, satellite-based measurements of solar output are available for only a few decades, indirect evidence of solar activity allows us to estimate past variations in solar energy output.[9] Such evidence suggests that solar forcing since preindustrial times amounts to about 0.3 W/m^2—enough to contribute somewhat to the observed global warming but far below what is needed to account for the warming of recent decades. However, there is some speculation that magnetic disturbances from the Sun can influence the flux of energetic particles impinging on Earth's atmosphere, which in turn affect stratospheric chemical processes and might thus indirectly alter the global energy balance. These speculations have led some to declare the warming of the past century to be wholly natural, but this notion is discounted by nearly all climatologists for two reasons: first, there is no demonstrated way in which solar energetic particles can have a large enough effect to account for the recent warming and, second, because it is unlikely that such solar magnetic events happened only in the past few decades and not over the past 1,000 years. But in the political world, scientific evidence cited by advocates of a solar explanation for recent climate change often is accorded equal credibility—until assessment groups such as the IPCC are convened to sort out such claims and to weigh their relative probabilities. That is why we report primarily the IPCC assessments rather than the claims of a few individual scientists.

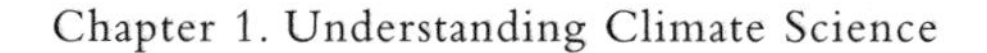

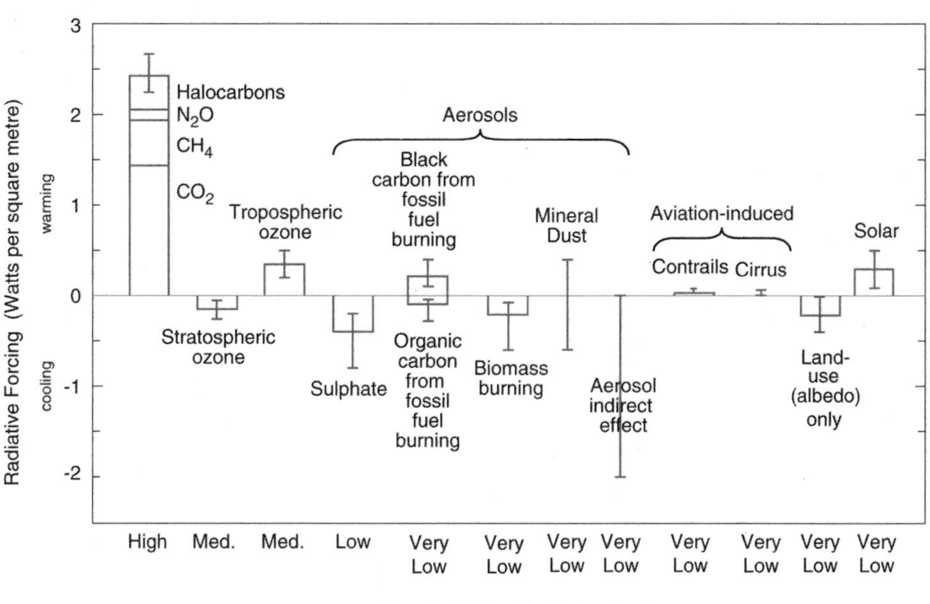

FIGURE 1.8. Radiative forcings caused by anthropogenic greenhouse gases, particulate emissions (aerosols), and other processes. Vertical bars indicate relative uncertainties, and the overall level of scientific understanding of and confidence in these processes is listed below the graph. (From IPCC, 2001.)

Radiative Forcing: The Overall Effect

Figure 1.8 summarizes our current understanding of radiative forcings caused by greenhouse gases, aerosols, land use changes, solar variability, and other effects since the start of the industrial era. The negative forcings from some of these anthropogenic changes might appear sufficient to offset the warming caused by anthropogenic greenhouse gases. This implication is misleading, however, because the effects of aerosols are short-lived and geographically localized compared with the long-term, global effects of the well-mixed greenhouse gases. The most advanced climate models, to be discussed shortly, are driven by a range of plausible assumptions for future emissions of all types and make it clear that the overall effect of human activity is almost certainly a net positive forcing.

Feedback Effects

Knowing the radiative forcing caused by changes in atmospheric constituents would be sufficient to project future climate if there were no additional climatic effects beyond the direct change in energy balance. But a change in climate caused by simple forcing can have significant effects on atmospheric, geological, oceanographic, biological, chemical, and even social processes. These effects, in turn, can further alter the climate. If that further alteration is in the same direction as its initial cause, then the effect is called a positive feedback. If the further alteration tends to counter the initial change, then it is a negative feedback. In reality, numerous feedback effects greatly complicate the full description of climate change. Here we list just a few to give a sense of their variety and complexity.

Ice-albedo feedback is an obvious and important feedback mechanism. Albedo is a planet's reflectance of incident sunlight. Figure 1.4 showed that Earth's albedo is about 0.31, meaning that 31 percent of incident sunlight is reflected back to space. A decrease in that number would mean more sunlight absorbed which would increase global temperature. One likely consequence of rising temperature is the melting of some ice and snow, which would eliminate a highly reflective surface and expose the darker land or water beneath the ice. The result is a decreased albedo, increased energy absorption, and additional heating. This is a positive feedback.

Rising temperature also results in increased evaporation of water from the oceans. That means more water vapor in the atmosphere. Because water vapor is itself a greenhouse gas, this effect results in still more warming and is thus a positive feedback. But increased water vapor in the atmosphere might mean more widespread cloudiness, which reflects sunlight and thus raises the albedo, resulting in less energy absorbed by the Earth–atmosphere system. The result is a negative feedback, tending to counter the initial warming. On the other hand, clouds also absorb outgoing infrared, resulting in a warming—a positive feedback. There are actually a number of processes associated with clouds, some of which produce warming and some cooling. These effects vary with the type of cloud, the location, and the season. Our limited understanding of cloud effects is one of the greatest sources of uncertainty in global climate sensitivity and thus in climate projections. However, the best estimates suggest that the overall effect of increased water vapor is a positive feedback that causes a temperature increase 50 percent higher than would occur in the absence of this feedback mechanism.[10]

Some feedbacks are biological. For example, increased atmospheric CO_2 stimulates plant growth, and plants in turn remove CO_2 from the atmosphere. This is a negative feedback. On the other hand, warmer soil temperatures stimulate microbial action that releases CO_2—a positive feedback effect. Drought and desertification resulting from climate change can alter the albedo of the land by replacing dark plant growth with lighter soil and sand. Greater reflection of sunlight results in cooling, so this is a negative feedback. But here, as so often with the climate system, the situation is even more complex. If sand is wet, as on a beach, then it is darker and therefore absorbs more sunlight than dry sand. Yet dry sand is hotter. The resolution of this conundrum is that the wet sand is cooler because of the cooling effects of evaporation, but the Earth is warmed by the wet sand because the evaporated water condenses in clouds elsewhere and puts the heat back into the overall system. Thus cooling or warming of the Earth–atmosphere system does not always imply cooling or warming of the Earth's surface at that location. Feedbacks can be a very complicated business.

There are even social feedbacks. For example, rising temperature causes more people to install air conditioners. The resulting increase in electrical consumption means more fossil fuel–generated atmospheric CO_2—again giving a positive feedback.

Accounting for all significant feedback effects entails not only identifying important feedback mechanisms but also developing a quantitative understanding of how those mechanisms work. That understanding often includes research at the boundaries of disciplines such as atmospheric chemistry and oceanography, biology and geology, even economics and sociology.

With positive feedback, there is a danger of runaway warming, whereby a modest initial warming triggers a positive feedback that results in additional warming. That, in turn, may increase the warming still further. This feedback could lead to extreme climate change. That is what has happened on Venus, where the thick, CO_2-rich atmosphere produced a runaway greenhouse effect that gives Venus its abnormally high surface temperature. Fortunately, we believe that the conceivable terrestrial feedbacks, at least under Earth's current conditions, are incapable of such dramatic effects. But that only means we aren't going to boil the oceans away; it doesn't preclude potentially disruptive climatic change.

Climate Modeling

Our earlier estimate that Earth's global average temperature in the absence of the greenhouse effect would be about –18°C was based on a simple climate model—a mathematical statement describing physical conditions that govern climate. In that case the statement was a single equation setting equal the temperature-dependent rate of energy loss and the rate of incoming solar energy. More generally, a climate model is a set of mathematical statements describing physical, biological, and chemical processes that determine climate. The ideal model would include all processes known to have climatological significance and would involve enough spatial and temporal detail to resolve phenomena occurring over limited geographic regions and in short times. Today's most comprehensive models approach this ideal but they still entail many compromises and approximations. Often less detailed models suffice, and in general the climate modeling enterprise involves comparisons between models with different levels of detail and sophistication. Computers are necessary to solve all but the simplest models.

What must go into a climate model depends on what one wants to learn from it. A few simple equations can give a decent estimate of the average global warming in response to specified greenhouse forcings. If we seek to model the long-term sequence of ice ages and interglacial warm periods (as shown in Fig. 1.6), our model must include explicitly the effects of all the important components of the climate system that act over timescales of a million years or so. These include atmosphere, oceans, the cryosphere (sea ice and glaciers), land surface and its changing biota, and long-term biogeochemical cycles as well as forcings from varying solar input associated with long-term variations in Earth's orbit and changes in the Sun itself. If we want to project climate over the next century, many of these long-term processes can be left out of our model. On the other hand, if we want to explore climate change on a regional basis or variations in climatic change from day to night, then we need models with more geographic and temporal detail. Computational limits impose trade offs between spatial and temporal scales.

This last point bears further emphasis in light of an unfortunately common misimpression among the general public. It is widely believed that meteorologists' inability to predict weather accurately beyond about 10 days bodes ill for any attempt at long-range climate projection. That misconception misses the differences of scale stressed in the preceding paragraph. In fact, it is impossible, even in principle, to predict credibly the small-scale details of local weather beyond about 10 days, and no amount of computing power or model sophistication is going to change that. This is because the atmosphere at small scales is an inherently chaotic system in which the slightest perturbation here today can make a huge difference in the weather a thousand miles away and a month hence. But large-scale climate shows little tendency to chaotic behavior (at least on decadal timescales), and appropriate models therefore can make reasonable climate projections decades or even centuries forward in time—provided, of course, that we have credible emission scenarios to drive the models.

A Hierarchy of Models

The simplest models involve just a few fundamental equations and a host of simplifying assumptions. For example, our basic global energy balance model treated Earth as a single point, with no atmosphere and no distinction between land and oceans. Simple models have the advantage that their predictions are easily understood on the basis of well-known physical laws. Furthermore, they produce results quickly and can, therefore, be used to test a wide range of assumptions by tweaking parameters of the model. In our simple energy balance model, for example, we could have studied the effect of different radiative forcings by subtracting a given forcing from the outgoing energy term to mimic the effect of infrared blockage.

More advanced are "multibox" models that treat land, ocean, and atmosphere as separate "boxes," and include flows of energy and matter between these boxes. Two-box models may ignore the land–ocean distinction and just treat Earth and its atmosphere separately. Three-box models handle all three components but do not distinguish different latitudes or altitudes. Still more sophisticated multibox models may break atmosphere and ocean into several layers or Earth into several latitude zones.

Most sophisticated are the large-scale computer models known as general circulation models (GCMs). These divide Earth's surface into a grid that, in today's highest-resolution models, measures just a few degrees of latitude and longitude on a side. At this scale, a model can represent with reasonable accuracy the actual shape of Earth's land masses. The atmosphere over and ocean below each surface cell are further divided into some 10–40 layers, making the basic unit of the model a small three-dimensional cell. Properties such as temperature, pressure, humidity, greenhouse gas concentrations, sunlight absorption, chemical activity, albedo, cloud cover, and biological activity are averaged within each cell. Equations based in physics, chemistry, and biology relate the various quantities within a cell, and other equations describe the transfer of energy and matter between adjacent cells. In some cases separate specialized models are developed for the atmosphere and the oceans and then linked together in a coupled atmosphere–ocean general circulation model (AOGCM).

GCMs are time-consuming and expensive to run, and their output can be difficult to interpret. Therefore, GCMs often are used to calibrate or to set empirical parameters (those not determined only from fundamental scientific principles) for simpler models that can then be used in specific studies. Thus the entire hierarchy of models becomes useful, indeed essential, for making progress in understanding climate.

Parameterization and Sub–Grid-Scale Effects

Even the best GCMs are limited to cell sizes roughly the size of a small country, such as Belgium. But climatically important phenomena occur on smaller scales.

Examples include clouds, which are far smaller than a typical grid cell, or the substantial thermal differences between cities and the surrounding countryside. Because all physical properties are averaged over a single grid cell, it is impossible to represent these phenomena explicitly within a model. But they can be treated implicitly.

Modelers use parametric representations, or parameterizations, in an attempt to include sub–grid-scale effects in their models. For example, a cell whose sky was half covered by fair-weather cumulus clouds might be parameterized by a uniform blockage of somewhat less than half the incident sunlight. Such a model manages not to ignore clouds altogether but doesn't quite handle them correctly. You can imagine that the effects of full sunlight penetrating to the ground in some small regions, while others are in full cloud shadow, might be different from those of a uniform light overcast, even with the same total energy reaching the ground. Developing and testing parameterizations that reliably incorporate sub–grid-scale effects is one of the most important and controversial tasks of climate modelers.

Transient Versus Equilibrium Models

Whether or not we manage to reduce anthropogenic greenhouse gas emissions, the atmospheric CO_2 concentration is likely to reach twice its preindustrial value (that is, CO_2 will reach some 560 parts per million) sometime in the present century. Although it may continue to rise well beyond that, a CO_2 concentration twice that of preindustrial times probably is the lowest level at which we have any hope of stabilizing atmospheric CO_2, barring a major breakthrough in low-cost, low–carbon-emitting energy technologies. For that reason, and because a doubling of atmospheric CO_2 from its preindustrial concentration provides a convenient benchmark, climate models often are run with doubled CO_2. The results can be summarized as a global average temperature rise for a doubling of CO_2, and this quantity is taken as a measure of the models' climate sensitivity. Most current models show a climate sensitivity of 1.5 to 4.5°C; that is, they predict a global average temperature rise of 1.5 to 4.5°C for a CO_2 concentration twice that of preindustrial times.

Until recently, most modeling groups did not have sufficient computer power to project future climate in response to the gradual increase in CO_2 concentration that will actually occur. Instead, they simply specified a doubled CO_2 concentration and solved their model equations once to determine the resulting climate. Physically, these equilibrium simulations give a projected climate that would result eventually if CO_2 were instantaneously doubled and then held fixed forever. In contrast, transient simulations solve the model equations over and over at successive times, allowing concentrations of greenhouse gases to evolve with time. The result is a more realistic projection of a changing climate. Transient simulations exhibit less immediate temperature rise because of the delay associated with the warming of the thermally massive oceans. In fact, the transient climate sensitivity—the warming at the instant CO_2 doubles during a transient calculation—typically is about half the equilibrium climate sensitivity (see Table 9.1 of IPCC, 2001a). That reduced rise can be deceptive because the full equilibrium warming must eventually occur, even if it is delayed for decades or more.

Transient simulations are essential in attempting to model climate records like that shown Fig. 1.1 in response, for example, to the CO_2 increase of Fig. 1.5. Recent advances in transient modeling have helped climatologists understand the role of anthropogenic gases in global warming by successfully reproducing the climate of the recent past in response to known anthropogenic and natural forcings.

Model Validation

How can modelers be more confident in their model results? How do they know that they have taken into account all climatologically significant processes and that they have satisfactorily parameterized processes whose scales are smaller than their models' grid cells? The answer lies in a variety of model validation techniques, most of which attempt to reproduce known climatic conditions in response to known forcings.

Major volcanic eruptions inject enough dust into the stratosphere to exert a global cooling influence that lasts several years. Such eruptions occur somewhat randomly, but typically once a decade or so, and they constitute natural experiments that can be used to test climate models. The last major eruption, of the Philippine volcano Mt. Pinatubo in 1991, was forecast by a number of climate modeling groups to cool the planet by several tenths of a degree Celsius. That is indeed what happened. Figure 1.9 shows a comparison between actual observed global temperature variations and those predicted by a climate model, for a period of 5 years after the Mt. Pinatubo eruption.[11] A few tenths of a degree is small enough that the observed variation might be a natural fluctuation. However, earlier eruptions including El Chichón in 1983 and Mt. Agung in 1963 were also followed by a marked global cooling of several tenths of a degree. Studying the climatic effects of a number of volcanic eruptions shows a clear and obvious correlation between major eruptions and subsequent global cooling.[12] Furthermore, a very simple calculation shows that the negative forcing of several watts per square meter produced by volcanic dust is consistent with the magni-

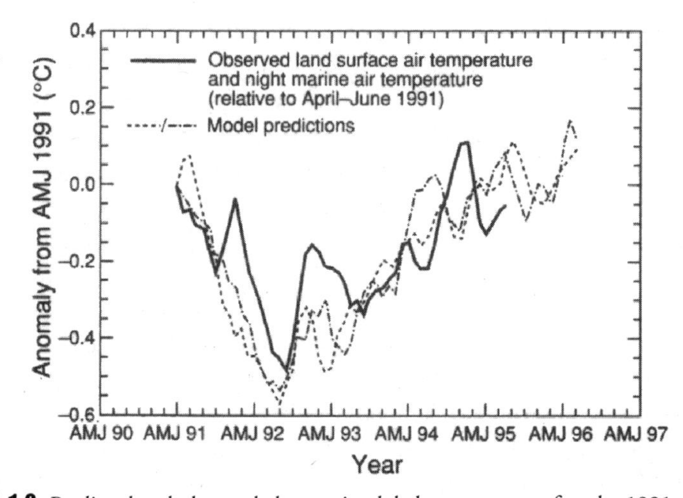

FIGURE 1.9. Predicted and observed changes in global temperature after the 1991 eruption of Mt. Pinatubo in the Philippines. Solid curve is derived from measured air temperatures over land and ocean surfaces. Broken curves represent climate model runs with slightly different initial conditions. In both cases the models included the effect of dust injected into the atmosphere by the volcanic eruption. (From Hansen et al., 1992, as adapted from IPCC, 2001a.)

tude of cooling after major volcanic eruptions. Taken together, all this evidence suggests that climate models do a reasonably good job of reproducing the climatic effects of volcanic eruptions.

Seasonality provides another natural experiment for testing climate models. Winter predictably follows summer, averaging some 15°C colder than summer in the Northern Hemisphere and 5°C colder in the Southern Hemisphere. (The Southern Hemisphere variation is smaller because a much larger portion of that hemisphere is water, whose high heat capacity moderates seasonal temperature variations.) Climate models do an excellent job reproducing the timing and magnitude of the seasonal temperature variations, although the absolute temperatures they predict may be off by several degrees in some regions of the world. The models are less good at reproducing other climatic variations, especially those involving precipitation and other aspects of the hydrologic cycle. Of course, reproducing the seasonal temperature cycle alone does not guarantee that models will describe accurately the climate variations resulting from other driving factors such as increasing anthropogenic greenhouse gas concentrations. However, the fact that GCMs reproduce seasonal variations so well is an assurance that the models' climate sensitivity is unlikely to be off by a factor of 10 or more, as some greenhouse contrarians assert.

Still another way to gain confidence in a model's future climate projections is to model past climates. Starting in 1860 with known climatic conditions, for example, can the model reproduce the temperature variation shown in Fig. 1.1? This approach not only provides some model validation but also helps modelers understand what physical processes may be significant in determining past climate trends. Figure 1.10 shows three different attempts, using the same basic climate model, to reproduce the historical temperature record of Fig. 1.1.[13] In the model runs of Fig. 1.10a, only estimates of solar variability and volcanic activity—purely natural forcings—were included in the model. The projected temperature variation, represented by a thick band indicating the degree of uncertainty in the model calculations, does not show an overall warming trend and clearly is a poor fit to the actual surface temperature record. The runs of Fig. 1.10b include only forcing caused by anthropogenic greenhouse gases and aerosols (e.g., the CO_2 record of Fig. 1.5, along with other known greenhouse gases and particulate emissions). This clearly does a much better job, especially in the late twentieth century, but deviates significantly from the historical record around midcentury. Finally, Fig. 1.10c shows the results from runs that include both natural and anthropogenic forcings. The fit is excellent, and it suggests that we can increase our confidence in this model's projections of future climate. Furthermore, the model runs of Fig. 1.10 taken together strongly suggest that the temperature rise of the past few decades is unlikely to be explained without invoking anthropogenic greenhouse gases as a significant causal factor. Thus the "experiments" of Fig. 1.10 illustrate one way of attempting to pry an anthropogenic climate signal from the natural climatic noise. In other words, Fig. 1.10 provides substantial circumstantial evidence of a discernible human influence on climate and supports the IPCC report's conclusion that "most of the warming observed over the last 50 years is attributable to human activities."[14]

Today's climate models provide geographic resolution down to the scale of a small country. Not only can they reproduce global temperature records, as shown in Fig. 1.10, but the best model results approach, although with less accuracy, the detailed geographic patterns of temperature, precipitation, and other climatic variables. These pattern-based comparisons of models and reality provide further confirmation of the models' essential validity.

No one model validation experiment alone is enough to give us high confidence in future climate projections. But considered together, results from the wide range of experiments probing the validity of climate models give considerable confidence that these models are treating the essential climate-determining processes with reasonable accuracy. Therefore, we can expect from them moderately realistic projections of future climate, given credible emission scenarios. That said, we still expect variations in the projections of different models. And

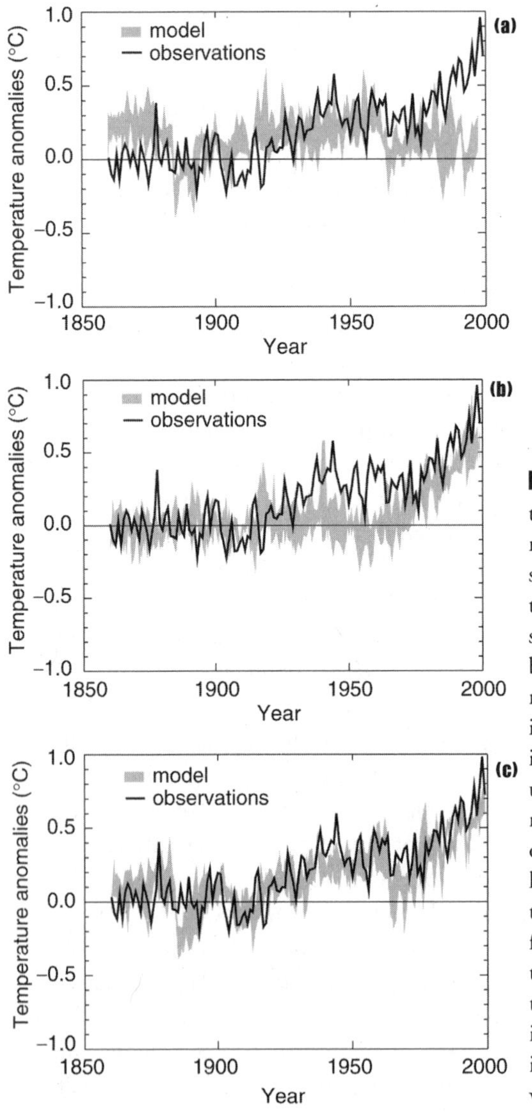

FIGURE 1.10. Attempts to model Earth's temperature from the 1860s using different model assumptions. In all three graphs, the solid curve is the observed surface temperature record of Fig. 1.1. Gray bands represent model projections. In each graph the bands encompass the results of four separate model runs. In (a), only natural forcings—volcanism and solar variability—are included. Clearly this simulation lacks the upward trend in the observed temperature record, suggesting that the temperature rise of the last century and a half is unlikely to have a purely natural explanation. Simulation (b), including only anthropogenic forcings, does much better, especially with the rapid temperature increase of the late twentieth century. Simulation (c), combining both natural and anthropogenic forcings, shows the best agreement with observations. (From IPCC, 2001a.)

because future greenhouse gas emissions depend on human behavior, future projections will differ depending on what assumptions modelers make about the human response to global warming. The uncertainties in projections of human behavior cause about as much spread in estimates of future warming as do uncertainties about the sensitivity of the climate system to radiative forcings. We probably will have to live with this frustrating situation for some time (see Chapter 2).

Consequences of Global Warming

The 1.5°C to 6°C global average temperature rise projected for the current century may seem modest, but as we noted, it could imply quite serious impacts. What might be the consequences? The most sophisticated climate models speak to a wide variety of possible impacts from global warming. Recall that a 6°C temperature drop means the difference between Earth's present climate and an ice age. Fortunately, it does not appear that a comparable rise will have consequences as devastating as two-kilometer thick ice sheets over populated areas of the Northern Hemisphere. But that doesn't mean the consequences of a few degrees' global warming will not be substantial and disruptive.

Global warming, obviously, means higher temperatures. But just how will the temperature rise be distributed in time and in space? We've been looking mostly at the global average temperature rise, characterized by a single number, but in fact global warming will vary substantially from one geographical region to another, and it will have different effects on night and day, winter and summer, land and sea.

Climate models provide rough consensus on many temperature-related projections. In general, projected temperature rises are greatest in the polar regions, and they affect the polar winter more dramatically than the summer. Similarly, nighttime temperatures are projected to rise more than daytime temperatures. Land temperatures are projected to rise more than oceans for the most part, influencing the patterns of monsoons and life-giving rains (and deadly floods) that they engender. Other obvious temperature-related consequences include increases in the maximum-observed temperatures and more hot days, increases in minimum temperatures and fewer cold days, and longer growing seasons owing to earlier last frosts and later first frosts. All these trends have already been seen in the climate change of the past few decades, and all are projected to continue through the present century. Climatologists' assessed confidence in these projections ranges from "likely" (two–thirds to 90 percent probability) to "very likely" (90 to 99 percent probability). Table 1.1 summarizes these and other effects of global warming, and gives the IPCC's quantitative estimates of the probability of each effect (see Chapter 2 for more explanation of what these probabilistic estimates really mean).[15]

The broadest impacts of direct temperature effects on human society are likely to be in agriculture and water supplies. However, health effects, including the spread of lowland tropical diseases vertically upward to plateaus and mountains and horizontally into temperate regions, may also be significant depending on the effectiveness of adaptive measures to reduce the threat. Natural ecosystems may also respond adversely to global warming. With temperatures chang-

TABLE 1.1. Projected Effects of Global Warming During the 21st Century

Projected Effect	*Probability Estimate*	*Examples of Projected Impacts with High Confidence of Occurrence (67–95% probability) in at Least Some Areas*
Higher maximum temperatures, more hot days and heat waves over nearly all land areas	Very likely (90–99%)	•Increased deaths and serious illness in older age groups and urban poor •Increased heat stress in livestock and wildlife •Shift in tourist destinations •Increased risk of damage to a number of crops •Increased electric cooling demand and reduced energy supply reliability
Higher minimum temperatures, fewer cold days, frost days and cold waves over nearly all land areas	Very likely (90–99%)	•Decreased cold-related human morbidity and mortality •Decreased risk of damage to a number of crops, and increased risk to others •Extended range and activity of some pest and disease vectors •Reduced heating energy demand
More intense precipitation events	Very likely (90–99%) over many areas	•Increased flood, landslide, avalanche, and mudslide damage •Increased soil erosion •Increased flood runoff increasing recharge of some floodplain aquifers •Increased pressure on government and private flood insurance systems and disaster relief
Increased summer drying over most mid-latitude continental interiors and associated risk of drought	Likely (67–90%)	•Decreased crop yields •Increased damage to building foundations caused by ground shrinkage •Decreased water resource quantity and quality •Increased risk of forest fire
Increase in tropical cyclone peak wind intensities, mean and peak precipitation intensities	Likely (67–90%) over some areas	•Increased risks to human life, risk of infectious disease epidemics and many other risks •Increased coastal erosion and damage to coastal buildings and infrastructure •Increased damage to coastal ecosystems such as coral reefs and mangroves
Intensified droughts and floods associated with El Niño events in many different regions	Likely (67–90%)	•Decreased agricultural and rangeland productivity in drought- and flood-prone regions •Decreased hydro-power potential in drought-prone regions

Projected Effect	*Probability Estimate*	*Examples of Projected Impacts with High Confidence of Occurrence (67–95% probability) in at Least Some Areas*
Increased Asian summer monsoon precipitation variability	Likely (67–90%)	•Increase in flood and drought magnitude and damages in temperate and tropical Asia
Increased intensity of mid-latitude storms	Uncertain (current models disagree)	•Increased risks to human life and health •Increased property and infrastructure losses •Increased damage to coastal ecosystems

Source: IPCC 2001b.

ing much more rapidly than in most natural sustained climatic shifts, temperature-sensitive plant species may find themselves unable to migrate fast enough to keep up with the changing climate. Even though their suitable habitat may shift only a few hundred miles, if plant species cannot reestablish themselves fast enough then they—and many animal species that depend on them—will go at least locally extinct. This is not just theory. Recent analyses of over 1,000 published studies have shown that, among other impacts, birds are laying eggs a few weeks earlier, butterflies are moving up mountains, and trees are blooming earlier in the spring and dropping their leaves later in the fall. In her capacity as a lead author for IPCC Working Group II, Terry Root led a group that combed recent literature to conclude that the most consistent explanation for these observed changes in environmental systems over the past few decades is global warming, and it appears that there is a discernible impact of regional climate change on wildlife and other environmental systems.[16] This opinion was first assessed and then echoed by the Working Group II, Third Assessment Report of the IPCC (2001b). Whether the regional climatic changes that seem to be driving these impacts are themselves manifestations of anthropogenic causation is more controversial. However, given that the responses observed are in about 80 percent of the cases in the direction expected with warming, Root and Schneider argue that global warming is the most consistent explanation.

Rising temperature also means rising sea level. A popular misconception holds that this is because of melting arctic ice. Actually, ice now floating on the oceans has almost no direct effect on sea level if it melts. Glaciers and the large ice sheets covering Greenland and Antarctica are a different story, as meltwater from these sources does increase sea level. But the bulk of sea-level rise observed to date or expected in the next century comes from the simple thermal expansion of seawater—the same process that drives up the liquid in a mercury or alcohol-based thermometer. Determining a global average level for the world ocean is difficult, but measurements suggest that sea level rose some 10–20 cm (4–8 inches) during the twentieth century. Climate models suggest that the rate of rise should increase as much as fourfold through the current century, resulting in a rise most likely near half a meter (about 20 inches). This may not seem like much, but it adds to the highest tides and to the surges associated with major storms (whose intensity is also expected to increase—see Table 1.1). Given that much of the world's population lives close to sea level, even a half-meter rise could have serious consequences in some regions, particularly those such as Bangladesh, which possess minimal resources and infrastructure to adapt to rising seas and higher storm surges. However, slow processes such as glacial melting would go on for many centuries, even after greenhouse gas emissions had long been replaced with non-emitting alternative energy systems. Thus, if humans use a substantial fraction of remaining fossil fuels and dump the greenhouses gases produced from their combustion into the atmosphere, then sea level is expected to go on rising, perhaps by several meters or more, over the thousand years that would follow the end of the fossil fuel era.[17]

Other weather-related projections include increased frequency of intense precipitation events, more heat waves in which the temperature remains at high levels for an extended time, fewer cold waves, more summer droughts, and more wet spells in winter. The intensity of tropical cyclones (hurricanes and typhoons) is likely to increase, although it is less clear whether the frequency or average locations of these storms will change. Hail and lightning are also likely to become more frequent. The large-scale Pacific Ocean fluctuation known as the El Niño/Southern Oscillation could become more persistent, which would have a substantial climatic impact on the Americas and Asia. All these projected changes will impact agriculture and may increase flooding and erosion, with concomitant effects on health and on the insurance industry. As shown in Table 1.1, confidence in this group of consequences ranges from medium (likelihood between one–third and two–thirds) to high (greater than two out of three chances). Keep in mind, however, that the probabilities given in Table 1.1 are not based on conventional statistical analysis because they refer to future events that do not follow past patterns—and obviously, the future hasn't occurred yet. Rather, these are *subjective* odds based on scientific judgment as sound as current understanding permits. Not surprisingly, that subjective element encourages some participants in the political process to attempt to discount these probability estimates (see Chapter 2 for more discussion on uncertainties and methods to deal with them).

Finally, there is the remote possibility of dramatic changes such as alterations in large-scale patterns of ocean circulation or the disintegration of the West Antarctic Ice Sheet. These could occur because the climate system is inherently nonlinear, meaning that a small change in some conditions can produce a disproportionately large change in others. A change in the Gulf Stream—part of the so-called ocean thermohaline circulation—could eventually—hundreds of years hence after anthropogenic greenhouse gases had dissipated—leaving northwest Europe with a chilly climate. Climate models predict with high confidence that the thermohaline circulation will weaken over the present century. But they also suggest, fortunately, that wholesale disruption is very unlikely at least before the year 2100. However, the models also warn that what humans do in the twenty-first century can precondition what the ocean currents will do in the twenty-second century and beyond. Potentially irreversible events could be built into the long-term planetary future even if those of us living in the twenty-first century are spared the experience of those effects.[18, 19] Similarly, recent studies suggest that the West Antarctic Ice Sheet is likely to remain stable for the foreseeable future, which is a very good thing because its breakup would result in a rise in sea level by some 6 meters (about 20 feet). But that "unlikely possibility" is not ruled out and looms as a potential threat that we need to check for periodically as we advance our understanding of the climate system and its potential for surprises.

We have given a brief description of the anticipated consequences of global warming in the present century. But even if we humans get our greenhouse gas emissions under control—not a likely occurrence in the near future—global temperature will continue to rise toward a new equilibrium value that will take at least many decades—more likely centuries—to become established. The effects of global warming, in particular sea-level rise, will almost certainly continue to increase beyond the end of the twenty-first century, and they may well become far more dramatic over the following centuries.

There is one final note on the issue of climatic impacts. In the above example of Bangladesh suffering from sea-level rises or more intense storms, we mentioned that adaptation would be difficult. This is much less so for a richer, more technologically advanced country such as the Netherlands. In fact, as is illustrated in Table 1.2 (in which IPCC 2001b authors summarize a comprehensive list of potential climate-change impacts for most regions of the world and economic sectors), a consensus is building in the scientific community that the damages that climatic changes might inflict on societies will depend in part on the adaptive capacities of those future societies, which in turn depend on their resource bases and technological and infrastructure capabilities.[20] This suggests, as Table 1.2 notes, that damages may be asymmetrically felt across the developed/developing country divide. The scenario where the northern rich countries

TABLE 1.2. Regional Adaptive Capacity, Vulnerability, and Key Concerns[a,b] (relevant sections of the Technical Summary of IPCC 2001b for each example are given in square brackets)

Region	
Africa	•Adaptive capacity of human systems in Africa is low due to lack of economic resources and technology, and vulnerability high as a result of heavy reliance on rain-fed agriculture, frequent droughts and floods, and poverty. [5.1.7] •Grain yields are projected to decrease for many scenarios, diminishing food security, particularly in small food-importing countries (*medium to high confidence*[c]). [5.1.2] •Major rivers of Africa are highly sensitive to climate variation; average runoff and water availability would decrease in Mediterranean and southern countries of Africa (*medium confidence*[c]). [5.1.1] •Extension of ranges of infectious disease vectors would adversely affect human health in Africa (*medium confidence*[c]). [5.1.4] •Desertification would be exacerbated by reductions in average annual rainfall, runoff, and soil moisture, especially in southern, North, and West Africa (*medium confidence*[c]). [5.1.6] •Increases in droughts, floods, and other extreme events would add to stresses on water resources, food security, human health, and infrastructures, and would constrain development in Africa (*high confidence*[c]). [5.1] •Significant extinctions of plant and animal species are projected and would impact rural livelihoods, tourism, and genetic resources (*medium confidence*[c]). [5.1.3] •Coastal settlements in, for example, the Gulf of Guinea, Senegal, Gambia, Egypt, and along the Southern–East African coast would be adversely impacted by sea-level rise through inundation and coastal erosion (*high confidence*[c]). [5.1.5]
Asia	•Adaptive capacity of human systems is low and vulnerability is high in the developing countries of Asia; the developed countries of Asia are more able to adapt and less vulnerable. [5.2.7] •Extreme events have increased in temperate and tropical Asia, including floods, droughts, forest fires, and tropical cyclones (*high confidence*[c]). [5.2.4] •Decreases in agricultural productivity and aquaculture due to thermal and water stress, sea-level rise, floods and droughts, and tropical cyclones would diminish food security in many countries of arid, tropical, and temperate Asia; agriculture would expand and increase in productivity in northern areas (*medium confidence*[c]). [5.2.1] •Runoff and water availability may decrease in arid and semi-arid Asia but increase in northern Asia (*medium confidence*[c]). [5.2.3] •Human health would be threatened by possible increased exposure to vector-borne infectious diseases and heat stress in parts of Asia (*medium confidence*[c]). [5.2.6]

Region	
Asia (*cont.*)	•Sea-level rise and an increase in the intensity of tropical cyclones would displace tens of millions of people in low-lying coastal areas of temperate and tropical Asia; increased intensity of rainfall would increase flood risks in temperate and tropical Asia (*high confidence*[c]). [5.2.5 and Table TS-8] •Climate change would increase energy demand, decrease tourism attraction, and influence transportation in some regions of Asia (*medium confidence*[c]). [5.2.4 and 5.2.7] •Climate change would exacerbate threats to biodiversity due to land-use and land-cover change and population pressure in Asia (*medium confidence*[c]). Sea-level rise would put ecological security at risk, including mangroves and coral reefs (*high confidence*[c]). [5.2.2] •Poleward movement of the southern boundary of the permafrost zones of Asia would result in a change of thermokarst and thermal erosion with negative impacts on social infrastructure and industries (*medium confidence*[c]). [5.2.2]
Australia & New Zealand	•Adaptive capacity of human systems is generally high, but there are groups in Australia and New Zealand, such as indigenous peoples in some regions, with low capacity to adapt and consequently high vulnerability. [5.3 and 5.3.5] •The net impact on some temperate crops of climate and CO_2 changes may initially be beneficial, but this balance is expected to become negative for some areas and crops with further climate change (*medium confidence*[c]). [5.3.3] •Water is likely to be a key issue (*high confidence*[c]) due to projected drying trends over much of the region and change to a more El Niño-like average state. [5.3 and 5.3.1] •Increases in the intensity of heavy rains and tropical cyclones (*medium confidence*[c]), and region-specific changes in the frequency of tropical cyclones, would alter the risks to life, property, and ecosystems from flooding, storm surges, and wind damage. [5.3.4] •Some species with restricted climatic niches and which are unable to migrate due to fragmentation of the landscape, soil differences, or topography could become endangered or extinct (*high confidence*[c]). Australian ecosystems that are particularly vulnerable to climate change include coral reefs, arid and semi-arid habitats in southwest and inland Australia, and Australian alpine systems. Freshwater wetlands in coastal zones in both Australia and New Zealand are vulnerable, and some New Zealand ecosystems are vulnerable to accelerated invasion by weeds. [5.3.2]
Europe	•Adaptive capacity is generally high in Europe for human systems; southern Europe and the European Arctic are more vulnerable than other parts of Europe. [5.4 and 5.4.6]

(*continues*)

TABLE 1.2. *Continued*

Region	
Europe (*cont.*)	•Summer runoff, water availability, and soil moisture are likely to decrease in southern Europe, and would widen the difference between the north and drought-prone south; increases are likely in winter in the north and south (*high confidence*[c]). [5.4.1] •Half of alpine glaciers and large permafrost areas could disappear by end of the 21st century (*medium confidence*[c]). [5.4.1] •River flood hazard will increase across much of Europe (*medium to high confidence*[c]); in coastal areas, the risk of flooding, erosion, and wetland loss will increase substantially with implications for human settlement, industry, tourism, agriculture, and coastal natural habitats. [5.4.1 and 5.4.4] •There will be some broadly positive effects on agriculture in northern Europe (*medium confidence*[c]); productivity will decrease in southern and eastern Europe (*medium confidence*[c]). [5.4.3] •Upward and northward shift of biotic zones will take place. Loss of important habitats (wetlands, tundra, isolated habitats) would threaten some species (*high confidence*[c]). [5.4.2] •Higher temperatures and heat waves may change traditional summer tourist destinations, and less reliable snow conditions may impact adversely on winter tourism (*medium confidence*[c]). [5.4.4]
Latin America	•Adaptive capacity of human systems in Latin America is low, particularly with respect to extreme climate events, and vulnerability is high. [5.5] •Loss and retreat of glaciers would adversely impact runoff and water supply in areas where glacier melt is an important water source (*high confidence*[c]). [5.5.1] •Floods and droughts would become more frequent with floods increasing sediment loads and degrade water quality in some areas (*high confidence*[c]). [5.5] •Increases in intensity of tropical cyclones would alter the risks to life, property, and ecosystems from heavy rain, flooding, storm surges, and wind damages (*high confidence*[c]). [5.5] •Yields of important crops are projected to decrease in many locations in Latin America, even when the effects of CO_2 are taken into account; subsistence farming in some regions of Latin America could be threatened (*high confidence*[c]). [5.5.4] •The geographical distribution of vector-borne infectious diseases would expand poleward and to higher elevations, and exposures to diseases such as malaria, dengue fever, and cholera will increase (*medium confidence*[c]). [5.5.5] •Coastal human settlements, productive activities, infrastructure, and mangrove ecosystems would be negatively affected by sea-level rise (*medium confidence*[c]). [5.5.3] •The rate of biodiversity loss would increase (*high confidence*[c]). [5.5.2]

Region	
North America	•Adaptive capacity of human systems is generally high and vulnerability low in North America, but some communities (e.g., indigenous peoples and those dependent on climate-sensitive resources) are more vulnerable; social, economic, and demographic trends are changing vulnerabilities in subregions. [5.6 and 5.6.1] •Some crops would benefit from modest warming accompanied by increasing CO_2, but effects would vary among crops and regions (*high confidence*[c]), including declines due to drought in some areas of Canada's Prairies and the U.S. Great Plains, potential increased food production in areas of Canada north of current production areas, and increased warm-temperate mixed forest production (*medium confidence*[c]). However, benefits for crops would decline at an increasing rate and possibly become a net loss with further warming (*medium confidence*[c]). [5.6.4] •Snowmelt-dominated watersheds in western North America will experience earlier spring peak flows (*high confidence*[c]), reductions in summer flows (*medium confidence*[c]), and reduced lake levels and outflows for the Great Lakes–St. Lawrence under most scenarios (*medium confidence*[c]); adaptive responses would offset some, but not all, of the impacts on water users and on aquatic ecosystems (*medium confidence*[c]). [5.6.2] •Unique natural ecosystems such as prairie wetlands, alpine tundra, and cold-water ecosystems will be at risk and effective adaptation is unlikely (*medium confidence*[c]). [5.6.5] •Sea-level rise would result in enhanced coastal erosion, coastal flooding, loss of coastal wetlands, and increased risk from storm surges, particularly in Florida and much of the U.S. Atlantic coast (*high confidence*[c]). [5.6.1] •Weather-related insured losses and public sector disaster relief payments in North America have been increasing; insurance sector planning has not yet systematically included climate change information, so there is potential for surprise (*high confidence*[c]). [5.6.1] •Vector-borne diseases—including malaria, dengue fever, and Lyme disease—may expand their ranges in North America; exacerbated air quality and heat stress morbidity and mortality would occur (*medium confidence*[c]); socioeconomic factors and public health measures would play a large role in determining the incidence and extent of health effects. [5.6.6]
Polar	•Natural systems in polar regions are highly vulnerable to climate change and current ecosystems have low adaptive capacity; technologically developed communities are likely to adapt readily to climate change, but some indigenous communities, in which traditional lifestyles are followed, have little capacity and few options for adaptation. [5.7] •Climate change in polar regions is expected to be among the largest and most rapid of any region on the Earth, and will cause major physical, ecological, sociological, and economic impacts, especially in the Arctic, Antarctic Peninsula, and Southern Ocean (*high confidence*[c]). [5.7]

TABLE 1.2. *Continued*

Region	
Polar (*cont.*)	•Changes in climate that have already taken place are manifested in the decrease in extent and thickness of Arctic sea ice, permafrost thawing, coastal erosion, changes in ice sheets and ice shelves, and altered distribution and abundance of species in polar regions (*high confidence*[c]). [5.7] •Some polar ecosystems may adapt through eventual replacement by migration of species and changing species composition, and possibly by eventual increases in overall productivity; ice edge systems that provide habitat for some species would be threatened (*medium confidence*[c]). [5.7] •Polar regions contain important drivers of climate change. Once triggered, they may continue for centuries, long after greenhouse gas concentrations are stabilized, and cause irreversible impacts on ice sheets, global ocean circulation, and sea-level rise (*medium confidence*[c]). [5.7]
Small Island States	•Adaptive capacity of human systems is generally low in small island states, and vulnerability high; small island states are likely to be among the countries most seriously impacted by climate change. [5.8] •The projected sea-level rise of 5 mm per year for the next 100 years would cause enhanced coastal erosion, loss of land and property, dislocation of people, increased risk from storm surges, reduced resilience of coastal ecosystems, saltwater intrusion into freshwater resources, and high resource costs to respond to and adapt to these changes (*high confidence*[c]). [5.8.2 and 5.8.5] •Islands with very limited water supplies are highly vulnerable to the impacts of climate change on the water balance (*high confidence*[c]). [5.8.4] •Coral reefs would be negatively affected by bleaching and by reduced calcification rates due to higher CO_2 levels (*medium confidence*[c]); mangrove, sea grass beds, and other coastal ecosystems and the associated biodiversity would be adversely affected by rising temperatures and accelerated sea-level rise (*medium confidence*[c]). [4.4 and 5.8.3] •Declines in coastal ecosystems would negatively impact reef fish and threaten reef fisheries, those who earn their livelihoods from reef fisheries, and those who rely on the fisheries as a significant food source (*medium confidence*[c]). [4.4 and 5.8.4] •Limited arable land and soil salinization makes agriculture of small island states, both for domestic food production and cash crop exports, highly vulnerable to climate change (*high confidence*[c]). [5.8.4] •Tourism, an important source of income and foreign exchange for many islands, would face severe disruption from climate change and sea-level rise (*high confidence*[c]). [5.8.5]

Source: IPCC 2001b.

[a] Because the available studies have not employed a common set of climate scenarios and methods, and because of uncertainties regarding the sensitivities and adaptability of natural and social systems, the assessment of regional vulnerabilities is necessarily qualitative.

[b] The regions listed in Table 2 are graphically depicted in Figure TS-2 of the Technical Summary of IPCC 2001b.

[c] These words represent collective estimates of confidence by authors of IPCC 2001b, based on observational evidence, modeling results, and theory: very high (95% or greater), high (67–95%), medium (33–67%), low (5–33%), and very low (5% or less).

---

get longer growing seasons and the poor tropical nations get more intense droughts and floods is clearly a situation ripe for increasing tensions in the world of the twenty-first century. Thus, not only is the climate-policy community faced with the need to estimate the impacts of a wide range of plausible climatic futures, but it must also estimate the relative adaptive capabilities of future societies so as to assess the equity implications of the consequences of slowing global warming. This in turn complicates the negotiations on solutions because many of the typically proposed mitigative activities could slow the economic growth rates of those very countries that need to build adaptive capabilities.[21] Yet, if these countries are allowed to emit unchecked amounts of greenhouse gases, the risks of severe impacts will increase. Therefore, the dilemma is to assess the range of possible outcomes as well as their costs and the distribution of those costs, and then to weigh those impacts versus the costs and benefits of a host of mitigation options carried out in various countries. All of this is played out against the historical background of large inequities in access to resources that make it difficult to achieve agreements that protect the global commons. It is our goal in this book to help you understand this complex interaction between political, economic, technological, and scientific issues as they relate to global climate change.

Is There Consensus on Global Warming?

The general public, especially in the United States, tends to think of global warming as a matter of intense and unsettled debate in the scientific community. A concerted effort by a handful of climate "contrarians" or "greenhouse skeptics"—scientists who do not share the views of most climate scientists—has kept the "debate" on global warming very much in public view.[22] The media, attempting to be fair to both sides has given the "contrarian" view publicity vastly disproportionate to its meager support in the community of climate scientists. Many policymakers also bring to their decisions a belief that prospects for global warming are murky, unsettled, and still very much a matter of debate—a belief reinforced by the dichotomous "debate" in the media between environmental activists proposing expensive sacrifices to avoid catastrophic climate change and those claiming that climate change would be beneficial, advocating that government stay out of all private matters, including their perceived "right" to dump wastes into the atmosphere without penalty.

Despite this ideologically-driven cacophony, there is a strong international consensus both on the basic science behind global climate change and on a broad range of future climate projections coming from modeling efforts. Why, then, is the public view—or at least the political debate in the U.S.—so out of step with mainstream science?

The Nature of Scientific Theories

Creationists attack Darwinian evolution because "it's just a theory." Critics still churn out counterproposals to Einstein's theory of relativity. And much of the public sees climate change in the same light: as just another scientific theory that might be right or might be wrong. The word *theory* is all too often an excuse to dismiss that with which one would like to disagree: "It's just a theory, so I don't have to accept it."

That attitude betrays a profound misunderstanding of the nature of scientific theories and scientific truth. A scientific theory is a coherent set of principles put forth to explain aspects of physical or biological or social reality. Decades of testing confirm a theory as providing the best available explanations for the phenomena at hand. It's always possible that an established theory may someday be proved wrong (or at least incomplete), but that possibility diminishes every time events in the real world live up the theory's predictions. Einstein's relativity, for example, is among the most solidly confirmed theories in science, tested not only in sophisticated astronomical observations and sensitive experiments but also in the workings of everyday devices from TV picture tubes to the Global Positioning Systems, neither of which would function correctly if relativity were wrong. Despite some gaps in the fossil record, Darwinian evolution remains the only consistent way science has found to understand the origin and demise of Earth's myriad species. Relativity and evolution may be "just theories," but they're so solidly confirmed that they've earned places in the canon of scientific truth. Likewise, gravity may be just a theory, but few would dare test it by jumping off the Empire State building.

The science at the basis of climate change has the same status. The essential idea—that Earth can maintain a constant temperature only if the rate at which energy reaches the planet equals the rate at which energy returns to outer space—is fundamental to the science of thermodynamics and was well established not only for Earth but for myriad other physical systems nearly two centuries ago. Measurements today confirm this idea of terrestrial energy balance to a high degree of precision. The role of greenhouse gases in that energy balance is also solidly established. We can measure the energy-absorbing properties of those gases in the laboratory, and field measurements provide accurate values for their atmospheric concentrations. The 33°C warming of Earth caused by natural greenhouse gases is well established and is further confirmed by our observations of the very different climates of Venus and Mars. The natural greenhouse effect is solidly established, and no reputable scientist would claim otherwise.

The public needs to recognize that established theories represent solidly confirmed bodies of scientific principles with broad explanatory powers and that, absent unlikely, radical new discoveries, such theories are the closest we can get to claiming we know the truth about physical reality. Many theories at the heart of modern science—including the thermodynamic basis of climate science and the theory of the greenhouse effect—fall into this category. They may be "just theories," but they're so solidly confirmed as to be universally accepted in the scientific community.

Does that mean there's no room for controversy about climate change? Of course not. It's one thing to accept a fundamental physical theory as rock-solid truth. It's quite another to affirm with high confidence the results of a complex computer model based on that theory but also depending on a host of other, more tenuous assumptions. Often our well-established theories are derived from very constrained and controllable situations, such as the fall of a particle in a gravitational field. But in the case of climate change, we are discussing a system of many interacting subcomponents. And, although we may be able to validate the behavior of many of the subcomponents via experiments and observations of the climate system, the interaction of all of them (that is, the behavior of the entire system) usually is not directly amenable to experimental confirmation. Furthermore, it is not possible even in principle to verify or to falsify a prediction for the year 2100—not before the fact, anyway. Thus, much of our confidence is based on the degree to which underlying principles are known for the major subcomponents of the system as a whole. This allows skeptics to cite out of context our poor understanding of a few subcomponents as proof that the whole system is poorly understood. Others do the opposite, singling out the best-verified components and neglecting the badly understood elements. That is why assessment teams of scientists from many disciplines and nations are summoned into activities such as the IPCC to try to provide a balanced perspective on the relative likelihood of various future events and their consequences. This is not "exact science" (itself an oxymoron) but the best representation of the state of the art. When the conclusions of such studies are juxtaposed against a few contrarian opinions in the name of "journalistic balance," the public and political process is muddied because few understand the very different relative credibility of these various claimants to state-of-the-art knowledge. Then it becomes incumbent on the citizen—whether in personal, corporate, or government capacity—to cut through this thicket of claims and counterclaims and use the literacy acquired in formal or lifelong learning activities to make sense of these controversies. Much of system science, of which climate change is a particularly important application, will always be murky. The fundamental principles are known and accepted, but the richness and complexity of nature coupled with imperfect knowledge of values, relationships, and processes make it impossible to predict accurately from first principles. Yet we can propose scenarios built on the best available science and provide meaningful estimates of our confidence in them.

Certainty and Uncertainty in Climate Science

Scientific "truth" is always a matter of probability. In the case of well-established theories such as evolution, relativity, thermodynamics, and the greenhouse effect, the probability that the theory is correct is so high as to constitute virtual certainty. But predictions and projections for complex systems may themselves be less certain. Again, the reasons are many and may include uncertainties in data that go into the calculations, uncertainties about the precise nature of physical processes, and uncertainties arising from approximations in the mathematical techniques used to solve complicated sets of equations. In climate science, examples of these uncertainties include, respectively, our imperfect knowledge of the global temperature record because of limited sampling sites and changes over time in instrumentation, urban growth, and other influences on temperature measurements; our limited understanding of physical processes in cloud formation and of the interaction of clouds with radiation; and the need for parameterization to handle mathematically processes (such as cloud formation) that occur on scales smaller than the numerical grids used in computer models.

However, the presence of uncertainty does not mean that such scientific results are speculative. Rather, it obligates the scientist to quantify just how uncertain a result may be, and it obligates any user of that result to take the stated uncertainty into account. In climate change studies, uncertainty manifests itself in the range of projected values for temperature, precipitation, and other climate variables. Uncertainty is further quantified by the confidence that the projected values will lie within the stated range.

A fundamental quantity in climate modeling is the sensitivity to a doubled atmospheric carbon dioxide concentration. The 1990, 1996, and 2001 IPCC reports all project a response to CO_2 doubling in the range 1.5–4.5°C. (Incidentally, the increased upper temperature range projected in the 2001 IPCC Working Group I report—6°C—results from an increased likelihood of higher greenhouse gas emissions, resulting in more than a doubling of preindustrial CO_2 levels by the end of the current century, combined with a decreased likelihood of offsetting aerosol cooling. The climate sensitivity estimates have not changed, but the CO_2 forcing has increased and the aerosol cooling decreased, hence the higher upper end of the projected temperature increase.) Where do these ranges in climate sensitivity come from?

The 2001 IPCC report lists 19 model runs, involving many different general circulation models, that show temperature increases (transient climate sensitivity) at the instant of doubled CO_2 ranging from 1.1°C to 3.1°C.[23] (These are transient simulations, which do not reflect the full amount of warming expected in a final equilibrium.) Expert opinion provides another measure of confidence in projections of global warming. A 1995 study by Morgan and Keith[24] elicited the subjective views of 16 leading climate scientists on the likely global response to doubled CO_2 concentration in equilibrium. All but one of the scientists gave their best estimates in the range 1.9–3.6°C. Each scientist (but one) also provided an interval in which he or she thought that the actual temperature change had a 90 percent chance of falling; the extent of these intervals ranged from 0.8°C to 8°C. None of this means we can say with absolute certainty that the twenty-first century will see a global temperature rise of several degrees. But when 15 different scientists and 19 distinct computer model runs suggest that a rise of this magnitude is likely, phrases such as "we're pretty sure" or "very likely" become appropriate ways of expressing confidence in projections of future climate. Chapter 2 explores further issues of uncertainty in relation to climate policy.

Scientific Consensus

What about that 1 scientist among the 16 experts whose estimate fell outside the 1.9–3.6°C range? That scientist suggested a best-estimate global temperature rise of only 0.3°C and was so confident as to be 90 percent certain that the actual rise would lie within a band only 0.7°C wide (see also the discussion in Chapter 2). Shouldn't his views be taken seriously? After all, scientific truth is not a matter of democratic vote. Might the 15 scientists be wrong and the single dissenter right? It's possible—but again, scientific truth is always a matter of probability, not absolute certainty. All 16 scientists in the Morgan and Keith survey share the same basic scientific knowledge, and it's the same Earth that they're all studying. Absent some overwhelmingly convincing reason to the contrary, it makes sense to weigh more strongly the views of the 15 scientists whose estimates are in general agreement.

Maybe the sample of scientists is biased. Had Morgan and Keith approached eight scientists who believe that we're due for significant global warming and eight who do not, their results would have been dramatically different. Would that have been a more balanced study? No, because that selection would not match the opinions of the scientific community in balanced proportion. Unfortunately, that is not what the public has always been led to believe. Instead, the public too often sees the debate over global warming as being between two factions with essentially equal scientific weight on both sides. That view is naïve both in its stark dichotomy and in its sense of equal weights. As we endeavor to inform the public about global climate change, it's crucial to set this point straight. First, climate scientists are not divided into two monolithic camps. The many scientists whose names appear as reviewers and contributors to the IPCC reports hold a range of views on the likely magnitude of future climate change, as the Morgan and Keith survey suggests. By 1995 these scientists were agreed that "the balance of evidence suggests a discernible human influence on global climate,"[25] and by 2001 they agreed that "most of the warming observed over the past 50 years is attributable to human activity."[26]

More importantly, there simply is no numerically substantial group of climate scientists whose views accord with the one dissenter in the Morgan and Keith survey—that is, who do not expect significant global warming in response to a doubling or greater increase of atmospheric carbon dioxide. What does exist is a small but vocal group that is visible out of all proportion to its numbers in the public debate over global warming. Often funded by the fossil fuel industry or by politically conservative think tanks, these scientists put forth the view that significant global warming is very unlikely or that limited warming will occur but will be beneficial. Unfortunately for scientific objectivity, they have been called in disproportionate numbers to testify at congressional hearings on climate change. They have also lent their names to slick, well-financed publications, Web sites, and video presentations that continue to leave the public with the impression of a balanced debate between equally tenable scientific positions. The amplified influence of these "greenhouse skeptics," and their close ties to the fossil-fuel industry, are well documented by journalist Ross Gelbspan in his book *The Heat Is On.*[27] More recent examples continue to crop up.[28]

So what should we teach the public about the nature of science, and climate science in particular? First, the basis of climate science—including the greenhouse effect—is firmly rooted in solidly proven scientific theories that are as close as we can get to scientific truth. Second, much of science is less certain than its fundamental theories, but that uncertainty can be quantified and may temper but not destroy our confidence in scientific projections. Third—and here science mixes with political reality—we need to convey the true nature of the scientific debate on the prospects for global climate change. That means exploring the idea of consensus in scientific communities and, in particular, revealing the substantial consensus that already exists in the climate science community.

Misconceptions

It's troubling enough that much of the public has only a vague understanding of climate science and of the nature of scientific debate and consensus on the subject. More troubling still are outright misconceptions that may be dangerous. In our efforts to educate the public and to implement sound policies toward climate change, we need to be aware of such widespread misconceptions and take explicit steps to eliminate them.

For much of the public—and even for a recent cabinet-level appointee in the U.S. with environmental responsibilities—ozone depletion is either synonymous or closely associated with global warming. This unfortunate confusion is confounded by the facts that both problems arise from anthropogenic gas emissions into the atmosphere and that both entered the public consciousness at about the same time. It doesn't help that one of the most visible environmental advocacy groups fighting for action against global warming was called Ozone Action. For the better informed, additional confusion arises because the two problems are related, albeit subtly: As we discussed earlier, ozone itself is a greenhouse gas, and the depletion of stratospheric ozone does affect Earth's energy balance.

But despite the public's confusion, ozone depletion and climate change contrast starkly not only in the scientific phenomena involved but also in light of attempts to solve each problem. Ozone depletion has been addressed by the most rapid and successful attack on an international environmental problem: the Montreal Protocol of 1987, which led to a worldwide ban on the most virulent ozone-depleting substances, the chlorinated fluorocarbons (CFCs). The solution to ozone depletion is being implemented (with full cooperation of most chemical manufacturers), and if the Montreal Protocol enjoys nearly full compliance, the problem of anthropogenic ozone depletion will be over in the roughly 50 years it will take for existing atmospheric CFCs to be removed naturally.

The status of international efforts to halt global climate change stands in dismal contrast (see Chapter 4). The most progressive international agreement on climate change, the Kyoto Protocol of 1997, takes some important steps but does not go nearly far enough to reduce anthropogenic global warming. And in the current political climate there appears to be no chance that the United States, the greatest single producer of greenhouse gases, will ratify the protocol.

Given that the problem of ozone depletion is essentially solved (albeit with a time delay), whereas there has been no effective progress on policies and measures to abate global climate change, public confusion of the two problems not only implies serious scientific misunderstanding but also carries the danger of public apathy toward urgently needed action on climate change.

To much of the public, carbon dioxide is just another of many pollutants produced by human activity, especially industry and transportation. Renewal of the Clean Air Act, tightening of automobile emission standards, inclusion of sport utility vehicles (SUVs) in automobile emission regulations, and lawsuits challenging older, dirtier power plants all sound like good news to a public that knows enough to recognize CO_2 as the main culprit in anthropogenic global climate change.

But this only highlights another misconception. In fact, CO_2 is not a pollutant, either in the legal sense or in the sense of an unwanted environmental contaminant produced inadvertently during combustion of fossil fuels. In fact, as the coal industry advertises, CO_2 is the raw material for photosynthesis, and the industry-supported Greening Earth Society advocates dumping more of it into the atmosphere as a "public service" to create a greener Earth. Unlike nitrogen oxides, carbon monoxide, sulfur oxides, ozone, and particulate matter, CO_2 is a necessary byproduct of fossil fuel combustion. Along with water vapor, it is what one wants to produce when burning fossil fuels. In that sense CO_2 is not a pollutant, and it is nonsensical to think of modifying the combustion process to eliminate CO_2 production. One might imagine sequestering combustion-produced CO_2 to keep it out of the atmosphere, but given some 20 pounds of CO_2 produced for every gallon of gasoline burned, that is a daunting and economically challenging prospect. But even that problem may succumb to technological solutions if hydrogen is extracted from fossil fuels, the carbon reinjected deep beneath Earth's surface, and the liberated hydrogen used in fuel cells to power cars and trucks. The key is at what cost this can be done and who pays, and the policy challenge is how to structure incentives to encourage our technological inventors to work on this problem.[29]

For a public that lumps CO_2 with "other" pollutants, there's a serious danger of complacency about CO_2 emissions in the face of tightening air quality regulations. Professional environmentalists have been heard to justify owning SUVs because they "meet California emission requirements." The vehicles in question may indeed be "clean" in that they emit few particulates or noxious gases, but a heavy SUV necessarily consumes more gasoline than a lighter car, and that gasoline produces CO_2 at the rate of some 20 pounds per gallon. No emission control technology can alter that figure. The gasoline-burning internal combustion engine is a Victorian industrial revolution technology, and to address the global warming issue more than superficially, we must reconsider the continued use—let alone expansion into the developing world—of this century-old technology. The only way to lower CO_2 emissions from cars and trucks is to burn less gasoline or to use another energy source altogether.

The confusion of CO_2 with other pollutants is a dangerous misconception because it leads to the complacent attitude that the CO_2 problem is coming under control. It isn't, and no amount of traditional pollution control will help (although building more fuel-efficient vehicles, power plants, and industrial boilers can reduce both traditional pollution and greenhouse gas emissions). Educating the public about climate change entails clearing up this glaring misconception.

An Informed Citizenry

This chapter began with the assertion that informed citizens and policymakers need a basic knowledge of climate science and climate policy to make intelligent policy decisions. So what's an informed citizen? First, it is one who understands the nature of science enough to appreciate that climate science is grounded in basic theories that are as close as we can get to scientific "truth" while recognizing that the projections of climate models are less certain but nevertheless carry a subjective but still expert-determined probability of being reasonably accurate. Second, an informed citizen is one who understands that the currently widespread view of a bipolar climate change debate between equally tenable scientific positions is simply incorrect and that most climate scientists are in a broad overall agreement that a significant global temperature increase is likely over the course of the twenty-first century. Third, an informed citizen understands the basic scientific ideas behind climate change projections, particularly energy balance, the greenhouse effect, and the nature and role of greenhouse gases. Finally, an informed citizen is aware of his or her own connection to the human processes that lead to climate change. Such a citizen is equipped to make intelligent value decisions about his or her own life choices as they influence climate and to participate in shaping the broader public response to the threat of climate change.

Notes

1. Intergovernmental Panel on Climate Change (IPCC), 2001a: *Climate Change 2001: The Scientific Basis*, Contribution of Working Group I to the Third Assessment Report of the IPCC, Houghton, J. T., Y. Ding, D. J. Griggs, M. Noguer, P. J. van der Linden, and D. Xiaosu (eds.) (Cambridge, England: Cambridge University Press), 881 pp.

2. Ibid.
3. Mann, M. E., R. S. Bradley, & M. K. Hughes, 1999: "Northern Hemisphere temperatures during the past millennium: Inferences, uncertainties, and limitations," *Geophysical Research Letters,* 26: 759.
4. Kiehl, J. T. and K. E. Trenberth, 1997: "Earth's annual global mean energy budget," *Bulletin of the American Meteorological Society,* 78: 197.
5. Neftel, A., H. Friedli, E. Moor, H. Lötscher, H. Oeschger, U. Siegenthaler, and B. Stauffer, 1994: "Historical CO_2 record from the Siple Station ice core," in *Trends: A Compendium of Data on Global Change* (Oak Ridge, TN: Carbon Dioxide Information Analysis Center, Oak Ridge National Laboratory); Keeling, C. D. and T. P. Whorf, 2000: "Atmospheric CO_2 records from sites in the SIO air sampling network," in *Trends: A Compendium of Data on Global Change* (Oak Ridge, TN: Carbon Dioxide Information Analysis Center, Oak Ridge National Laboratory, U.S. Department of Energy).
6. Petit, J. R., D. Raynaud, C. Lorius, J. Jouzel, G. Delaygue, N. I. Barkov, and V. M. Kotlyakov, 2000: "Historical isotopic temperature record from the Vostok ice core" in *Trends: A Compendium of Data on Global Change* (Oak Ridge, TN: Carbon Dioxide Information Analysis Center, Oak Ridge National Laboratory, U.S. Department of Energy); Jouzel, J., C. Lorius, J. R. Petit, C. Genthon, N. I. Barkov, V. M. Kotlyakov, and V. M. Petrov, 1987: "Vostok ice core: a continuous isotope temperature record over the last climatic cycle (160,000 years)," *Nature,* 329: 403; CDIAC, 2001: *Trends: A Compendium of Data on Global Change* (Oak Ridge, TN: Carbon Dioxide Information Analysis Center, Oak Ridge National Laboratory, U.S. Department of Energy; http://cdiac.esd.ornl.gov/trends/trends.htm).
7. IPCC, 2001a, op. cit.
8. Jacobson, M. Z., 2001: "A physically based treatment of elemental carbon: Implications for global direct forcing of aerosols," *Geophysical Research Letters,* 27(2): 217–220.
9. Hoyt, D. V. and K. H. Schatten, 1997: *The Role of the Sun in Climate Change* (New York: Oxford University Press).
10. Harvey, L. and D. Danny, 2000: *Global Warming: The Hard Science* (Englewood Cliffs, NJ: Prentice Hall).
11. Hansen, J., A. Lacis, R. Ruedy, and M. Sato, 1992: "Potential climate impact of Mount Pinatubo eruption," *Geophysical Research Letters,* 19: 215. IPCC, 1995, op. cit.
12. Mass, C. and S. H. Schneider, 1977: "Influence of sunspots and volcanic dust on long-term temperature records inferred by statistical investigations," *Journal of Atmospheric Science,* 34: 12.
13. IPCC, 2001a, op. cit.
14. IPCC, 2001a, op. cit., p. 10.
15. IPCC, 2001b: *Climate Change 2001: Impacts, Adaptation, and Vulnerability,* contribution of Working Group II to the Third Assessment Report of the IPCC, McCarthy, J. J., O. F. Canziani, N. A. Leary, D. J. Dokken, K. S. White (eds.) (Cambridge, England: Cambridge University Press), 1032 pp.
16. See Root, T. L. and Schneider, S. H., 2002: "Climate Change: Overview and Implications for Wildlife," in Schneider, S. H. and T. L. Root (eds.), *Wildlife Responses to*

Climate Change: North American Case Studies, National Wildlife Federation, Washington, DC: Island Press, pp. 1–55, for a full description of the methods used.
17. IPCC, 2002c: *Climate Change 2001: Synthesis Report*, contribution of Working Groups I, II, and III to the Third Assessment Report of the IPCC, Watson, R. T., et al. (eds.) (Cambridge, England: Cambridge University Press), 397 pp.
18. Rahmstorff, S., 1999: "Shifting Seas in the Greenhouse," *Nature* 399: 523–24.
19. Schneider, S. H. and S. L. Thompson, 2000: "A simple climate model used in economic studies of global change," in S. J. DeCanio, R. B. Howarth, A. H. Sanstad, S. H. Schneider, and S. L. Thompson, *New Directions in the Economics and Integrated Assessment of Global Change* (Washington, DC: Pew Center on Global Climate Change), 59–80.
20. IPCC, 2001b, op. cit.
21. IPCC, 2001c: *Climate Change 2001: Mitigation*, contribution of Working Group III to the Third Assessment Report of the IPCC, Metz, B., O. Davidson, R. Swart, and J. Pan (eds.) (Cambridge, England: Cambridge University Press), 752 pp.
22. Gelbspan, R., 2000: *The Heat Is On: The Climate Crisis, the Cover-Up, the Prescription* (New York: Perseus Books), 288 pp.
23. IPCC, 2001a, op. cit., Table 9.1.
24. Morgan, M. G. & D. W. Keith, 1995: "Subjective judgments by climate experts," *Environmental Science & Technology,* 29: 468A.
25. IPCC, 1996: *Climate Change 1995: The Science of Climate Change*, J. T. Houghton et al. (eds.) (Cambridge, England: Cambridge University Press).
26. IPCC, 2001a, op. cit.
27. Gelbspan, 2000, op. cit.
28. Greenhouse "contrarian" skepticism resurfaced in Lomborg, B. 2001: *The Skeptical Environmentalist: Measuring the Real State of the World,* (Cambridge, England: Cambridge University Press), 515 pp, and was sharply critiqued by Schneider, S. H. 2002: "Global Warming: Neglecting the Complexities," *Scientific American,* January 2002, 62–65, and others in that same issue.
29. Schneider, S. H., 2001: "Earth systems engineering and management," *Nature,* 409: 417–421. (See Chapter 20, this volume.)

CHAPTER 2

Uncertainty and Climate Change Policy

Stephen H. Schneider and Kristin Kuntz-Duriseti

Given modern satellite technology, it is ironic that the thinning of the ozone layer over the South Pole in the late 1970s went undetected for years. The satellite instrumentation did not fail; rather, the computer programs written to analyze the vast volumes of satellite data were instructed to reject measurements that diverged sharply from expected normal conditions. Amazingly, the rejected values were called to no one's attention. Noticing outliers in ground-based records of ultraviolet (UV) radiation reaching the earth's surface at a British station on the coast of Antarctica,[1] incredulous British scientists plotted the data by hand. To the surprise of all, they discovered a steady decrease in the ozone in the Southern Hemisphere springtime from the mid-1970s to the mid-1980s. This unexpected phenomenon immediately triggered a reprogramming of the U.S. computers to analyze all data points and revealed a deep hole in the ozone over the Antarctic continent, which was growing in intensity over time and drifting over nearby oceans and continents. This example shows that sometimes the knowable remains undetected because of the assumptions that frame the question or methods of analysis.

It is almost a tautology to note that unexpected global changes such as the development of the hole in the ozone layer are inherently difficult to predict. It is equally noninformative to suggest that other climate "surprises" can arise in the future. Yet despite the difficulty in forecasting climate change and its consequences, it remains imperative to address the wide uncertainties in our understanding of climate change and its effects. Global change science and policymaking will continue to deal with uncertainty and surprise. Therefore, more systematic analysis of surprise issues and more formal and consistent methods of incorporating uncertainty into global change assessments will become increasingly necessary.

Significant uncertainties plague projections of climate change and its consequences. The extent of the human influence on the environment is unprecedented: Human-induced climate change is projected to occur at a very rapid rate; natural habitat is fragmented by agriculture, settlements, and other development activities; exotic species are imported across natural barriers; and we assault our environment with a host of chemical agents.[2] For these reasons it is essential to understand not only how much climate change is likely but also how to characterize and analyze the effects of climate change.

The combination of increasing population and increasing energy consumption per capita is expected to contribute to increasing CO_2 and sulfate emissions over the twenty-first century. However, projections of the extent and effect of the increase are very uncertain.[3] Central estimates of emissions suggest a doubling of current CO_2 concentrations by the mid–twenty-first century, leading to projected warming of more than 1°C to nearly 6°C by the end of the twenty-first century.[4] Warming at the upper end of this range is even more likely beyond a doubling of CO_2, which is likely to occur during the twenty-second century in most scenarios. Although warming at the low end of the uncertainty range could still have significant implications for species adaptation, warming of 5°C or more could have catastrophic effects on natural and human ecosystems, including serious coastal flooding, collapse of thermohaline circulation (THC) in the Atlantic Ocean (i.e., changes in the Gulf Stream currents), or nonlinear responses of ecosystems.[5] The market value cost of these impacts could easily run into many tens of billions of dollars annually[6] to perhaps as much as trillions of dollars by the late twenty-first century.[7]

Policymakers struggle with the need to make decisions that have far-reaching and often irreversible effects on both environment and society with sparse and imprecise information. Not surprisingly, efforts to incorporate uncertainty into decision making enter the negotiating parlance through catchphrases such as "the precautionary principle," "adaptive environmental management," "the preventive paradigm," and "principles of stewardship."[8] The shift toward prevention in environmental policy "implies an acceptance of the inherent limitations of the anticipatory knowledge on which decisions about environmental [problems] are based."[9]

Uncertainty or, more generally, debate about the level of certainty needed to reach a firm conclusion is a perennial issue in science. The difficulties of explaining uncertainty have become increasingly salient as society seeks policy advice to deal with global environmental change. How can science be most useful to society when evidence is incomplete or ambiguous, the subjective judgments of experts about the likelihood of outcomes vary, and policymakers seek guidance and justification for courses of action that could cause significant environmental and societal changes? How can scientists improve their characterization of uncertainties so that areas of slight disagreement are not perceived as major scientific disputes, as occurs all too often in media or political debates? Finally, how can policymakers synthesize this information and formulate policy? In short, how can the full spectrum of the scientific content of public policy debates be assessed fairly and openly?

Decision Making Under Uncertainty

The term *uncertainty* implies anything from confidence just short of certainty to informed guesses or speculation. Lack of information obviously results in uncertainty, but often disagreement about what is known or even knowable is a source of uncertainty (Box 2.1). Some categories of uncertainty are quantifiable, yet other kinds cannot be expressed readily in terms of probabilities. Uncertainties arise from such factors as linguistic imprecision, statistical variation, measurement error, variability, approximation, subjective judgment, and disagreement. These problems are compounded by the global scale of climate change, but local scales of impacts, long time lags between forcing and response, low-frequency climate variability that exceeds the length of most instrumental records, and the impossibility of before-the-fact experimental controls. Moreover, because climate change and other complex sociotechnical policy issues are not just scientific topics but also matters of public debate, it is important to recognize that even good data and thoughtful analysis may be insufficient to resolve some uncertainties associated with the different standards of evidence and degrees of risk aversion or acceptance that people participating in this debate may hold.

In dealing with uncertainty in science or the policy arena policymakers typically consider two options: bound the uncertainty or reduce the effects of uncertainty. The first option is to reduce the uncertainty through data collection, research, modeling, simulation, and so forth. This effort is characteristic of normal scientific study. The objective is to overcome the uncertainty—to make known the unknown. However, the daunting uncertainty surrounding global environmental change and the need to make decisions before the uncertainty is resolved make the first option difficult to achieve. That leaves policymakers an alternative: to manage uncertainty rather than master it. Thus, the second option is to integrate uncertainty into policymaking.

The emphasis on managing uncertainty rather than mastering it can be traced to work on resilience in ecology, most notably by Holling.[11] Resilience is the ability to recover from a disturbance without compromising the overall health of the system.

BOX 2.1. Examples of Sources of Uncertainty[10]

Problems with Data

- Missing components or errors in the data
- "Noise" in the data associated with biased or incomplete observations
- Random sampling error and biases (nonrepresentativeness) in a sample

Problems with Models

- Known processes but unknown functional relationships or errors in the structure of the model
- Known structure but unknown or erroneous values of some important parameters
- Known historical data and model structure but reasons to believe that the parameters or model structure will change over time
- Uncertainty about the predictability (e.g., chaotic or stochastic behavior) of the system or effect
- Uncertainties introduced by approximation techniques used to solve a set of equations that characterize the model

Other Sources of Uncertainty

- Ambiguously defined concepts and terminology
- Inappropriate spatial or temporal units
- Inappropriateness or lack of confidence in underlying assumptions
- Uncertainty caused by projections of human behavior (e.g., future consumption patterns or technological change), which is distinct from uncertainty from "natural" sources (e.g., climate sensitivity, chaos)

The fields of mathematics, statistics, and physics independently and concurrently developed methods to deal with uncertainty. These methods offer many powerful ways to conceptualize, quantify, and manage uncertainty, including frequentist probability distributions, subjective probability and belief statements of Bayesian statistics, and even a method for quantifying ignorance.[12] Addressing other aspects of uncertainty, fuzzy set logic offers an alternative to classic set theory for situations in which the definitions of set membership are vague, ambiguous, or nonexclusive.[13] More recently, researchers have proposed chaos theory and complexification theory to focus on expecting the unexpected in models and theory.[14]

Risk Assessment

One method for incorporating uncertainty is to perform an expected value analysis. The expected value is simply the sum across all possible outcomes of the product of the probability of an outcome and the value (cost or benefit) of that outcome. Typically, modelers postulate two outcomes: a low-probability, high-damage case and a high-probability, low-damage case. However, this method is fraught with problems when applied to study climate change. First, expected value calculations assume risk neutrality and thus neglect any consideration of risk aversion, especially with respect to low-probability, catastrophic outcomes. A gambling analogy clarifies this concept. Suppose you are offered the following gamble: a 50 percent chance of winning $100 and a 50 percent chance of losing $100. The expected value, or average outcome, of this gamble is zero. Would you opt to take the gamble? If you were risk neutral, you would be indifferent between the two options. If you were risk averse, you would forgo the gamble, but if you were risk accepting, you would take your chances on the gamble. Risk neutrality implies indifference between receiving for certain the expected value across outcomes and accepting the single outcome from a one-time gamble across all possible outcomes. Risk aversion implies a preference for receiving the expected value over facing the gamble; in technical terms, the utility of the expected *value* of the gamble is greater than the expected *utility* of the gamble (Fig. 2.1). The difference between the expected value and the expected utility is the amount forgone to avoid facing the gamble—in other words, the risk premium.[15]

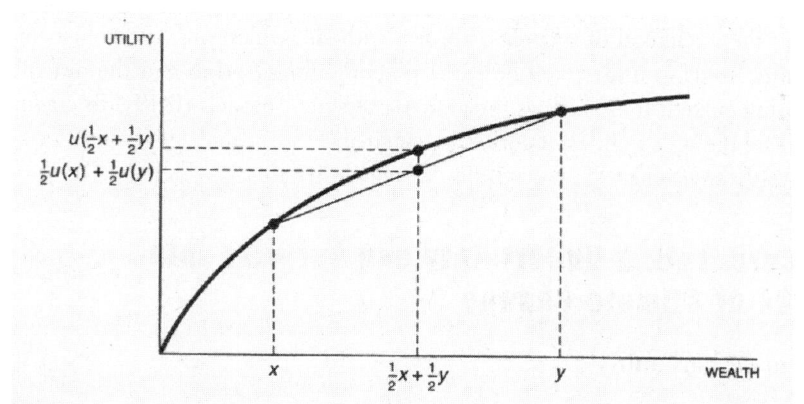

FIGURE 2.1. The expected utility of the gamble is 1/2U(X) + 1/2U(Y). The utility of the expected value of the gamble is U(1/2X + 1/2Y). In the risk averse case depicted the utility of the expected value is higher than the expected utility of the gamble. *Source:* Varian, 1992.

Imaginable Surprise

Strictly speaking, a surprise is an unanticipated outcome; by definition it is an unexpected event. Potential climate change and, more broadly, global environmental change are replete with this kind of surprise because of the enormous complexities of the processes and relationships involved (such as coupled ocean, atmosphere, and terrestrial systems) and our insufficient understanding of them.[16]

The Intergovernmental Panel on Climate Change (IPCC) Second Assessment Report (SAR),[17] defines "surprise" as rapid, nonlinear responses of the climatic system to anthropogenic forcing, such as the collapse of the "conveyor" belt circulation in the North Atlantic Ocean[18] or rapid deglaciation of polar ice sheets.

Unfortunately, most climate change assessments rarely consider low-probability, high-consequence events. Instead, assessments primarily consider scenarios that supposedly "bracket the uncertainty" rather than explicitly integrate unlikely events. Not even considered in the standard paradigm are structural changes in political or economic systems or changes in public consciousness regarding environmental issues. Although researchers recognize the wide range of uncertainty surrounding global climate change, their analyses are essentially surprise free.

Extreme events that are not truly unexpected are better defined as imaginable abrupt events. And for some surprises, although the outcome is unknown, it is possible to identify imaginable conditions for surprise to occur. For example, as the rate of change of CO_2 concentrations is one imaginable condition for surprise, the system would be less rapidly forced if decision makers chose to slow down the rate at which human activities modify the atmosphere. This would lower the likelihood of surprises. To deal with such questions, the policy community needs to understand both the potential for surprises and the difficulty of using current tools such as integrated assessment models (IAMs) to credibly evaluate the probabilities of currently imaginable "surprises," let alone those not currently envisioned.[19]

Incorporating Uncertainty and Surprise into IAMs of Climate Change

Climate Variability

A critical assumption of the standard assessment paradigm is whether the probability of climate extremes, such as droughts, floods, and super-hurricanes, will remain unchanged or will change with the mean change in climate according to unchanged variability distributions. As Mearns et al.[20] have shown, however, changes in the daily temperature variance or the autocorrelation of daily weather extremes can significantly reduce or dramatically exacerbate the vulnerability of agriculture, ecosystems, or other climate extreme–sensitive components of the environment to global warming. How such variability measures might change as the climatic mean changes is highly uncertain, although an increase in the number of extreme events with global warming is expected.[21] Variability in precipitation, most notably from an increase in high-intensity rainfall, is expected to increase. Karl and Knight[22] have observed that about half of the 8 percent increase in precipitation in the United States since 1910 occurred in the most damaging heavy downpours. In addition, the El Niño–Southern Oscillation (ENSO) could well continue the trend of the past two decades and become a more recurrent or stronger phenomenon, which will increase climate variability.[23]

Projections for storms (tropical cyclones, mid-latitude storms, tornadoes, and severe storms) are more controversial. Based on the assumption that wet bulb temperatures (a measure of the humidity of the atmosphere; the higher the humidity, the higher the temperature that a wet, i.e., evaporating, thermometer would measure) translate into larger potential energy for severe weather, recent studies have examined the link between temperature and extreme weather. Arguing that "storm activity may be more dependent on daily minimum temperatures than on daily maximums," Dessens[24] notes a positive correlation between nighttime temperatures and the frequency of severe storms (specifically hail storms in France) and projects a 40 percent increase in hail damage from a 1°C increase in mean minimum temperatures. Reeve and Toumi[25] note a link between lightning and temperature and predict a 40 percent ± 14 percent increase in lightning for a 1°C increase in wet bulb temperature. Currently, the climate record is too noisy to detect a clear signal of increased hurricane intensities, but the theoretical understanding of the driving forces behind hurricanes strongly suggests that peak intensities should be higher in a warmer world.[26] Although it is not possible to determine with high confidence, given current data and methods, the possibility of increased climate extremes from human disturbances is not remote.

Transient Effects of Climate Change

Standard assessments model responses to a one-time doubling of CO_2 and analyze the effects once the system reaches equilibrium. Clearly, what happens along the path to a new equilibrium is of interest as well, especially in the event of abrupt change. For example, resultant environmental or societal impacts are likely to be quite different from those that would occur with smoother, slower changes. The long-term impact of climate change may not be predictable solely from a single steady-state outcome but may well depend on the characteristics of the transient path. In other words, the outcome may be path dependent. Any exercise that neglects surprises or assumes transitivity of the earth system (i.e., a path-independent response) is indeed questionable and should carry a clear warning to users of the fundamental assumptions implicit in the technique dependent on steady state results. Furthermore, rapid transients and nonlinear events are likely to affect not only the mean values of key climate indicators but also higher statistical moments, such as variability, of the climate (e.g., week-to-week variability, seasonal highs and lows, and day-to-night temperature differences).

Rate of Forcing Matters

Even the most comprehensive coupled-system models are likely to have unanticipated results when forced to change very rapidly by external disturbances such as CO_2 and aerosols. Indeed, some of the transient coupled atmosphere–ocean models run out for hundreds of years exhibit dramatic change in the basic climate state.[27] Stocker and Schmittner[28] argue that rapid alterations in oceanic currents could be induced by faster rates of climate change. For very rapid increases in CO_2 concentrations, Thompson and Schneider[29] simulate a reversal of the equator-to-pole temperature difference in the Southern Hemisphere over the century immediately during and after the rapid buildup of CO_2. Slower increases in CO_2 would not create such a surprise. More recent research by Schneider and Thompson[30] suggests that factors contributing to a collapse of the THC in the North Atlantic Ocean include changes in the climate sensitivity, the overturning rate of the THC (i.e., how quickly cold, salty waters sink), the CO_2 stabilization level, and the rate of increase of CO_2 concentrations (the former two are uncertain biogeophysical factors and the latter two are social factors dependent on human decisions). Mastrandrea and Schneider show that the combination of these factors and the discount rate critically affect the "optimal" rate of CO_2 mitigation.[31] Furthermore, Schneider and Thompson demonstrate that abrupt and discontinuous environmental change can occur even when climate forcings are smooth.

Simulations by Schneider and Thompson[32] and Mastrandrea and Schneider[33] suggest that actions taken in the short term may have serious long-term, abrupt, potentially irreversible consequences. Mastrandrea and Schneider[34] demonstrate for some scenarios that only low discount rates stimulate sufficient controls on CO_2 emissions to prevent a circulation collapse, which implies that myopic policymakers may implement weak short-term climate policies that build into the long-term future unexpected, major changes in climatic conditions. To develop a climate policy that will lower the risk of climate catastrophes, policymakers need to consider consequences of climate change beyond the twenty-first century, including very uncertain but highly consequential events such as a THC collapse.

Estimating Climate Damages

A critical issue in climate change policy is costing climatic impacts, particularly when the possibility for nonlinearities, surprises, and irreversible events is allowed. The assumptions made when carrying out such estimations largely explain why different authors obtain different policy conclusions.

Subjective probability assessments of potential climate change impacts provide a crude metric for assigning dollar values to certain aspects of ecosystem services. We can anticipate costs associated with global change and place a preliminary value on some of the ecosystem services that could be affected. One way to assess the costs of climate change is to evaluate the historic losses from extreme climatic events, such as floods, droughts, and hurricanes.[35] Catastrophic floods and droughts are cautiously projected to increase in both frequency and intensity with a warmer climate and the influence of human activities such as urbanization, deforestation, aquifer depletion, groundwater contamination, and poor irrigation practices.[36] The financial service sector has taken particular note of the potential losses from climate change. Losses from weather-related disasters in the 1990s were eight times higher than in the 1960s. Although there is no clear evidence that hurricane frequency has changed over the past few decades (or will change in the next few decades), there is overwhelming data that damage from such storms has increased astronomically. Attribution of this trend to changes in socioeconomic factors (e.g., economic growth, population growth and other demographic changes, or increased penetration of insurance coverage) or to an increase in the occurrence or intensity of extreme weather events as a result of global climate change is uncertain and controversial. (Compare Vellinga et al.,[37] which acknowledges both social and climatic influences and recognizes the difficulty in attribution, to Pielke and Landsea,[38] which dismisses any effects of climate change.) Damage assessment is one possible way in which we can relate the cost of more inland and coastal flooding, droughts, and possible intensification of hurricanes to the value of preventing the disruption of climate stability.[39]

An assumption in cost–benefit calculations in the standard assessment paradigm is that "nature" is either constant or irrelevant. Because "nature" is beyond the purview of the market, cost–benefit analyses ignore its nonmarket value. For example, ecological services[40] such as pest control and waste recycling are omitted from most assessment calculations. Implicitly, this assumes that the economic value of ecological services is negligible or will remain unchanged with human disturbances. Recent assessments of the value of ecosystem services acknowledge the tremendous public good provided, not to mention the recreational and aesthetic value. For example, a cost assessment study in New York discovered that paying residents and farmers to reduce toxic discharges and other environmental disruptions to protect the Catskills, which provide a natural water purification service, produced a significant savings (on the order of billions of dollars) over building a new water treatment plant. Furthermore, it is highly likely that communities of species will be disrupted, especially if climate change occurs in the middle to upper range projected.[41]

The Discount Rate

Discounting plays a crucial role in the economics of climate change, yet it is a highly uncertain parameter. Discounting is a method of aggregating costs and benefits over a long time horizon by summing across future time periods net costs (or benefits) that have been multiplied by a discount rate, typically greater than zero. If the discount rate equals zero, then each time period is valued equally (case of infinite patience). If the discount rate is infinite, then only the current period is valued (case of extreme myopia). The discount rate chosen in assessment models is critical because abatement costs typically are incurred in the near term, but the brunt of climate damages is realized primarily in the long term. Thus, if the future is sufficiently discounted, present abatement costs will outweigh discounted future climate damages because discount rates will eventually reduce future damage costs to negligible present values.

Consider a climate impact that would cost $1 billion 200 years from now. A discount rate of 5 percent per year would make the present value of that future cost equal to $58,000. At a discount rate of 10 percent per year, the present value would be only $5. Changes in this parameter largely explain why some authors,[42] using large discount rates, conclude that CO_2 emission increases could be socially beneficial whereas others,[43] using low or zero discount rates, justify substantial emission reductions, even when using similar damage functions.[44]

It might seem that the appropriate discount rate is a matter of empirical determination, but the conflict involves a serious normative debate about how to value the welfare of future generations relative to current ones. Moreover, it requires that the current generation estimate what kinds of goods and services future generations will value (e.g., what trade-offs they will want to make between extra material wealth and greater loss of environmental services). Much of the debate centers around different interpretations of the normative implications of the choice of the discount rate.[45]

The descriptive approach chooses a discount rate based on observed market interest rates to ensure that investments are made in the most profitable projects. Supporters of this approach often argue that using a market-based discount rate is the most efficient way to allocate scarce resources used for competing priorities, of which one is mitigating the effects of climate change.

The prescriptive approach emphasizes that the choice of discount rate entails a choice on how the future should be valued. Proponents of intergenerational equity often argue that it is difficult to argue that the welfare of future generations should be discounted simply because they exist in the future.

Although these two approaches are the most common in IAMs of climate change, alternative discount methods have been proposed. There is empirical evidence to suggest that people exhibit hyperbolic discounting, in which discount rates decline over time with higher-than-market discount rates in the short run and lower discount rates over the long term.[46] This behavior is consistent with a common finding that "human response to a change in a stimulus is inversely proportional to the pre-existing stimulus."[47] Hyperbolic discounting can be derived from both the descriptive and the prescriptive approach, and is obtained when discount rates fall over time. This can be modeled in IAMs with a logarithmic discount factor[48] or by assuming that per capita income grows logistically over the next century; because the discount rate is proportional to growth rates, declining discount rates are obtained.[49]

Furthermore, if climate change is severe, such that future income falls rather than grows—growth is assumed in almost all IAMs—then the discount rate can be negative, provided that the rate of time preference is sufficiently low.[50] In this case, future welfare should be valued more than the present. The complexity in the discounting issue stems not only from uncertainty in how to calculate the value of the future once a discount rate is specified but also from uncertainty over whether any particular choice is appropriate for alternative value systems.

Agency

The predominant approach to the discounting problem is based on an infinitely lived representative agent (ILA) who maximizes utility from a future welfare stream subject to economic and environmental conditions, usually assumed to be known. The ILA framework imposes strong assumptions regarding intergenerational fairness.[51] An alternative modeling paradigm, the overlapping generations model (OLG), differentiates between individual time preference and intergenerational equity (the distinction is suppressed in the ILA model) and endogenizes the choice of discount rate.[52] A distinctive characteristic of OLG models (unlike ILA models in most IAMs) is that the OLG framework explicitly models the existence of generations who work and save when young and consume savings, or "dissave," when old. Thus, the two modeling frameworks represent different conceptions of intergenerational equity. The policy recommendations derived from the OGM differ fundamentally from those of the ILA model, including higher carbon emission abatement (however, Manne and Stephan[53] show that under certain restrictions, the results from the ILA and the OGM models concur).

Natural Variability Masks Trends and Delays Adaptation

One of the major differences in estimates of climatic impacts across different studies is how the impact assessment model treats adaptation of natural and human systems to climate change. For example, it has often been assumed that agriculture is the most vulnerable economic market sector to climate change. For decades agronomists have calculated potential changes in crop yields from various climate change scenarios, suggesting that some regions now too hot would sustain heavy losses from warming, whereas others, now too cold, could gain.[54] Rosenberg[55] has long argued that such agricultural impact studies implicitly invoke the "dumb farmer assumption." That is, they neglect the fact that farmers do adapt to changing market, technology, and climatic conditions. For example, Mendelsohn et al.[56] use cross-sectional analyses to estimate empirically the adaptation responses of real farmers to changes in climate (e.g., changes in crop yields and land rent values) by simply comparing land use activities in warm places such as the U.S. Southeast and colder places such as the Northeast as a proxy for how temperature changes might affect these segments of the economy. Agricultural economists such as John Reilly[57] argue that such adaptations will dramatically reduce the climate impact costs to market sectors such as farming, transportation, coastal protection, and energy use. However, ecologists and some social scientists often dispute this optimism. Haneman[58] notes that Mendelsohn et al. confound the normative statement that public policy *should* encourage efficient adaptation with the positive statement that adaptation *will be* efficient: "It is a large leap to go from the observation that there will be *some* adaptation to the inference that there will be *perfectly efficient* adaptation." Furthermore, Schneider[59] objects that the statistical analysis Mendelsohn et al. use ignores time-evolving or transient changes in temperature and other variables, not to mention surprises. In essence, they assume perfect substitutability for changes at one place over time with changes across space at the same time. Assuming a high level of adaptation neglects such real-world problems as resistance to trying unfamiliar practices, problems with new technologies, unexpected pest outbreaks,[60] or the high degree of natural variability of weather.

The high natural variability of climate probably will mask any slowly evolving anthropogenically induced climate trends, either real or forecasted. Furthermore, adaptation is likely to be a reaction to an already changed climate rather than a preemptive response to anticipated or projected climate change. Therefore, adaptations to slowly evolving trends embedded in a noisy background of inherent variability are likely to be delayed by decades behind the slowly evolving global change trends.[61] Moreover, were agents to mistake background variability for trend or vice versa, the possibility arises of adaptation following the wrong set of climatic cues. In particular, agents might be more influenced by regional anomalies of the recent past in projecting future trends. They may be unaware of the likelihood that very recent anomalous experience in one region may well be largely uncorrelated with slowly building long-term trends at a global scale or may be part of a transient response that will reverse later on. In addition, unwarranted complacency may result from the inability to foresee nonlinear events.

Passive Versus Anticipatory Adaptation

Schneider and Thompson,[62] in an intercomparison of climate change, ozone depletion, and acid rain problems, differentiate passive adaptation (e.g., buying more water rights to offset impacts of a drying climate) from anticipatory adaptation. They suggest, as a hedging strategy, investing in a vigorous research and development program for low-carbon energy systems in anticipation of the possibility of needing to reduce CO_2 emissions in the decades ahead. The idea is that it would be cheaper to switch to systems that were better developed as a result of such anticipatory investments. Such proactive forms of adaptation (e.g., building a dam a few meters higher in anticipation of an altered future climate) have been prominent in most subsequent formal assessments of anthropogenic climate change.[63] Nearly all modern integrated assessments explicitly[64] or implicitly[65] attempt to incorporate (mostly passive) adaptation. Although these studies should be applauded for attempting to recognize and evaluate the implications of adaptive responses on the impact costs of climate change scenarios, serious problems with data, theory, and method remain. In particular, analyses must incorporate a wide range of assumptions,[66] and both costs and benefits of climate change scenarios should be presented in the form of statistical distributions based on a wide range of subjective probability estimates of each step in the assessment process.[67]

Guidance on Uncertainties

Attempts to achieve more consistency in assessing and reporting on uncertainties are just beginning to receive increasing attention. However, the scientific complexity of the climate change issue and the need for information that is useful for policy formulation present a large challenge to researchers and policymakers alike; both groups must work together toward improved communication of uncertainties. The research community must also bear in mind that readers often assume for themselves what they think the authors believe to be the distribution of probabilities when the authors do not specify it themselves. For example, integrated assessment specialists may have to assign probabilities to alternative outcomes (even if only qualitatively specified by natural scientists) because many integrated assessment tools require estimates of the likelihood of a range of events to calculate efficient policy responses. Moss and Schneider[68] argue that it is more rational for experts to provide their best estimates of probability distributions and possible outliers than to have novice users make their own determinations. In particular, a guidance paper on uncertainties commissioned by the IPCC[69] recommends developing an estimate of a probability distribution based on the documented ranges and distributions in the literature, including sources of information on the key causes of uncertainty. An assessment should include a measure of the central tendency (if appropriate) of the distribution as well as a characterization of the end points of the range of outcomes and possible outliers—i.e., the likelihood of outcomes beyond the end points of the range. Truncating the estimated probability distribution should be avoided because this narrows the range of outcomes described and excludes outliers that may include "surprises" and does not convey to potential users a representation of the full range of uncertainty associated with the estimate. It is inappropriate to combine different distributions into one summary distribution if this obscures differences between two (or more) schools of thought. Representing the full distribution has important implications regarding the extent to which the analysis accurately conveys uncertainties.

A projected range is a quantifiable range of uncertainty situated within a population of possible futures that cannot be fully identified (nominated as "knowable" and "unknowable" uncertainties by Morgan et al.).[70] The limits of this total range of uncertainty are unknown but may be estimated subjectively.[71] The inner range represents a well-calibrated range of uncertainty based on doc

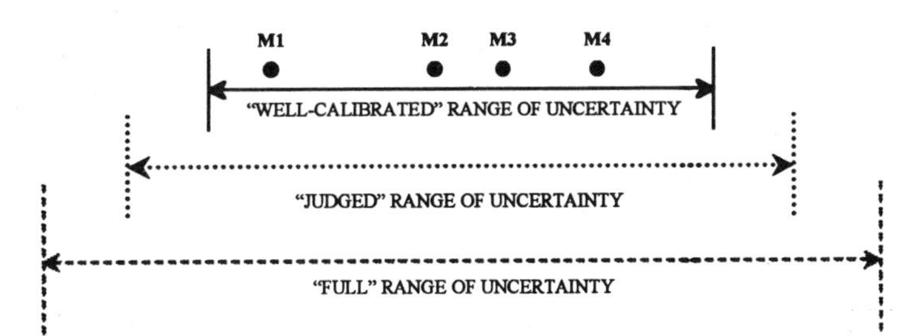

FIGURE 2.2. Schematic depiction of the relationship between "well-calibrated" scenarios, the wider range of "judged" uncertainty that might be elicited through decision analytic survey techniques, and the "full" range of uncertainty, which is drawn wider to represent overconfidence in human judgments. M1 to M4 represent scenarios produced by four models (e.g., globally averaged temperature increases from an equilibrium response to doubled CO_2 concentrations). This lies within a "full" range of uncertainty that is not fully identified, much less directly quantified by existing theoretical or empirical evidence. (Modified from Jones, 2000.)

umented literature. The wider range of uncertainty represents a "judged" range of uncertainty based on expert judgments, which may not encompass the full range of uncertainty given the possibility of cognitive biases such as overconfidence (Fig. 2.2). New information, particularly reliable and comprehensive empirical data, may eventually narrow the range of uncertainty by falsifying certain outlier values.

Aggregation and the Cascade of Uncertainty

A single aggregated damage function or a best-guess climate sensitivity estimate is a very restricted representation of the wide range of beliefs available in the literature or among lead authors, particularly because these estimates rely on a causal chain that includes several different processes. The resultant aggregate distribution might have very different characteristics than the various distributions that make up the links of the chain of causality.[72] Thus, poorly managed projected ranges in impact assessment may inadvertently propagate uncertainty. The process whereby uncertainty accumulates throughout the process of climate change prediction and impact assessment has been variously described as a cascade of uncertainty[73] or an uncertainty explosion.[74] The cascade of uncertainty implied by coupling the separate probability distributions for emissions, biogeochemical cycle calculations needed to calculate radiative forcing, climate sensitivity, climate impacts, and valuation of such impacts in climate damage functions has yet to be produced in the literature.[75] If an assessment is continued through to economic and social outcomes, even larger ranges of uncertainty can be accumulated (Fig. 2.3).

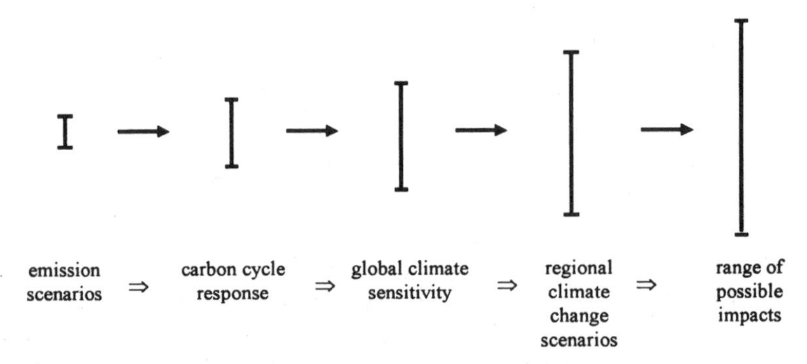

FIGURE 2.3. Range of major uncertainties typical in impact assessments showing the "uncertainty explosion" as these ranges are multiplied to encompass a comprehensive range of future consequences, including physical, economic, social, and political impacts and policy responses. (Modified after Jones, 2000 and the "cascading pyramid of uncertainties" in Schneider, 1983.)

This cascade of uncertainty produces a range of possible outcomes rather than best guesses.

Using Probability Distributions to Evaluate Climate Damage

Many recommendations for modest controls are based on point estimate values, that is, results that are derived from a series of best guesses. This point estimate method fails to account for the wide range of plausible values for many parameters. Similarly, output from a single model run does not display all the information available, nor does it offer sufficient information to provide the insights needed for well-informed policy decisions. Clearly, the use of probabilistic information, even if subjective, provides a much more representative picture of the broad views of the experts and a fairer representation of costs, which, in turn, allows better potential policy insights. The characterization and range of uncertainties of the information provided by decision analysis tools must be made explicit and transparent to policymakers.[76] Policymaking in the business, health, and security sectors often is based on hedging against low-probability but high-consequence outcomes. Thus, any climate policy analysis that represents best guess point values or limited ranges of outcomes limits the ability of policymakers to make strategic hedges against such risky outlier events. The end result of any set of integrated assessment modeling exercises will be the subjective choice of a decision maker,[77] but a more comprehensive analysis with uncertainties in all major components explicitly categorized and displayed should lead to a better-informed choice.[78]

Morgan and Keith[79] and Nordhaus[80] tap the knowledgeable opinions of what they believe to be representative groups of scientists from physical, biological, and social sciences on two separate questions: the climate science itself and policy-relevant impact assessment. In the Morgan and Keith study, 16 scientists were interviewed to elicit their subjective probability estimates for a number of factors, including the climate sensitivity factor (i.e., the increase in global mean temperature for a doubling of CO_2). The Morgan and Keith survey shows that although there is a wide divergence of opinion, nearly all scientists assign some probability of negligible outcomes and some probability of highly serious outcomes (Fig. 2.4).

Nordhaus[81] conducted a survey of conventional economists, environmental economists, atmospheric scientists, and ecologists to assess expert opinion on estimated climate damages. Interestingly, the survey reveals a striking cultural divide between natural and social scientists in the study. The most striking difference in the study is that conventional economists believe that even extreme climate change (i.e., 6°C warming by 2105) would not impose severe economic losses. Natural scientists' estimates of the economic impact of extreme climate change

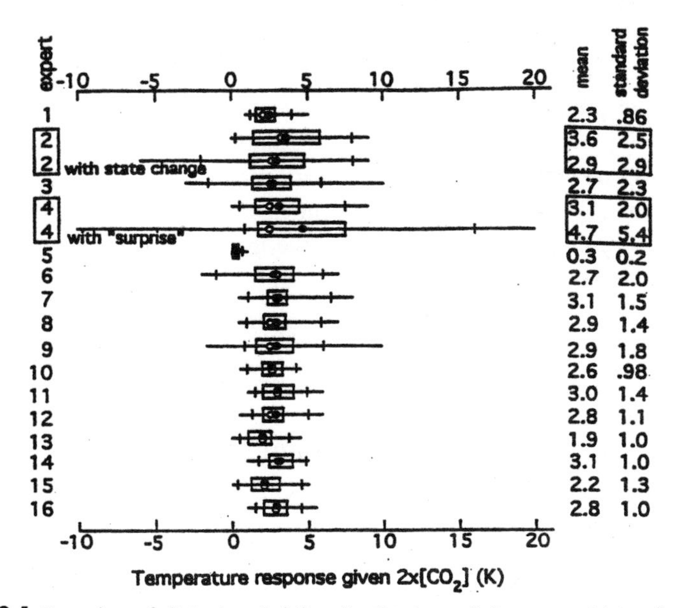

FIGURE 2.4. Box plots of elicited probability distributions of climate sensitivity, the change in globally averaged surface temperature for a doubling of CO_2 (2x[CO_2] forcing). Horizontal line denotes range from minimum (1%) to maximum (99%) assessed possible values. Vertical tick marks indicate locations of lower (5) and upper (95) percentiles. Box indicates interval spanned by 50% confidence interval. Solid dot is the mean and open dot is the median. The two columns of numbers on right hand side of the figure report values of mean and standard deviation of the distributions. (From Morgan & Keith, 1995.)

are 20 to 30 times higher than conventional economists'.[82] Despite the magnitude in difference of damage estimates between economists and ecologists, the shape of the damage estimate curve was similar. The respondents indicate accelerating costs with higher climate changes. Most respondents, economists and natural scientists alike, offer right-skewed subjective probability distributions. That is, most of the respondents consider the probability of severe climate damage ("nasty surprises") to be higher than the probability of moderate benefits ("pleasant surprises"). Roughgarden and Schneider[83] demonstrate that adopting such right-skewed probability distributions into integrated assessment models produces optimal carbon taxes several times higher than point estimates. The long, heavy tails of the skewed distribution (which Roughgarden and Schneider label "surprise") pull the median and means of the distribution away from the mode. Figure 2.5 shows this right skewness clearly for the Nordhaus survey.

We will not easily reconcile the optimistic and pessimistic views of these specialists with different training, traditions, and world views. One thing that is clear from the Morgan and Keith and the Nordhaus studies is that most knowledgeable experts from a variety of fields admit to a wide range of plausible outcomes in the area of climate change, including both mild and catastrophic outcomes. This condition is ripe for misinterpretation by those who are unfamiliar with the wide range of probabilities most scientists attach to climate change issues. The wide range of probabilities follows from recognition of the many uncertainties in data and assumptions still inherent in climate models, climatic impact models, economic models, or their synthesis via integrated assessment models.[84] In a highly interdisciplinary enterprise such as the integrated assessment of climate change, it is necessary to include a wide range of possible outcomes along with a

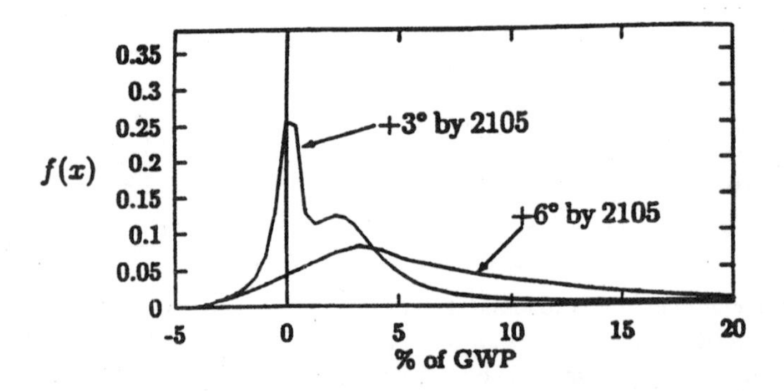

FIGURE 2.5. Probability distributions ($f(x)$) of climate damages as a percentage of gross world product (market and non-market components combined) from an expert survey in which respondents were asked to estimate 10th, 50th, and 90th percentiles for the two climate change scenarios shown. (From Roughgarden & Schneider, 1999. Data from Nordhaus, 1994a.)

representative sample of the subjective probabilities that knowledgeable assessment groups believe accompany each of those possible outcomes.

In essence, the "bottom line" of estimating climatic impacts is that extremely optimistic and pessimistic projections are the two lowest-probability outcomes (see Fig. 2.5) and that most knowledgeable scientists and economists consider there to be a significant chance of climatic damage to both natural and social systems. Under conditions of persistent uncertainty it is not surprising that most formal climatic impact assessments have called for cautious but positive steps to slow down the rate at which humans modify the climatic system and to make natural and social systems more resilient to whatever changes eventually materialize.[85]

Using Scenarios to Develop a Plausible Range of Outcomes

The IPCC commissioned a Special Report on Emission Scenarios (SRES)[86] both to broaden assessments to include a range of outcomes and to focus analysis on a coherent set of scenario outcomes to facilitate comparison. The scenarios concentrate on assumptions about economic growth, technological developments, and population growth, arguably the three most critical variables affecting the uncertainty over future climate change and policy options. To the extent possible, the Third Assessment Report (TAR)[87] has referred to the SRES to inform and guide the assessment. Box 2.2 describes the baseline SRES scenarios; Fig. 2.6 demonstrates how the SRES scenarios have been used to evaluate projected temperature changes.[88] However, IPCC did not assign subjective

BOX 2.2. The Emission Scenarios of the Special Report on Emission Scenarios (SRES)

A1: The A1 storyline and scenario family describes a future world of very rapid economic growth, global population that peaks in mid-century and declines thereafter, and the rapid introduction of new and more efficient technologies. Major underlying themes are convergence between regions, capacity building, and increased cultural and social interactions, with a substantial reduction in regional differences in per capita income. The A1 scenario family develops into three groups that describe alternative directions of technological change in the energy system. The three A1 groups are distinguished by their technological emphasis: fossil intensive (A1FI), non–fossil energy sources (A1T), or a balance across all sources (A1B) (where balance is defined as not relying too heavily on one particular energy source, on the assumption that similar improvement rates apply to all energy supply and end use technologies).

BOX 2.2. *Continued*

A2: The A2 storyline and scenario family describes a very heterogeneous world. The underlying theme is self-reliance and preservation of local identities. Fertility patterns across regions converge very slowly, which results in continuously increasing population. Economic development is primarily regionally oriented, and per capita economic growth and technological change are more fragmented and slower than in other storylines.

B1: The B1 storyline and scenario family describes a convergent world with the same global population (which peaks in midcentury and declines thereafter) as in the A1 storyline but with rapid change in economic structures toward a service and information economy, with reductions in material intensity and the introduction of clean and resource-efficient technologies. The emphasis is on global solutions to economic, social, and environmental sustainability, including improved equity but without additional climate initiatives.

B2: The B2 storyline and scenario family describes a world in which the emphasis is on local solutions to economic, social, and environmental sustainability. It is a world with continuously increasing global population (at a rate lower than in A2), intermediate levels of economic development, and less rapid and more diverse technological change than in the B1 and A1 storylines. Although the scenario is also oriented toward environmental protection and social equity, it focuses on local and regional levels.

An illustrative scenario was chosen for each of the six scenario groups A1B, A1FI, A1T, A2, B1, and B2 represented in Fig. 2.6. The SRES authors consider the scenarios equally sound, which offers no guidance on which scenarios are more or less likely. A subjective probability assessment of the likelihood of the scenarios would offer policymakers a useful characterization of which scenarios may entail dangerous outcomes.

The SRES scenarios do not include additional climate initiatives, which means that no scenarios are included that explicitly assume implementation of the United Nations Framework Convention on Climate Change or the emission targets of the Kyoto Protocol or any next generation agreements.

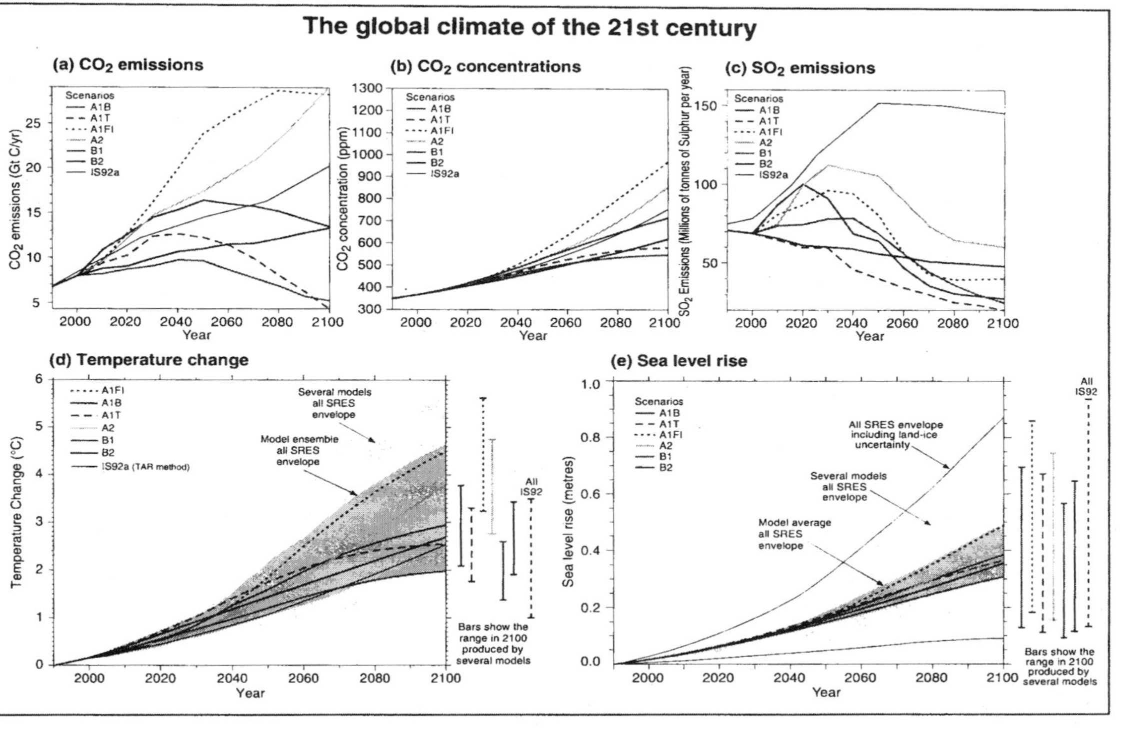

FIGURE 2.6. The global climate of the twenty-first century will depend on natural changes and the response of the climate system to human activities. Climate models project the response of many climate variables—such as increases in global surface temperature and sea level—to various scenarios of greenhouse gas and other human-related emissions. (a) CO_2 emissions of the six illustrative SRES scenarios, summarized Box 2.2, along with IS92a for comparison purposes with the SAR. (b) Projected CO_2 concentrations. (c) Anthropogenic SO_2 emissions. Emissions of other gases and other aerosols were included in the temperature change model but are not shown in the figure. (d), (e) The projected temperature and sea level responses, respectively. The "several models all SRES envelope" in (d) and (e) shows the temperature and sea level rise, respectively, for the simple model when tuned to a number of complex models with a range of climate sensitivities. The "all SRES envelopes" refer to the full range of 35 SRES scenarios. The "model average all SRES envelope" shows the average from these models for the range of scenarios. Note that the warming and sea level rise from these emissions would continue well beyond 2100. Also note that this range does not allow for uncertainty relating to ice dynamic changes in the West Antarctic ice sheet, nor does it account for uncertainties in projecting nonsulfate aerosols and greenhouse gas concentrations. (From IPCC, 2001a, Working Group I Summary for Policymakers, available online at http://www.ipcc.ch.)

probabilities to the SRES scenarios or to various climate model uncertainties, making it difficult for policymakers to compare risks or evaluate tradeoffs.[89]

Policy Implications

What Are Some Actions to Consider?

Given an uncertain environment with respect to our knowledge about the science of climate change, the impacts of climate change, and the effects of policy actions, what are reasonable policy options to mitigate or adapt to climate change? We suggest several options that collectively or separately will help to manage this uncertainty while assessing and addressing climate change.

Focus on Win–Win Strategies

Paramount is the need to pursue a climate policy with significant "co-benefits" that address other policy objectives. Despite the widespread agreement that at least some climatic change is inevitable, that major change is quite possible, and that most of the world will experience net effects that are more likely to be negative than positive, particularly if global warming is allowed to increase beyond a few degrees (which is likely to occur after the mid–twenty-first century if no policies are undertaken to mitigate emissions), many more pressing concerns critical to human health and well-being are competing for attention. Many countries are struggling to raise literacy rates, lower death rates, increase life expectancy, provide employment for burgeoning populations, and reduce local air and water pollution, which pose imminent health hazards to their citizens and environments. These demands are concrete, imminent, and vital to human welfare. In contrast, costs imposed by climate change often are diffuse, delayed, and intangible. Uncertainty about the consequences of climate change only exacerbates the problem. Slowing climate change is simply a low priority for many countries, even if it would be efficient to do so. It is unfortunate that less developed countries, in particular, place a low priority on the abatement of global climate change despite the fact that nearly all impact assessments suggest that it is these very countries that are most vulnerable to climatic change.[90] Furthermore, climate change probably will exacerbate these existing stresses. Policy responses to climate change, including both mitigation and adaptation, are more likely to succeed if they are linked to or integrated with policies designed to address nonclimatic stresses. Part of the assessment should include not only weighing competing risks and priorities against the costs of climate policy options but also considering how policies to address competing objectives may complement each other.

Understandably, policymakers place an emphasis on identifying no-regrets policies (measures that demonstrate positive net benefits) and co-benefits of climate policies (secondary benefits from climate policy options that also meet other policy objectives, such as reducing air and water pollution), both of which suggest linkages to other policy objectives. In addition to direct effects on problems other than greenhouse gas emissions, such as reducing air and water pollution, co-benefits of climate change policies may also include indirect effects on transportation, agriculture, land use practices, employment, and fuel security. Co-benefits may be experienced in the other direction as well; climate change mitigation may be an ancillary benefit of other policies. For example, a low greenhouse gas emission scenario could result from a sustainable development policy. Forest preservation is a particularly important, contemporary example of how accounting for co-benefits affects the value of policy options.

By current estimates, tropical deforestation accounts for 20 to 30 percent of carbon emissions. Clearly, protecting primary forests is a preferred global climate policy. However, conflicts between global, local, and national interests can undermine support for conservation. For example, the opportunity costs of the economic alternatives to the Masoala National Park Integrated Conservation and Development Program reveal that at the national level, industrial logging was the preferred option, despite the tremendous benefits of the conservation program to the local community (conservation yields local benefits greater than the slash-and-burn alternative).[91] It behooves the international community to support conservation efforts because of the tremendous *global* economic (and intrinsic) value of these forests. Paying national constituencies to preserve the Masoala forests would safeguard a valuable carbon sink at a low cost. Note that this is a value in addition—sometimes called a "double dividend"—to protecting biodiversity and ecosystem services. Despite the tremendous uncertainty regarding climate change and its implications for human welfare, all parties to the climate negotiations should recognize that potential damages to a global commons such as the earth's climate are not mere ideological rhetoric. Policies to mitigate the effects of climate change are not cost free, but we should emphasize win–win solutions in which economic efficiency, cost-effectiveness, equity in the distributional impacts, and environmental protection can coexist.[92] Emphasizing the co-benefits of climate policy for other policy priorities can promote multiple objectives and secure support for mitigating climate change.[93]

Sensitivity Studies Are Essential

It is unlikely that all important uncertainties in either climatic or social and environmental impact assessment models will be resolved to the satisfaction of most of the scientific community in the near future. However, this does not imply that model results are uninformative. On the contrary, sensitivity analyses in which various policy-driven alternative radiative forcing assumptions are made and the consequences of these assumptions compared can offer insights into the potential effectiveness of such policies in terms of their differential climatic effects and impacts.[94] Even though absolute accuracy is not likely to be claimed for the foreseeable future, greater precision concerning the sensitivity of the physical and biological subsystems of the earth can be obtained via carefully planned and executed sensitivity studies across a hierarchy of models.

Validation and Testing Are Needed

Although it may be impractical, if not theoretically impossible, to validate the precise future course of climate given the uncertainties that remain in scenarios of emissions and land use changes, internal dynamics, and surprise events, many of the basic features of the coupled physical and biological subsystems of the earth already can be simulated. Testing models against each other when driven by the same sets of climate scenarios, testing the overall simulation skill of models against empirical observations, testing model parameterizations against high-resolution process models or data sets, testing models against proxy data of paleoclimatic changes, and testing the sensitivity of models to anthropogenic radiative forcings by computing their sensitivity to natural radiative forcings (e.g., seasonal radiative forcing, volcanic dust forcing, orbital element variation forcings, meltwater-induced rapid ocean current changes) make up a necessary set of validation-oriented exercises that all modelers should agree to perform. Impact assessment models should also be subjected to an analogous set of validation protocols (e.g., testing model projections against actual storm damage) to increase the credibility of their results. Similarly, economic models can be tested to see how they perform when simulating such shocks as the OPEC oil embargoes or the free trade agreements implementation. Further analysis should focus on systematically extending and evaluating existing assessment models to gauge the range of outcomes and their sensitivity to a variety of specification assumptions.

Finally, the most complex and difficult testing challenge is to fashion methods to test the behavior of emergent properties of coupled physical, biological, and social scientific submodels because the behavior of such highly integrated socioecological models is what most matches the complexity of the world we live in. The best suggestion we can offer here is that a hierarchy of models of increasing complexity be compared first against each other and then against data at as many scales as possible.[95] As the hierarchy is expanded and more testing protocols implemented, the confidence of the scientific community in the credibility of such modeling of the dynamics of the socioecological system will increase.

Incorporate Subjective Probability Assessment

In addition to standard simulation modeling exercises in which various parameters are specified or varied over an uncertainty range, formal decision analytic techniques can be used to provide a more consistent set of values for uncertain model parameters or functional relationships.[96] The embedding of subjective probability distributions into climatic models is just beginning[97] but may become an important element of integrated assessment modeling in future generations of model building.[98]

Provide for "Rolling Reassessment"

Changes in environmental and societal systems and our understanding of them will certainly occur over the next few decades. Under these circumstances, flexible management of global commons such as the earth's climate seems necessary to incorporate new discoveries. Therefore, a series of assessments of climatic effects, related impacts, and policy options to prevent potentially dangerous impacts will be needed periodically—perhaps every five years, as IPCC has chosen for the repeat period of its major Assessment Reports, which consider climatic effects, impacts, and policy issues. Whatever policy instruments are used (either mitigative or adaptive) must be flexible enough to respond quickly and cost-effectively to the evolving science that will emerge from this rolling reassessment process.

Some politicians are reluctant to revisit politically contentious issues every five years or so and prefer to "solve" them once and for all. Although that is a politically more palatable strategy for some, it is certain to be less efficient than flexible management given the high probability that new information will reduce some risks currently believed to be potentially serious and elevate others not now perceived as dangerous. Learning to live with changing assessments and flexible management instruments will be a hallmark of environmental debates in the twenty-first century.

Consider Surprises and Irreversibility

Given the many uncertainties that still attend most aspects of the climate change debate, priority should be given to the aspects that could exhibit irreversible damages (e.g., extinction of species whose already-shrinking habitat is further stressed by rapid climatic changes) or for which imaginable "surprises"[99] have been identified (e.g., changes in oceanic currents caused by rapid increases in greenhouse gases).[100] For these reasons, management of climatic risks must be considered well in advance of more certain knowledge of climatic effects and impacts.

Promote Environmentally Friendly Technologies

Schneider and Goulder[101] show that current policy actions, such as imposing a moderate carbon tax, are urgently needed to induce the technological innovations assumed in economic cost-effectiveness studies. In other words, policy actions to help induce technological changes (e.g., through research and development or "learning by doing") are needed now to promote cost-effective abatement in the decades ahead.[102]

Controversy will remain, of course, because total emissions are the product of world population size, per capita economic output, and the activities that produce that economic output. Technological innovations to reduce emissions are less controversial than social policies, which affect factors such as population and economic growth or consumption patterns. Thus, incentives for technology development and deployment are likely to be the focus of climate policy for the immediate future. Social factors eventually will need to be considered if very large human impacts on the environment are to be averted.[103]

Consider Carbon Management Alternatives

Two broad classes of carbon management can be distinguished. The first includes attempts to manipulate natural biogeochemical processes of carbon removal—so-called carbon sinks—such as adding iron to the oceans to enhance uptake of carbon by the resulting blooms of phytoplankton, planting vast forests of fast-growing trees to sequester carbon, or altering agricultural practices to increase carbon storage in soils.[104] The second kind of carbon management stresses prevention of carbon emissions that otherwise would have been directly injected into the atmosphere, including preservation of primary forests that otherwise might have been cut down (also helping to preserve biodiversity); industrial processing to increase the hydrogen content and remove carbon from fuels such as coal or methane, injecting the carbon into (hopefully stable) reservoirs for long-term storage; and using less carbon-intensive energy supply systems and improving energy efficiency. Keith[105] suggests that the dividing line between geoengineering (carbon management through deliberate modification of biogeochemical cycles) and mitigation (carbon management through prevention of carbon emission release to the atmosphere) occurs when the technology acts by counterbalancing an anthropogenic forcing rather than by reducing it.

Conclusions

We have argued that nonlinearities and the likelihood of rapid, unanticipated events ("surprises") require that integrated assessment methods use a wide range of estimates for key parameters or structural formulations and that, when possible, results be cast in probabilistic terms rather than central tendencies because the latter mask the wide range of policy relevant results. We have also argued that the underlying structural assumptions and parameter ranges be explicit to make the conclusions as transparent as possible. For example, although it is often acknowledged that a wide range of uncertainty accompanies estimates of climate damages from scenarios of anthropogenic climatic change (because of uncertainties in adaptation capacity, synergistic impacts, and so on), it is less common[106] to have a comparably wide set of estimates for mitigation costs of carbon policies (e.g., a carbon tax being a common analytic benchmark). Yet the tighter range of mitigation cost estimates occurs in part because standard costing methods make common assumptions about the lack of preexisting market failures or do not explicitly account for the possibility of climate policy–induced technological changes reducing mitigation cost estimates.[107]

Moreover, in view of the wide range of plausible climatic change scenarios available in the literature—including a growing number of rapid non-linear change projections—it is important for costing analyses to consider many such scenarios, including the implications of rapid changes in emissions triggering nonlinear climatic changes with potentially significant implications for costing.[108]

In short, the key is transparency of assumptions and the use of as wide a range of eventualities (and their attendant probabilities) as possible to help decision makers become aware of the arguments for flexibility of policy options.

Acknowledgments

This chapter is modified from Schneider, Turner, and Morehouse Garriga; Schneider, 2002b;[109] and Schneider et al., 2000.[110] We gratefully acknowledge the comments by Robert van der Zwan.

We also gratefully acknowledge the partial support of the Winslow Foundation, which has made this contribution possible.

Kristin Kuntz–Duriseti is in the department of political science at the University of Michigan and is also associated with the Institute for International Studies and Biological Sciences at Stanford University. She gratefully acknowledges support from the Winslow Foundation for part of this effort.

Notes

1. Farman, J. C., B. G. Gardiner, and J. D. Shanklin, 1985: "Larger losses of total ozone in Antarctica reveal seasonal ClO_x/NO_x interaction," *Nature,* 315: 207–210.
2. For example, Root, T. L. and S. H. Schneider, 1993: "Can large-scale climatic models be linked with multiscale ecological studies?" *Conservation Biology,* 7 (2): 256–270.
3. Intergovernmental Panel on Climate Change (IPCC), 2000: *Special Report: Emissions Scenarios (SRES),* Nakicenovic, N., et al. (eds.), A Special Report of Working Group III of the IPCC (Cambridge, England: Cambridge University Press), 599 pp., available online: http://www.ipcc.ch.
4. IPCC, 1996a: *Climate Change 1995: The Science of Climate Change,* contribution of Working Group I to the Second Assessment Report of the IPCC, Houghton, J. T., L. G. Meira Filho, B. A. Callander, N. Harris, A. Kattenberg, and K. Maskell (eds.) (Cambridge, England: Cambridge University Press), 572 pp.; IPCC, 2001a: *Climate Change 2001: The Scientific Basis,* contribution of Working Group I to the Third Assessment Report of the IPCC, Houghton, J. T., Y. Ding, D. J. Griggs, M. Noguer, P. J. van der Linden, and D. Xiaosu (eds.) (Cambridge, England: Cambridge University Press), 881 pp.
5. IPCC, 2001b: *Climate Change 2001: Impacts, Adaptations, and Vulnerability,* contribution of Working Group II to the Third Assessment Report of the IPCC, McCarthy, J. J., O. F. Canziani, N. A. Leary, D. J. Dokken, K. S. White (eds.) (Cambridge, England: Cambridge University Press), 1032 pp.
6. Smith, J. and D. Tirpak (eds.), 1988: *The Potential Effects of Global Climate Change on the United States: Draft Report to Congress.* Vols. 1 and 2. U.S. Environmental Protection Agency, Office of Policy Planning and Evaluation, Office of Research and Development (Washington, DC: U.S. Government Printing Office); IPCC, 1996b: *Climate Change 1995. Impacts, Adaptations and Mitigation of Climate Change: Scientific–Technical Analyses,* contribution of Working Group II to the Second Assessment Report of the Intergovernmental Panel on Climate Change. R. T. Watson, M. C. Zinyowera, and R. H. Moss (eds.) (Cambridge, England: Cambridge University Press), 878 pp; IPCC, 2001b.
7. Nordhaus, W. D., 1994a: "Expert opinion on climatic change," *American Scientist* (January issue), 82: 45–52.
8. For example, Brown, P., 1997: "Stewardship of climate," an editorial for *Climatic Change,* 37 (2): 329–334.
9. Wynne, B., 1992: "Uncertainty and environmental learning: reconceiving science and policy in the preventive paradigm," *Global Environmental Change,* 20: 111–127; quote on p. 111.
10. From Moss, R. H. and S. H. Schneider, 2000: "Uncertainties in the IPCC TAR: recommendation to lead authors for more consistent assessment and reporting," in R. Pachauri, T. Taniguchi, and K. Tanaka (eds.), *Third Assessment Report: Cross Cutting Issues Guidance Papers* (Geneva: World Meteorological Organisation), pp. 33–51. Available on request from the Global Industrial and Social Progress Institute at http://www.gispri.or.jp.

11. Holling, C. S., 1973: "Resilience and stability of ecological systems," *Annual Review of Ecology and Systematics,* 4: 1–23; Holling, C. S., 1986: "The resilience of terrestrial ecosystems: local surprise and global change," in W. C. Clark and R. E. Munn (eds.), *Sustainable Development of the Biosphere* (Cambridge, England: Cambridge University Press for the International Institute for Applied Systems Analysis), pp. 292–317.
12. For example, see Ayyub, B. M., M. M. Gupta, and L. N. Kanal (eds.), 1992: *Analysis and Management of Uncertainty: Theory and Applications* (Amsterdam: North-Holland Publishers), 428 pp; Tonn, B., 1991: "Research policy and review 34: the development of ideas of uncertainty representation," *Environment and Planning,* 23: 783–812; Yager, R. R., 1992: "Decision making under Dempster–Shafer uncertainty," *International Journal of General Systems,* 20: 233–245.
13. For example, Zadeh, L. A., 1965: "Fuzzy sets," *Information and Control,* 8 (3): 338–353; Zadeh, L. A., 1990: "The birth and evolution of fuzzy logic," *International Journal of General Systems,* 17: 95–105.
14. For example, Casti, J. L., 1994: *Complexification: Explaining a Paradoxical World Through the Science of Surprise* (New York, Harper Collins), 320 pp.
15. See any undergraduate microeconomics textbook for a fuller exposition, e.g., Varian, H., 1992: *Intermediate Microeconomics,* 3rd ed. (New York: W. W. Norton), 506 pp.
16. For example, Chapter 1 and Summary for Policymakers, IPCC, 1996a; Casti, 1994; Darmstadter, J. and M. A. Toman (eds.), 1993: *Assessing Surprises and Nonlinearities in Greenhouse Warming. Proceedings of an Interdisciplinary Workshop* (Washington, DC: Resources for the Future); Broecker, W. S., 1994: "Massive iceberg discharges as triggers for global climate change," *Nature,* 372 (1): 421–424.
17. See IPCC, 1996a, p 7.
18. Rahmstorf, S., 2000: "The thermohaline ocean circulation: a system with dangerous thresholds? An editorial comment," *Climatic Change,* 46 (3): 247–256.
19. For discussions and a review of the literature, see Schneider, S. H., B. L. Turner, and H. Morehouse Garriga, 1998: "Imaginable surprise in global change science," *Journal of Risk Research,* 1 (2): 165–185.
20. Mearns, L. O., R. W. Katz, and S. H. Schneider, 1984: "Extreme high temperature events: changes in their probabilities and changes in mean temperature," *Journal of Climate and Applied Meteorology,* 23: 1601–1613.
21. See IPCC, 2001a.
22. Karl, T. R., and R. W. Knight, 1998: "Secular trends of precipitation amount, frequency, and intensity in the U.S.A.," *Bulletin of the American Meteorological Society,* 79 (2): 231–241.
23. See IPCC, 2001b.
24. Dessens, J., 1995: "Severe convective weather in the context of a nighttime global warming," *Geophysical Research Letters,* 22 (10): 1241–1244.
25. Reeve, N. and R. Toumi, 1999: "Lightning activity as an indicator of climate change," *Quarterly Journal of the Royal Meteorological Society,* 125: 893–903.
26. Emanuel, K. A., 1987: "The dependence of hurricane intensity on climate," *Nature,* 326: 483–485; Walsh, K. and A. B. Pittock, 1998: "Potential changes in tropical

storms, hurricanes, and extreme rainfall events as a result of climate change," *Climatic Change,* 39: 199–213; IPCC, 2001b.

27. Several authors examine possible radical changes in global ocean currents, such as Manabe S., and R. J. Stouffer, 1993: "Century scale-effects of increased atmospheric CO_2 on the ocean–atmosphere system," *Nature,* 364: 215–218; Haywood, J. M., R. J. Stouffer, R. T. Wetherald, S. Manabe, and V. Ramaswamy, 1997: "Transient response of a coupled model to estimated changes in greenhouse gas and sulfate concentrations," *Geophysical Research Letters,* 24 (11): 1335–1338; Rahmstorf, S., 1999: "Shifting seas in the greenhouse?" *Nature,* 399: 523–524.
28. Stocker, T. F. and A. Schmittner, 1997. "Influence of CO_2 emission rates on the stability of the thermohaline circulation," *Nature,* 388: 862–864.
29. Thompson, S. L. and S. H. Schneider, 1982: "CO_2 and climate: the importance of realistic geography in estimating the transient response," *Science,* 217: 1031–1033.
30. Schneider, S. H. and S. L. Thompson, 2000. "A simple climate model used in economic studies of global change," in S. J. DeCanio, R. B. Howarth, A. H. Sanstad, S. H. Schneider, and S. L. Thompson, *New Directions in the Economics and Integrated Assessment of Global Climate Change* (Washington, DC: The Pew Center on Global Climate Change), pp. 59–80.
31. Mastrandrea, M. and S. H. Schneider, 2001: "Integrated assessment of abrupt climatic changes," *Climate Policy,* 1: 433–449.
32. Schneider and Thompson, 2000.
33. Mastrandrea and Schneider, 2001.
34. Ibid.
35. Alexander, S. E., S. H. Schneider, and K. Lagerquist, 1997: "The interaction of climate and life," in G. C. Daily (ed.), *Nature's Services: Societal Dependence on Natural Ecosystems* (Washington, DC: Island Press), pp. 71–92.
36. IPCC, 1996a, 2001b.
37. Vellinga, P., E. Mills, G. Berz, S. Huq, L. Kozak, J. Palutikof, B. Schanzenbächer, S. Shida, and G. Soler, 2001. "Insurance and other financial services," in IPCC 2001b, pp. 417–450.
38. Pielke, R. A., Jr., and C. W. Landsea, 1998: "Normalized hurricane damages in the United States, 1925–97," *Weather Forecasting,* 13: 351–361.
39. For example, see Alexander et al., 1997; Titus, J. and V. Narayanan, 1996: "The risk of sea level rise: A Delphic Monte Carlo analysis in which twenty researchers specify subjective probability distributions for model coefficients within their respective areas of expertise," *Climatic Change,* 33 (2): 151–212; Yohe, G., J. Neumann, P. Marshall, and H. Ameden, 1996: "The economic cost of greenhouse induced sea level rise for developed property in the United States," *Climatic Change,* 32 (4): 387–410; Yohe, G., 1989: "The cost of not holding back the sea: economic vulnerability," *Ocean and Shoreline Management,* 15: 233–255.
40. Daily, G. C., 1997: *Nature's Services: Societal Dependence on Natural Ecosystems* (Washington, DC: Island Press), 392 pp.
41. Root, T. L. and S. H. Schneider, 1993: "Can large-scale climatic models be linked with multiscale ecological studies?" *Conservation Biology,* 7 (2): 256–270; Root, T. L. and S. H. Schneider, 2002: "Climate change: Overview and implications for

wildlife," in Schneider, S. H. and T. L. Roots (eds.), *Wildlife Responses to Climate Change: North American Case Studies,* National Wildlife Federation, (Washington, DC: Island Press), pp 1–55.

42. Nordhaus, W. D., 1994b: *Managing the Global Commons: the Economics of Climate Change* (Cambridge, MA: MIT Press); Manne, A. S., R. Mendelsohn, and R. G. Richels, 1995: "MERGE: a model for evaluating regional and global effects of GHG reduction policies," *Energy Policy,* 23: 17–34.
43. Cline, W., 1992: *The Economics of Global Warming* (Washington, DC: Institute of International Economics), 399 pp; Azar, C. and T. Sterner, 1996: "Discounting and distributional considerations in the context of climate change," *Ecological Economics,* 19: 169–185; Hasselmann, K., S. Hasselmann, R. Giering, V. Ocaña, and H. von Storch, 1997: "Sensitivity study of optimal CO_2 emission paths using a simplified structural integrated assessment model (SIAM)," *Climatic Change,* 37 (2): 345–386; Schultz P. A. and J. F. Kasting, 1997: "Optimal reductions in CO_2-emissions," *Energy Policy,* 25: 491–500; Mastrandrea and Schneider, 2001; Lind, R. C. (ed.), 1982: *Discounting for Time and Risk in Energy Policy* (Washington, DC: Resources for the Future), pp. 257–271.
44. For an overview of the discounting debate, see discussion in Portney, P. R. and J. P. Weyant (eds.), 1999: *Discounting and Intergenerational Equity* (Washington, DC: Resources for the Future), 186 pp.
45. Arrow, K. J., W. Cline, K. G. Mäler, M. Munasinghe, R. Squitieri, and J. Stiglitz, 1996: "Intertemporal equity, discounting and economic efficiency," in J. P. Bruce, H. Lee, and E. F. Haites (eds.), *Climate Change 1995: Economic and Social Dimensions of Climate Change. Second Assessment of the Intergovernmental Panel on Climate Change* (Cambridge, England: Cambridge University Press), pp. 125–144; Nordhaus, W. D., 1997: "Discounting in economics and climate change" (editorial), *Climatic Change,* 37 (2): 315–328; Markandya, A. and K. Halsnaes, 2000: "Costing methodologies," in R. Pachauri, T. Taniguchi, and K. Tanaka (eds.), *The Third Assessment Report: Cross Cutting Issues Guidance Papers* (Geneva: World Meteorological Organization), pp. 15–31. Available on request from the Global Industrial and Social Progress Institute at http://www.gispri.or.jp.
46. Ainslie, G., 1991: "Derivation of 'rational' economic behaviour from hyperbolic discount curves," *American Economic Association Papers and Proceedings,* 81 (2): 334–340.
47. Heal, G., 1997. "Discounting and climate change, an editorial comment," *Climatic Change,* 37 (2): 335–343, quote on page 339.
48. Ibid.
49. Azar, C. and T. Sterner, 1996; Portney and Weyant, 1999.
50. Azar, C. and O. Johansson, 1996: *Uncertainty and Climate Change—or the Economics of Twin Peaks,* presented at the annual conference of European Environmental and Resource Economists, June 1996, Lisbon.
51. Solow, R. M., 1974a: "Intergenerational equity and exhaustible resources," *Review of Economic Studies* (Symposium), 29–45; Solow, R. M., 1974b. "The economics of resources or the resources of economics," *American Economic Review,* 64 (2): 1–14; Lind, 1982; Richard Howarth, personal communication, 2000.

52. Howarth, R. B., 2000: "Climate change and international fairness," in S. J. DeCanio, R. B. Howarth, A. H. Sanstad, S. H. Schneider, and S. L. Thompson, *New Directions in the Economics and Integrated Assessment of Global Climate Change* (Washington, DC: The Pew Center on Global Climate Change), pp. 43–58; Gerlagh, R. and B. C. C. van der Zwaan, 2001: "The effects of aging and an environmental trust fund in an overlapping generations model on carbon emission reductions," *Ecological Economics,* 36 (2): 311–326.
53. Manne, A. S. and G. Stephan, 1999: "Climate-change policies and intergenerational rate-of-return differentials," *Energy Policy,* 27 (6): 309–316.
54. For example, see references in Rosenzweig, C. and M. Parry, 1994: "Potential impact of climate change on world food supply," *Nature,* 367: 133–138; Smith and Tirpak, 1988.
55. Rosenberg, N. J. and M. J. Scott, 1994: "Implications of policies to prevent climate change for future food security," *Global Environmental Change,* 4: 49–62.
56. Mendelsohn, R., W. Nordhaus, and D. Shaw, 1994: "The impact of global warming on agriculture: a Ricardian analysis," *The American Economic Review,* 84 (4): 753–771.
57. Reilly, J., W. Baethgen, F. E. Chege, S. C. van de Geijn, L. Erda, A. Iglesias, G. Kenny, D. Patterson, J. Rogasik, R. Rötter, C. Rosenzweig, W. Sombroek, J. Westbrook, D. Bachelet, M. Brklacich, U. Dämmgen, M. Howden, R. J. V. Joyce, P. D. Lingren, D. Schimmelpfennig, U. Singh, O. Sirotenko, and E. Wheaton, 1996. "Agriculture in a changing climate: impacts and adaptation," in R. T. Watson, M. C. Zinyowera, and R. H. Moss (eds.), Intergovernmental Panel on Climate Change (IPCC), 1996: *Climate Change 1995. Impacts, Adaptations and Mitigation of Climate Change: Scientific–Technical Analyses,* contribution of Working Group II to the Second Assessment Report of the IPCC (Cambridge, England: Cambridge University Press), pp. 427–467.
58. Haneman, W. M., 2000: "Adaptation and its measurement. An editorial comment," *Climatic Change,* 45 (3–4): 571–581; quote on p. 575.
59. Schneider, S. H., 1997: "Integrated assessment modeling of global climate change: transparent rational tool for policy making or opaque screen hiding value-laden assumptions?" *Environmental Modeling and Assessment,* 2 (4): 229–248.
60. Ehrlich, P. R., A. H. Ehrlich, and G. Daily, 1995: *The Stork and the Plow* (New York: Putnam).
61. For example, Kaiser, H. M., S. Riha, D. Wilks, D. Rossiter, and R. Sampath, 1993: "A farm-level analysis of economic and agronomic impacts of gradual climate warming," *American Journal of Agricultural Economics,* 75: 387–398; Schneider, S. H., 1996: "The future of climate: potential for interaction and surprises," in T. E. Downing (ed.), *Climate Change and World Food Security* (Heidelberg: Springer-Verlag), NATO ASI Series 137: 77–113; Morgan, G. and H. Dowlatabadi, 1996: "Learning from integrated assessment of climate change," *Climatic Change,* 34 (3–4): 337–368; Kolstad, C. D., D. L. Kelly, and G. Mitchell, 1999: *Adjustment Costs from Environmental Change Induced by Incomplete Information and Learning,* Department of Economics, UCSB Working Paper; Schneider, S. H., W. E. Easter-

ling, and L. Mearns, 2000: "Adaptation: Sensitivity to natural variability, agent assumptions and dynamic climate changes," *Climatic Change,* 45 (1): 203–221.

62. Schneider, S. H. and S. L. Thompson, 1985. "Future changes in the atmosphere," in R. Repetto (ed.), *The Global Possible* (New Haven, CT: Yale University Press), pp. 397–430.
63. For example, National Academy of Sciences (NAS), 1991: *Policy Implications of Greenhouse Warming: Mitigation, Adaptation, and the Science Base.* Panel on Policy Implications of Greenhouse Warming, Committee on Science, Engineering, and Public Policy (Washington, DC: National Academy Press).
64. For example, Rosenberg, N. J. (ed.), 1993: "Towards an integrated impact assessment of climate change: The MINK study," *Climatic Change* (Special Issue), 24: 1–173; Rosenzweig and Parry, 1994; Reilly et al., 1996.
65. For example, Mendelsohn et al., 1994; Mendelsohn, R., W. Nordhaus, and D. Shaw, 1996: "Climate impacts on aggregate farm value: accounting for adaptation," *Agricultural and Forest Meteorology,* 80: 55–66; Mendelsohn, R., W. Morrison, M. Schlesinger, and N. Andronova, 2000: "Country-specific market impacts of climate change," *Climatic Change,* 45 (3–4): 553–569.
66. For example, as recommended by Carter, T. R., M. L. Parry, H. Harasawa, and S. Nishioka, 1994: *IPCC Technical Guidelines for Assessing Climate Change Impacts and Adaptations. Summary for Policy Makers and a Technical Summary* (London: Department of Geography, University College London, and the Center for Global Environmental Research, National Institute for Environmental Studies, Japan), 59 pp.
67. For example, as advocated by Yohe, G. W., 1991: "Uncertainty, climate change, and the economic value of information: an economic methodology for evaluating the timing and relative efficacy of alternative response to climate change with application to protecting developed property from greenhouse induced sea level rise," *Policy Science,* 24 (3): 245–269; Morgan and Dowlatabadi, 1996; Schneider, 1997.
68. Moss and Schneider, 2000; see also Moss, R. H., 2000: "Cost estimation and uncertainty," *Pacific and Asian Journal of Energy,* 10 (1): 43–62.
69. Moss and Schneider, 2000.
70. Morgan, M. G., M. Henrion, and M. Small, 1990: *Uncertainty: A Guide to Dealing with Uncertainty in Quantitative Risk and Policy Analysis* (New York: Cambridge University Press).
71. Morgan, M. G. and D. W. Keith, 1995: "Subjective judgments by climate experts," *Environmental Science and Technology,* 29: 468A–476A; Nordhaus, 1994a.
72. Modified from Jones, R. N., 2000: "Managing uncertainty in climate change projections–Issues for impact assessment, an Editorial Comment," *Climate Change,* 45 (3–4): 403–419.
73. Schneider, S. H., 1983: "CO_2, climate and society: a brief overview," in R. S. Chen, E. Boulding, and S. H. Schneider (eds.), *Social Science Research and Climate Change: An Interdisciplinary Appraisal* (Boston: D. Reidel), pp. 9–15.
74. Henderson-Sellers, A., 1993: "An Antipodean climate of uncertainty," *Climatic Change,* 25: 203–224.
75. See Table 2 in Schneider, 1997.

76. Moss, R. and S. H. Schneider, 1997: "Characterizing and communicating scientific uncertainty: moving ahead from the IPCC second assessment," in S. J. Hassol and J. Katzenberger (eds.), *Elements of Change 1996* (Aspen, CO: Aspen Global Change Institute), pp. 90–135; Moss and Schneider, 2000.
77. Schneider, 1997.
78. For example, Titus and Narayanan, 1996, combine climate models with expert subjective opinion to derive a statistical distribution for future sea level rise, and Morgan and Dowlatabadi, 1996, present a probability distribution comparing CO_2 emission abatement costs with averted climate damages.
79. Morgan and Keith, 1995.
80. Nordhaus, 1994a.
81. Ibid.
82. Ibid.; Roughgarden, T. and S. H. Schneider, 1999: "Climate change policy: quantifying uncertainties for damages and optimal carbon taxes," *Energy Policy,* 27 (7): 415–429.
83. Ibid.
84. Schneider, 1997.
85. For example, NAS, 1991.
86. IPCC, 2000.
87. IPCC, 2000a, 2000b; Intergovernmental Panel of Climatic Change (IPCC), 2001c: *Climate Change 2001: Mitigation,* contribution of Working Group III to the Third Assessment Report of the IPCC on Climate Change, Metz, B., O. Davidson, R. Swart, and J. Pan (eds.) (Cambridge, England: Cambridge University Press), 752 pp.
88. IPCC, 2001a.
89. For example, see Schneider, S. H., 2001a: "What is 'dangerous' climate change?" *Nature,* 411 (May 3): 17–19, and the rebuttal by Grübler, A. and N. Nakicenovic, 2001: "Identifying dangers in an uncertain climate," *Nature,* 412 (July 5): 15, and the perspective by Pittock, A. B., R. N. Jones, C. D. Mitchell, 2001: "Probabilities will help us plan for climate change," *Nature,* 413, p. 249. See also the later rebuttal: Schneider, S. H., 2002: Editorial Comment: "Can we estimate the likelihood of climatic changes at 2100?" *Climatic Change,* 52 (4): 441–451.
90. For example, Rosenzweig and Parry, 1994; IPCC, 2001b.
91. Kremen, C., J. O. Niles, M. G. Dalton, G. C. Daily, P. R. Ehrlich, J. P. Fay, D. Grewal, and R. P. Guillery, 2000: "Economic incentives for rain forest conservation across scales," *Science,* 288: 1828–1832.
92. Schneider, S. H., 1998: "The climate for greenhouse policy in the U.S. and the incorporation of uncertainties into integrated assessments," *Energy and Environment,* 9 (4): 425–440; IPCC, 2001c.
93. IPCC, 2001b, 2001c.
94. Nordhaus, 1994b.
95. Root, T. L. and S. H. Schneider, 1995: "Ecology and climate: research strategies and implications," *Science,* 269: 331–341.
96. Moss and Schneider, 1997, 2000.
97. For example, Titus and Narayanan, 1996; West, J. J., and H. Dowlatabadi, 1999. "Assessing economic impacts of sea level rise," in T. E. Downing, A. A. Olsthoorn,

and R. Tol (eds.), *Climate Change and Risk* (London: Routledge), pp. 205–220; West, J. J., M. J. Small, and H. Dowlatabadi, 2001: "Storms, investor decisions and the economic impacts of sea level rise," *Climatic Change,* 48 (2–3): 317–342; Roughgarden and Schneider, 1999.

98. For example, see the discussion of the hierarchy of integrated assessment models in Schneider, 1997.
99. Schneider et al., 1998.
100. For example, Broecker, W. S., 1997: "Thermohaline circulation, the Achilles heel of our climate system: will man-made CO_2 upset the current balance?" *Science,* 278: 1582–1588; Rahmstorf, S., 1997: "Risk of sea-change in the Atlantic," *Nature,* 388: 825–826.
101. Schneider S. H. and L. H. Goulder, 1997: "Achieving low-cost emissions targets," *Nature,* 389: 13–14.
102. See also Hoffert, M. I., K. Caldeira, A. K. Jain, L. D. D. Harvey, E. F. Haites, S. D. Potter, M. E. Schlesinger, S. H. Schneider, R. G. Watts, T. M. L. Wigley, and D. J. Wuebbles, 1998. "Energy implications of future stabilization of atmospheric CO_2 content," *Nature,* 395: 881–884.
103. For example, Yang, C. and S. H. Schneider, 1998: "Global carbon dioxide emissions scenarios: sensitivity to social and technological factors in three regions," *Mitigation and Adaptation Strategies for Global Change,* 2: 373–404.
104. IPCC, 2000.
105. Keith, D., 2001: "Box 1, Geoengineering," p. 420, accompanying Schneider, S. H., 2001b: "Insight Feature: Earth Systems Engineering and Management," *Nature,* 409: 417–421. See also Chapter 20, this volume.
106. For example, see Moss and Schneider, 1997.
107. For example, see Azar, C., 1996: *Technological Change and the Long-Run Cost of Reducing CO_2 Emissions.* Working Paper, INSEAD, France; Goulder, L. H., S. H. Schneider, 1999: Induced technological change and the attractiveness of CO_2 emissions abatement policies. *Resources and Energy Economics,* 21, 211–253; Grubb, M., M. Ha-Duong, T. Chapuis, 1994: "Optimizing climate change abatement responses: On inertia and induced technology development," in N. Nakicenovic, W. D. Nofdhaus, R. Richels, F. L. Tóth (eds.), *Integrative Assessment of Mitigation, Impacts, and Adaptations to Climate Change* (Laxenburg, Austria: International Institute for Applied Systems Analysis), 513–534; Brugg, M., J. Koehler, and D. Anderson, 1995: *Induced technical change: Evidence and implications for energy-environmental modeling and policy,* Report to the Deputy Secretary-General, OECD; Repetto, R., D. Austin, 1997: *The costs of climate protection: A guide for the perplexed* (Washington, DC: World Resources Institute), 60 pp.
108. Mastrandrea and Schneider, 2001.
109. Schneider, S. H., 2002b: "Modeling climate change impacts and their related uncertainties," in Cooper, R. N. and R. Layard (eds.) *What the Future Holds: Insights from Social Sciences,* (MIT Press, Cambridge, Mass., London), 123–155.
110. Schneider, S. H., K. Kuntz-Duriseti, and C. Azar, 2000: "Costing non-linearities, surprises and irreversible events," *Pacific and Asian Journal of Energy,* 10 (1): 81–106.

CHAPTER 3

Regional Impact Assessments: A Case Study of California

Eleanor G. Turman

In recent years, a number of studies and workshops have investigated the potential impacts of global warming on the state of California. The results of their analysis point to the following changes over the next century:

- Higher temperatures mean more precipitation will fall as rain and less as snow, decreasing the snowpack and shifting the spring runoff from the Sierra Nevada and Cascade ranges to earlier in the year;
- Thermal expansion of the oceans will cause sea-levels to rise, resulting in the inundation of low-lying coastal areas and wetlands, including those in the San Francisco Bay and the Sacramento/San Joaquin Delta areas;
- Weather variability will increase as more frequent and more intense El Niño events caused by the warming of the Pacific occur;
- Changed hydraulic and El Niño patterns will trigger more frequent and intense floods, mudslides, extreme tides, and convective storms;
- Wildfires will become more frequent, severe, and widespread as temperatures rise and the summer dry season lengthens;
- Forest growth patterns will shift northward and upslope as temperatures rise, with some species gaining and others loosing ranges; deserts will give way to grasslands; savannas will replace shrubland which will replace forests in some areas; and
- Many plant and animal species will become regionally or totally extinct, as isolated ecosystems disappear and new ecosystems develop that consist of species aggregations different from those of today.

As a result of these environmental changes, the following impacts on the human population of California are likely:

- Agricultural environments will shift geographically as temperatures rise and precipitation patterns change, causing losses of crop acreage in some areas and increases in others;
- As a result of increased spring runoff and seawater inundation of low-lying areas, California's summer water supply will decrease unless major water supply infrastructure investments and/or demand reduction measures are undertaken.
- Competition for water among agricultural, urban, and environmental interests will increase as demand grows due to higher summer temperatures;
- Electricity demand will increase (beyond what is already forescast due to expected population growth) as the energy supply from all existing sources shrinks both within and outside California;
- Air and water pollution, and associated health problems and crop damage, could worsen;
- Human morbidity and mortality is likely to increase due to higher summer temperatures, vector-borne diseases, and (potentially) increased air pollution; and
- Prices for water, electricity, wood, fuels, farm products, and their derivatives will in all likelihood increase.

Regional Climate Studies

Much effort has gone into developing general circulation models (GCMs) that simulate global atmospheric circulation and interaction with the earth's land surface and oceans. These models are used to assess the potential impacts of greenhouse gas emissions on the world's climate. GCMs predict an increase in average global surface temperature of 1.5°C to 4.5°C (2.5°F to 8°F) by 2100 that is now widely accepted in the scientific community.[1] This range is based on a projected doubling of CO_2 during this century. Recently, the 2001 IPCC Working Group I raised the upper bound to 6°C based on a projection of greater greenhouse gas emissions.[2] GCMs indicate that warming will be greatest over continental land masses at middle to high latitudes because melting snow and ice will make these surfaces less reflective to sunlight, allowing them to absorb more heat. A recent assessment of climate change impacts on the United States using GCMs projects temperature increases of 3°C to 5°C (5°F to 9°F) in the next 100 years.[3]

GCMs are useful in projecting long-term hemispheric or continental climate changes. But because they operate on a coarse scale (several degrees of latitude and longitude) and ignore the effects of regionally important features such as mountain ranges, they fall short when it comes to predicting regional climate changes. This is particularly true when the region's topography is highly variable, as is the case for California. Recently, progress has been made in assessing regional sensitivity to global climate change using finer-scale regional climate models along with landscape-scale hydrology and vegetation and ecosystem models, often in combination with GCMs. These models forecast temperature changes for the western United States of 2°C (4°F) by the 2030s and 4.5–6°C (8–11°F) by the 2090s[4,5] (Fig. 3.1). Temperature increases will vary seasonally in the West, particularly in California, where statewide average temperature increases on the order of 3°C (5°F) in winter and 1°C (2°F) in summer are expected to occur by 2030–2050.[6]

GCM forecasts of mean annual global precipitation differ substantially, and regional forecasts are even more uncertain. Figure 3.1 shows the predictions of two widely reviewed climate models for precipitation increases in the western United States by 2095. They differ by 40 percent. Although the percentage of precipitation increase remains uncertain, it is clear that major frontal storms will bring more rain to the western edges of continents.[7] This means increased rainfall over the California coast, particularly during winter. Several models predict increased precipitation for California because of warming of the eastern Pacific and southward movement of the storm-generating Aleutian Low.[8–11] Changes in precipitation are likely to vary across the state. Drier summer conditions will predominate in the southwestern half of the state, with wetter conditions occurring in the Sierra Nevada and Klamath mountains. A 25 percent increase (or more) in winter precipitation over the California coast and the western Sierra is likely, but areas east of the Cascades and the Sierras and in the southeastern part of the state probably will be drier.

Recently, GCMs capable of modeling both atmospheric and oceanic circula-

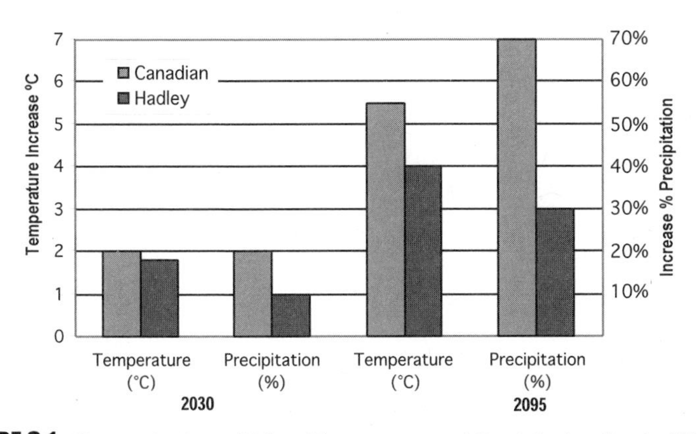

FIGURE 3.1. Changes in Annual Mean Temperature and Precipitation for the West Projected by the Hadley and Canadian Models Compared to 1961–90 Base Period.

tion have been used to study how global warming may affect the El Niño–Southern Oscillation (ENSO). They point toward more frequent and intense El Niños caused by warming of the eastern Pacific.[12–15] If this proves true, Californians can expect an increase in winter storms and therefore higher winter precipitation in some years. The frequency and intensity of convective storms (thunderstorms) probably will increase in warmer areas, such as California's southern deserts and the Sierra Nevada Mountains.[16]

GCMs also predict that global sea level will rise 6–37.5 inches on average by the year 2100 because of thermal expansion of the oceans and continental ice melt.[17–19] Many studies assume rises on the high end of this range (1 meter is typical) for California, although a recent study puts the California increase somewhat lower, predicting only 8–12 inches by 2100.[20] This is still very significant because an 8- to 12-inch rise is two to three times the observed increase over the past 150 years.[21]

Hydrologic Effects and Water Resources

Higher temperatures mean changes in precipitation patterns that will have a dramatic impact on California's water resources and hydrology. Most researchers agree that global warming will change the form, timing, intensity, and distribution of precipitation in very significant ways, whether or not there is any change in the overall amount of precipitation.[22–26] This has profound implications for California's surface water supply.

Available surface water depends largely on how much snow falls and how long it is stored in the Sierra Nevada and Cascade mountain ranges each winter. Higher temperatures mean that more precipitation will fall as rain and less as snow.[27] Growing evidence suggests that the rain-to-snow ratio is already increasing as a consequence of global warming. Several studies indicate that annual stream flow from the Sierras in the fall and winter has increased during the second half of the twentieth century, and spring flow has decreased.[28,29] In general, snow lines in California's mountains will rise 500 vertical feet for each degree centigrade increase in atmospheric temperature.[30] Therefore, a future 3°C increase will raise the snowline 1,500 feet. Higher temperatures also mean earlier snowpack melt.[31–33] Assuming a 3°C temperature increase, the combined effect of less snow and earlier snowpack melt will reduce the amount of water stored in the snowpack by an estimated 33 percent in an average year.[34] Such changes in the snowpack will cause a shift in the timing of water runoff from the mountains toward the winter and early spring.[35] Runoff will increase in winter and early spring and decline in late spring and early summer.[36,37] This probably will be the case even if there is no change in the amount of precipitation.[38]

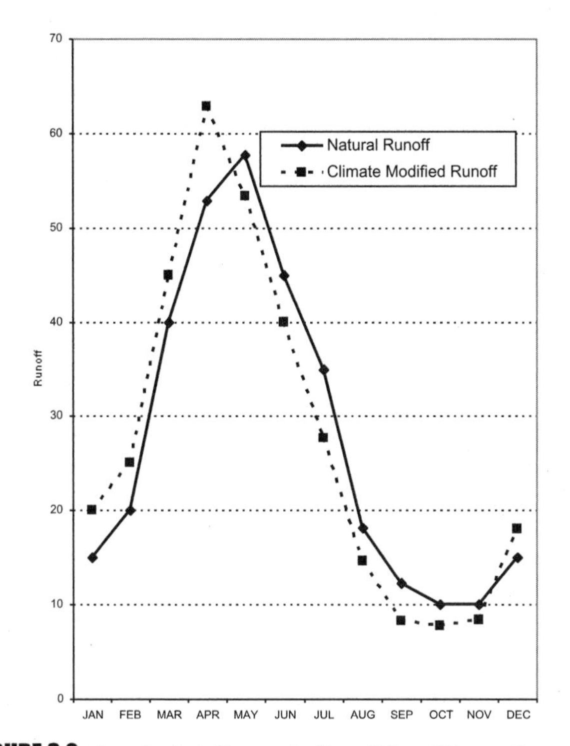

FIGURE 3.2. Hypothetical Changes in Runoff for a Western Snowmelt Basin.

Should winter precipitation double by the end of the century (as some models predict), runoff will increase three-fifths by the 2030s and double by the 2090s.[39] Figure 3.2 illustrates the shift in timing of the spring runoff.

Currently, winter runoff from Sierra Nevada and Cascade rivers is captured and stored in reservoirs. These reservoirs, together with a lengthy canal system, make water available throughout the year to urban population centers and agricultural areas statewide. This elaborate system moves water from northern California, where two-thirds of the state's surface water originates, to southern California, which has 70 percent of the population and 80 percent of the water demand.[40] The reservoirs are also used to prevent flooding of agricultural lands and urban areas that would otherwise occur as a result of natural variations in the amount of runoff.

Changes in the timing and amount of runoff will alter the frequency, timing, and severity of floods. Today, more than 75 percent of California's communities are built on floodplains or Special Flood Hazard Areas.[41] Continued population growth probably will mean more development in floodplains. Increased spring runoff will expand flood-prone acreage to include many communities that are not yet at risk. Further urbanization, which increases surface runoff during storms, will exacerbate the problem. An increase in wildfires caused by higher summer temperatures and intensified winds is another likely consequence of global warming in California.[42] Wildfires denude watersheds and increase soil erosion. This increases the amount of sediment deposited in California's streams and rivers, thereby reducing their flood-carrying capacity and placing an even heavier burden on the reservoir system.

The water delivery system was constructed under the federally funded Central Valley Project and the California State Water Project in the mid-twentieth century. It was engineered based on runoff patterns that had been observed over the preceding hundred years. Changes in the spring runoff patterns are likely to reduce summer water supplies unless major changes are made to California's water delivery system or demand is reduced.[43] This is true even if annual precipitation increases because the increase will occur in the winter. Existing reservoirs will not have enough capacity to store the increased winter volume for use during the summer and, at the same time, prevent flooding unless current operating rules change, and even that may be insufficient.[44]

Construction of additional infrastructure may be necessary to capture the increased winter and spring runoff. However, there are significant environmental and cost impediments to such construction.[45,46] In the absence of new infrastructures, existing reservoirs would need to be maintained at lower levels during the winter. This could reduce statewide water supply in the summer by an estimated 7–20 percent.[47] Alternative flood control measures could be implemented so that more of the winter runoff could be stored in reservoirs. Development could be restricted in floodplains, for example. Rivers could be restored to their natural state so that floods would spread out along their length and concentrated flooding downstream would be avoided. Californians could also adopt conservation technologies (e.g., low-flush toilets and better landscaping practices) to improve the efficiency with which municipalities and industry use the reduced summer water supply.[48]

Without infrastructure or water management policy changes, the surface water supply will also be adversely affected by another major consequence of global warming: sea level rise. The Sacramento and San Joaquin Delta is the major source of California's water, providing 65 percent of the state's total water supply and 45 percent of its drinking water.[49] Today, water is released from upstream reservoirs to maintain delta outflow at the level needed to prevent saltwater intrusion into pumping stations that supply freshwater to the rest of California. Increased seawater intrusion caused by higher sea levels, together with lower dry season runoff, would significantly degrade water quality in the delta.

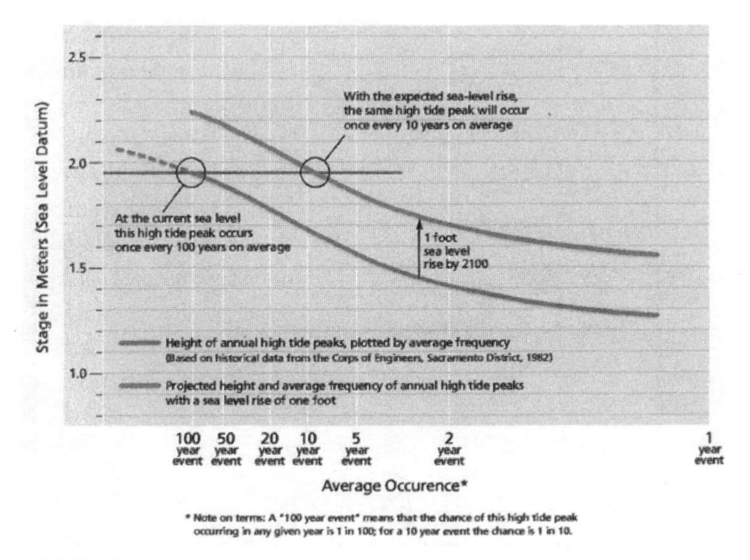

FIGURE 3.3. Sea-Level Rise and Delta Flooding.

Twice as much water would have to be released from reservoirs to maintain water quality if sea level rises by 1 meter.[50]

Today, a significant portion of land in the delta lies as much as 25 feet below sea level.[51] It is protected from inundation by a system of levees, sumps, and pumps. If sea level rises 1 foot, today's 100-year high tide mark will become the 10-year high mark (Fig. 3.3).[52] A 1-meter sea level rise, together with increased winter runoff from the Sacramento and San Joaquin rivers, probably would overtax the current levee system.[53] A large inland lake with fresh to brackish water would then replace the existing delta.

Furthermore, a 1-meter rise in sea level will cause seawater to inundate the San Francisco Bay. Figure 3.4 illustrates what could happen to the bay's wetlands during the twenty-first century. If the median estimate of the likely rate of sea level change caused by global warming is assumed together with a continuation of the historic rate of sea level rise in the bay caused by local subsidence, sea level will rise 1.0 meter by 2050 and 1.9 meters by 2100. This will result in a loss of 57.7 percent of tidal flat habitat by 2050 and 62.1 percent by 2100.[54]

Today, surface water supplies approximately 60 percent and groundwater supplies 40 percent of the water used for all purposes in California.[55] In many areas, the use of groundwater could be increased to compensate for a reduction in surface water during the summer. However, groundwater quality will be compromised in areas along the heavily populated coastal strip if aquifers are

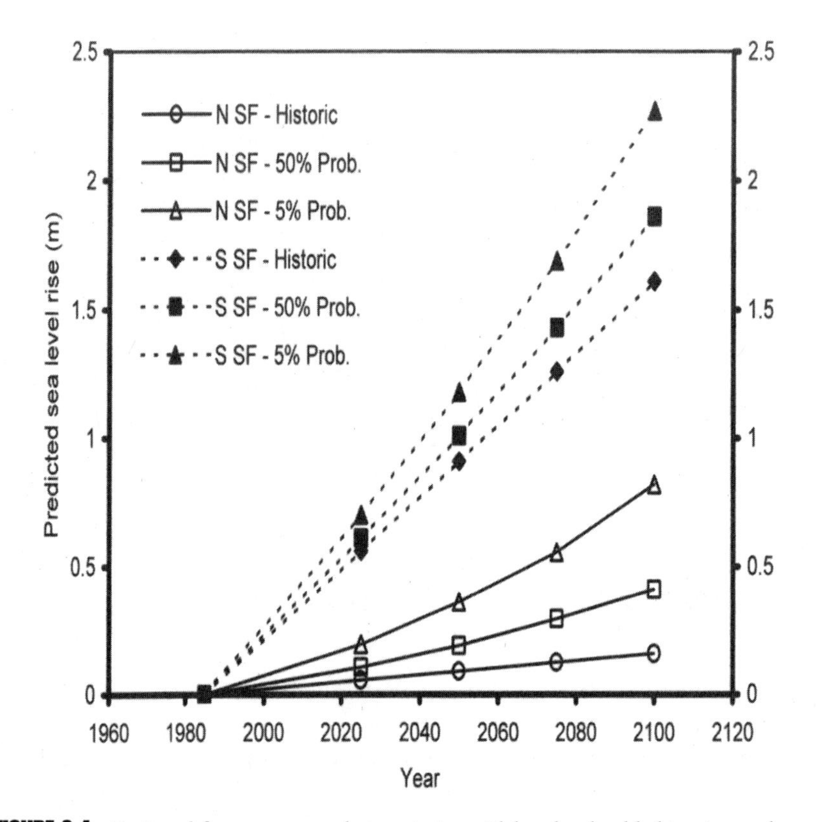

FIGURE 3.4. Projected future percent changes in intertidal and upland habitat in northern and southern San Francisco Bay under three sea level rise scenarios.

supplies 40 percent of the water used for all purposes in California.[55] In many areas, the use of groundwater could be increased to compensate for a reduction in surface water during the summer. However, groundwater quality will be compromised in areas along the heavily populated coastal strip if aquifers are inundated by seawater. Hazardous waste sites and landfills along the coast could be flooded and contaminate aquifers, further reducing the groundwater supply.

More reservoirs and levees probably would be needed to ensure an adequate surface water supply in the summer.[56] The entire water delivery system will have to be reengineered to maintain water supply and flood control at current levels. However, major construction projects of this sort take 30–50 years to plan and build. Voters must approve them. Serious disruptions in the water supply system may have to occur to create the political will to undertake a major new construction project.

Agriculture and Ranching

Experts disagree on the combined impact changes in temperature, CO_2 concentrations, and precipitation will have on California's crops. On one hand, higher temperatures and drier soils in summer would tend to reduce crop yields. On the other hand, a longer growing season resulting from higher winter temperatures probably would increase crop yields. Higher CO_2 levels would be good for crops because CO_2 stimulates plant growth and improves the efficiency with which plants use water.[57] More winter precipitation would increase crop yields if the soil does not become saturated. Flooding and waterlogging of crops would occur if the soil could not absorb additional moisture. This would reduce crop yields. One study predicts net gains in key California crops on the order of 5–30 percent (depending on the crop) when projected changes in temperature, precipitation, and CO_2 levels are taken into account.[58] Another considers all these factors and predicts an overall decline in California crop yields.[59]

In any event, California's farmers will have to adapt quickly to the changing climate by switching to crop varieties that better match new temperatures, CO_2 levels, and precipitation patterns. Unfortunately, growers of certain crops that are vital to California's current economy may not be able to switch easily to new varieties. For example, grapevines and fruit trees take years to mature.

Experts agree that climate change will shift the locations where specific crops can be grown successfully. A 3°C increase in summer temperature could shift crops currently grown in the Central Valley 200 miles north.[60] Today's Imperial Valley crops would become better suited to the San Joaquin Valley. Crops currently farmed in the San Joaquin would move to the Sacramento Valley. The largest gains in agricultural acreage could be in the San Joaquin Delta region. A temperature increase would make this area more suitable for many key crops, assuming the levee system is successfully maintained. As discussed earlier, however, the highly productive delta islands would be lost to rising sea levels if the levee system were to fail.

Climate change could also cause a shift of another kind. The range of agricultural pest populations could shift north because of higher temperatures.[61] Some insects that cannot tolerate today's colder temperatures in key agricultural areas such as the San Joaquin Valley might survive quite well as these areas warm.

The precipitation changes discussed previously could make it economically difficult or impossible to continue farming some areas of the state that rely heavily on irrigation. Today, the Central Valley Project and the State Water Project deliver water to semiarid agricultural land throughout the Central Valley. A reduction in the amount of surface water available for summer irrigation could raise the cost of farming substantially. Groundwater could fill the gap where it is available. However, sharp reductions in irrigated acreage can be expected in areas where groundwater is not available in sufficient quantity. This includes the southern San Joaquin Valley, a large area west of Sacramento, and some Sierra foothills.[62]

Because it relies heavily on irrigation, California agriculture consumes 80 percent of the managed water supply (surface and groundwater) available statewide.[63] In the absence of climate change, agricultural demand for water is not expected to increase. In fact, it probably would decrease as competition for water between agriculture and rapidly growing municipalities forces farmers to adopt new water efficiency practices. However, if the climate changes as expected, higher temperatures will cause more evaporation, which will dry the soil. Crops grown in drier soil need more water. More efficient water use or a shift to crops that need less moisture may not offset the increased need for water caused by higher California temperatures.[64]

Some computer models indicate a shift in the timing of the rainy season, with the rain starting later in the fall and lasting later into the spring.[65] Greater variability in the timing of rainfall is also predicted. In fact, global warming is expected to increase weather variation generally.[66,67] This poses a significant problem for farmers who rely on accurate weather forecasts when making planting and harvesting decisions. Agriculture will become an increasingly risky business if the weather becomes more volatile and unpredictable.

Ranching could benefit from a longer grazing season and increased forage production caused by higher temperature combined with increased precipitation. However, increased flooding and higher risk of animal diseases could offset these advantages. The California dairy industry suffered heavy losses to bacterial infections during the 1998 El Niño.[68]

Forestry

Global biome models, which are used to analyze changes in vegetation distribution, predict a shift of Northern Hemisphere forests poleward and to higher elevations.[69] In California, tree species such as ponderosa pine that are able to tolerate longer dry seasons will increase the size of their overall range. Species such as the Douglas fir that are more drought sensitive will disappear from the coastal lowlands but survive at higher elevations. Because most tree species migrate at slow rates, it may take several centuries before these shifts are completed and new forests are securely established. During the transition period, tree dieoffs will provide additional fuel for wildfires, increasing their frequency and intensity. Higher temperatures and associated changes in precipitation will place additional stress on many tree species and increase their susceptibility to insect infestations and disease. This has happened in the past. During the 1976–77 drought, for example, there were significant tree dieoffs because of bark beetle infestation.[70]

Increased CO_2 will stimulate forest growth in areas with abundant water and soil nutrients such as the Sierra Nevada Mountains, which could be a boon for commercial forestry.[71] On the other hand, a study that considered the Sierra Nevada forests that existed 9,000 years ago, when the temperature is believed to have been 1°C to 3°C warmer than it is today, suggests a different outcome.[72] At that time, the forest cover and tree density were lower. This provides evidence that a longer and more intense dry season could reduce California forest density, especially where pine and fir species dominate. In areas such as northwestern California that are poorer in mineral nutrients and more susceptible to drought, higher temperatures would most certainly mean reduced forest productivity.[73]

Natural Ecosystems

Figure 3.5 illustrates changes in California's vegetation distribution that are predicted over the next 100 years.[74] Forests will move upslope as temperatures climb. The land they occupy today will be taken over by arid shrubs. Grassy savannas will replace shrubland in the western foothills of the Sierras and the coastal range. California's southeastern deserts will give way to grasslands and shrublands because precipitation over these areas will increase because of more frequent and intense convective storms.[75]

If this happens, some plant and animal species will be able to shift their ranges better than others. Some long-lived trees, including coastal redwoods and giant sequoias, could persist in their current locations for many centuries if conditions remain within tolerable limits for adult trees. However, seedlings and saplings might have difficulty surviving in more extreme temperature or drought conditions. This raises the possibility that these popular species will become extinct in California if they cannot successfully migrate to new ranges where conditions are more favorable to reproduction.[76]

In general, it is unlikely that entire ecosystems could move as a unit. Animals that do not migrate could die off in the southern portions of their current ranges as temperatures increase. Consequently, the ecosystems of the future probably will consist of a different mix of species.[77]

Migration may be problematic for species or entire ecosystems that are isolated by natural phenomena or human development.[78] Differences in soil chemistry in adjacent areas, topographic features such as mountains, and microclimates can hinder dispersal of seeds and insect pollinators. Highways, farms, and housing as well as commercial developments are increasingly becoming barriers

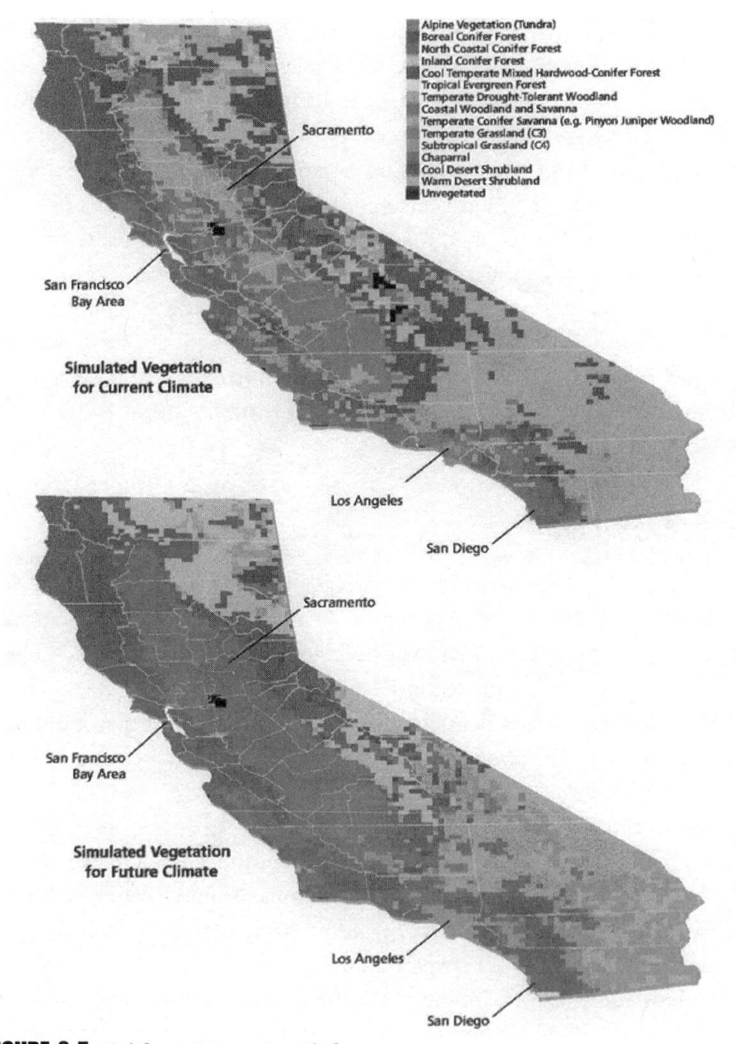

FIGURE 3.5. California Vegetation Shifts.

opportunities such as the Channel Islands. The serpentine outcrops of northern California and the vernal pools in the Central Valley, both home to rare species, are so isolated that natural migration may be impossible.[80]

The length of the dry season and its temperature intensity are major factors that determine where particular plant species grow. As temperatures climb, vegetation that tolerates or prefers warmer summer conditions will invade habitat currently occupied by plants that find it harder to adapt to higher temperatures.

The length of the dry season and its temperature intensity are major factors that determine where particular plant species grow. As temperatures climb, vegetation that tolerates or prefers warmer summer conditions will invade habitat currently occupied by plants that find it harder to adapt to higher temperatures. Where opportunities for migration exist, some species may decline as their habitat is invaded by neighboring species. Some endangered alpine plants, for example, could be crowded out as temperature increases cause subalpine species to move to higher elevations. In many areas, nonnative species can move in after climate change and associated disturbances such as wildfire, pests, and disease weaken native organisms.[81]

Increased air pollution caused by global warming would also have adverse effects on some plant communities. Ozone is strongly damaging to coastal sage, which today dominates coastal areas south of San Francisco to the Mexican border.[82] If convective storms, which readily pick up and transport pollutants, increase in frequency and intensity, as some studies predict, more acid rain will fall on the forests and lakes of the Sierras.[83] It has been estimated that 20–50 percent of current habitat could become unsuitable for the plant and animal species that inhabit California today.[84]

Throughout most of California, soil moisture levels could decrease substantially in the dry season as a result of warming.[85–88] Higher temperatures mean more evaporation, which dries the soil. Increased winter precipitation could offset this in regions such as the desert southeast, where the soil can absorb the additional moisture. But in much of California, today's levels of winter precipitation already saturate the soil. Additional winter precipitation probably would add to runoff rather than soil moisture. Plants that can withstand drier soil conditions in the summer would be favored over those needing more moisture. Thus, global warming could trigger range expansion for the former and a loss of range for the latter.

Higher temperatures, which dry out vegetation and stir up winds, could mean an increase in wildfire frequency and intensity in the dry season. Some fire behavior models used in combination with climate models point to higher losses to wildfire, particularly in the grasslands and shrublands of California's coasts and foothills.[89] More frequent wildfires could also trigger more frequent mudslides and erosion of mountainous terrain. This in turn could increase sediment buildup and adversely affect the clarity of estuarine water and the productivity of fisheries. But it should be emphasized that much uncertainty remains because wildfires typically occur during extreme weather conditions that are not well predicted by current models.

Although alpine lake evaporation could increase because of higher temperatures, simulations show that moisture loss to evaporation will be more than offset by an increase in annual precipitation over the Sierras.[90] Net moisture (precipitation minus evaporation) over lakes probably would be higher than it is today. Therefore, the quantity of lake water is unlikely to be adversely affected by global warming. Quality is a different matter. The amount of pollutants deposited in alpine lakes could increase to the point where acidification occurs as a consequence of a higher rain-to-snow precipitation ratio in the Sierras.[91] This is because acids and other chemicals are rapidly flushed out of the snowpack that feeds the lakes during the early spring snowmelt. Faster snowmelt and higher chemical loading (both projected consequences of higher temperatures and acid rain from more frequent and intense convective storms) could significantly change lake chemistry.

Simulations also show that warmer winters may reduce or eliminate turnover in some alpine lakes.[92] This would reduce biological productivity, which depends on the mixing and redistribution of nutrients that occurs during lake turnover. However, this trend could be at least partially offset by a longer plant-growing season resulting from higher temperatures, which would enhance productivity. One study predicts that, taking all these factors into account, productivity will actually increase.[93] Rising temperatures could mean an increase in algae production, which would decrease dissolved oxygen and block light from reaching lower levels. More frequent fires, which cause more nutrients to be deposited in lakes, may also stimulate algal growth. All these factors are likely to dramatically change the fish species composition in California's lakes.

Saline lakes such as Mono Lake could see fluctuations in plant and animal productivity as a result of global warming.[94] Salinity will increase if lake levels drop during longer and hotter dry seasons. Aquatic organisms will expend more energy to regulate the amount of salt in their tissues and less on growth and reproduction. Sudden increases in winter lake level caused by more frequent and intense El Niño storms and the increased runoff they bring probably will cause persistent chemical stratification, which also reduces productivity. But productivity can rise or fall as runoff amounts fluctuate.

The mix of fish species in California rivers, streams, and estuaries is likely to change as a result of global warming.[95] A reduced flow in late spring and summer and warmer temperatures would favor species that are adapted to warm water. Species that cannot tolerate the higher temperatures will decline. This includes many commercially important saltwater species (including Chinook salmon, striped bass, American shad, and steelhead rainbow trout) that breed in freshwater. Freshwater runoff in winter from the Sierras could carry higher concentrations of nutrients, as well as pollutants, to California's coastal bays and estuaries if more precipitation falls as rain than as snow. Increased runoff caused by more frequent and intense El Niño storms would transport more sediment as well. California's streams will carry higher concentrations of pollutants because of changes in the snowpack and increased acid rain. The poorer water quality combined with a reduction in stream flow in the late spring and summer could harm aquatic organisms and wildlife using the streams for food and water. Wetlands and the migrating birds that use them would also be adversely affected by the higher concentrations of pollutants.

California's estuaries and bays could be inundated by saltwater as a result of global warming. Specifically, a 1-meter rise in sea level by 2100 could result in the tidal marshes of the San Francisco Bay being inundated around 2040, according to one study.[96] Drier springs (caused by the shift in timing of runoff from the Sierras discussed previously) would tend to further increase bay salinity because there would be less freshwater available to dilute the saltwater. Higher salinity could result in a decline in freshwater and brackish plant species and a shift to more salt-tolerant plants. This would reduce waterfowl food and cover and potentially reduce waterfowl populations severely. Marine fish species could increase in the Suisun and San Pablo bays as salinity increases, and freshwater and anadromous (saltwater fish that spawn in freshwater) species would decrease.

Wetlands and beaches along California's coastal areas could be lost as sea level rises, particularly if people construct bulkheads and other barriers to protect property.[97] This could reduce or even eliminate turtles, birds, and marine mammals that rely on these areas for breeding and raising young.[98]

Global warming may raise the temperature of the California Current.[99] This could change the mix of marine life along California's shores. A population decline in zooplankton associated with warming of the California Current has already been observed,[100] as has an increase in southern animal species at the expense of native northern species in Monterey Bay.[101] A strengthened California Current could also cool the California coastline during the summer and strengthen onshore breezes. These changes would favor some species over others and thus tend to alter the current species mix. Some seabirds (such as sooty shearwaters) are already declining in response to the warming of the California Current.[102]

Air Quality and Health

Global warming could have both good and bad effects on California's air quality. Higher temperatures could increase ozone levels in urban population centers and the Central Valley, where ozone is already problematic. The number of people-hours exceeding Environmental Protection Agency (EPA) pollution standards could triple in the San Francisco Bay area, and ozone concentrations in August in the Central Valley could rise by as much as 20 percent in response to a 4°C temperature increase.[103] The projected increase results from temperature-induced acceleration in the chemical reactions that create ozone and from increased emission expected from higher electricity usage during the summer. But higher temperatures could also increase air circulation. Changes in air currents could shift pollutants away from populated areas and thus have a positive effect on the overall health of the urban populace. The overall impact of climate change on California air quality therefore is uncertain.

Generally, increased illness and mortality among the older population caused by stroke and heart disease can be expected with higher summer temperatures.[104] Premature births and perinatal deaths can also be expected to increase in the hotter summers. Increased air pollution would aggravate lung conditions such as asthma and emphysema and put those with heart disease at greater risk. It would also increase the production of airborne plant aeroallergens, thereby intensifying seasonal allergies. Additional death and injuries (both physical and psychological) and an increase in diseases attributable to unsafe water can be expected if climate change increases the frequency and severity of natural disasters such as floods, extreme storms, landslides, and wildfire.[105]

A warmer and wetter climate would favor insect pests such as mosquitoes, mites, and ticks and potentially increase the incidence of the diseases they carry (plague, typhus, malaria, yellow fever, dengue fever, and encephalitis) and extend their range northward. Heavier and more frequent El Niño rains could increase the frequency of rodent population explosions that precede Hantavirus outbreaks, for example.[106] However, many factors (e.g., public health policy, standard of living) influence the transmission dynamics of these diseases as well. Further study is needed to determine what impact climate change will have on vector-borne diseases in California.[107] If climate variability increases, current watershed protection and sewage management systems may be inadequate to control contamination of water and food by microbial agents that are transported via rainfall runoff. This could result in an increase in disease.[108]

Energy Production

Higher temperatures will increase demand for electricity for residential and commercial air-conditioning in the summer beyond the already large increase caused by continued population growth. A 3°C mean temperature gain could increase annual demand for electricity in California a further 1.4–2.5 percent by 2050. Peak demand could increase 2.9–6.7 percent.[109] Increased air-conditioning will also be needed for agriculture (e.g., to cool poultry operations or freeze produce). Additional energy will be needed to pump groundwater to agricultural fields as higher temperatures reduce surface water supply and increase crop water demand in the summer.

At the same time, higher temperatures are likely to reduce the amount of power generated by California's hydroelectric power plants during the summer.[110] As the overall surface water supply is reduced, competition between various users for water stored in California reservoirs could increase. As previously discussed, reservoir storage for hydroelectric power production may be sacrificed to meet demands for flood control in the wet season and to preserve wildlife habitat and fisheries in the dry season. This would make it more difficult for hydroelectric power generating systems to meet increased peak summer demands.

California purchases significant amounts of electricity from other states in the Pacific Northwest and Southwest. It is expected that the Pacific Northwest will experience precipitation changes and reductions in hydroelectric power production similar to those predicted for California.[111] This will reduce the amount of hydroelectric power available for import to California. In the Southwest, global warming is expected to reduce hydroelectric supplies as well.[112] A switch to coal-based electric supplies that are readily available in the Southwest is possible but unlikely if federal policies to reduce greenhouse gases are put in place.

Global warming could improve the economics of alternative, nonpolluting energy sources in California. A doubling of CO_2 could increase agricultural biomass in California as much as 50 percent, assuming an adequate water supply.[113] This raises the possibility of biomass residuals as an additional energy source. Global warming may also intensify offshore winds in California during the summer,[114] providing an opportunity for increasing power supplied by wind turbines.

Quality of Life

The aesthetic value of California's scenic areas could diminish as a result of climate change. More extensive storm damage and greater coastal erosion from higher storm surges can be expected to result from sea level rise combined with more severe storms. Flooding of coastal areas is likely to increase as well. California's national and state parks could suffer from more frequent and extensive forest fires. Trees stressed by climate change could succumb at a higher rate to forest diseases and pest infestations such as pine bark beetles. As a result, park wildlife that depends on a healthy forest may decline both in overall numbers and in diversity.

Global warming could mean less water for outdoor use. This would adversely affect businesses such as landscaping, nurseries, car washes, and theme parks.

Higher temperatures could shorten California's ski season. The ski industry would also be negatively affected by the reduced snowpack. Loss of wetlands in areas such as the northern side of the Suisun Bay would adversely affect the hunting and fishing industry.

The California economy will be strongly affected by climate change. Higher prices for water, electricity, wood, fuels, farm products, and goods produced from these primary goods can be expected.[115] Higher insurance rates and higher costs associated with increased property damage caused by more frequent national disasters are also likely. The San Francisco and Oakland port facilities are central to California's internal and external trade, and they are sensitive to sea level change and changes in freshwater runoff. A rise in sea level could reduce the need to dredge navigation channels but could also make existing piers unusable.[116] There could be significant changes in demographics as California absorbs immigrants from areas around the globe that have been inundated by a rising sea or whose economies have been devastated by regional climate changes that undermine local agriculture or industries.

Conclusions

Global warming will produce significant changes over the next century in the hydrology and weather of California. Runoff from snowpack-driven streams in the Sierras and Cascades will increase in winter and early spring. Seawater will inundate low-lying areas. Wildfires, floods, and landslides will become more frequent and severe. El Niño events will become more frequent and intense. As a consequence, California's ecosystems will shift geographically, and the species mix within them will change. Agricultural environments will also shift as crop acreage and crop yields increase in some areas and decline in others. More water will be needed in summer to meet the competing needs of agriculture, urban populations, industry, and natural ecosystems. At the same time, summer water supplies will decrease unless significant changes are made in the water delivery infrastructure or water conservation policies are enacted. Air and water pollution, weather-related death and disease, and the cost of living in California will increase.

Some of the studies that yield these conclusions address various actions that could be taken to alleviate the most detrimental effects.[117–121] They suggest various adaptations, such as pricing water to encourage conservation, creating corridors for species migration, and developing heat- and drought-resistant crops. These studies also identify areas where more research is needed to address the remaining uncertainties about the specific effects of climate change on California. Much work remains to be done to formulate and implement specific adaptation strategies that will allow California to cope effectively with the consequences of the coming climate change.

Notes

1. Houghton, J. T., L. G. Meira Filho, B. A. Callander, N. Harris, A. Kattenburg, and K. Maskell (eds.), 1996: *Climate Change 1995: The Science of Climate Change* (Cambridge, England: Cambridge University Press).
2. Intergovernmental Panel on Climate Change (IPCC), 2001a: *Third Assessment Report of Working Group I: The Science of Climate Change,* Houghton, J. T., Y. Ding, D. J. Griggs, M. Noguer, P. J. van Der Linden, and D. Xiaosu (eds.) (Cambridge, England: Cambridge University Press), 944 pp.
3. National Assessment Synthesis Team, 2000: *Climate Change Impacts on the United States: The Potential Consequences of Climate Variability and Change* (Washington, DC: U.S. Global Change Research Program).
4. Kattenberg, A., F. Glorgi, H. Grassl, G. A. Meehl, J. F. B. Mitchell, R. J. Stouffer, T. Tokioka, A. J. Weaver, T. M., and L. Wigley, 1996: *Climate Models: Projections of Future Climate in Climate Change 1995: Science of Climate Change,* J. Houghton, Y. Ding, D. J. Griggs, M. Noguer, P. J. van Der Linden, and D. Xiaosu (eds.) (Cambridge, England: Cambridge University Press).
5. National Assessment Synthesis Team, 2000.
6. Field, C. B., G. D. Daily, F. W. Davis, S. Gaines, P. A. Matson, J. Melack, and N. L. Miller, 1999: *Confronting Climate Change in California: Ecological Impacts on the Golden State* (Cambridge, MA: The Union of Concerned Scientists, and Washington, DC: Ecological Society of America).
7. Ibid.
8. Giorgi, F., C. S. Brodeur, and G. T. Bates, 1994: "Regional climate change scenarios over the United States produced with a nested regional climate model," *Journal of Climate,* 7: 375–399.
9. Mock, C. J., 1996: "Climatic controls and spatial variations of precipitation in the western United States," *Journal of Climate,* 9: 1111–1125.
10. National Assessment Synthesis Team, 2000.
11. Stamm, J. F. and A. Gettleman, 1995: "Simulation of the effect of doubled atmos pheric CO_2 on the climate of northern and central California," *Climatic Change,* 30: 295–325.
12. Collins, M., 2000: "The El Niño–Southern Oscillation in the second Hadley centre coupled model and its response to greenhouse warming," *Journal of Climate,* 13: 1299–1312.
13. Meehl, G. A. and W. M. Washington, 1999: "Climate oscillations: genesis and evolution of the 1997–1998 El Niño," *Science,* 283: 950–954.
14. National Assessment Synthesis Team, 2000.
15. Timmerman, A. 1999: "Increased El Niño frequency in climate model forced by future greenhouse warming," *Nature,* 398: 694–697.
16. Dessens, J., 1995: "Severe convective weather in the context of a nighttime global warming," *Geophysical Research Letters,* 22: 1241–1244.

17. Houghton et al., 1996.
18. National Assessment Synthesis Team, 2000.
19. Titus, J. G. and V. Narayanan. 1996. "The risk of sea level rise," *Climate Change,* 33: 151–212.
20. Field et al., 1999.
21. Warrick, R. A., C. Le Provost, M. F. Meier, J. Oerlemans, and P. L. Woodworth, 1995: "Changes in Sea Level," in *Climate Change 1995: Science of Climate Change,* J. Houghton et al. (eds.) (Cambridge, England: Cambridge University Press), 359–405.
22. California Energy Commission, 1991: *Global Climate Change Potential Impacts & Policy Recommendation,* Vol. II (Sacramento: California Energy Commission).
23. Field et al., 1999.
24. Knox, J. B. and A. Foley, 1992: *Global Climate Change and California: Potential Impacts and Responses* (Los Angeles: University of California Press), 228 pp.
25. Melack J. M., J. Dozier, C. R. Goldman, D. Greenland, A. M. Milner, and R. J. Naiman, 1997: "Effects of climate change on inland waters of the Pacific coastal mountains and Western Great Basin of North America," *Hydrological Processes,* 11: 971–2231.
26. National Assessment Synthesis Team, 2000.
27. Wolock, D. M. and G. J. McCabe, 1999: "Simulated effects of climate change on mean annual runoff in the conterminous United States," *Journal of the American Water Resources Association,* 35 (6): 1341–1350.
28. Aguado, E., D. Cayan, L. Riddle, and M. Roos, 1992: "Climatic fluctuations and the timing of West Coast streamflow," *Journal of Climate,* 5: 1468–1483.
29. Pupacko, A., 1993: "Variations in northern Sierra Nevada streamflow: implications of climate change," *Water Resources Bulletin,* 26: 69–86.
30. Gleick, P. H., 1987: "Regional hydrologic consequences of increases in atmospheric CO_2 and other trace gases," *Climatic Change,* 10: 137–159.
31. Gleick, P. H. and E. L. Chalecki, 1999: "The impacts of climate changes for water resources of the Colorado and Sacramento–San Joaquin river basins," *Journal of the American Water Resources Association,* 35 (6): 1429–1441.
32. Jeton, A. E., M. D. Dettinger, and J. L. Smith, 1996: *Potential Effects of Climate Change on Streamflow, Eastern and Western Slopes of the Sierra Nevada, California and Nevada.* USGS Water-Resources Investigations Report 95-4260 (Washington, DC.: Government Printing Office).
33. Melack et al., 1997.
34. Knox, 1992.
35. Jeton et al., 1996.
36. Gleick, P. H., 1987: "The development and testing of a water balance model for climate impact assessment: modeling the Sacramento Basin," *Water Resources Research,* 23: 1049–1061.
37. Gleick, P. H., et al., 1999.
38. Lettenmaier, D. P. and T. Y. Gan, 1990: "Hydrologic sensitivities of the Sacramento–San Joaquin River Basin, California, due to global warming," *Water Resources Research,* 26: 69–86.

39. Wolock and McCabe, 1999.
40. Smith, J. B. and D. A. Tirpak, 1990: *The Potential Effects of Global Climate Change on the United States* (Washington, DC: Environmental Protection Agency).
41. California Energy Commission, 1991.
42. Torn, M. S., E. Mills, and J. Fried, 1998: *Will Climate Change Spark More Wildfire Damage?* LBNL-42592 (Berkeley, CA: Lawrence Berkeley National Laboratory).
43. Gleick, 1987.
44. Lettenmaier, D. P. and D. P. Sheer, 1991: "Climatic sensitivity of California water resources," *Journal of Water Resources Planning and Management,* 117: 108–125.
45. Knox, 1992.
46. National Assessment Synthesis Team, 2000.
47. Smith and Tirpak, 1990.
48. Wong, A. K., L. Owens-Viani, A. Steding, P. H. Gleick, D. Haasz, R. Wilkinson, M. Fidell, and S. Gomez, 1999: *Sustainable Use of Water: California Success Stories* (Oakland, CA: Pacific Institute for Studies in Development, Environment, and Security).
49. Knox, 1992.
50. Smith and Tirpak, 1990.
51. National Assessment Synthesis Team, 2000.
52. Field et al., 1999.
53. California Energy Commission, 1991.
54. Galbraith, H., R. Jones, R. Park, J. Clough, S. Herrod-Julius, B. Harrington, and G. Page, *Global Climate Change and Sea Level Rise: Potential Losses of Intertidal Habitat for Shorebirds. Waterbirds,* in press.
55. Regional Impacts Assessment Initiative, October 1997, Western Region of NIGEC, http://nigec.ucdavis.edu/westgec/news/impact.html.
56. Smith and Tirpak, 1990.
57. Kimball, B. A., 1983: "Carbon dioxide and agricultural yield: An assemblage and analysis of 430 prior observations," *Agronomy Journal,* 75: 779–788.
58. Smith and Tirpak, 1990.
59. Rosenzweig, C. and D. Hillel, 1998: *Climate Change and the Global Harvest* (New York: Oxford University Press), p. 324.
60. Knox, 1992.
61. Watson, M. C., M. C. Zinyowera, and R. H. Moss, 1997: *The Regional Impacts of Climate Change: An Assessment of Vulnerability* (Cambridge, England: Cambridge University Press), 527 pp.
62. Smith and Tirpak, 1990.
63. Regional Impacts Assessment Initiative, October 1997.
64. Knox, 1992.
65. Lettenmaier and Gan, 1990.
66. Field et al., 1999.
67. National Assessment Synthesis Team, 2000.
68. Ibid.
69. Leverenz, J. W. and D. J. Lev, 1987: "Effects of carbon dioxide-induced climate

changes on the natural ranges of six major commercial tree species in the western United States," in *The Greenhouse Effect: Climate Change, and U.S. Forests* (Washington, D. C.: The Conservation Foundation), 123–155.

70. Smith and Tirpak, 1990.
71. VEMAP, 1995: "Vegetation/ecosystem modeling and analysis project: comparing biogeography and biogeochemistry models in a continental-scale study of terrestrial ecosystem responses to climate change and CO_2 doubling," *Global Biogeochemical Cycles,* 9: 407–438.
72. Smith and Tirpak, 1990.
73. Dickinson, R. E., R. M. Errico, F. Giorgi, and G. T. Bates, 1989: "A regional climate model for the western United States," *Climatic Change,* 15: 383–422.
74. Field et al., 1999.
75. National Assessment Synthesis Team, 2000.
76. Field et al., 1999.
77. Ibid.
78. Ibid.
79. Smith and Tirpak, 1990.
80. Field et al., 1999.
81. Ibid.
82. Westman, W. E., 1991: "Measuring realized niche spaces: climatic response of chaparral and coastal sage scrub," *Ecology,* 72 (5): 1678–1684.
83. Collins, M., 2000.
84. Knox, 1992.
85. Field et al., 1999.
86. Gleick, P. H., 1987.
87. Gleick, P. H., 1987.
88. Lettenmaier and Gan, 1990.
89. Torn, M. S., E. Mills, and J. Fried, 1998.
90. Thompson, R. S., S. W. Hostetler, P. J. Bartlein, and K. H. Anderson, 1998: *A Strategy for Assessing Potential Future Changes in Climate, Hydrology, and Vegetation in the Western United States,* U.S. Geological Survey Circular 1153, Washington, DC.
91. Wolford, R. A. and R. C. Bales: 1996: "Hydrochemical modeling of Emerald Lake watershed, Sierra Nevada, California: Sensitivity of stream chemistry to changes in fluxes and model parameters," *Limnology and Oceanography,* 41: 947–954.
92. Thompson et al., 1998.
93. Melack et al., 1997.
94. Ibid.
95. Smith and Tirpak, 1990.
96. Ibid.
97. National Assessment Synthesis Team, 2000.
98. Peters, R. and J. Darling, 1985: "The greenhouse effect and nature reserves," *BioScience,* 35: 11.
99. Chelton, D. B., P. A. Bernal, and J. A. McGowan, 1982: "Large-scale interannual physical and biological interaction in the California current," *Journal of Marine Research,* 40: 1095–2215.

100. *Changes in Coastal Marine Communities off California Show Intimate Linkage to Global Climate Change, 1999.* Western Region of NIGEC, http://nigec.ucdavis.edu/westgec/news/article3.html.
101. Field et al., 1999.
102. Ibid.
103. Smith and Tirpak, 1990.
104. Ibid.
105. National Assessment Synthesis Team, 2000.
106. Field et al., 1999.
107. National Assessment Synthesis Team, 2000.
108. Ibid.
109. California Energy Commission, 1991.
110. National Assessment Synthesis Team, 2000.
111. Ibid.
112. Ibid.
113. Knox, 1992.
114. Ibid.
115. California Energy Commission, 1991.
116. Smith and Tirpak, 1990.
117. California Energy Commission, 1991.
118. Field et al., 1999.
119. Knox, 1992.
120. Regional Impacts Assessment Initiative, October 1997.
121. Smith and Tirpak, 1990.

PART II

ECONOMIC ANALYSIS

CHAPTER 4

International Approaches to Reducing Greenhouse Gas Emissions

Lawrence H. Goulder and Brian M. Nadreau

Climate change is a global problem, and dealing successfully with this problem will require the efforts of many nations. Although some climate policies can be implemented unilaterally, international coordination of national efforts is crucial to addressing climate change in the most effective and equitable manner.

In recent years, policy analysts have proposed a number of international approaches for confronting climate change. These policies include different proposals for burden sharing, that is, the international division of responsibilities for dealing with climate change. The policies also include mechanisms for coordinating the various national efforts to reduce emissions of greenhouse gases (GHGs). Among the coordination devices are flexibility mechanisms that give nations additional options for addressing climate change. Flexibility mechanisms can reduce the global cost of achieving reductions in greenhouse gases relative to the cost that would apply if countries acted in a unilateral, uncoordinated fashion. The cost savings from additional flexibility can be significant: Studies indicate that international flexibility mechanisms can reduce by more than 50 percent the costs of achieving certain global targets for reducing emissions.[1]

The centerpiece for recent international policy discussions is the Kyoto Protocol, an international agreement formulated in December 1997. Under the protocol, industrialized nations commit themselves to national targets (or ceilings) for their emissions of greenhouse gases; these targets would need to be reached by the commitment period of 2008–2012. The protocol embraces a number of flexibility mechanisms, including a system of international emissions permit trading and various credits for the international transfer of clean (low-carbon) technologies. Although these mechanisms have many adherents, some policy analysts prefer alternatives such as an international carbon tax.

This chapter examines a number of possible international policies to address global climate change. It considers the approaches contained in the Kyoto Protocol and other approaches as well. Our focus is broader than the policies covered under the Kyoto Protocol partly because at present it is unclear whether the protocol will be ratified by the legislatures of enough nations to be implemented as an international policy.[2] A second and possibly more important reason for our broader focus is that the protocol almost certainly will not be the last international initiative to deal with climate change. It is likely to be the first in a series of international efforts along these lines. Therefore, it makes sense to consider a range of leading policy alternatives, not just the ones articulated as part of the protocol.

National Costs of Reducing Greenhouse Gas Emissions

As discussed in Chapter 1, the proximate cause of climate change is the atmospheric buildup of heat-trapping greenhouse gases. To mitigate future climate change, nations must reduce this buildup. A direct way to do this is to reduce emissions of greenhouse gases, although expanding carbon sinks can also help address this problem.

There may be some opportunities for nations to reduce emissions at no cost; this is the no-regrets situation that politicians love. But most studies indicate that large-scale reductions in such emissions will entail costs.[3] Industrialized nations currently are highly dependent on fossil fuels, combustion of which is a principal source of atmospheric carbon dioxide (CO_2). Reducing CO_2 emissions generally entails reducing the use of coal and other fossil fuels. Such reductions, in turn, entail the use of alternative industrial processes that in many cases are more expensive than the existing processes. New technologies that allow cheap production with reduced input of fossil fuels may be developed, but such development often is expensive. Thus, there may be national costs of reducing emissions even if the channel for such reductions is the advent of a "carbon-free" technology that proves useful for a given industry.

Figure 4.1a shows a typical relationship between CO_2 abatement and national cost. The figure indicates that national costs increase with the extent of abatement. The figure also shows that costs may be zero or below zero within a range but eventually become positive. Costs rise at an increasing rate, as indicated by the fact that the slope of the total cost curve gets steeper.

A central concept in evaluating climate change policy is the marginal cost of emission reductions. The marginal cost is the cost of a given increment to the

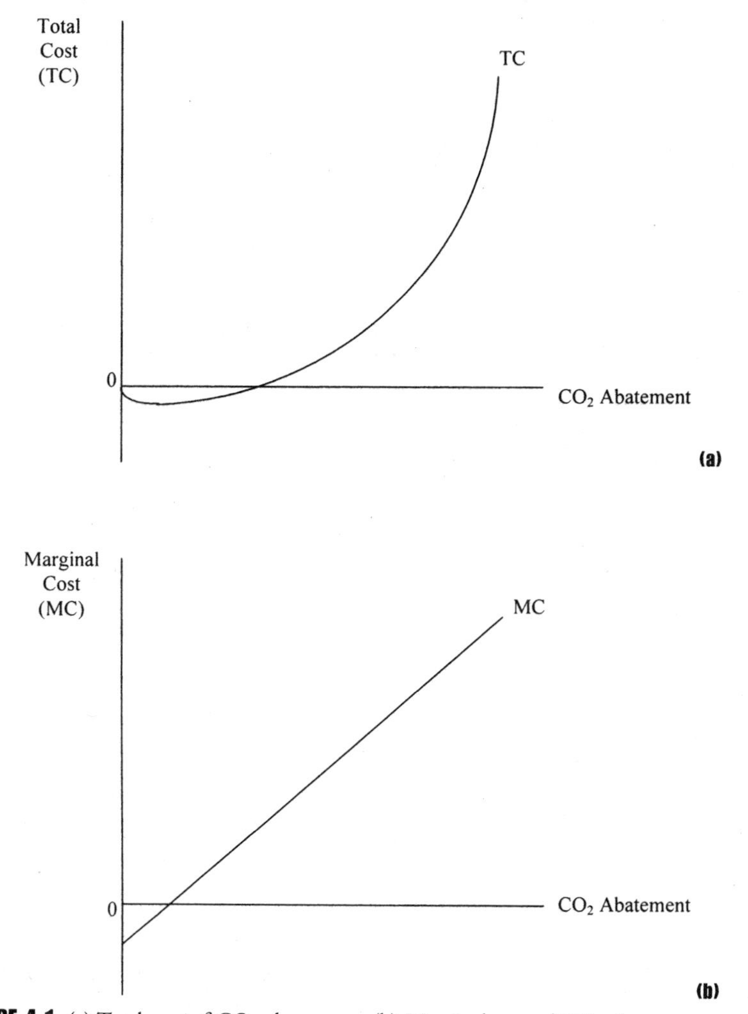

FIGURE 4.1. (a) Total cost of CO_2 abatement. (b) Marginal cost of CO_2 abatement.

amount of abatement (e.g., the cost associated with increasing CO_2 reduction from 5 to 6 tons, from 30 to 31 tons, or from 150 to 151 tons). Marginal cost is represented by the slope of the total cost curve. Figure 4.1b shows the marginal costs corresponding to the various levels of abatement in Fig. 4.1a. In this figure, at any given time it is relatively cheap to reduce the first units of CO_2 and much more costly to make incremental reductions beyond that (at the same point in time).

In Fig. 4.1b, the marginal costs increase with the amount of abatement. This corresponds to the fact that the slope of the total cost curve in Fig. 4.1a

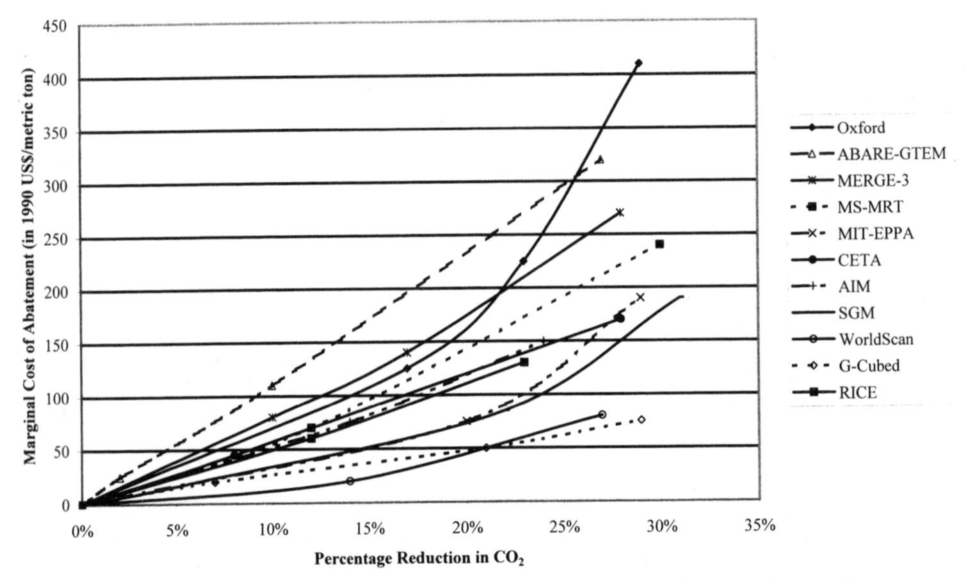

FIGURE 4.2. Marginal cost of CO_2 abatement in the United States for 2010. (From Weyant and Hill, 1999. Marginal cost data estimated from Fig. 10(a), p. xxxvii.)

increases as the amount of abatement gets larger. Rising marginal costs are consistent with the idea that it's relatively easy to remove the first units of carbon from the economy, but removing additional units becomes increasingly difficult. An example of rising marginal costs can be found in the electric power industry. In the United States, carbon emissions can be reduced to some extent by substituting natural gas in the generation of electric power. Natural gas has much lower carbon content per unit of energy than does coal, so substituting natural gas for coal reduces emissions. This is a relatively inexpensive way to reduce emissions. However, further emission reductions could be more costly, necessitating perhaps the replacement of existing power plants with new plants.

Sophisticated computer models have been developed to measure the marginal costs of reducing CO_2 emissions in various countries. These models take account of substitution possibilities in industrial, commercial, and household activities throughout the economy. Figure 4.2 shows the marginal costs of emission reductions in the United States, as predicted for the year 2010 by 11 models that participated in a study by the Stanford University Energy Modeling Forum. The figure shows significant variation in the marginal cost projections from the various models. However, each model projects rising marginal costs. Projections for other industrialized nations follow a similar pattern.[4]

The Importance of the National Targets: The Problem of Burden Sharing

The costs of abatement depend on the stringency of the national target: Reducing emissions by 8 percent will cost more than reducing emissions by 4 percent. One of the most difficult political challenges is to achieve an international consensus on burden sharing, or allocating responsibility for the global reduction in emissions between nations.

Decisions about burden sharing (or national targets) can be based on a number of different country characteristics, including current or past emissions, population, and gross domestic product (GDP). Each of these characteristics implies differences in the distribution of responsibilities for emission reductions (as well as differences in the gains from permit trades). Table 4.1 gives a glimpse

TABLE 4.1. Implications of Alternative Allocation Criteria

Country	*Percentage Share of World Industrial CO_2 Emissions, 1992*	*Percentage Share of World GDP, 1993*	*Percentage Share of World Population, 1993*
United States	21.85	27.08	4.69
China	11.94	1.84	21.42
Russian Federation	9.41	1.43	2.70
Japan	4.89	18.23	2.26
Germany	3.93	8.27	1.47
India	3.44	0.98	16.33
United Kingdom	2.53	3.54	1.05
Canada	1.83	2.07	0.52
Italy	1.83	4.29	1.04
France	1.62	5.42	1.05
Poland	1.53	0.37	0.70
Mexico	1.49	1.49	1.64
Australia	1.20	1.25	0.32
Spain	1.00	2.07	0.72
Brazil	0.97	1.92	2.84
Indonesia	0.83	0.63	3.40
Netherlands	0.62	1.34	0.28
Czech Republic	0.61	0.14	0.19
Romania	0.55	0.11	0.41
Sweden	0.25	0.72	0.16

Sources: World Resources Institute, 1996: *World Resources 1996–1997* (New York: Oxford University Press); World Bank, 1995: *World Development Report 1995* (New York: Oxford University Press).

of how much difference it would make to the selected countries if allocation of national targets were based on GDP, population, or emissions. For some countries, it makes a huge difference. For example, China's share of world population is more than 20 percent, but its share of world GDP is less than 2 percent. The United States accounts for more than a quarter of global GDP but for less than 5 percent of its population. Thus, China would bear a much smaller share of the global responsibility for emission reductions if national targets were based on population rather than GDP. The situation is the opposite for the United States.

Pointing out the political difficulties of reaching international agreements on burden sharing, Schelling[5] and Cooper[6] argue against the use of national targets. They claim that efforts to reach international agreements on quantity limits (national targets) ultimately will be fruitless, in part because of vast uncertainties about the ultimate costs of achieving given quantity targets and in part because nations fundamentally disagree on the principles that should apply to determine relative commitments. They maintain that, instead, nations should negotiate the use of instruments (e.g., carbon taxes, technology subsidies) rather than the emission reduction targets. Although the Kyoto Protocol has embraced national targets, the use of such targets remains highly controversial. An alternative policy—the international carbon tax—avoids national targets. We discuss this alternative later in this chapter.

Under the Kyoto Protocol, each industrialized nation would be required to reduce its greenhouse gas emissions by a certain percentage relative to its emissions in 1990.[7] Each of these countries would have to achieve this reduction by 2008 and maintain the reduction through the compliance period 2008–2012. The greenhouse gases are all expressed in terms of carbon equivalents. That is, a unit of each gas is meant to imply the same contribution to the greenhouse effect.[8] The United States, in particular, would need to reduce its greenhouse gas emissions by 7 percent relative to its 1990 emissions. This constitutes an emission reduction of about 30 percent relative to annual business-as-usual projections during the period 2008–2012. Thus, the far-right portions of the curves in Fig. 4.2 seem relevant to the commitments the United States would face under the Kyoto Protocol.

Reducing Costs Through International Trades in Emission Permits

The projections in Fig. 4.2 assume that there are no international flexibility mechanisms, such as provisions for trading rights to emit greenhouse gases. Under these conditions, when nations meet their national targets, they will usu-

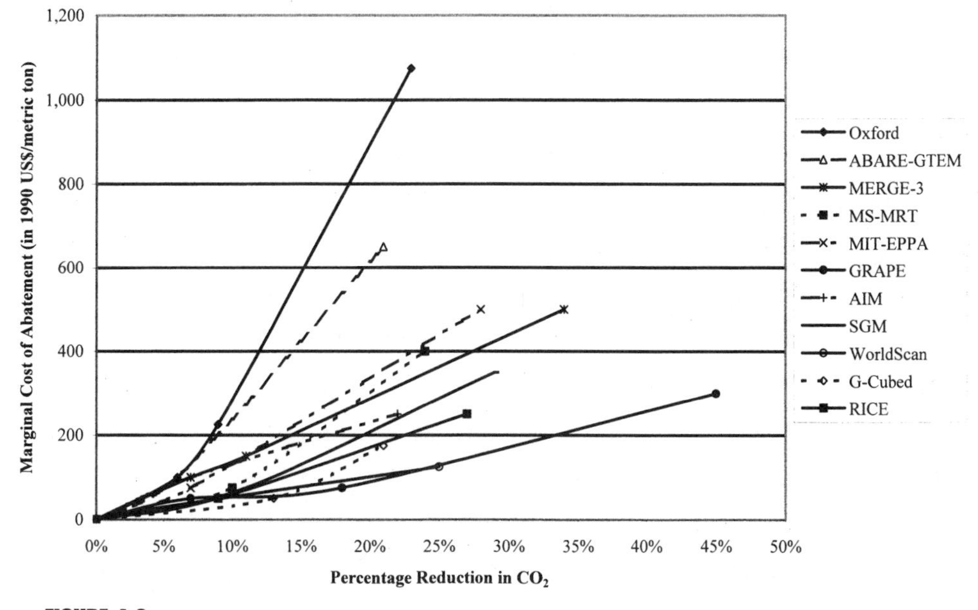

FIGURE 4.3. Marginal cost of CO_2 abatement in Japan for 2010. (From Weyant and Hill, 1999. Marginal cost data estimated from Fig. 10(c), p. xxxix.)

ally have different emission reduction costs for the last, or marginal, unit of reduction. This can be seen with the following example. Consider the marginal costs that the United States and Japan would face if they needed to reach their Kyoto targets without international trades in emission rights. The emission reductions required under the protocol are approximately 30 percent and 24 percent for the United States and Japan, respectively. If we focus on the results in Fig. 4.2 for the Multi-Sector–Multi-Region Trade (MS-MRT) model, in the United States the marginal cost associated with a 30 percent reduction is \$240 per metric ton; that is, the last increment of GHG reduction necessary to bring the overall reduction to 30 percent costs \$240. In contrast, Fig. 4.3 indicates that, according to the same model, the marginal cost associated with Japan meeting its required 24 percent reduction is about \$400 per metric ton. The lower marginal cost in the United States reflects fact that the United States currently uses a lot of fossil fuels in its production processes, whereas Japan tends to use less. Achieving emission reductions is less costly for the United States than for Japan because there are more opportunities in the United States for cheaply switching to processes less dependent on fossil fuels.

Gains from Trade

Consider what happens if a particular flexibility mechanism is introduced: a system allowing international trades in CO_2 emission rights, such as the system authorized under Article 17 of the Kyoto Protocol.[9] Specifically, assume that the United States is obliged to reduce its emissions to a level of 1,243 million metric tons per year (during the years 2008–2012) and that Japan is required to reduce its emissions to a level of 290 million metric tons per year during that period. A system of international trades in greenhouse gas emission rights would involve the following. First, each country would give out carbon emission permits to firms, with each permit entitling the recipient firm to a unit of emissions, say 1 metric ton. In the United States, 1,243 million permits would be given out, so that the total allowable emissions would be 1,243 million metric tons, the same as the national target. Similarly, Japanese firms would receive from their government a total of 290 million permits. Recall that the marginal cost of reducing emissions was much higher in Japan than in the United States. If there were no trades of permits between the United States and Japan, then at least one Japanese firm would incur a cost of $400 to reduce its emissions to the level associated with the number of permits owned.

However, if firms are offered the opportunity to trade permits internationally, the costs of reducing emissions can be brought down, and all parties involved in the trade can benefit. For example, suppose the Japanese firm that faces marginal abatement costs of $400 buys a permit from a U.S. firm. If the Japanese firm can buy a permit for anything less than $400, it will be better off. Because it would own one more permit, it would be entitled to emit one more unit of CO_2. This means the firm would avoid $400 in abatement costs. The net savings to the firm from this trade is thus $400 minus the cost of the permit.

U.S. firms can also benefit from the trade. Suppose the U.S. firm faces marginal abatement costs of $240. If it can sell the permit for anything above $240, it will benefit from the trade. Selling the permit obligates the U.S. firm to reduce its emissions by one more unit (because it owns one less permit). This means the firm will suffer a cost of $240. But the firm will make money on the deal if it sells the permit for more than this cost. Let's assume that the firms agree on a price of $300 for the permit. In this case, the Japanese firm benefits by paying $300 to avoid a cost of $400, thereby gaining $400 – $300 = $100. The American firm benefits by receiving the $300 permit price minus the $240 needed to abate the additional unit of emissions, thereby gaining $300 – $240 = $60. Thus, the purchase and sale of permits at prices between the marginal costs of different countries can make firms in both countries better off than they would be with no such flexibility.

International trades in permits also reduce the world's cost of reaching given emission targets. When Japan buys permits from the United States, it compels the United States to engage in additional emission abatement (because fewer permits are owned by U.S. firms) and enables Japan to emit a bit more. In essence, the trades promote emission reductions by firms that can achieve the reductions most cheaply. Because the total number of permits in circulation has not changed, there is no change in total emissions: Only the distribution of the emissions around the globe changes.

Figure 4.4 heuristically depicts the impact of emission trading on the abatement achieved in the United States and Japan. In the absence of trading, the United States would need to achieve abatement level A_1, and Japan would need to achieve level A_1^*. By purchasing permits from the United States, Japan can move to abatement level A_2^*. The sale of permits obliges the United States to expand abatement to A_2. After purchases and sales have moved the United States and Japan to A_2 and A_2^*, there is no further potential for gains from trade because marginal abatement costs (MC) are the same. In theory, the ultimate price of permits (p) will be equal to the marginal abatement costs of each country (which, we have just seen, will be equal to each other when there is no further potential for gains from trade).[10]

We have illustrated the potential savings from trades by focusing on trades between the United States and Japan. But the same principles apply when many countries are simultaneously involved in trades. In general, countries with high marginal costs of emission reductions will tend to purchase permits from coun-

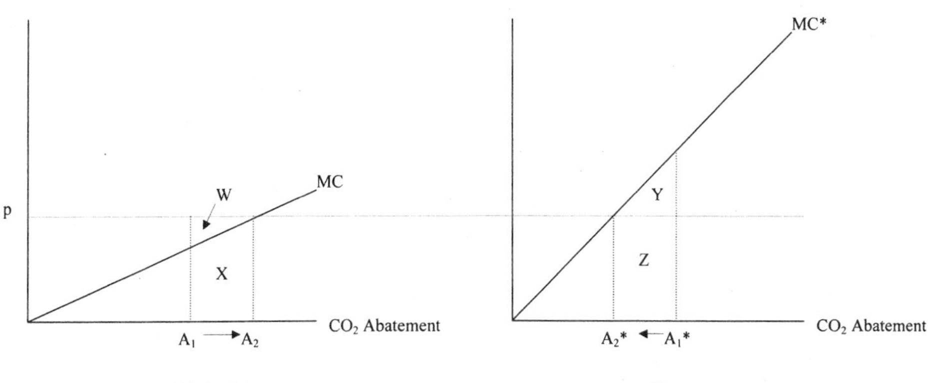

FIGURE 4.4. Abatement under emission permit trading.

tries with low marginal costs of emission reductions. For the high–marginal cost countries, the price of the permit is lower than the cost savings associated with the ability to generate more emissions. For the low–marginal cost countries, the receipt from the sale of permits more than compensates for the obligation to reduce emissions a bit more. Trades are attractive not only because they lower costs to participating countries but also because nations as a whole achieve emission reductions at lower cost. Thus, a virtue of emission permit trading is cost-effectiveness: the ability to achieve a given target at low cost. An international agreement involving tradable emission permits is considered more cost-effective than one without trades because emission reduction goals can be achieved at lower cost with trades than without.

Trades can yield cost savings whenever there are differences between potential traders in the marginal costs of emission reduction. Trades help iron out these differences. In our example with the United States and Japan, the sale of permits by U.S. firms forces the United States to move outward (to the right) along its marginal cost curve, so U.S. marginal costs rise. In contrast, the purchase of permits causes Japan to move to the left on its marginal cost curve, thus reaching a portion of the curve with lower marginal costs. In theory, permit trades continue until there no longer are any differences in the marginal costs between potential traders.

Computer models have examined the potential of international permit trading to reduce costs. Table 4.2 indicates the cost savings from such trades, based on 4 of the 11 models included in a recent study by the Stanford Energy Modeling Forum and reported in *The Energy Journal.* The results indicate significant reductions in the cost (as described by year 2010 GDP loss in 1990 U.S. dollars) of meeting the Kyoto targets in the presence of trading. Cost savings for the United States for example, range from roughly $16 billion to approximately $94 billion when a no-trading scenario is compared with trading between Annex B (i.e., industrialized) nations only. The cost savings are from $31 billion to $153 billion when the no-trading scenario is compared with a global trading system.[11] Global trading offers the greatest opportunities to exploit gains from trade, so the cost savings potential is highest in this case.

These computer analyses tend to assume that the emission trading system functions well, i.e., that nations meet their obligations and that firms comply with the rules by emitting no more than the amount allowed by the number of permits they possess. In fact, such compliance is unlikely to occur unless there is an effective international enforcement mechanism. The enforcement agency needs to be able to detect violations of the rules and to impose stiff penalties for violations. Developing the international institutions for such enforcement is no easy task, yet doing so is crucial for emission trading to work.[12]

TABLE 4.2. Cost of Meeting Kyoto Protocol Targets in the Presence and Absence of Permit Trading

Model	No Trading	Annex B Trading		Global Trading	
	GDP Loss*	GDP Loss*	Percentage Cost Saving Relative to No Trading	GDP Loss*	Percentage Cost Saving Relative to No Trading
MS-MRT					
United States	181	87	52%	28	85%
Japan	78	14	82%	2	97%
European Union	60	10	83%	4	93%
Oxford					
United States	158	90	43%	49	69%
Japan	83	23	72%	14	83%
European Union	195	60	69%	45	77%
MERGE3					
United States	90	42	53%	18	80%
Japan	34	8	76%	N/A	N/A
European Union	105	50	52%	20	81%
G-Cubed					
United States	36	20	44%	5	86%
Japan	23	19	17%	5	78%
European Union	170	60	65%	25	85%

Source: Weyant, J. P. and J. N. Hill, 1999: "Introduction and Overview," *The Energy Journal,* Kyoto Special Issue. GDP losses are estimated from Fig. 9, pp. xxxiii–xxxiv.

•MS-MRT (Multi-Sector–Multi-Region Trade Model), developed by Charles River Associates and the University of Colorado.
•Oxford, developed by Oxford Economic Forecasting.
•MERGE3 (Model for Evaluating Regional and Global Effects of GHG Reductions Policies), developed by Stanford University and the Electric Power Research Institute.
•G-Cubed (Global General Equilibrium Growth Model), developed by Australian National University, University of Texas, and the U.S. Environmental Protection Agency.

For more information, please see the special issue of *The Energy Journal* cited above.

*In billions of 1990 U.S. dollars.

The Significance of the Breadth of the Trading System

The costs of achieving given global reductions in GHG emissions also depend on the breadth of the international agreement—that is, on the number of countries that take part in international trades in emission permits. When more countries participate, there is greater potential to exploit gains from trade by concentrating the abatement among enterprises than can reduce emissions most cheaply. The results in Table 4.2 illustrate this principle. Under global trading (with all countries involved), the gains from trade are significantly larger than when only the Annex B countries are involved in the trading system.

The breadth of the agreement also affects the directions of trades. If only the Annex B countries are involved and if emission targets are based on their 1990 emissions, then eastern European nations and the nations of the former Soviet Union would be the primary sellers of permits. The reason is that the carbon emissions in these countries have declined by about 40 percent because of the nearly 50 percent decline in their GDPs since 1990. Because permits are assumed to be distributed based on 1990 emissions, these countries will have more permits than actual emissions and will thus be willing to sell a large share of their permits. As indicated in Table 4.3, the central forecast of the Energy Information Administration of the U.S. Department of Energy indicates that emissions from the former Soviet Union will not reach 1990 levels until after 2020. A crucial and highly controversial issue is whether special provision should be made in emission allocations to account for the significant reductions in emissions by the former Soviet Union and eastern Europe since 1990. In most analyses, the United States is perceived to have low costs of emission abatement, and most simulation studies indicate that under an emission trading system involving only the Annex B countries, the United States will be the main seller of permits other than the former Soviet Union.

If developing countries are involved in trades, however, the direction of trading is likely to differ. A typical scenario for a broader agreement stipulates that the initial allocation for less developed countries (LDCs) is simply their projected baseline emissions, that is, the level of emissions in the absence of regulation. The LDCs would be able to sell permits if they reduced emissions below this baseline. Under almost every projection of trades under a broader agreement, China would be a major player. China would be a huge seller of permits for two reasons. First, China's carbon emissions are projected to grow rapidly over the next several decades. As shown in Table 4.3, China's emissions are projected to grow at an average annual rate of 4.5 percent between 1999 and 2020. Over this time period, China's share of world trade emissions is expected to rise from 11 percent to 17 percent, and the ratio of its emissions to those of the

TABLE 4.3. Historical and Projected Carbon Emissions by Region (million metric tons carbon equivalent)

	History			Projections				
Region/Country	*1990*	*1998*	*1999*	*2005*	*2010*	*2015*	*2020*	*Average Annual Percentage Change 1999–2020*
Industrialized Countries								
North America	1,555	1,742	1,762	1,972	2,119	2,271	2,424	1.5
United States	1,345	1,495	1,511	1,690	1,809	1,928	2,041	1.4
Canada	126	146	150	158	165	173	180	0.9
Mexico	84	101	101	124	145	170	203	3.4
Western Europe	930	947	940	1,005	1,040	1,076	1,123	0.9
Industrialized Asia	357	412	422	447	460	479	497	0.8
Japan	269	300	307	324	330	342	353	0.7
Australasia	88	112	115	123	130	137	144	1.1
Total industrialized	2,842	3,101	3,124	3,424	3,619	3,826	4,044	1.3
Eastern Europe and Former Soviet Union								
Former Soviet Union	1,036	599	607	665	712	795	857	1.7
Eastern Europe	301	217	203	221	227	233	237	0.8
Total Eastern Europe and and Former Soviet Union	1,337	816	810	886	939	1,028	1,094	1.4
Developing countries								
Developing Asia	1,054	1,435	1,361	1,751	2,137	2,563	3,012	3.9
China	617	765	669	889	1,131	1,398	1,683	4.5
India	153	231	242	300	351	411	475	3.3
Other Asia	284	439	450	562	655	754	854	3.1
Middle East	231	325	330	378	451	531	627	3.1
Africa	179	216	218	262	294	334	373	2.6
Central and South America	178	246	249	312	394	492	611	4.4
Total developing	1,642	2,222	2,158	2,703	3,276	3,920	4,623	3.7
TOTAL WORLD	5,821	6,139	6,092	7,013	7,834	8,774	9,761	2.3

Source: Energy Information Agency (EIA), 2001: *International Energy Outlook 2001* (Washington, DC: United States Department of Energy), Table A10.

The U.S. numbers include carbon emissions attributable to renewable energy sources.

United States is projected to increase from 2/5 to 4/5. Under any agreement that confers permits to LDCs on the basis of (unconstrained) baseline projections, China would obtain a very large share of the global total. Second, studies show that China's energy use (in particular, its use of coal) is highly inefficient, implying that significant reductions could be made at very low cost. Thus, most simulation studies show China with a very large number of permits that could be sold at a profit because its abatement costs would be well below those of most industrialized nations.

Under a broader agreement, other LDCs would also be permit sellers, along with the former Soviet Union and eastern European countries. Thus, under typical assumptions for the global allocations of permits, a broader agreement effectively amounts to an arrangement whereby the more industrialized nations pay states of the former Soviet Union and developing nations to reduce CO_2 emissions. This situation has led some analysts to question the fairness of emission trading. We shall address the issue of fairness and several other criticisms of a tradable permits system shortly, but first we will examine one last aspect of the design of an international emission trading system: the banking and borrowing of permits across different time periods.

Banking and Borrowing

Thus far, we have focused mainly on static issues in the design of an international emission trading system, but climate change is a dynamic problem that will take many years to combat. Adding an analysis that considers economic effects over time is necessary. Many analysts have endorsed introducing intertemporal flexibility in international emission trades through banking and borrowing provisions. Under banking, a nation that reduces emissions below the level implied by its holdings of emission allowances can bank the difference—that is, apply the difference to future abatement obligations. Similarly, under borrowing, a nation can exceed the level of emissions implied by its current allowance in a given year by borrowing on future emission allowances.

In theory, the added flexibility provided by these banking and borrowing provisions enables nations and firms to lower the overall discounted costs of achieving given emission reductions. There is fairly strong empirical evidence that banking can generate large cost savings. Banking already seems to play an important role in the performance of the SO_2 trading program under the 1990 Clean Air Act Amendments, and it was important to the success of the lead rights trading program of the 1980s.[13,14] In the climate change context, numerical simulations by Manne and Richels[15] indicate that intertemporal emission trading—or "when" flexibility—is nearly as important as the "where" flexibility in enabling nations or firms to trade permits with each other.[16]

In the Manne–Richels analysis, borrowing leads to substantial reductions in the present value of abatement costs. Because national targets are tighter in the short run (i.e., the present value cost of meeting earlier targets is greater than the present value cost of meeting subsequent ones) it is financially prudent for nations to borrow from their future permit holdings. In theory, nations would borrow to the point at which the present value costs are equal across all time periods because doing so minimizes the aggregate present value cost of emission reductions for all periods.

Two potential difficulties arise in connection with intertemporal flexibility. First, the time profile of firms' abatement can affect long-run concentrations of CO_2. There is a natural removal rate of CO_2 from the atmosphere. Consider two alternative abatement paths that involve the same cumulative abatement. The abatement path that involves less abatement in the near term will exploit the natural removal process more and therefore contribute less to the concentration of CO_2 over the long term. Thus, borrowing reinforces efforts to reduce CO_2 concentrations, and banking weakens such efforts. In these ways, banking and borrowing can affect concentrations and impacts on climate.[17]

Second, nations that do not plan to comply with their emission quotas can disguise this intention through continual borrowing. Because of the difficulty of enforcing an international agreement, this could be a serious source of tension.

Some Criticisms of International Emission Trading Systems

We have emphasized a key virtue of international trading in CO_2 emission rights: the ability to lower the overall costs of achieving given overall targets for emission reductions. Notwithstanding this attraction, many participants in policy debates are uneasy with the prospect of emission trading. Note that under emission trading, nations that purchase additional permits can avoid reducing their own emissions by as much as they would have in the absence of trades. Some participants in international discussions have raised an ethical objection to emission trading, asserting that it is immoral to grant nations the ability to buy their way out of emission reductions on their own soil. This criticism has been raised by representatives from LDCs and by representatives of some European countries.[18]

Others have criticized emission trading in light of its implications for the speed of transition to a low-carbon or carbon-free economy. Some participants in the international discussions have argued that for nations that purchase emission permits, the ability to meet national targets through purchases of permits slows down the transition away from dependence on fossil energy and that a faster transition is preferable.

Motivated by the idea that nations have an obligation to reduce emissions on their own soil, some policymakers have recommended that a principle of supplementarity be followed in international agreements. Under the supplementarity principle, whatever emission reductions a nation achieves through purchases of emission permits must supplement a certain level of domestically achieved reductions.[19] In the Conference of Parties held at the Hague in November 2000, European representatives advocated adherence to the supplementarity principle, arguing that nations should not be able to meet the bulk of their required reductions by purchasing permits.[20] To enforce the supplementarity principle, policymakers have launched proposals under which each industrialized nation must achieve a certain fraction of its national emission reduction target through domestic reductions. This effectively puts a limit on the number of emission permits a nation can purchase.

From a strictly financial point of view, it is hard to justify restrictions on emission trading. Such restrictions reduce the abilities of both purchasers and sellers to benefit financially, and they raise the global economic cost of meeting given targets. Buyers benefit from trades because the ability to generate additional emissions is worth more than the cost of the additional permits purchased. Sellers benefit because the revenues from the sale of permits exceed the costs of additional abatement. Thus, restrictions on trades hurt potential buyers and sellers and raise the global economic costs.

The "transition" argument—that restrictions on emission trading are necessary to avoid unacceptable slowdowns in the transition away from a fossil-based economy—does not seem fully convincing. It is likely that such restrictions would induce faster development of low-carbon or carbon-free technologies in countries that otherwise would meet most of their national commitments through purchases of permits. However, these same restrictions also would imply a more prolonged reliance on fossil fuels by nations that otherwise would sell more permits. It is not clear whether, for the world as a whole, such restrictions would promote a faster weaning from fossil fuels.

Still, some critics of trades argue that financial gain, global economic cost, and the speed of transition away from a carbon-dependent economy are not the only relevant considerations. They maintain that trades produce a moral cost that overrides the financial gains to participating parties. There is no simple answer to this point. Is it immoral to shift the international locus of abatement to places where abatement can be achieved at the lowest cost? And if so, does this justify restricting or eliminating the possibility for trade when doing so eliminates financial benefits to potential traders? These are questions that must be addressed in the policy debate and, ultimately, through the political process.

An Alternative: The International Carbon Tax

So far we have focused on international approaches involving national emission targets, perhaps accompanied by provisions for emission permit trading. A key alternative to these approaches is an international carbon tax. Under this policy, every participating country would impose a tax on its emissions of carbon dioxide. The effective tax rate on emissions would be the same for all countries. The simplest form of the international carbon tax would impose the tax on suppliers of fossil fuels—coal, crude oil, and natural gas—with the tax based on the carbon content of each fuel. This tax would cause the prices of fossil fuels to rise, which in turn would cause the prices of various fuel-based products to go up because the amount of carbon dioxide released from combustion of refined fuel products (e.g., gasoline) is proportional to the carbon content of such products. By imposing a tax on fossil fuels in proportion to their carbon content, governments effectively impose a tax on carbon dioxide emissions.

Under the international carbon tax, the higher prices of carbon-based products provide incentives for industrial, commercial, and residential users to reduce their demands for such products. The higher the carbon tax rate, the greater the reduction that would be achieved. Studies indicate that to achieve the 7 percent reduction in CO_2 emissions stipulated under the Kyoto Protocol, the United States would need to impose a carbon tax of $50–$150 per ton. This would cause coal prices to rise by more than 100 percent and would cause oil and natural gas prices to rise by 35–40 percent. The price of gasoline would rise by 12–14 percent. Under an international carbon tax policy, there would be no need for internationally traded emission permits. In principle, if the tax is introduced on an international basis, then, in all countries, users of carbon-based products will reduce their demands until the marginal cost of doing so is just equal to the benefit from doing so. This benefit is the value of the avoided tax. If the tax rate is the same in all countries, then the benefit from reducing demands will be same in all countries, and the marginal costs of reducing carbon use will be equated across countries as well. Thus, in theory, the tax automatically causes marginal costs of emission reduction to be the same across participating countries, so there are no opportunities for gains from trade in emission permits.

These ideas are expressed by Fig. 4.5. Like Fig. 4.4, this figure shows the marginal costs of abatement in the United States and Japan. Here the interna-

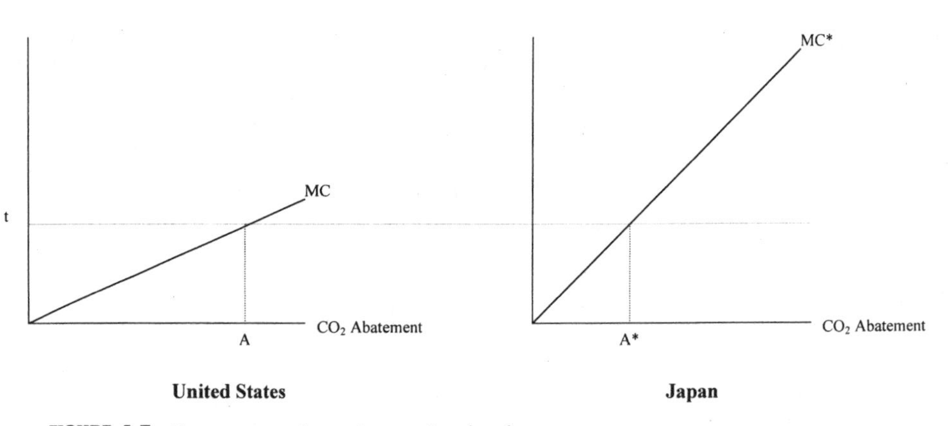

FIGURE 4.5. Abatement under an international carbon tax.

tional carbon tax is at a rate of *t*. Firms in the United States and Japan will pursue abatement until the marginal cost of abatement is equal to the tax rate. If firms pursued less than that amount of abatement, then the potential tax savings (given by the tax rate) for an additional unit of abatement would exceed the cost of abatement. Because abating the unit of emissions is cheaper than paying the tax on it, firms would reduce their emissions further to avoid the tax. Similarly, if firms pursued more than the amount of abatement indicated by A and A^*, the cost of the last units of abatement would exceed the tax savings associated with those units. Firms therefore would be better off paying the tax on those units than abating then. Thus, costs are minimized by abating up to the point where the marginal cost of abatement equals the tax rate. Note that when the United States and Japan are at A and A^*, the marginal costs of abatement are the same for the two countries, so there are no potential gains from permit trading.

Suppose that the international carbon tax is equal to the value of p in Fig. 4.4. In this case, Japan's abatement level A^* in Fig. 4.5 is exactly the same as its abatement level A_2^* in Fig. 4.4 under emission trading. But Japan is worse off under the carbon tax than under emissions trading. (Can you explain why?[21]) Therefore, although the emissions under both policies are the same, the distributional impacts are different.

Evaluating the International Carbon Tax

Three issues are relevant in comparing an international carbon tax to a tradable permit system: the ability to address distributional concerns, differences in the locus of uncertainty, and differences in transaction costs. We discuss each of these areas in turn.

Distributional Concerns and the Need for Explicit Transfers

We have seen that under a system involving national targets, the relative cost of emission reductions depends heavily on the stringency of a country's target: The greater the reduction required, the higher the cost. Although the global cost of meeting a given global reduction might be similar under two policies in which national targets differed, the distribution of this cost across various countries could differ dramatically.[22] Thus, the allocation of national targets affects the distribution of costs among nations and can in principle be used to address distributional concerns.

In contrast, a (uniform) international carbon tax by itself is not flexible in its distributional impact. If the international distribution of the burden of this carbon tax were politically unacceptable, a set of explicit international monetary transfers would be necessary to alter the distributional impacts. One way to accomplish such international transfers would be through an international agency that collects carbon tax revenues and then spends them in agreed-upon ways.

There might be stiff political opposition to entrusting an international agency with fund transfers of the magnitudes that would be anticipated, and the transparency of the transfers itself could generate controversy. Based on these considerations, some have argued that the need for explicit international transfers is a basic disadvantage of an international carbon tax relative to a system of national targets with emissions permit trading.[23]

Differences in the Locus of Uncertainty

A system of national targets makes it clear what the total emissions by participating countries will be. This total is simply the sum of the reductions called for by each of the participating countries. This is the case whether or not the system of national targets is accompanied by provisions for international trades in emission permits. Allowing for emission permit trading does not affect the emission total because (assuming full compliance by participating countries) the total number of permits in circulation is simply the sum of allowable emissions under the original national targets. Although permit trading has no effect on the total emissions, it does change where the emissions are generated. Thus, under a system involving national targets (whether or not this is accompanied by permit trading), there is relatively little uncertainty about the quantity of emissions. What is uncertain is the marginal cost of achieving the emission reductions necessary to realize this quantity.

Under an international carbon tax, the locus of uncertainty is different. In principle, the tax makes it clear what the marginal cost of reducing emissions will be; it is simply the carbon tax rate. For example, if the carbon tax is $25 per ton of CO_2 emissions, then producers will find it advantageous to reduce demands for carbon-based fuels until the marginal cost of doing so is equal to the avoided tax payment of $25 per ton. What is uncertain under the carbon tax is the quantity of emissions that will be generated. This depends on how much producers can reduce their use of carbon-based fuels before the marginal cost of doing so reaches the carbon tax rate.

Which uncertainty is worse? There is no easy answer. Some scientists and environmentalists are concerned that allowing emissions and concentrations of CO_2 to cross a critical threshold could have calamitous effects on climate. These people might prefer the approach involving national targets, assuming that the targets imply total emissions and concentrations below the threshold value. To these analysts, an international carbon tax is too risky because it does not indicate in advance what level of emissions will result. On the other hand, many business groups express concern about the uncertainties inherent in the national targets approach, pointing out that such an approach does not make clear what it will cost to reach the targets. These groups would prefer the international carbon tax because the tax rate indicates the maximal marginal costs of abatement. These differing viewpoints about uncertainty continue to be expressed in debates about climate policy.[24]

Differences in Transaction Costs

Transaction costs are the costs of finding buyers and sellers in a market and the costs of coordinating transactions. Once implemented, a carbon tax might involve small transaction costs, particularly if the tax is applied at the source (at the mine mouth for coal, at the wellhead for oil and natural gas). Transaction costs would be low in this case because the tax could be incorporated in the price of the fuel, just as sales tax is added to the cost of consumer goods. This does not make it more difficult to obtain the good or trade the good and therefore would not cause significant transaction costs. Transaction costs associated with permit trades might be more significant, although much depends on the particular institutional arrangement involved. Although a well-functioning international market in which permits are traded could involve low transaction costs, regulations and other barriers to trade could increase the cost of finding a trading partner and thus discourage advantageous trades. If these costs are significant, the overall attractiveness of an emission trading system will be reduced.

A Hybrid Approach: National Targets with a Safety Valve

We have thus far focused on an international emission permit trading system and an international carbon tax separately. What would happen if we combined these approaches?

As mentioned, under a system with national emission targets, it is not clear in advance what the marginal costs of abatement (or equilibrium permit price) will be. In contrast, under the international carbon tax, one cannot tell in advance what the emission levels will be. Some analysts have proposed a hybrid approach that they claim combines the best features of the national targets and carbon tax systems.[25]

The hybrid approach is a national target system with a safety valve. Each participating nation would agree to a national target for emissions. Within each nation, emission permits would be allocated to firms, with the total number of permits equal to the national target. Permits would be bought and sold on the international market. However, once the price of permits reaches a given trigger price, governments in each country would offer to sell additional permits to firms at that price. This prevents the permit price from exceeding the trigger price; this is the safety valve.

An attraction of this hybrid approach is that it puts a ceiling on—and thus reduces uncertainty about—the marginal costs of abatement. In addition, because national costs under this policy are determined largely through the allocation of national targets, it avoids the explicit financial transfers that might have to accompany an international carbon tax. At the same time, the hybrid approach introduces uncertainty about the equilibrium level of emissions. Once governments begin to sell additional permits, the global quantity of permits is increased beyond the quantity associated with the sum of the national targets. Thus, this policy allows global emissions to rise beyond the amount initially prescribed under the national targets.

Environmental groups in the United States have already expressed strong criticism of this scheme on the grounds that it leaves open the possibility that emissions will not be reduced substantially; producers could purchase a large number of additional permits if the trigger price is set too low. Indeed, in late October 1997 several prominent U.S. environmental groups wrote an open letter to the Clinton Administration condemning the hybrid system on these grounds.

For the safety valve approach to work, it must be implemented by all countries that participate in emission trading. To see why, suppose just one country—say, France—introduced the safety valve. If this were the case, then as soon as the permit price reached the ceiling price, all potential purchasers of permits would buy permits from France because the French government was promising to sell at the ceiling price and other potential sellers would require higher prices. This would be great for the French Treasury—it is as if France were levying a tax on emissions from all over the world. But many countries are likely to be unhappy with this outcome, which redistributes global wealth to France. This problem can be overcome only if all governments agree to sell permits at the ceiling price.

Therefore, the safety valve must be implemented internationally. An effective safety valve mechanism also must include careful attention to how the revenues from permit sales are used. If such revenues are controlled by national governments, the effectiveness of the international mechanism might be jeopardized. For example, a given nation could use these revenues to finance subsidies to producers of fossil fuel. This obviously would undermine the purposes of the international agreement. Recognizing this potential problem, Kopp, Morgenstern, and Pizer (2000) urge that the revenues from the additional permit sales should go directly to an international fund rather than to national governments. The revenues in this fund would be used to finance emission-reduction projects in various countries, based on a reverse auction.[26] In theory, such an auction would ensure that the revenues are devoted to projects that complement rather than undermine the objectives of reducing greenhouse gas emissions.

Project-Based Mechanisms

The approaches we have discussed so far focus on putting a price on greenhouse gas emissions, but there is another set of flexibility mechanisms that can be used to reduce the overall cost of emission abatement. These mechanisms are based on private investment in projects that lead to emission reductions in other countries, hence the name *project-based mechanisms.* Two such mechanisms are included in the Kyoto Protocol: joint implementation and the Clean Development Mechanism.

Joint Implementation

Joint implementation (JI), described and endorsed under Article 6 of the Kyoto Protocol, is an example of a project-based mechanism for reducing the global costs of achieving emission reductions. Like emission permit trading, it operates with a backdrop of national commitments to reduce emissions to specified targets. However, in the case of JI, permits are not exchanged. Rather, a firm in one nation (Country A) sponsors a project—for example, the construction of a modern electricity-generating plant—to reduce emissions in another nation (Country B). Both parties indicate in a contract how much emissions will be reduced in the latter country as a result of the project. Suppose the emission reduction is *X.* Based on this contract, the firm in Country A gets credit for *X* units of emission reductions, and the total allowable emissions in Country B are reduced by *X.* Both parties benefit. For firms in Country A, it is cheaper to finance the project in Country B than to reduce emissions on its own (or buy additional emission permits). For Country B, gaining technical and financial assistance from the Country A firm is worth the price of having to reduce emissions by *X* more units.

The ultimate basis for cost savings under JI is the same as under emission trading: The policy encourages emission reductions where they can be implemented most cheaply. Both parties in the JI transaction benefit, and the global costs of achieving given emission targets are reduced.

JI can be introduced in parallel with emission permit trading. If both mechanisms are in place, a given firm holding emission permits can choose among a number of options. If the firm finds it exceptionally costly to reduce emissions to the level dictated by its current holdings of emission permits, it can purchase additional permits at the market price. Alternatively, if it can sponsor a JI project in another country at low cost (that is, at lower cost than the cost of achieving comparable emission reductions and lower than the cost of purchasing permits), it will be financially more attractive for the firm to take this option.

The Clean Development Mechanism

In Article 12, the Kyoto Protocol articulates and endorses another flexibility mechanism: the Clean Development Mechanism (CDM). This mechanism is similar to JI in certain respects. Like JI, the CDM gives emission reduction credit to firms in an industrialized nation for the reductions that occur as a result of its financing of an investment project in another country. However, unlike JI, the CDM gives credit for emission reducing projects in countries that do not face national caps. Under the Kyoto Protocol, it is the developing countries that do not face such caps. Therefore, in the context of the protocol, the CDM is a means for developed nations to receive emission reductions credits by financing projects in developing nations.

The CDM has been embraced as an important vehicle for involving developing countries in the global effort to reduce greenhouse gas emissions. In principle, both the investing firm and the developing country benefit from a CDM project that is voluntarily agreed upon by both parties: The investing firm benefits from the emission reduction credits, and the developing nation benefits from the transfer of technology and the increased sustainability of its economy.

However, the CDM faces some difficult problems that do not appear under JI or emission permit trading. The most important seems to be that of ensuring that CDM projects reduce emissions in the country where the project is undertaken. To see how this problem arises, compare what happens under a JI program with the situation under the CDM. When a JI project is undertaken, the country receiving financial assistance has its allowable national emission total reduced by the emission reduction identified with the JI project. This is the price that the recipient country pays for the project's financing. Likewise, the firm that finances the project is entitled to increase its emissions by the same amount; this is its compensation for financing the JI project. The changes in entitlements in the financing country and the country receiving assistance offset each other, so there is no change in global emissions.

Ideally, a similar situation would arise under the CDM: The presence of a CDM project would cause emissions in the developing country to be lower than they would otherwise be, exactly offsetting the financing firm's entitlement to generate more emissions. However, because there are no national caps on the developing country's emissions, it is difficult to know whether developing country emissions are reduced by the CDM project. Furthermore, it is possible that the CDM project would have been undertaken anyway, so that the CDM project has no real impact on national emissions. This is sometimes called the problem of additionality: The emission reductions under a CDM project are supposed to be additional to any emission reductions that otherwise would have occurred.

Some analysts suggest that these problems can be overcome if sector-specific baselines are established for developing countries.[27] Such baselines would indicate total emissions that the developing country could generate in connection with a given industrial sector (e.g., the electricity-generating sector). Under this plan, a precondition for a developing country's involvement in CDM would be the establishment of sector baselines. If a developing nation wanted to accept a CDM project in a particular sector, its emissions from that sector would have to be reduced below the sector baseline by an amount equal to the emission reduction associated with the CDM project. In other words, there must not be an increase in emissions elsewhere in the sector that offsets the reduction from the CDM project.

To many analysts, sector baselines make the CDM a viable and effective option. However, others express skepticism about this mechanism. They argue that the CDM cannot be counted on to yield global emission reductions without specific baselines and express doubts as to whether sector-specific baselines are workable or politically acceptable.

Another design issue, which could offset the baseline problem, is the extent to which CDM projects are regulated and approved by an international governing body. Although institutional oversight may help prevent the baseline problem, it will increase the cost of undertaking a CDM project, thereby reducing the cost-effectiveness of CDM: Cheap CDM projects that would have been undertaken in the absence of institutional oversight will no longer be viable. The reduced cost-effectiveness that accompanies stricter oversight has led some analysts to argue against such oversight, despite the baseline problem.[28]

In theory, the CDM benefits all parties that participate in it. The decision by a developing country to participate, in particular, suggests that the value of the technological assistance exceeds the developing nation's own sacrifice in undertaking the project. However, several developing country representatives and analysts from developed nations are skeptical about the benefits of CDM. Analysts have raised two major concerns:

- If decision makers do not consider all the national sacrifices involved, or if the financial transfers are not put to good use, participation could fail to benefit the developing country, they argue. The benefits of the CDM to developing countries could be offset by misuse of financial transfers or by weakening the bargaining position of the developing country in future negotiations that might include emission targets for developing countries.[29]
- If developing countries will eventually have to face binding national targets in the future, the CDM could increase developing countries' cost of meeting those targets because the CDM projects "may end up using most of the cheapest mitigation opportunities."[30] This "low-hanging fruit" problem could be addressed by giving developing nations the option to buy back the emission reductions credits from the sponsoring Annex B country at a prespecified price.[31]

Coverage Issues

Two further issues deserve emphasis in the design of international mechanisms to address climate change. We have already alluded to these issues in our previous analysis, but a more explicit treatment is necessary given their importance. These issues address the coverage of an international agreement and focus on which nations should be included and which gases should be controlled.

National Coverage: How Broad Should an Agreement Be?

The Kyoto Protocol includes binding commitments by the industrialized (Annex B) countries to reduce their greenhouse gas emissions. The developing countries have no binding commitments. In general, a system of national targets with emission permit trading becomes more cost-effective as it becomes broader, encompassing more countries. A broader trading system yields more opportunities for gains from trade. Thus, it would be more cost-effective to include developing countries within a system of tradable emission permits.

Participation by LDCs is ultimately crucial because their baseline emissions are projected to grow very rapidly over the next few decades. The growth in baseline emissions is particularly striking for China and India. No serious global reductions in the growth of GHG concentrations can take place without significant contributions by China, India, and other LDCs.[32]

Although the ultimate importance of LDC participation is clear, thorny issues remain as to the timing and extent of LDC participation. The timing issue remains a key source of controversy as nations consider whether to ratify the Kyoto Protocol. Some analysts argue that it makes most sense to push from the beginning for a broad agreement with LDC participation instead of starting with an agreement involving Annex B countries alone that might later be expanded to bring in the LDCs.[33]

Jacoby, Prinn, and Schmalensee[34] provide two main arguments to support the "start out broad" approach:

- Under a narrow agreement, carbon "leakage" will undo much of the desired global emission reduction. If the Annex B countries act alone, their reductions in carbon use will tend to decrease demand for fossil fuels, thereby depressing world prices. This will increase non–Annex B countries' demand for such fuels and lead to higher consumption of fossil fuels outside the Annex B countries. Reduced pretax fuel prices also will lower prices for goods and services produced with fossil fuels, particularly carbon-intensive goods such as refined petroleum products. This will stimulate higher consumption of these goods and services outside the Annex B countries, again potentially offsetting the carbon reductions by the Annex B countries. Bernstein et al.[35] argue that carbon leakage could be a significant problem under the Kyoto Protocol and estimate the ratio of the increase in carbon emissions in non–Annex B countries to the reduction of emissions in Annex B countries in 2010 to be around 16 percent. Their analysis suggests that the carbon leakage ratio will increase through time as well.[36]
- A narrow agreement promotes an increase in the share of global fossil fuel production and consumption by nonparticipating countries. This can expand dependence on fossil fuels by these countries, which could make it more difficult to attain their future participation in global carbon abatement efforts.

These authors argue that, in the short term, the breadth of an international agreement is far more important than its depth. A broad agreement involving modest emission reductions is viewed as preferable to a narrow one with more substantial emissions cuts.

However, other analysts argue that it is more realistic to start out with a narrower agreement and bring the LDCs into the fold later on. The Kyoto Protocol clearly is more consistent with this latter approach. Opponents of immediate LDC commitments raise two major arguments:

- Industrialized nations largely have caused the problem of elevated GHG concentrations. From this notion two distinct conclusions often are drawn:

  LDC participation should take place later. Brazil, China, India, Indonesia, and other LDCs should not face emission limits until their emissions reach parity (in terms of total tons) with those of the major industrialized nations.

  Emission trading would be unfair to LDCs. As mentioned earlier, some LDC representatives find it onerous that industrialized nations could buy their way out of abatement—in effect, paying for the privilege of maintaining high consumption of fossil fuels. Moral considerations indicate that industrialized nations should not buy their way out, even if emission trading could lower the financial costs of achieving given emission reductions and even if it could be financially advantageous to LDCs.

- Binding commitments could harm the LDCs in ways that are not acknowledged by advocates from industrialized nations. Stavins and Weiner indicate that LDC representatives at times express a fear of "carbon colonialism" and a belief that industrialized nations would use an international emission permit trading system strategically to their own advantage.[37]

Any attempts to involve LDCs in international agreements must confront these arguments and objections.

Pollution Coverage: What Pollutants or Activities Should Be Covered by an International Agreement?

Most discussions of international abatement efforts center on carbon dioxide (CO_2), the most important anthropogenic contributor to radiative forcing. However, the most cost-effective abatement policy would be one that causes marginal costs of reducing GHGs (in carbon equivalents) to be equated across all types of greenhouse gases. Thus, other things equal, it would enhance cost-effectiveness to include other GHGs in an international agreement. A 1999 study by Reilly et al.[38] estimates that the total annual abatement cost in 2010 under the Kyoto Protocol targets could be more than 60 percent higher when only CO_2 is regulated than under an agreement that includes other GHGs and carbon sinks.

However, the other GHGs are more difficult to monitor than CO_2. A detailed analysis by Victor (1991) concluded that monitoring problems effectively prevented including any GHGs other than CO_2 in an international agreement.[39] Although including other GHGs has become more feasible with advances in monitoring technologies, the Kyoto Protocol is still inexact in its treatment of the breadth of gases covered. The protocol refers to a basket of gases rather than focusing solely on CO_2 but gives little insight into how these gases are to be monitored. The potential monitoring problems with respect to greenhouse gases other than CO_2 raise questions as to whether Kyoto can be implemented effectively.

Another problem is how to establish the exchange ratios for different greenhouse gases.[40] These ratios determine how emission reductions of different gases can be added up or converted to carbon equivalents. Suppose a given nation reduces methane emissions by 4 million tons. How many units of carbon dioxide emissions is this equivalent to? In principle, the different gases should be exchanged according to their impact on the environment and human welfare. If a ton of methane does twice as much damage as a ton of carbon dioxide, a nation should get twice as much credit for a ton of methane reduced as it gets for each ton of CO_2 reduced. But the relative impact of a ton of CO_2 and a ton of methane can vary at different points in time because of different residence times in the atmosphere and because of interactions between the stocks of different gases. Moreover, there are huge uncertainties as to the damages produced by the different gases. These complications pose significant challenges to the task of coming up with exchange ratios across greenhouse gases.[41]

Thus, there are significant difficulties in extending the coverage emission agreements to incorporate gases other than CO_2.[42]

A parallel breadth question applies to carbon sinks: Should an international agreement give credit to carbon sequestration through afforestation? The Kyoto Protocol includes carbon sequestration as a means of reducing national emissions, expressing countries' obligations in terms of net emissions (carbon emissions minus the absorption of carbon attributable to expanded sinks).

Carbon sequestration raises a baseline problem analogous to that which arises under the CDM: Some forest expansion would occur even in the absence of an international agreement, and it is difficult to determine which expansion is linked to the agreement.

The sequestration issue was a sticking point at the Conference of Parties meeting at the Hague in November 2000. Representatives from the United States viewed sequestration credits as essential to meeting the U.S. emission targets as mandated by the Kyoto Protocol. However, European representatives and environmental groups believed that the American proposals would illegitimately relax U.S. obligations to cut back on greenhouse gas emissions and would prevent the protocol from achieving its environmental goals. The impasse on sequestration credit ultimately dominated the discussion at the conference and was never resolved satisfactorily.[43]

Recently, researchers have explored the possibility of chemical sequestration of carbon, which involves capturing CO_2 after it is generated by the combustion of fossil fuels or refined fuels. This sequestration is still in the experimental stage, but at some point it might make sense to consider incorporating it in international arrangements to control atmospheric CO_2.[44]

In the near term, however, the most promising approach for international agreements may be to focus on CO_2 emission abatement because monitoring is less of a problem for this gas: Emissions can fairly reliably be linked to domestic consumption of fossil fuels.[45] In subsequent agreements, the breadth of GHGs and of GHG-related activities (e.g., emission abatement, sequestration) could be expanded as the scope of the problems of monitoring and baseline definition become clearer.

Conclusions

A decade ago there were no broad-based discussions of international approaches to the problem of global climate change. Now, with the Kyoto Protocol as the centerpiece, there is a great deal of discussion worldwide of a great many policy options. Substantial uncertainties remain as to what will be the shape of an international agreement on global climate policy. Indeed, given that few nations have ratified the Kyoto Protocol since its adoption in 1997, and given the inability of the parties to agree on the implementation details of the protocol, it is possible that no agreement will be reached and implemented for quite a while, if at all.

The Kyoto Protocol focuses on a system of national targets combined with various flexibility mechanisms. In this chapter we have described and evaluated the various potential features of this system. In addition, we have explored significant alternatives to the Kyoto approaches, including an international carbon tax and a hybrid policy that combines features of national targets and emission trading with elements of an international carbon tax.

It is possible to become overwhelmed by the divergence of opinion on the relative merits of different policy options and the associated uncertainties about what the future will bring. However, the wider scope and increased sophistication of the international discussion of policy options are encouraging signs.

Three points deserve emphasis. First, although the Kyoto Protocol does not include binding emission targets for LDCs, the participation of LDCs ultimately will be crucial for any serious global reductions in GHG concentrations. The LDCs are likely to resist committing to significant sacrifices in the near future. Therefore, it may be fruitful to pursue a policy that engages the LDCs in a legally binding way without committing them to significant near-term emission reductions. One such policy would include the LDCs in an emission trading system while making the emission allocations to LDCs sufficient to avoid requiring any significant reductions in the short term. (For example, the allocation to LDCs could reflect the projected unconstrained baseline for these countries.) Even if this has little or no effect on near-term emissions by LDCs, it involves LDCs in the global abatement process, which may enhance the prospects for serious LDC emission reductions in the more distant future.

A second point is the importance of flexibility in global efforts to address the prospect of climate change. There are great uncertainties about current and future benefits and costs of reducing GHGs. This gives value to flexibility. It seems sensible to design policies that can adjust as scientific information and nations' circumstances change. The most attractive international approach would allow periodic adjustments concerning future global emission targets, changes in the range of GHGs and abatement activities it embraces, and modifications to the obligations of various countries.[46]

A final point is that it is useful to keep in mind the likely time frame of the climate change problem. If an international agreement is reached and implemented in the near term—whether it be the Kyoto Protocol or some other agreement—it is likely to be just a first step in a series of international efforts stretching over decades. One potential goal of current policies might be to lay the groundwork for future policies by helping develop appropriate legal and fiscal institutions for dealing with the problem of climate change—institutions that will remain useful even as the specific policies change.

Acknowledgments

We thank Ray Kopp, Stephen Schneider, Robert Shackleton, and Rob Stavins for helpful suggestions. We are especially grateful to Michael Toman for extensive and insightful comments on an earlier draft.

Notes

1. For a review of 13 computer modeling studies of abatement costs for industrialized nations, see Weyant, J. P. and J. N. Hill, 1999: "Introduction and overview," *The Energy Journal,* Kyoto Special Issue: vii–xliv.
2. As of December 11, 2001, although 84 parties had signed, only 46 had ratified or acceded to the treaty. Article 25 of the Kyoto Protocol states that the protocol will become international law when at least 55 of the parties to the convention, incorporating Annex I countries that account for no less than 55 percent of the total carbon dioxide emissions for 1990, have ratified the treaty (UN Framework Convention on Climate Change, 2002: "The Convention and Kyoto Protocol," http://unfccc.int/resource/convkp.html. Accessed February 4, 2002).
3. For instance, all 11 of the models reporting U.S. data in the 1999 Kyoto Special

Issue of *The Energy Journal* predict positive and increasing costs to emission reductions beyond approximately a 1 percent reduction from the 2010 U.S. baseline projection (Weyant and Hill, 1999, Fig. 10(a), p. xxxvii).

4. Weyant and Hill, 1999.
5. Schelling, T. C., 1992: "Some economics of global warming," *American Economic Review,* 82 (March): 1–14.
6. Cooper, R. N.,1998: "Toward a real global warming treaty," *Foreign Affairs,* 77 (March/April): 66–79.
7. More precisely, Annex B nations are required to reduce their emissions. Annex B countries are defined in Annex B of the protocol and include the OECD and the nations of the former Soviet Union and eastern Europe.
8. The conversion of one gas to another (in this case to carbon dioxide) is not a simple task. Differences in the residence times of various gases and uncertainties about the damage of different gases make determining the "exchange rates" between gases difficult. We discuss this issue more fully later in the chapter.
9. The interaction between international institutions that facilitate the permit trading system and domestic actions taken by national governments can have a significant effect on the benefits of an international emission permit trading system. For a discussion of these issues and their effect on the cost savings of flexibility mechanisms, see Hahn, R. W. and R. N. Stavins, 1995: "Trading in greenhouse permits: a critical examination of design and implementation issues," in H. Lee, ed., *Shaping National Responses to Climate Change* (Washington, DC: Island Press); and Hahn, R. W. and R. N. Stavins, 1999: "What has Kyoto wrought? The real architecture of international tradable permit markets," Discussion paper 99-30 (Washington, DC: Resources for the Future, http://www.rff.org/disc_papers/PDF_files/9930.pdf).
10. Fig. 4.4 also indicates the gains from trade. U.S. firms collectively gain area W. This is the revenue from sale of permits ($W + X$) minus the cost of additional abatement (X). Japanese firms collectively gain area Y. This is the benefit ($Y + Z$) associated with avoiding the need to extend abatement beyond A_2^* to A_1^*, minus the cost of purchasing additional permits (Z). These national gains correspond to global savings.
11. Numbers are estimated from graphs appearing in Weyant and Hill, 1999, "Introduction and overview," *The Energy Journal,* Kyoto Special Issue: pp. xxxiii–xxxiv (Fig. 9).
12. For an insightful discussion of the enforcement issue and other institutional issues associated with international climate change policy agreements, see Victor, D. G., 2001: *The Collapse of the Kyoto Protocol: Economics, Politics and the Struggle to Slow Global Warming* (Princeton, NJ: Princeton University Press), 160 pp.
13. Ellerman, A. D., R. Schmalensee, P. L. Joskow, J. P. Montero, and E. M. Bailey, 1997: "Sulfur dioxide emissions trading under Title IV of the 1990 Clean Air Act amendments: evaluation of compliance costs and allowance market performance," MIT Center for Energy and Environmental Policy Research, May.
14. Kerr, S. and D. Mare, 1997: "Efficient regulation through tradeable permit markets: the United States lead phasedown," Working Paper 96-06, Department of Agricultural and Resource Economics, University of Maryland, College Park, January.
15. Manne, A. and R. Richels, 1997: "On stabilizing CO_2 concentrations: cost-effective emission reduction strategies," Working Paper, Stanford University, April.

16. Manne and Richels, 1997, compare the costs of a fixed-quota system (with no provision for international or intertemporal trading) with systems involving international trading and intertemporal trading. Allowing intertemporal trading (banking and borrowing) lowers costs by about as much as allowing international trading.
17. In the preceding discussion we have implicitly assumed that impacts are based on the stock of greenhouse gases; however, impacts on climate can depend both on the stock of greenhouse gases and on the rate of change of the stock. For a discussion and critique of this assumption and other climate change modeling assumptions, see Schneider, S. H., W. E. Easterling, and L. O. Mearns, 1997: "Adaptation: sensitivity to natural variability, agent assumptions and dynamic climate changes," *Climatic Change,* 45 (1): 203–222; and Schneider, S. H., 2000: "Integrated assessment modeling of global climate change: transparent rational tool for policy making or opaque screen hiding value-laden assumptions?" *Environmental Modeling and Assessment,* 2 (4): 229–248.
18. This and other criticisms of international emissions trading are described in Part 1 of Kopp, R., M. Toman, and M. Cazorla, 1998: "International emissions trading and the Clean Development Mechanism," Climate Issue Brief No. 13, Resources for the Future, Washington, DC, October, http://www.rff.org/issue_briefs/PDF_files/ccbrf13.pdf.
19. The language of the Kyoto Protocol is inexact on the issue of supplementarity. Article 17 states, "Any such trading shall be supplemental to domestic actions for the purpose of meeting quantified emission limitation and reduction commitments under [Article 3]." However, the criteria for determining whether a purchase is "supplemental" to domestic actions presumably are left to the political process.
20. Anderson, J. W., 2000: "Why the climate change conference failed: an analysis," Weathervane Feature, Resources for the Future, Washington, DC, December 4, http://www.weathervane.rff.org/negtable/COP6/analysis_anderson.htm.
21. Under both emission trading and the international carbon tax, Japan must pay abatement costs equal to the area under its marginal cost curve MC^* from the origin to A^* (which is equal to A^*_2 in Fig. 4.4). But under emission trading, Japan enjoys a net gain of Y from trading (see Note 10). Thus, it is better off under emission trading. The United States is worse off under emission trading than under the international carbon tax. Note that if the initial allocations of permits had been different, with Japan receiving fewer (less than A^*_2) permits initially and the United States receiving more (more than A_2) permits initially, the situation would be reversed: In this case the emission trading scenario is better for the United States and worse for Japan than the international carbon tax scenario. Thus, the initial allocation of permits importantly influences national costs both absolutely and relative to an equivalent international carbon tax.
22. In theory, under a perfectly functioning emission trading system, the global costs of achieving a given global reduction will be the same regardless of the distribution of national targets.
23. See Weiner, J. B., 2000: "Policy design for international greenhouse gas control," RFF Climate Issues Brief No. 6, revised July, p. 6. Others worry that nations could

dilute the effectiveness of an international carbon tax in various ways. For example, if the international tax rate is $25 per ton, a country could offset its impact by introducing a domestic subsidy to coal production. To prevent abuses of this sort, an international carbon tax agreement must take notice of the full set of tax policies introduced in the participating nations.

24. Several studies have formally analyzed the relative strengths of quantity-based (like national targets) and price-based (like the carbon tax) approaches to environmental regulation. Highly technical, general treatments are offered in Weitzman, M. L., 1974: "Prices vs. quantities," *Review of Economic Studies,* 41 (4): 477–491; and Stavins, R. N., 1996: "Correlated uncertainty and policy instrument choice," *Journal of Environmental Economics and Management,* 30 (2): 218–232. For a less technical general discussion, see Kolstad, C. D., 2000: *Environmental Economics* (New York: Oxford University Press). For an analysis in the context of climate change policy, see Newell, R. G. and W. A. Pizer, 2000: "Regulating stock externalities under uncertainty," Discussion Paper 99-10-REV, Resources for the Future, Washington, DC, May, http://www.rff.org/CFDOCS/disc_papers/PDF_files/9910rev.pdf; and Pizer, W. A., 1997: "Prices vs. quantities revisited: the case of climate change," Discussion Paper 98-02, Resources for the Future, Washington, DC, October, http://www.rff.org/CFDOCS/disc_papers/PDF_files/9802.pdf. Pizer finds that a price-based policy offers expected efficiency gains that are approximately five times greater than those of a quantity-based policy.
25. See Pizer, 1997; McKibbin, W. J. and P. J. Wilcoxen, 1997: "A better way to slow global climate change," Brookings Policy Brief No. 17, The Brookings Institution, Washington, DC, June, http://www.brook.edu/comm/policybriefs/pb017/pb17.htm; Hahn, R. W., 1997: "Climate change: economics, politics, and policy," Working Paper, American Enterprise Institute, Washington DC, October; and Kopp, R., R. Morgenstern and W. Pizer, 2000: "Limiting cost, assuring effort, and encouraging ratification: compliance under the Kyoto Protocol," paper presented at CIRED/RFF Workshop on Compliance and Supplementarity in the Kyoto Framework, June 26–27, Paris, http://www.weathervane.rff.org/features/parisconf0721/KMP-RFF-CIRED.pdf. For a more technical discussion of the benefits of hybrid approaches over target and tax systems, see Roberts, M. J. and M. Spence, 1976: "Effluent charges and licenses under uncertainty," *Journal of Public Economics,* 5 (3–4): 193–208. As indicated by Roberts and Spence, a hybrid policy of the type described here dominates a pure price-based instrument (such as a carbon tax) or pure quantity-based instrument (such as a carbon quota) on average, assuming there are no implementation costs, no transaction costs, and no other real-world complications. Using an extended version of William Nordhaus's Dynamic Integrated Climate-Economy (DICE) model, Pizer, 1997, indicates that this hybrid approach produces slightly larger expected efficiency gains than a carbon tax and significantly larger expected gains than a carbon quota.
26. Kopp, Morgenstern, and Pizer, 2000.
27. See Hargrave, T., N. Helme, and I. Puhl, 1998: "Options for simplifying baseline setting for joint implementation and Clean Development Mechanism projects," Center for Clean Air Policy, Washington, DC, November, http://www.ccap.org.

28. For an excellent discussion of this and other design issues related to CDM, see Toman, M., 2000: "Establishing and operating the Clean Development Mechanism," Climate Issue Brief No. 22, Resources for the Future, Washington, DC, September, http://www.rff.org/issue_briefs/PDF_files/ccbrf22_Toman.pdf.
29. van t' Veld, K., 2000: "Incentives for developing country participation in CDM," Weathervane at the Negotiating Table Article, Resources for the Future, Washington, DC, November, http://www.weathervane.rff.org/negtable/cop6/cop6%5Fvant veld.htm.
30. Cullet, P., 1999: "Equity and flexibility mechanisms in the climate change regime: conceptual and practical issues," *Review of European Community and International Environmental Law,* 8 (2) :177.
31. van t' Veld, 2000.
32. In this connection, Jacoby, Prinn, and Schmalensee, 1998, point out that if the Annex B nations were solely responsible for achieving the IPCC-endorsed global emission path that leads to a 550 parts per million by volume concentration target, their emissions would have to become negative by the middle of the next century. (Jacoby, H. D., R. G. Prinn, and R. Schmalensee, 1998: "Kyoto's unfinished business," *Foreign Affairs,* 77 [July/August]: 54–66.)
33. See Jacoby, Prinn, and Schmalensee, 1999, and Stavins, R. N., 1997: "Policy instruments for climate change: how can national governments address a global problem?" *The University of Chicago Legal Forum,* 1997: 293–329.
34. Jacoby, Prinn, and Schmalensee, 1999.
35. Bernstein, P. M., W. D. Montgomery, and T. F. Rutherford, 1999: "Global impacts of the Kyoto agreement: results from the MS-MRT model," *Resource and Energy Economics,* 21 (3–4): 375–413.
36. Ibid.
37. Stavins, 1997; Weiner, J. B., 1997: "Policy design for international greenhouse gas control," Climate Issue Brief No. 6, Resources for the Future, Washington, DC, September.
38. Reilly, J., R. Prinn, J Harnisch, J. Fitzmaurice, H. Jacoby, D. Kicklighter, J. Melillo, P. Stone, A. Sokolov, and C. Wang, 1999: "Multi-gas assessment of the Kyoto Protocol," *Nature,* 401 (6735, October): 549–555.
39. Victor, D. G., 1991: "Limits of market-based strategies for slowing global warming: the case of tradeable permits," *Policy Sciences,* 24: 199–222.
40. The Kyoto Protocol uses global warming potentials (GWPs) to convert emissions into carbon equivalents. However, GWPs are linear approximations of a nonlinear climate system and therefore are inexact. The accuracy of the GWP approximation approach has been called into question recently, and there is no clear-cut answer to whether GWPs are an accurate enough index of warming potential. See Smith, S. J. and T. M. L. Wigley, 2000a: "Global warming potentials: 1. climate implications of emissions reductions," *Climatic Change,* 44 (4): 445–457; and Smith, S. J. and T. M. L. Wigley, 2000b: "Global warming potentials: 2. accuracy," *Climatic Change,* 44 (4): 459–469.
41. For a discussion of these issues, see Hammitt, J. K., A. K. Jain, J. L. Adams, and D. J. Wuebbies, 1996: "A welfare-based index for assessing environmental effects of

greenhouse-gas emissions," *Nature,* 381 (May 23): 301–303; Bradford, D. and K. Keller, 2000: "Global warming potentials: a cost-effectiveness approach," Working Paper, Princeton University, August; and Lashof, D., 2000: "The use of global warming potentials in the Kyoto Protocol: an editorial comment," *Climatic Change,* 44: 423–425.

42. The discussion here is on converting greenhouse gases under policies involving national targets. The same challenges arise in attempts to expand an international emission tax to include other greenhouse gases in addition to CO_2.
43. Anderson, 2000.
44. For information on the current state of research into carbon sequestration, see U.S. Department of Energy, 1999: *Carbon Sequestration Research and Development.* Available online at http://www.ornl.gov/carbon_sequestration/.
45. Some adjustments would be needed for the use of fossil fuels as petrochemical feedstocks because in this use the fuels are not combusted and no CO_2 is generated. Such uses represent less than 5 percent of total consumption of fossil fuels in the United States. For a useful discussion of monitoring and enforcement issues arising in the context of global climate policy, see Hahn, 1997.
46. Stressing the importance of flexibility, Hahn, 1997, recommends that industrialized nations adopt an agreement involving several different approaches to emission reduction, including both a modest domestic carbon tax and tradable permits. He claims that this would provide more information than any single approach and would help nurture the development of appropriate institutions necessary for dealing effectively with climate change.

CHAPTER 5

Designing Global Climate Regulation

Jonathan Baert Wiener

In response to scientific research suggesting that global climate change is a serious prospect, political negotiations have sought to establish an international regulatory policy to constrain greenhouse gas emissions. Major new treaties—the 1992 Framework Convention on Climate Change and the 1997 Kyoto Protocol—have been negotiated. But identifying the problem is not the same as crafting the solution. Climate science is a necessary but not sufficient basis for climate policy. It remains crucial, and often not simple, to design the regulatory system best suited to addressing global climate change.[1]

U.S. Supreme Court Justice Stephen Breyer, a scholar of regulatory design, observed two decades ago that "mismatches" defeat many well-intentioned regulatory programs and that regulatory systems should match the social and environmental systems they regulate.[2] More specifically, regulatory programs should employ cost-effective tools, foster creativity in achieving solutions, and match the scale of the ecosystems and spillover effects they are meant to govern. But actual legal responses have too often created mismatches with social and environmental systems, such as regulatory programs that are unduly narrow and inflexible, resulting in excessive costs or even perverse increases in environmental harm.

Global climate change plainly illustrates this problem. Climate change is complex on many dimensions, frustrating simple and hasty regulatory responses. The challenge is to design a regulatory system that matches these complex realities and thereby accomplishes cost-effective advances in global environmental protection. At least three kinds of complexity confront regulatory design for global climate change: causal complexity, spatial complexity, and temporal complexity.

Causal complexity denotes the diverse interconnected factors that drive climate change. Multiple greenhouse gases (GHGs) are affected by almost every human activity, including industry, transportation, agriculture, and forest management. The sources and sinks of the multiple GHGs are numerous and widespread. Policies aimed at only one of these causal factors, such as one GHG, can unintentionally exacerbate other causal factors. If global climate regulation is to be effective, it must address these complex causal factors comprehensively; it must match the causal scope of the problem. Yet there are persistent pressures to design regulatory regimes narrowly.

Spatial complexity involves the great breadth and diversity of GHG sources and sinks in almost every country. Regulating a global problem is difficult because the institutions of governance are not matched to the spatial scale of the problem. Geographically narrow policies limited to one country or region may induce emitting activities to relocate to other areas. But establishing global environmental regulations is more difficult than instituting national ones; without a global government (in part for some good reasons), global regulations must be made by the cooperation of numerous national governments. Across the planet, countries have diverse economies, social norms, political institutions, and interests. This spatial diversity makes a single uniform regulatory approach unwise and makes global cooperation on coordinated regulatory policies difficult. Furthermore, whereas national law typically is imposed by some form of majority vote, at the international level each nation is sovereign and is bound only by the treaties to which it consents. Compliance with national pollution control laws can be compelled, but participation in international treaties cannot be compelled and must instead be attracted. Effective climate regulation must therefore deal with global scale, global diversity, and the global legal framework.

Temporal complexity refers to the dynamic character over time of the climate, human activities and technologies, and our understanding of these systems. Climate policy cannot be made once and for all; it must be updated to adapt to changing circumstances and knowledge. Yet designing a dynamically adaptive regulatory regime is difficult: We never have full knowledge of the future, investors want predictable rules, early decisions about emissions and investments may endure for many years, and political planning horizons may not match environmental time horizons. And even a climate policy without repeated adaptation must decide how to allocate abatement efforts over time.

Nevertheless, through careful analysis an effective and efficient global regulatory regime for climate change can be constructed. A comprehensive, incentive-based, and adaptive regulatory design can be matched to the casual, spatial and temporal complexities of global climate change.

Causal Complexity and Comprehensive Scope[3]

The Scope of Environmental Regulation

How comprehensive should environmental regulation be? When faced with a problem, how much of it should we try to tackle? The essence of the environment is its interconnectedness. But the complexities of policymaking often push decision makers toward narrow, piecemeal solutions that address one obvious symptom or cause of an environmental problem. Advocates of narrow solutions claim that limited, incremental steps are easier to accomplish than broader, comprehensive approaches.[4]

Piecemeal regulatory strategies, however, may ignore the full scope of a problem, miss lower-cost options to achieve better results, and produce unintended side effects.[5] A broader, more comprehensive approach takes into account the complex nature of environmental issues. It attempts to match the regulatory design to the complex environmental system being regulated.

Discussions about global climate change policy in the late 1980s centered on reducing the amount of carbon dioxide (CO_2) emitted from the energy sector because CO_2 was the most plentiful greenhouse gas, and the energy sector was the largest source of CO_2. The initial negotiating positions of major countries proposed a treaty calling for cuts in energy sector CO_2.

But at the same time, scientists were demonstrating to policymakers that CO_2 was only one of several important GHGs. First, although the volume of CO_2 emitted far exceeds that of other GHGs, each CO_2 molecule is a weak absorber of infrared radiation (heat). Other GHGs, such as methane (CH_4) and nitrous oxide (N_2O), are important contributors to global warming potential because despite their smaller volume of emissions, they are roughly 20 and 300 times more potent per unit, respectively, than CO_2 at retaining heat in the atmosphere over time. Thus CO_2 was estimated to be responsible for only about one-half[6] of the global warming potential of anthropogenic GHG emissions in the 1980s.

Second, the relative influence of CH_4 and N_2O was expected to increase in the future. GHGs absorb infrared radiation in wavelengths specific to each gas. As the concentration of CO_2 in the atmosphere has risen, more and more of the infrared radiation at the wavelength blocked by CO_2 molecules is already being absorbed. Because of this saturation effect, additional emissions of abundant atmospheric gases such as CO_2 will have decreasing marginal impacts relative to those of less abundant gases such as methane. Thus, narrowly targeting CO_2 and omitting the other salient GHGs would limit the effectiveness of the regulatory regime in averting climate change.

Advantages of the Comprehensive Approach to Climate Change

Environmental Advantages

Taking a comprehensive approach to climate policy has several significant advantages. First, it is environmentally superior. Piecemeal approaches ignore important sources of the problem and thus neglect important opportunities to solve it. Moreover, they tend to be self-defeating because efforts to solve one aspect of a problem intensify other, neglected aspects. The history of pollution control in the United States offers an example. Our federal environmental statutes have focused on one medium at a time: separate laws for air, water, and land. Restrictions on one medium have induced disposal into other media.[7] Like squeezing one end of a balloon, this approach shifts the problems elsewhere and delays attainment of the primary goal: a cleaner environment. An integrated approach would control pollution more comprehensively and effectively.[8]

Similarly, focusing solely on energy sector CO_2 would induce perverse shifts in emissions. For example, controlling energy sector CO_2 alone would invite fuel switching from coal to natural gas because burning coal emits about twice as much CO_2 per unit of energy produced as does natural gas. But natural gas is almost pure methane (CH_4), and methane is roughly 20 times more potent than CO_2 per mass at causing global warming. As little as 6 percent fugitive methane emissions from natural gas systems would be enough to fully offset the CO_2-related benefits of this fuel switching.[9] In the United States, natural gas systems rarely release more than 2 percent of their methane, but in Europe the methane leakage rate has been much higher, often exceeding 6 percent, especially in Russia, where much of the natural gas to replace European coal would come from. Thus a CO_2-only policy in Europe could yield a net increase in the contribution to global warming.[10]

Another example involves replacing fossil fuels with biomass fuels, such as ethanol made from corn. At first glance such a policy seems attractive because it would reduce energy sector CO_2 emissions. The CO_2 emissions from burning the fossil fuels would be reduced or eliminated, and the CO_2 emissions from burning the biomass fuels would, one might presume, be at least partly offset by the sequestration of that same CO_2 from the atmosphere by the corn as it grew. But the story is not that simple. Focusing only on energy sector CO_2 neglects three important emission categories. First, the CO_2 emissions from the ancillary agricultural operations needed to farm the corn, manufacture fertilizer, irrigate the land, and convert the corn into fuel probably would be large.[11] Second, growing corn uses large quantities of nitrogen fertilizer, which release nitrous oxide (N_2O), a GHG almost 300 times more potent per mass than CO_2. Third, if the corn is grown on cleared forest lands, the carbon liberated from the forest ecosystem (trees, plants, and soils) when cleared and the lesser ability of the corn field to sequester carbon as compared to the forest must be counted as well. Together, these three side effects could make biomass fuel much less attractive, and possibly even perverse, as a climate protection strategy.

The solution to these perverse shifts is not to abandon climate protection but to include all the major GHGs (including methane and nitrous oxide, and others, as well as CO_2) and all sectors (including agriculture and forests as well as energy). A comprehensive approach defines performance and measures results in terms of the full impacts of any policy intervention on climate change, thus preventing perverse shifts across GHGs and sectors.

A comprehensive approach would also give sources the incentive to find ways to reduce all of these GHGs in all sectors. For example, under a comprehensive regulatory design, Russia and other countries with leaky natural gas systems would have a greater incentive to invest in closing methane leaks. And sources would invest in conserving and expanding forests to sequester carbon, potentially aiding biodiversity as well as climate protection.[12]

Economic Advantages

There are also economic advantages associated with the comprehensive approach. Allowing a wider array of control options reduces the cost of achieving the overall objective. By allowing countries to choose which GHGs they reduce in which sectors, the comprehensive approach gives them the opportunity to make the most cost-effective reductions. Because there is so much variety in GHG limitation opportunities across nations, the comprehensive approach would yield large cost savings compared with a piecemeal approach that fixes limits for CO_2 alone or for each gas separately. A comprehensive approach would regulate the net CO_2-equivalent emissions from each country, not the specifics of how it was achieved, thereby protecting the climate at lower cost. For example, the U.S. Department of Energy estimated that meeting a U.S. emission target of 20 percent below 1990 levels by the year 2010 by comprehensively addressing all GHGs, instead of just energy sector CO_2 alone, would reduce costs by 75 percent; adding the option of sink enhancement would cost 90 percent less than the energy sector CO_2 policy.[13] Similarly, a World Bank study found that India could reduce its costs 80 percent by controlling all GHGs instead of energy sector CO_2 alone.[14] The most recent and thorough study confirms these results worldwide. Using an integrated assessment model of the world economy, a research team at the Massachusetts Institute of Technology (MIT) found that a comprehensive approach to all GHGs and sectors reduces the global costs of meeting the Kyoto Protocol targets by at least 60 percent.[15] The MIT study also noted that the multigas approach could be more effective at protecting the climate than the CO_2-only approach, both because the relative global warming impact of the non-CO_2 gases is expected to increase in the future and because the ability of CO_2 to fertilize plant growth and hence stimulate carbon storage means that CO_2 creates a negative feedback on global warming that the other gases do not. A new study by National Aeronautics and Space Administration (NASA) climate scientist James Hansen and colleagues offers further support for the comprehensive approach, showing that control of non-CO_2 GHGs (including methane and dark soot) would be cost-effective and would yield significant side benefits to human health by reducing local air pollutants.[16]

Innovation

By rewarding efforts in a wider array of gases and sectors, the comprehensive approach also provides better incentives for innovation in abatement strategies. Focusing narrowly on a specific sector or gas misses the chance to stimulate new approaches that have not yet been identified. The comprehensive approach also offers the flexibility to change tactics as our understanding of technologies and climate impacts evolves.

Fairness

The comprehensive approach establishes a more equitable position for all nations at the regulatory negotiation table. Because of the differences across countries in opportunities to control sources and expand sinks and differences in their economic status, a piecemeal policy inevitably favors some nations while disproportionately burdening others. The comprehensive approach allows each country to choose its best mix of policies, dealing more even-handedly with countries of widely different internal economic and social configurations.

Participation

The cost and fairness advantages of the comprehensive approach have another benefit. As will be discussed in more detail later in this chapter, attracting participation in international climate policy by a large number of countries is critical. Because climate change and regulatory actions to address it affect each nation differently, their own best policy responses will vary. No single, narrow regulatory tactic will be attractive to all of the world's countries; flexible approaches will have wider appeal. Policy instruments that are less costly, individually and collectively, will stand a greater chance of being acceptable to all parties and attracting their participation in the treaty.

Progress on the Comprehensive Approach

The climate treaties have made progress in adopting the comprehensive approach to addressing all major GHGs in all sectors, and including sinks as well as sources. The United States proposed the comprehensive approach in 1990,[17] and that approach was adopted in the Framework Convention on Climate Change (FCCC) signed at the Rio Earth Summit in 1992. Article 3 of the FCCC endorses the comprehensive approach, and Article 4 states that parties shall reduce emissions of all GHGs and enhance GHG sinks.

The Kyoto Protocol, signed in 1997, maintained the comprehensive approach. It specifically included six GHG classes in its quantitative emission targets: carbon dioxide (CO_2), methane (CH_4), nitrous oxide (N_2O), hydrochlorofluorocarbons (HCFCs) and hydrofluorocarbons (HFCs), perfluorocarbons (PFCs), and sulfur hexafluoride (SF_6). It also gives credit for sink expansion. The Kyoto Protocol requires countries to attain levels of net GHG emission reductions, weighted by the Global Warming Potential (GWP) index according to their relative contribution to global warming, and does not specify separate limitations for each gas. This comprehensive approach offers each country the flexibility to reduce the sum of its GHG emissions in the most cost-effective way it chooses while requiring countries to monitor and manage all the salient GHGs. The Bonn and Marrakech accords on implementing Kyoto reinforced the comprehensive approach in almost all respects, although they did impose quantitative ceilings on the use of sinks by each country. These limits may increase the costs of achieving the Kyoto targets.

Concerns have been raised about the administrative practicality of a multigas approach, including that emissions of some gases might be difficult to monitor and that the GWP index used to compare the heat-trapping ability of the different GHGs is imperfect. Some critics proposed that a narrow regulatory mechanism (addressing only CO_2) be devised initially and then expanded stepwise into a more comprehensive instrument (addressing multiple GHGs) later on. But this strategy is flawed. First, it would initially forfeit the environmental and economic advantages of the comprehensive approach: It would invite perverse shifts, and it would cost much more. These benefits of comprehensiveness vastly outweigh its administrative costs. Second, the intended stepwise expansion probably would be delayed or thwarted: The countries and interest groups least burdened by the initial narrow design would become entrenched in their favored positions and would resist expansion to a more comprehensive approach later. Third, this piecemeal strategy would fail to provide the incentives for innovation in the monitoring and abatement methods for non-CO_2

gases that eventually would be needed to run an effective comprehensive program. Moreover, the comprehensive approach is not impractical.[18] The measurement of non-CO_2 gases and nonenergy sectors, even if initially difficult, would improve in response to policy incentives. And such measurement is necessary even under a CO_2-only policy if we are to evaluate the true effectiveness of the policy in protecting the climate; ignoring the non-CO_2 gases does not make them go away. The GWP index is not perfect, but it is more accurate than ignoring the non-CO_2 gases (implicitly assigning them an index weight of zero). The treaties expressly contemplate improving the GWP index over time in response to new science. In sum, the comprehensive approach is a practical and advantageous design for effective and efficient climate policy.

Spatial Complexity, Participation, and Instrument Choice[19]

GHG emissions could be regulated in several ways, such as technology requirements, emission taxes, subsidies for abatement, maximum emission levels, or tradable emission allowances. This question of instrument choice has long been a central theme of environmental law, policy, and economics. And it has taken center stage in the international negotiations on the FCCC and the Kyoto Protocol (and the 2001 Bonn/Marrakech accord on implementing Kyoto). The FCCC adopted an informal version of allowance trading called joint implementation (JI) (Article 4(2)(a)). The Kyoto Protocol retained JI (Article 6) and added a formal system of tradable allowances (Article 17) as well as a new informal trading system for the sale of emission reduction credits by developing countries, called the Clean Development Mechanism (CDM) (Article 12). Were these the best choices? The answer relates to the spatial complexity of global climate, global economic activity, and global regulation.

Spatial Complexity

Global Impacts

A primary challenge of global environmental problems is that they have global impacts. Each country's GHG emissions create global environmental spillover effects, or externalities.[20] The atmosphere is being treated as an open-access commons that anyone can use as a disposal site for GHGs.[21] Prevention of these global externalities (i.e., climate protection) is a global public good because it is nonexcludable: Once an improved climate is provided, it is impossible to exclude anyone from enjoying its benefits; abatement of emissions at any one location generates benefits enjoyed by people around the world. As a result, any individual country is likely to receive only a small fraction of the benefits of its own abatement efforts.

If GHG abatement is costly, countries will prefer to avoid the costs of abatement while enjoying the shared benefit of others' efforts, trying to take a free ride on others' abatement.[22] Collective abatement action would bring greater net gains to all the participants, but fear of free riding by others can lead each country to hesitate to act. Thus, the global nature of the climate problem means that individual countries will tend to invest in less abatement than would be desirable from a collective global point of view. A central challenge for global regulatory design is to choose instruments that help overcome free riding and facilitate collective action.[23]

Global Sources

Overcoming free riding in the provision of public goods is never easy, even at the local level, but doing so in the global context is even more difficult. The sources of GHG emissions are spread all around the planet, so climate policy must have nearly global coverage to be effective. The U.S. and China are the world's top emitters, and developing countries are expected to increase their GHG-emitting activities rapidly over the next few decades.[24] A spatially limited policy that covers only industrialized countries, or omits China and the U.S., would omit a major fraction of global emissions and fail to forestall adverse climate change.

Worse, a policy that restricts emissions only in some countries could induce emission sources to shift or "leak" to unregulated countries through both industry relocation and changing world commodity prices. Such leakage has several undesirable consequences. First, it at least partly offsets the environmental effectiveness of the policy. Second, the economies of the initially unregulated nations receiving the leakage become more GHG-intensive as a result of the leakage so that later participation in the regulatory treaty becomes even more costly and unappealing to them.[25] Third, even if actual leakage is small, fear of leakage can be a potent political obstacle to treaty participation. For example, in 1997 the U.S. Senate voted 95–0 not to ratify a climate treaty that exempted the developing countries.[26]

Local Diversity

A further complexity is that sources and impacts vary widely around the world. There is significant local diversity in the costs and benefits of abatement and in social and legal systems. The costs of abatement vary because differences in technology, available substitutes, and economic structures make avoiding future emissions much less (or more) costly in some places than others. One study found a 50-fold difference in GHG abatement costs just within the membership of the European Union (EU).[27] The range of variation in global abatement costs is likely to be even greater than that.

Meanwhile, the benefits of preventing global environmental change also vary. Even though climate protection is a global public good, its benefits would vary regionally. Island nations and countries with low-lying coastal areas are at greater risk from sea level rise and so stand to see greater benefits from averting global warming. Wind and precipitation patterns may change so that some areas will experience drier weather and others wetter weather. Host ranges for vegetation and pests may shift. Poorer countries with agrarian or coastal economies and little social safety net may be physically more vulnerable to these changing patterns than are wealthier countries. But wealthier countries, even if physically less vulnerable to climate change, typically place a higher priority on long-term global environmental protection than do poorer countries for whom more local and more immediate problems—such as hunger and infectious disease—are more pressing. Thus, with the exception of poor island and coastal nations, it is largely the wealthier countries that press for long-term climate protection.[28]

Some countries, perhaps including China and Russia, might even believe they stand to gain from climate change, on the view that they will enjoy greater agricultural yields in currently cold areas if temperatures rise. A recent synthesis of global climate change impacts on key end points—agriculture, forestry, water resources, energy consumption, sea level rise, ecosystems, and human health—indicates that some initial warming (1°C) and CO_2 fertilization may help agriculture and human health in some areas (including the Organisation for Economic Co-operation and Development [OECD], Russia, and China), for a near-term gain of 1–3 percent of GDP. However, this climate change will have adverse impacts in poorer areas (especially Africa and southeast Asia, which would lose 1–4 percent of GDP), and the impacts of greater warming will become adverse worldwide over the longer term, including losses of 1–2 percent in OECD countries and 4–9 percent in Russia and developing countries (but not in China, which exhibits persistent gains from climate change of about 2 percent of GDP).[29] Therefore, China and perhaps Russia (initially) may not just be free riders (players for whom cooperative action is beneficial but who would rather let others bear the cost) but may be "cooperative losers": players for whom climate change is benign (or not seriously adverse) and for whom cooperative action to prevent climate change is costly and who therefore dislike cooperative prevention efforts.[30] Because these countries are also large GHG emitters, successful climate regulation must include these countries. But attracting participation by cooperative losers is even more difficult than overcoming free riding.

Participation and Voting Rules

This spatial complexity would not make so much difference in the choice of regulatory instrument if global regulation could simply be imposed on all emitters worldwide by one rational benevolent dictator. That imaginary world of welfare-maximizing despotism is the dream of some, the nightmare of others, and the routine assumption of most economic models of regulation.[31]

In reality, the voting rule for policy adoption ranges along a spectrum from rule by one (autocracy) to rule by all (unanimity). In autocracy, a single decision maker makes the law, and all are bound regardless of their consent. In democracies, legislation embodies a version of majority rule: A majority of consent is sufficient to adopt a law that then binds all, including those (up to 49 percent) who dissented from the adoption of this law.[32] By contrast, the voting rule for international treaties is consent: Treaties bind only those who agree to be bound.[33] Unlike autocracy and majority rule, under the consent voting rule regulation cannot be imposed on dissenters. Note that consent is not quite the same as unanimity. The latter requires the consent of every voter for a law to become binding on any voter, whereas the former does not. Under consent, the law is binding on those who do consent, even if others demur. Under unanimity, each voter can veto the entire law; under consent, each voter can only choose not to participate herself.

In practice, the real international voting rule for global climate treaties is consent,[34] tinged with aspects of both coercion and unanimity. Overlaid on the basic rule of consent to treaties are some coercive pressures, such as military force and trade sanctions.[35] But military force is rarely used to secure adoption of environmental treaties (although disputes over fisheries have recently come close to naval combat), and the use of trade sanctions to penalize treaty non-participants may be limited by the General Agreement on Tariffs and Trade (GATT) and World Trade Organization (WTO) free trade disciplines. Shaming[36] and interest group pressures[37] are elements of a country's calculus of whether to consent. Meanwhile, the tradition of seeking consensus in treaty negotiations[38] and the need to avoid emission leakage by covering all major players tend to place the consent-based voting rule for international climate treaties fairly near to the unanimity end of the spectrum.

The voting rule of consent has fundamental implications for participation and, in turn, for the choice among regulatory instruments.[39] In general, national consent to a treaty requires a positive national net benefit compared to not joining.[40] Unless a country views joining a treaty as favoring its interests, it is highly unlikely to join. Of course, *net benefit* and *interest* are to be construed broadly, including considerations of fairness and reputation as well as economic, environmental, social, political, and other concerns. In economic terms, treaties must satisfy not just Kaldor–Hicks efficiency (aggregate net benefits) but also the more stringent test of actual Pareto improvement (individual net benefits for each participant).[41] International treaties thus are adopted by a voting rule much more analogous to marketplace contracts than to national legislation.[42] And this consent voting rule, together with the problems of free riders, leakage, and cooperative losers, makes collective action more difficult to organize than under coercive voting rules such as majority rule.

Global Instrument Choice

Most analyses of regulation assume autocracy: If one rational person could pick the best regulatory instrument, which would she choose? This section begins by reviewing the analysis under autocracy and then examines how this choice is different when the voting rule is consent.

The Regulator's Toolbox

The instruments available to the regulator include technology requirements, emission taxes, subsidies for abatement, performance standards, and tradable emission allowances. A broad distinction can be drawn between basing regulation on conduct and basing it on outcomes.

Conduct-based instruments specify how firms shall act, in the hope that improved conduct will reduce pollution. For example, a conduct-based instrument might dictate specific technologies that firms must install or specific fuels that firms must use to limit emissions. In contrast, outcome-based instruments (also called incentive-based) seek to achieve a certain degree of environmental protection but allow firms to choose how they will meet that goal. They internalize externalities by "reconstituting" flawed markets, using incentives that motivate firms to adjust their own behavior by taking account of the environmental impacts they had previously neglected.[43] Two basic types of incentive-based instruments are price-based and quantity-based.[44] Price-based instruments set a price for emitting, and firms then decide what quantity of emissions to generate in light of having to pay this price. Price-based instruments include taxes on emissions and subsidies for abatement. Quantity-based instruments set a total quantity of acceptable emissions and then allocate entitlements to emit. Quantity-based instruments include fixed performance standards (i.e., an emission limit for each source) and tradable emission allowances (i.e., emission limits for each source, which sources can buy or sell). Once the total quantity of emissions is chosen and allowances adding up to that total are assigned (or once an emission tax is set), each source then decides how much to emit in light of having to buy an additional allowance (or pay the tax) and in light of the opportunity to earn the market price to sell an extra allowance.

The Analysis Under Autocracy

There is no universal best regulatory instrument; the choice among them depends on several contextual factors, including their environmental effectiveness and their cost in achieving any given level of protection.[45] Still, under the standard assumption of autocracy—that the law is imposed by a single rational actor—three presumptions have emerged in the literature on instrument choice. These three presumptions are that incentive instruments are superior to conduct instruments, taxes and tradable allowances are superior to subsidies for abatement, and taxes often are superior to tradable allowances. After briefly describing these presumptions in the world of autocracy, we can examine their validity in a world of consent.

Incentives Versus Conduct

First, incentive-based instruments such as taxes and tradable allowances generally are more cost-effective than conduct instruments or fixed performance standards. Uniform standards require all firms to do the same thing regardless of cost. If abatement costs vary across sources—as they do for GHGs—then cost-effectiveness can be improved by using a regulatory mechanism that obtains more abatement from the lower-cost abaters. Both emission taxes and tradable allowances achieve overall environmental protection at lower total cost by inducing lower-cost firms to abate more and higher-cost firms to abate less. In the United States, allowance trading programs have proven to be far more cost-effective than conduct rules or fixed performance standards, cutting costs by roughly half.[46] For example, the SO_2 emission trading system adopted in the 1990 Clean Air Act amendments to reduce acid rain has achieved a dramatic reduction in SO_2 emissions at roughly half the cost of the prior uniform approach.[47] Because GHG abatement costs vary a great deal across countries, the cost savings for global GHG emission trading (compared with fixed national targets) are predicted to be large (30 to 70 percent).[48]

Second, incentive instruments are more effective in stimulating dynamic innovation. Technology requirements provide no incentive for the firm to invest in improved abatement methods beyond what has been mandated. Performance standards provide a modest incentive for innovation. Taxes and trading give sources the strongest continuous motivation to improve abatement methods, which enables the source to sell allowances or pay lower taxes.[49]

Third, incentive instruments need not involve undue administrative costs.

Technology standards require detailed engineering choices and monitoring of devices installed. Incentive methods must determine the tax rate or number of allowances and monitor actual emissions. Monitoring emissions can be costly (especially for dispersed sources), but monitoring the technology in place at a source does not measure environmental impact. Monitoring actual emissions can be worthwhile if it improves environmental effectiveness. Moreover, the social cost savings and enhanced innovation under incentive instruments would often dwarf their administrative costs.

Fourth, incentive instruments can be designed to promote fairness. There is concern that efficiency-enhancing policies (such as emission trading) might be unfair to poorer communities and developing countries.[50] Developing countries worry that global environmental law may be a form of eco-imperialism. They want developed countries to take the lead in controlling GHG emissions. It would be unfair to make poorer countries worse off in the effort to correct a problem caused by and of primary concern to wealthier industrialized countries. Technology standards, performance standards, and emission taxes could be regressive. But global tradable allowances could be structured to achieve fairness for poorer societies by giving them valuable headroom in their initial assignment of allowances. This would enable poorer countries to grow economically by emitting somewhat more GHGs (perhaps up to or even over their business-as-usual forecast) or by earning substantial revenues from selling a valuable new asset—the tradable allowances—to wealthier sources facing higher abatement costs. This system would benefit poorer societies by giving them a substantial revenue stream.[51] It would also oblige richer countries to take the lead by financing global emission reductions (in a way that is also cost saving). The basic logic of voluntary exchange (market trading) means that allowance sales would not occur unless both parties felt better off. On the other hand, insisting that industrialized countries control their emissions entirely at home would be unfair to developing countries because it would deprive developing countries of the allowance sale revenue stream. It would be like insisting that rich people must spend their money only in rich neighborhoods.

Fifth, incentive mechanisms do not represent immoral means of achieving environmental protection. Critics worry that translating environmental protection into market prices and commodities may debase its moral value.[52] But insofar as environmental degradation stems from the failure of markets to take account of environmental impacts, the problem is not that the environment is too important to leave to markets but rather that the environment is too important to leave out of markets. Nor do tradable allowances amount to a special "license to pollute." Conduct instruments and fixed performance standards amount to a license to pollute for free once the technology has been installed or the performance standard achieved. Taxes and tradable emission allowances, by contrast, force the source to pay for every unit of emissions, either by paying the tax or by foregoing the revenue from the sale of the allowance. Furthermore, if causing additional pollution is the immoral act and if incentive instruments are more cost-effective and innovation-enhancing, then the moralist who opposes incentive instruments is committing an immoral act.

Taxes and Trading Versus Subsidies

The second presumption is that subsidies for abatement are inefficient. Subsidies for abatement can act like emission taxes at the margin: For each source, declining to abate means forfeiting the subsidy, which is equivalent to paying a tax of the same amount. But whereas taxes also charge the source for all its unabated emissions and thereby raise the average cost of doing business in that industry, subsidies pay the source for abatement and thereby reduce the average cost of doing business in that industry. This attracts investment to the emitting sector and could increase total emissions even if the subsidy reduced emissions at individual plants.[53] The subsidy payment may be seen as insurance against the social cost of the emitting activity and thus lead to its increase.[54] Sources may also increase pollution to secure larger subsidies for abatement.[55]

Taxes Versus Trading

The third presumption of the standard analysis is that taxes are preferred to tradable allowances. In theory, these instruments can produce identical results. Taxes set the price of emitting and allow the quantity of emissions to vary, whereas allowances set the aggregate quantity of emissions and allow the price of emitting to vary. If the actor adopting these instruments (our assumed rational autocrat) knows firms' costs with certainty, she can use either instrument to achieve the same result: If she issues *Q* allowances, the market price *P* for each allowance to emit 1 ton of pollutant will be equal to the tax of *P* that she would set to achieve the same *Q* amount of emissions.

But if the decision maker is uncertain about firms' costs, then these instruments diverge. A tax set at *P* might achieve *Q* emissions, but if firms' true costs are higher than expected, this tax will yield more than *Q* emissions (firms will pay the tax rather than abate). Issuing *Q* allowances might achieve a market price of *P* for each allowance, but if firms' true costs are higher than expected, this policy will yield a higher price for allowances. Thus the tax prevents cost escalation (firms will not pay more than the tax) but lets emissions vary, whereas the allowance system prevents emission escalation (there is a finite number of allowances) but lets costs vary. Under uncertainty, the choice between these instruments depends on one's relative concern about cost escalation versus emission escalation (i.e., on the relative steepness of the marginal cost of abatement versus the marginal benefit of abatement).[56] One study found that, given significant uncertainty about true abatement costs and assuming a very flat marginal benefit curve (i.e., assuming that escalating emissions would have only very gradual impact on the damages from climate change), a GHG tax would yield roughly five times greater net benefits than would a system of tradable emission allowances.[57]

Tax regimes and auctioning allowances also have the advantage of raising revenues that can be used to reduce previously existing taxes. Often, these preexisting taxes act as a disincentive to something good, such as labor or investment. Revenue-raising GHG abatement policies can also reduce those distortionary taxes, yielding a "double-dividend."[58]

A system of tradable allowances, like any market, also faces other challenges. One is market power: A few large allowance sellers (e.g., Russia or China) could try to charge excessive monopoly prices.[59] This is a particularly knotty problem at the international level, where there is no antitrust law. Another problem for a GHG allowance market is transaction costs.[60] The costs of finding trading partners, negotiating deals, monitoring and enforcing performance, and insuring against nonperformance can hinder efficient transactions. Formal allowance trading seeks to reduce transaction costs by making allowances fungible and fostering risk diversification and market transparency. But informal allowance trading such as JI and the CDM may face high transaction costs.

Taxes, however, face their own difficulties. First, if emission escalation is a more serious concern than cost escalation (the converse of the assumption described earlier), then allowances are superior to taxes under uncertainty.[61] Second, whereas allowance markets can face market power and transaction costs, taxes can face high administrative costs to calculate and collect the tax and to audit and enforce against taxpayers. Third, raising revenue may become more important to tax officials than the environmental purpose of the tax, leading them to set the tax too low to discourage GHG emissions. Fourth, there is the question of which country, countries, or organization would collect GHG taxes and distribute tax revenues, a particularly sensitive issue on the international front. Fifth, as discussed earlier, GHG taxes could be unfairly regressive to poorer countries. Sixth, as discussed below, under the consent voting rule taxes may not attract adequate participation by countries.

The Analysis Under Consent

The foregoing assumes autocracy. As discussed earlier, real global regulation occurs under a voting rule of consent: No country can be bound to a treaty except by its agreement, which in turn depends on its perceived national net benefit. Basing global instrument choice on the assumption of autocracy may therefore lead to serious errors.

At the international level, participation must be attracted, not coerced. Free riding must be overcome. Cooperative losers (countries that perceive a national net cost from preventing global warming) must be persuaded to participate by some inducement other than global environmental protection itself, such as side payments sufficient to overcome their foregone gains from warming plus their abatement costs.[62]

Participation Efficiency

Attracting participation yields benefits but can be costly. The benefits of participation include greater coverage of globally dispersed emissions, reduced free riding, reduced cross-border emission leakage, and a wider array of abatement opportunities. The costs of securing participation include the out-of-pocket costs of side payments and the perverse incentives of subsidizing abatement (discussed earlier).[63] The best regulatory instrument under consent must therefore strive to satisfy a criterion that is not relevant under autocracy: "participation efficiency."[64] Participation efficiency is the ability to attract participation at least cost. The most participation-efficient regulatory instrument would minimize the sum of the costs of nonparticipation plus the costs of securing participation. Equivalently, it would maximize the difference between the benefits of securing participation and the costs of securing participation.

The less coercive the voting rule, the more participation efficiency matters in selecting among regulatory instruments. Under majority rule, some participation-efficient inducements are needed to gain the majority needed to adopt a law. After that, coercive power exists over remaining dissenters. Under consent, every important cooperative loser must be paid to play.

Comparing Instruments

Under autocracy, as discussed above, the standard conclusion is that taxes are the superior instrument. But under consent, the relative merits of alternative regulatory instruments depend significantly on their participation efficiency.

First, direct subsidies for abatement, in the form of a cash payment to nonbeneficiary countries, would be one way to provide the compensation needed to attract participation.[65] Unfortunately, subsidies for abatement generate perverse incentives for increased aggregate emissions.[66] There is also the possibility that some countries would posture as cooperative losers to demand side payments via threatened or actual increases in GHG emissions, potentially decreasing the degree of cooperation enough to result in higher total emissions.[67]

Second, participation might be coerced through threats of trade sanctions.[68]

Loss of trading partners could induce free riders and even cooperative losers to participate because of the fear that noncooperation would be more costly than cooperation.[69] Although this approach avoids the perverse incentive problem of subsidies, several other problems would arise. Threats of trade sanctions may not be credible because they would impose high costs on both sides of the trade barriers. Trade sanctions may also distort trade, impair global economic efficiency, and spur a retaliatory trade war. Trade sanctions often are ineffective because they strengthen the target government's domestic political case for resistance to foreign meddling.[70] Trade sanctions can also injure the target country's economy so much that compliance would become more difficult or impossible, thwarting the goal of inducing environmental protection.[71] Finally, trade sanctions imposed by wealthy countries against poorer countries cut against principles of fairness.[72]

Third, GHG taxes might be used. But because taxes impose the highest costs on sources, they probably will induce the greatest rate of nonparticipation. GHG taxes probably would attract the fewest cooperative losers, leading to significant leakage and a failure to reduce global emissions. Perhaps a tax paired with side payments could succeed. But to attract participation, the side payments would have to be large enough to ensure positive national net benefits, compensating for abatement costs, foregone environmental benefits to cooperative losers, and the burden of the tax on residual unabated emissions. Such a side payment would undercut the ability of the tax to reduce emissions in recipient countries. The side payment could not be a "lump sum" (a single one-time payment unrelated to the country's marginal costs) because the side payment would have to repay the country for every incremental dollar of burden incurred as a result of the tax, or else the policy would not be attractive on net (Pareto improving) to the recipient country and would not attract the country's participation.

Fourth, one could use quantity-based instruments. Fixed-quantity targets (performance standards) for each country, on their own, would incur high nonparticipation costs. Large and growing cooperative losers would simply decline to be bound. This has been the predictable experience under the FCCC and Kyoto Protocol: Large and growing developing countries, including China, India, and Brazil, have declined to adopt quantitative emission limitations.

Coupling fixed quantity targets with a direct payment to cooperative losers could help secure those countries' participation. This was the approach taken in the Montreal Protocol to phase out CFCs: Its Multilateral Fund was created to secure participation by China and India. Such side payments would still generate perverse incentives, but now—in contrast to the cases of direct subsidies and taxes plus side payments—the fixed-quantity limits would constrain the perverse incentives from increasing aggregate emissions. This is a distinct advantage of quantity limits over taxes under the consent voting rule, where side payments are necessary.

But fixed-quantity limits would not be cost-effective because they would not allow emission reductions to be accomplished wherever abatement costs are lowest. An even better design for quantity-based instruments would be to use tradable allowances, reducing costs dramatically. The side payments could then be embedded in the allowance trading system itself. In this "cap-and-trade" system, poorer countries with large emissions such as Russia, China, India, and Brazil would be assigned extra allowances as a side payment to attract their participation. These headroom allowances would be a new asset that poorer countries could sell to earn profits in the allowance trading market. Wealthier countries would thereby finance abatement (and a lower-GHG economic growth path) in poorer countries by buying headroom allowances. This cap-and-trade system would attract participation through in-kind side payments while constraining the perverse incentives of those side payments by securing the adoption of quantity caps on participating countries.[73] This was the strategy used in the Kyoto Protocol to engage Russia's participation: Russia was assigned headroom allowances in exchange for its agreement to join the treaty. Without these extra allowances, Russia might well have stayed out of the treaty, impairing its effectiveness. This approach might also be used to attract participation by China and other major developing countries.

A critical step in this cap-and-trade approach is the initial allocation of emission allowances. Of course, the negotiations will be difficult, as with any burden-sharing negotiation. Some critics have asserted that negotiating the assignment of GHG emission allowances would be so difficult that the system would never get off the ground.[74] But this concern applies to any regulatory instrument because all forms of regulation impose varying burdens on those regulated and because all forms of regulation under the consent voting rule entail a burden-sharing negotiation. The real question is the relative difficulty of negotiating the initial assignment using the alternative instruments, given the consent framework.[75] In that context, tradable allowances would ease the problem of initial negotiations. As Coase taught, the lower the impediments to subsequent reallocations of entitlements among the parties, the less the initial assignment binds.[76] Technology standards, fixed quantity limits, and taxes provide no flexibility for subsequent reallocations of entitlements. But allowance trading makes postagreement reallocations possible, thereby reducing the initial assignment impasse.

To summarize, under the voting rule of consent that governs global climate treaties, participation efficiency is crucial. A way must be found to pay reluctant sources to participate while also inhibiting the perverse incentives that these payments create. The best instrument for achieving this result is a system of international tradable emission allowances, with headroom allowances allocated to cooperative losers. It secures broad participation and enables cost-effective flexibility in the spatial location of abatement but caps total emissions and thereby constrains the perverse environmental effects of subsidizing abatement.

Compliance

Compliance is a general problem of any regulatory system. But it figures prominently in criticisms of international environmental regulation because it is more troublesome under the consent voting rule, where countries—even after agreeing to participate—cannot be compelled to comply but must be attracted by the continuing desirability of participation. Critics often charge that ensuring compliance with international emission trading would be difficult. Yet the problem of compliance is not unique to allowance trading; all regulatory instruments require monitoring and enforcement. The key question is the relative ability of the instruments to maintain compliance, given the voting rule of consent. The criticisms of weak enforcement systems are really criticisms of the weak ability of the international system to deal with any nation-states' noncompliance with any treaty obligations.

Noncompliance is a partial version of free riding. Once free riding is overcome—once countries are attracted to participate by the net gains they perceive from joining the treaty—then "compliance comes free of charge."[77] Therefore, there are good reasons to expect allowance trading to be superior to alternative regulatory instruments at inducing compliance. First, the improved cost-effectiveness (30–70 percent lower abatement costs) under allowance trading makes participation less costly and thus lowers the incentive to free ride or cheat. Second, the assignment of headroom allowances attracts participation by erstwhile noncooperators, and the prospect of continuing to sell allowances over time provides a strong disincentive to cheat. Third, a system of allowance trading furnishes useful enforcement tools, including the ability to debit a violator's allowance account and to exclude the violator from the allowance market. Fourth, a tradable allowance system is likely to nurture domestic political constituencies—allowance sellers, allowance buyers, abatement investors, brokers, and environmentalists—who would pressure their governments to comply with emission limits so as not to have their allowances devalued or their market access hindered.[78]

Meanwhile, the actual effectiveness of internationally agreed GHG taxes or technology standards would be extremely difficult to ensure. In response to a GHG tax or technology standard, countries would have strong incentives to adjust their internal tax and subsidy policies to counteract the effect of the international policy on domestic industries. This "fiscal cushioning" would undermine the effect of the tax or technology standard on actual emissions.[79] Thus, a country could be in technical compliance with the tax or technology standard, but its fiscal cushioning countermoves could vitiate the environmental effectiveness of these instruments. It would be difficult for international authorities to detect and block these detailed domestic fiscal games. By contrast, the effectiveness of international allowance trading would be simpler to monitor. Under a quantity instrument, participants need not monitor all the domestic tactics being practiced in each country. Instead, they need only monitor the nation's aggregate emissions and compare them with the country's allowed total (its cap or allowance holdings). This real environmental effectiveness—as opposed to apparent compliance—would be easier to monitor than would the intricacies of domestic implementation under a global tax or technology standard.

Assessing the Kyoto Protocol

In terms of spatial complexity and participation efficiency, the Kyoto Protocol gets things about half right. On the bright side, it adopts a quantity constraint on emissions, eschewing technology standards and emission taxes, and it authorizes emission trading (in Article 17) to enhance cost-effectiveness rather than adopting fixed performance standards. Moreover, it makes some use of allowance allocations to secure participation. It allocates the burden of emission reductions among nations roughly in proportion to national wealth, which as discussed earlier is a rough proxy for national perceived benefits of climate protection. And it assigns headroom allowances to Russia—a move that some observers have criticized as ineptitude and dubbed "hot air" but can be better understood as a very rational and necessary form of compensation to secure Russia's participation in the treaty. Russia's agreement to emission controls was by no means guaranteed, and without headroom allowances it might well have stayed out of the treaty, squandering many low-cost abatement options and inviting significant leakage.

But this cap-and-trade regime is only a half-step in the right direction because the Kyoto Protocol omits the developing countries from this regime. China, India, Brazil, Indonesia, and other developing countries have no obligations to limit their emissions under the treaty. Their growing emissions will render the treaty increasingly ineffective. The prospects for emission leakage from capped industrialized countries to uncapped developing countries are serious. Under the consent voting rule (and also for reasons of distributional fairness), side payments will be needed to attract their participation.

The Kyoto Protocol tries to address developing country abatement by introducing a new and well-intentioned device—the Clean Development Mechanism (CDM) created in Article 12—through which industrialized country sources could purchase emission reduction credits from developing countries. The CDM does promise significant abatement at low cost and the possibility of introducing lower-emitting technologies into developing countries before they become dependent on high-emission growth paths. These are important advantages.

But the CDM could have a perverse impact on global emissions and could undermine future efforts to bring developing countries into the cap-and-trade regime. First, because CDM seller countries are not subject to national quantity caps, the CDM transactions amount to pure subsidies for abatement. As discussed earlier, this regulatory instrument is disfavored because it induces perverse increases in the total size of the emitting sector. By reducing the relative cost of operating emitting enterprises in developing countries, the CDM will attract investment to those industries (accelerate leakage) and thus could be of limited effectiveness or even expand total emissions. (Moreover, because there are no national quantity caps on developing countries, CDM abatement investments might be offset by unseen increases in emissions elsewhere in the same country.)

Second, the opportunity to sell CDM credits could discourage uncapped developing countries from joining the cap regime. Recall that it is the prospect of selling headroom allowances that provides the pivotal incentive for developing countries to participate in the cap-and-trade system. But if those countries can earn just as much by selling CDM credits without a cap, why should they accept caps? And if they don't join the cap regime, increased net leakage may render the entire treaty futile or worse. One way to address this problem would be to discount CDM credits (or "certify" them at less than the claimed tons of abatement) to reflect their lesser effectiveness in achieving global abatement. This would lower their attractiveness and push more countries toward agreeing to caps to take advantage of more lucrative formal trading.[80]

Third, the CDM may be a battleground for political and market power. It is constituted under Article 12 as a discrete entity governed by an executive board. This apparently centralized organization could exert control over the market in CDM credits.

Thus, the Kyoto Protocol makes some progress in the use of allowance trading to secure efficient participation but fails to engage developing countries in the cap-and-trade system. For that reason the U.S. Senate announced its unanimous opposition to the treaty, and the Clinton administration never submitted the treaty to the Senate for ratification. In 2001 the Bush administration announced that it would not pursue the Kyoto Protocol but did not propose an alternative.

The accords reached at Bonn and Marrakech in 2001 to implement Kyoto omitted both the developing countries and the United States, portending limited effectiveness in reducing global emissions. They also retained some restrictions on emission trading, including a "reserve" requirement to limit allowance selling and quantitative limits on credit for sink expansion. The cost savings expected from emission trading in theory must be reestimated with the actual Bonn/Marrakech restrictions in place. To be environmentally effective (as well as less costly), the accords should be revised to include major developing countries in a fully flexible cap-and-trade system on terms beneficial to all through the assignment of headroom allowances.

Although the events of 2001 seemed to sacrifice broad participation, they might set the stage for an even better result: joint accession by both the United States and China.[81] Politically the United States will not join targets without China (as made clear by the Bush administration and by the Senate's 95–0 vote against joining a climate treaty that omits the major developing countries). And China will not join targets without the United States (because it will not act unless the wealthy industrialized countries act first). So both will have to join for either to join. Moreover, the current parties to Kyoto will want the United States and China to join simultaneously. If one joins without the other, it will distort allowance prices in the emission trading market: Prices will go way up if the United States (a large net demander) joins alone and way down if China (a large net supplier) joins alone. The EU and Japan will not want prices to rise sharply, and Russia will not want prices to fall sharply.

Thus, perhaps unintentionally, the initially awkward result in 2001 may pave the way for joint accession by the United States and China. If not, the Kyoto accord will amount to very little. Without the world's largest emitters participating, it will not affect global emissions or concentrations much at all. Thus joint accession by the United States and China may be the only plausible future for the climate treaties. And this reality in turn gives the United States and China significant leverage to negotiate for a sound global regime that improves on Marrakech, Bonn, and Kyoto through full global emissions trading. The real difficulty in this scenario will not be the United States; it will be China. The United States faces both costs and benefits from joining. But China may well perceive only costs because many forecasts of the impacts of global warming, as noted earlier, suggest that China would on balance benefit from a warmer world. China will have to be paid to play. The best way to compensate China for joining the abatement regime will be through assignments of headroom allowances that China can then sell, as was done in Kyoto to engage Russia.

Temporal Complexity and Dynamic Adaptation

Perhaps the most vexing form of complexity confronting climate policy is temporal: Things change over time. The environment changes, so climate change may turn out to be more or less serious (or different in kind) than we now envision. The economy changes in ways that may ease or exacerbate abatement costs. Temporal complexity implies two challenges: optimally allocating abatement efforts over time and adapting climate policy as conditions and knowledge evolve. Compared with causal and spatial complexity, temporal complexity has received the least attention in the actual climate change treaty negotiations.

Optimal Allocation of Abatement over Time

Any given level of climate protection may be achieved with different allocations of abatement over time. These different time paths of emission reduction will imply different costs and benefits. Earlier reductions may protect the climate more because they prevent the buildup of gases that would reside in the atmosphere for decades thereafter. But later reductions may cost less because they ease the turnover of capital investments, allow the development of new technologies, and spend scarce resources later rather than sooner.

One strategy to optimize abatement over time is to set emission targets not for single years but for multiyear aggregates such as 10-year emission budgets for each country. Such multiyear targets (or extended commitment periods) give each country flexibility in the timing of abatement, thereby reducing the costs of compliance because different countries may have different expectations for the turnover of capital stock, acquisition of new technologies, and social discount rates. Temporal flexibility through multiyear budgets is conceptually similar to the spatial flexibility afforded by tradable allowances: Because abatement costs vary across the relevant dimension (temporal or spatial), flexibility improves cost-effectiveness. A more embracing version of temporal flexibility would authorize banking of extra early emission reductions for application to subsequent emission limitations, and perhaps borrowing against later limitations by promising to achieve extra abatement later to make up for earlier excess emissions. (If the climate benefits more from emission reductions achieved earlier than later, then banking should earn and borrowing should be charged an "interest rate" that renders equivalent the abatement occurring at the different times.)

Second, targets could be announced at least 10 years in advance of their effective dates, or take effect 10 years after the treaty enters into force. Major investments in capital and innovation often take longer than 5 years to turn over, so a longer time horizon would provide early signals that enable more cost-effective changes in technology. Targets set too close to the present will be harder to achieve, perhaps impossible, and will invite repeated deferral in a process that makes the initial targets lack credibility and inculcates public cynicism about the regulatory regime. A similar cycle of unrealistic targets followed by deferral and cynicism has characterized several major U.S. environmental laws, such as the national ambient air quality standards (NAAQS) under the Clean Air Act amendments of 1970, 1977, and 1990 and the best technology standards under the Clean Water Act amendments of 1972, 1977, and 1987. On the other hand, a downside of setting targets for many years hence is that they may fail to motivate changes in businesses' investments, and they may lack credibility because there is so much time available to debate and revise them. Perhaps a middle course is to set not a single target for one out-year or period but a continuous schedule of emission limits, beginning with small or no reductions and tightening over time. This approach was successful in the lead phasedown from the 1970s through late 1980s and was approximated in the acid rain title of the 1990 Clean Air Act and the Montreal Protocol on ozone-depleting substances.

Third, the time path of emission limitations can be optimized in light of the benefits and costs of climate protection. The FCCC states in Article 2 that its objective is the stabilization of atmospheric GHG concentrations at a level that will avoid dangerous anthropogenic interference with the climate. (No such level has yet been defined or agreed upon.) Such a stabilization objective can be achieved through many different time paths of abatement, some of which are much less costly than others. In particular, delaying abatement for several decades and then reducing emissions more sharply can significantly reduce the cost of stabilization by allowing for capital turnover, new technologies, and discounting.[82] On the other hand, if one takes account of the damages resulting from climate change as it occurs (instead of pegging a single level at which to stabilize concentrations), then the optimal time path of abatement is different. Hammitt[83] compares the emission reductions implied by the least-cost path to stabilize atmospheric GHG concentrations at designated levels with the emission reductions implied by the optimal (net benefits maximizing) path to prevent climate change (based on several assumptions about benefits and costs). He finds that the optimal path involves more stringent near-term emission reductions below the business as usual (BAU) emission forecast than does the least-cost path to stabilize atmospheric GHG concentrations at 750, 650, or even 550 ppm by the period 2100–2150.[84] The reason is that the optimal path takes into account the damages from near-term emissions, whereas the least-cost path to stabilize concentrations does not. Thus the optimal path in Hammitt's analysis calls for some near-term emission reductions—roughly 3 percent below BAU by

2010, 5 percent below BAU by 2025, and 20 percent below BAU by 2100—whereas the least-cost stabilization path for hitting 750, 650, or 550 ppm calls for near-term emissions essentially unchanged from BAU until around 2070, 2050, and 2010, respectively, and then much steeper declines in emissions thereafter (beginning about 2025 in the case of the 550 ppm target, for example). The optimal path exhibits a more smoothly but slowly rising emission profile that is about 2–5 percent below the least-cost stabilization profile in the near term (through about 2025) but eventually exceeds the least-cost stabilization emission profile after 2107, 2069, and 2024, respectively, for stabilization at 750, 650, and 550 ppm.[85] Hammitt's approach, which minimizes overall costs (both economic and environmental), is conceptually preferable to the least-cost stabilization strategy, which minimizes only economic costs to achieve an arbitrarily chosen stabilization level.[86]

As Hammitt notes, one would need to start building the institutional structure for climate policy some time before the dates at which emission reductions would be expected, in order to send credible policy signals that will in turn stimulate the needed shifts in investments, practices, and technologies. To achieve Hammitt's optimal path of 3 percent below BAU in 2010, 5 percent below BAU in 2025, and 20 percent below BAU in 2100, one would need to begin constructing and implementing the institutional design well before 2010—that is, roughly, now.

Adaptation of Policy over Time

Temporal complexity also means that the level of protection initially set may later seem erroneous and need to be updated as conditions and knowledge have changed. The direction of our likely errors is highly debatable: Are we acting too hastily or not fast enough? Some say that temporal complexity counsels against adopting quantity limits on emissions and in favor of more gradual institution building and research;[87] others say that temporal complexity counsels in favor of adopting more stringent limits now to prevent even greater harms than we now foresee.[88]

A central lesson of temporal complexity is the value of adaptation over time. "Adaptive management" has become a popular idea but an elusive reality. Designing an adaptive regulatory regime is difficult because knowledge is always changing, but investors want predictable rules, and the establishment of rules itself invites investments that entrench opposition to subsequent changes in those rules. The challenge is to design regulatory institutions that are able to evolve as conditions and understandings change yet are not so mercurial that they upset investors' expectations and undermine their own credibility.

Several steps toward an adaptive approach are desirable. First, governments should continue investing in scientific and economic research as regulations are imposed and reassess regulations regularly in light of the latest expert advice. The role of the Intergovernmental Panel on Climate Change (IPCC) and of national research programs therefore will continue to be crucial. All regulatory institutions, at every scale, need to be geared toward learning and updating.

Second, the iterative negotiating sessions held under the FCCC and Kyoto Protocol—roughly one or two Conferences of the Parties each year—can be seen as fostering the regime's adaptive capacity. Through this process, parties debate new emission targets every few years, keeping options open rather than trying to adopt a permanent set of emission limits once and for all. On the other hand, this process of sequential decisions creates uncertainty about future targets and may be at odds with the objective of setting a schedule of continuous emission limitations over many years so that investments respond accordingly and cost-effectively. Sequential target-setting should be undertaken transparently so that investors have advance signals of likely next steps.[89]

Third, policy should be based on an evaluation of multiple plausible scenarios rather than the choice of a single best scenario. Adaptive management is particularly valuable in cases such as climate change that involve fundamental uncertainty about how the system works.[90] Our current forecasts may not only be off a bit but may rely on models that do not even describe reality. One hedge against this uncertainty is to base policy on a collage of several plausible but conceptually different models and to update this collage over time, with predictions weighted by experts' relative confidence in the different models.

Fourth, in the face of such uncertainty, policy should at least begin by instituting measures that would be desirable under all of these scenarios. These could include reducing subsidies for energy use, reforming incentives for forest clearing, supporting basic research into low-GHG energy systems, improving the capacity for technology diffusion and application in developing countries, reducing emissions of air pollutants in ways that both protect human health and help prevent climate change, and making social and environmental systems more resilient against climate changes. At the same time, some measures will be warranted on grounds of climate protection alone, even in the face of significant uncertainty.

Assessing the Kyoto Protocol

The FCCC and Kyoto Protocol have done little to address temporal complexity. Kyoto allowed some temporal flexibility by setting targets as average emissions over a 5-year commitment period, 2008–2012. But even greater temporal efficiencies could have been achieved through a longer commitment period (such as 10 years) and through expressly authorizing both banking of early reductions and borrowing against later limitations (with an interest rate reflecting the time value of abatement). Kyoto did not give any credit for emission reductions before 2008 (except, oddly, for CDM projects) and did not allow borrowing. Banking and borrowing make the most sense as early departures below and above a continuous emission reduction schedule, whereas Kyoto set a single commitment period target, and negotiations on a second commitment period target have not yet begun.

Regarding the time to achieve targets, Kyoto announced its targets in 1997 for an effective date beginning 11 years into the future. Eleven years might seem like a long time, but the practical realities of treaty negotiations and energy system investments suggest that a longer time between announcement and effective date could have been prudent. By the time the Kyoto process neared even initial ratification it was already 2001, with entry into force expected no earlier than late 2002, making the 2008 effective date seem too near to achieve substantial emission cuts without major costs.

Kyoto also set targets that depart significantly from both the least-cost stabilization path and Hammitt's illustrative optimal path. The Kyoto Protocol called for emission reductions by industrialized countries of about 5 percent below 1990 levels by 2012, which corresponds to a U.S. reduction of about 30 percent below BAU in 2012 and a reduction in all industrialized countries' emissions of roughly 15–20 percent below BAU by 2012. Thus the Kyoto Protocol appears to require (at least for industrialized countries) much sharper near-term emission reductions than those required by either Hammitt's optimal path (which requires global emissions to be 3 percent below BAU by 2010, 5 percent below BAU by 2025, and 20 percent below BAU by 2100) or the least-cost path to stabilizing concentrations at 750, 650, or 550 ppm (all of which require essentially zero reduction below BAU through 2025 but steeper reductions later).[91] More fundamentally, Hammitt's analysis suggests that the stabilization objective enshrined in the FCCC is not the best goal for climate policy, even if achieved at least cost, because it neglects the continuous impacts of GHG accumulation over time. Analyses of optimal climate policy must do a better job of accounting for damages over time and nonlinear climatic effects.[92]

As to adaptive management, the Kyoto process involves iterated negotiation of targets, with regular scientific input from the IPCC. This sequential process of adjustment could be helpful in adapting to new information. But the IPCC has not done enough to advise the treaty negotiators on the optimal time path of abatement. The Kyoto process may well result in repeated updating of its emission targets, but those updates may not reflect a considered evaluation of the optimal temporal path for abatement.

Conclusion

Global climate policy is deeply complex. This chapter has examined three kinds of complexity—causal, spatial, and temporal—and three corresponding innovations in the design of the regulatory regime for climate change. First, the comprehensive approach would protect the environment more effectively (avoiding perverse cross-gas shifts) and at perhaps 60 percent lower cost than a piecemeal approach. Second, international allowance trading would cost perhaps 70 percent less than fixed national caps, and, under the consent voting rule that prevails at the global level, would be more participation-efficient than alternative regulatory instruments. Participation is crucial to global success; it has been neglected in the Kyoto and Bonn/Marrakech agreements, but well-designed global allowance trading holds the promise of engaging both the United States and China in the future. Third, optimal time paths and adaptive management would enable climate policy to be flexible as technologies, environmental conditions, and our knowledge all change over time.

This is not to say that these approaches are perfect, nor that other regulatory approaches do not have their strengths in other contexts. The administrative costs of the comprehensive approach could become unreasonable if its scope were expanded indefinitely. The presumptive advantage of tradable allowances could diminish if cooperative losers were unimportant to global emissions or if abatement cost uncertainties were so large that containing those costs through taxes (or through a price ceiling on allowances) became a higher priority than participation efficiency and containing climate damages. Optimal temporal policies could raise questions about the credibility of long-term commitments by governments. Nonetheless, the advantages of these three policy designs appear to far outweigh their administrative difficulties.

The phenomena of causal, spatial, and temporal complexity will continue to challenge and intrigue those who design global climate policy. The Kyoto Protocol and the Bonn/Marrakech accord have made good progress on comprehensive coverage and on emission trading among industrialized countries, but they have limited sinks, have made meager headway in the effort to secure broad global participation, and have only begun to address optimal temporal policy design. Thus there is much work remaining in the design of successful global climate policy.

Acknowledgments

The author thanks John Barton, Aimee Christensen, James Hammitt, John O. Niles, Stephen Schneider, and Richard Stewart for comments on prior drafts and David Halsing for editorial assistance.

Notes

1. This chapter examines the design of the regulatory mechanism, not its goals; that is, it discusses the means or instruments of protection, not the ends or level of protection that should be sought. Although the two questions ultimately are interrelated, deciding how to reduce emissions can usefully be analyzed as a distinct matter from deciding how much to reduce emissions. See Bohm, P. and C. S. Russell, 1995: "Comparative analysis of alternative policy instruments," in A. V. Kneese and J. L. Sweeney (eds.) *Handbook of Natural Resource and Energy Economics* (New York: North-Holland/Elsevier), 395, 397 ("Choice of policy goal and choice of instrument or implementation system are essentially separable problems.").
2. Breyer, S. G., 1982: *Regulation and Its Reform* (Cambridge, MA: Harvard University Press).
3. This section draws on Wiener, J. B. 1995: "Protecting the global environment," in J. D. Graham and J. B. Wiener (eds.), *Risk vs. Risk: Tradeoffs in Protecting Health and the Environment* (Cambridge, MA: Harvard University Press); Stewart, R. B. and J. B. Wiener, 1992: "The comprehensive approach to global climate policy: issues of design and practicality," *Arizona Journal of International & Comparative Law,* 9:83; and Stewart, R. B. and J. B. Wiener, 1990: "A comprehensive approach to climate change," *American Enterprise,* 1 (6): 75.
4. The classic case for narrow incrementalism is Lindblom, C. E., 1959: "The science of 'muddling through,'" *Public Administration Review,* 19: 79.
5. Wiener, J. B., 1998: "Managing the iatrogenic risks of risk management," *Risk: Health, Safety & Environment,* 9: 39.
6. Houghton, J. T., G. J. Jenkins, and J. J. Ephraums (eds.), 1992: *IPCC First Assessment Report, The Science of Climate Change* (Cambridge, England: Cambridge University Press).
7. Harrington, W., 1989: *Acid Rain: A Primer* (Washington, DC: Resources for the Future).
8. Guruswamy, L., 1991: "The case for integrated pollution control," *Law & Contemporary Problems,* 54: 41.
9. Rodhe, H., 1990: "A comparison of the contribution of various gases to the greenhouse effect," *Science,* 248 (June 8): 1217–1219.
10. Wiener, 1995, pp. 209–212.
11. Schlesinger, W. H., 1999: "Carbon sequestration in soils," *Science,* 284 (June 25): 2095.
12. "Potentially" because, although conserving forests would protect biodiversity, new afforestation projects to sequester carbon might replace biodiverse mature forests

with monoculture plantation forests. See Wiener, 1995, 218–219. Meanwhile, some GHG emissions could aid forests: CO_2 emissions help fertilize plant photosynthesis, a beneficial effect that the other GHGs do not offer. See Wiener, 1995, 214–218; DeLucia, E. H., et al., "Net primary production of a forest ecosystem with experimental CO_2 enrichment," *Science,* 284 (May 14): 1177–1179. Thus, to be fully environmentally comprehensive, a climate policy would need to be broadened or accompanied by biodiversity protections and by a gas comparison index reflecting GHGs' full ecosystem impacts. See Stewart and Wiener, 1990, 1992.

13. Bradley, R., E. Watts, and E. Williams, 1991: *Limiting Net Greenhouse Emissions in the United States,* Volume II: *Energy Responses* 8.10–8.12 (Washington, DC: U.S. Department of Energy, Office of Environmental Analysis).
14. World Bank, 1992: *World Development Report 1992: Development and the Environment,* Box 8.6 (Washington, DC: The World Bank).
15. Reilly, J., R. Prinn, J. Harnisch, J. Fitzmaurice, H. Jacoby, D. Kicklighter, J. Melillo, P. Stone, A. Sokolov, and C. Wang, 1999: "Multi-gas assessment of the Kyoto Protocol," *Nature,* 401 (October 7): 549–555.
16. Hansen, J., M. Sato, R. Ruedy, A. Lacis, and V. Oinas, 2000: "Global warming in the twenty-first century: an alternative scenario," *Proceedings of the National Academy of Sciences,* 97: 9875.
17. Stewart and Wiener, 1990, 1992. The comprehensive approach was staunchly advocated by the United States through both the Bush and Clinton administrations. See Revken, A. C., 2000: "Study proposes new strategy to stem global warming," *New York Times,* Aug. 19, p. A13 (reporting that the White House supports the comprehensive approach embodied in the Kyoto Protocol as the "best approach to slowing warming" because it addressed "all of the greenhouse gases . . . largely because of the insistence by American negotiators.").
18. For further discussion see Stewart and Wiener, 1992.
19. This section draws on Wiener, J. B., 1999: "Global environmental regulation: instrument choice in legal context," *Yale Law Journal,* 108: 677–800.
20. Externalities are the effects of an economic transaction not faced by the parties to the transaction. Economists view externalities as a source of inefficiency because the actors involved in the transaction are making decisions without considering their full social consequences. Economists therefore seek ways to internalize externalities into market decisions, such as through regulatory instruments. Because regulations themselves can pose costs, the presence of an externality is a necessary but not sufficient condition for regulating.
21. Hardin, G., 1968: "The tragedy of the commons," *Science,* 162: 1243.
22. Axelrod, R., 1984: *The Evolution of Cooperation* (New York: Basic Books) pp. 7–9.
23. Keohane, R. O., 1984: *After Hegemony* (Princeton: Princeton University Press).
24. Council of Economic Advisors, 1998: *Economic Report of the President* (Washington, DC: Government Printing Office) pp. 170–172 (forecasting that GHG emissions from developing countries will exceed those from industrialized countries by the year 2030). To be sure, such forecasts are uncertain. See U.S. Energy Information Administration, 2001: *International Energy Outlook* (Washington, DC: U.S. Department of Energy) (forecasting China's energy sector CO_2 emissions to grow

from roughly 650 million tons of carbon in 1999 to somewhere between 1,115 and 2,059 million tons in 2020).

25. Schmalensee, R., 1998: "Greenhouse policy architectures and institutions," in W. D. Nordhaus (ed.), *Economics and Policy Issues in Climate Change* (Washington, DC: Resources for the Future), pp. 137, 146.
26. S. Res. 98, 143 *Congressional Record* S8113-05 (July 25, 1997). The Clinton administration decided not even to submit the Kyoto Protocol to the Senate for ratification until "meaningful participation" by developing countries had been secured.
27. Barrett, S., 1992: "Reaching a CO_2 emission limitation agreement for the community: implications for equity and cost-effectiveness," *European Economics,* 1:3, 16.
28. Baumol, W. J., 1974: "Environmental protection and income distribution," in H. H. Hochman and G. E. Peterson (eds.), *Redistribution Through Public Choice* (New York: Columbia University Press), p. 93 (observing that the demand for environmental protection typically rises with income).
29. Tol, R., 2001a: "Estimates of the damage costs of climate change, part I: benchmark estimates," *Environment & Resource Economics* (Dordrecht: Kluwer Academic Publishers); Tol, R., 2001b: "Estimates of the damage costs of climate change, part II: dynamic estimates," *Environment & Resource Economics* (Dordrecht: Kluwer Academic Publishers). Tol's synthesis does not account for other adverse impacts, such as fishery losses, extreme weather events, and the possibility of catastrophic changes in ocean currents or other critical natural systems.
30. Aronson, A. L., 1993: "From 'cooperator's loss' to cooperative gain: negotiating greenhouse gas abatement," *Yale Law Journal,* 102: 2143, 2150–2151.
31. Wiener, "Global environmental regulation," 1999, pp. 701–704 (showing the ubiquity of the assumption of coercive fiat in discussions of regulatory instrument choice). As James Buchanan put it in his Nobel Prize address, economists have long been "proffering advice as if they were employed by a benevolent despot." Buchanan, J. N., 1987: "The constitution of economic policy," *American Economic Review,* 77: 243.
32. This is an approximation. Real national law in the United States involves majority or super-majority votes in more than one legislative chamber, plus signature by the executive and review by the courts. And majority rule can be limited, for example, by constitutional protection of free speech or compensation to those from whom the majority takes property.
33. A classic statement is Lord McNair, 1961: *The Law of Treaties* (Oxford: Clarendon Press), p. 162: "No State can be bound by any treaty provisions unless it has given its assent."
34. Hoel, M. and K. Schneider, 1997: "Incentives to participate in an international environmental agreement," *Environmental & Resource Economics,* 9: 153, 165–167.
35. Cameron, J., 1994: "The GATT and the environment," in P. Sands (ed.), *Greening International Law* (New York: New Press), pp. 106–116.
36. Chayes, A. and A. Handler Chayes, 1995: *The New Sovereignty* (Cambridge: Harvard University Press), p. 27.
37. Keohane, R. O. and J. Nye, 1972: *Transnational Relations and International Governance* (Boston: Little Brown and Company); Lee, H., 1995: *Shaping National*

Responses to Climate Change (Washington, DC: Island Press), p. 14 ("de facto transnational coalitions" often have "enormous influence" on international diplomacy).

38. Humphreys, D., 1996: *Forest Politics* (London: Earthscan), p. 162.
39. In this discussion I take the voting rules as given and make no normative comment on their relative desirability. Cf. Wiener, J. B., 1999: "On the political economy of global environmental regulation," *Georgetown Law Journal,* 87: 749 (discussing normative implications of alternative voting rules for international environmental law).
40. Keohane, 1984, p. 104; Wiener, "Global environmental regulation," 1999, pp. 735–747. The question for each country is whether joining would be better than not joining. A country need not view the entire treaty as a net gain compared with the complete absence of any treaty (the situation before anyone had joined the treaty). The fact that some countries have already formed the treaty could impose net costs on the remaining countries. See Gruber, L., 2000: *Ruling the World* (Princeton: Princeton University Press). Even if a country would prefer a world with no treaty at all, it may still prefer joining the treaty to staying out if there are further costs to staying out (such as trade sanctions) or new benefits to joining (such as side payments).
41. Wiener, "Global environmental regulation," 1999, pp. 743–755.
42. Buchanan, J. and G. Tullock, 1962: *The Calculus of Consent* (Ann Arbor: University of Michigan Press), p. 113 (analogizing consent voting rules to contracts); Keohane, R. O., 1983: "The demand for international regimes," in S. D. Krasner (ed.), *International Regimes* (Ithaca, NY: Cornell University Press), pp. 141, 146–149, 152 (arguing that "international regimes . . . are more like contracts. . . . In general, we expect states to join those regimes in which they expect the benefits of membership to outweigh the costs").
43. Stewart, R. B., 1986: "Reconstitutive law," *Maryland Law Review,* 46: 86, 92.
44. Hahn, R. W. and R. N. Stavins, 1991: "Incentive-based environmental regulation: a new era from an old idea?" *Ecology Law Quarterly,* 18: 1. A third type of incentive instrument is information disclosure.
45. Stavins, R. N., 2000: "Introduction," in P. R. Portney and R. N. Stavins (eds.), *Public Policies for Environmental Protection,* 2nd ed (Washington, DC: Resources for the Future). Improving cost-effectiveness is valuable because it saves resources that can be used for other important social goals, including greater investments in environmental protection. See Baumol, W. J. and W. E. Oates, 1988: *The Theory of Environmental Policy,* 2nd ed. (Cambridge, England: Cambridge University Press).
46. Hahn, R. W. and G. L. Hester, 1989: "Marketable permits: lessons for theory and practice," *Ecology Law Quarterly,* 16: 387.
47. Joskow, P. L., R. Schmalensee, and E. Baily. 1998: "The market for sulfur dioxide emissions," *American Economic Review,* 88: 669.
48. See Manne, A. S. and R. G. Richels, 2000: "The Kyoto Protocol: A Cost effective strategy for meeting environmental objectives?" in C. Carraro (ed.), *Efficiency and Equity in Climate Change* (Amsterdam: Kluwer Academic Publishers), p. 59; Weyant, J. P. and J. Hill, 1999: "Introduction and overview," *The Energy Journal,* Special Issue vii; Bohm, P., 1997: "Joint implementation as emission quota trade:

an experiment among four Nordic countries," *Nord,* 1997: 4, Nordic Council of Ministers, Copenhagen; Manne, A. and R. Richels, 1996: "The Berlin mandate: the costs of meeting post-2000 targets and timetables," *Energy Policy,* 24: 205; Burniaux, J.-M., J. P. Martin, G. Nicoletti, and J. Oliveira-Martins. 1992: "The costs of reducing CO_2 emissions: evidence from GREEN," OECD Economic Department Working Paper No. 115.

49. Jaffe, A. and R. N. Stavins, 1995: "Dynamic incentives of environmental regulation: the effects of alternative policy instruments on technology diffusion," *Journal of Environmental Economics and Management,* 29: S-43.
50. See Okun, A. M., 1975: *Equality and Efficiency: The Big Tradeoff* (Washington, DC: The Brookings Institution); Kelman, S., 1981: *What Price Incentives? Economists and the Environment* (Boston: Auburn Publishing Co.), pp. 84–86. A related concern is that trading allowances of toxic emissions might yield local hotspots, but this concern is minimal in the case of CO_2 and other GHGs with few or no local health impacts.
51. Oliveira-Martins, J., J. M. Burniaux, J. P. Martin, and G. Nicoletti. 1992: "The costs of reducing CO_2 emissions: a comparison of carbon tax curves with GREEN," OECD Economics Department Working Paper No. 118 (estimating developing country revenues at $10 billion per year initially and increasing over several decades to $100 billion per year under global emission trading).
52. Sandel, M. J., 1997: "It's immoral to buy the right to pollute," *New York Times,* December 15, p. A23 (op-ed).
53. See Baumol and Oates, 1988, pp. 211–222; Oates, W. E. 1990: "Economics, economists, and environmental policy," *Eastern Economic Journal,* 16: 289, 290; Kohn, R. E., 1992: "When subsidies for pollution abatement increase total emissions," *Southern Economic Journal,* 58: 77.
54. Posner, R. A., 1992: *Economic Analysis of Law,* 4th ed. (Boston: Little Brown) p. 64.
55. Hoel and Schneider, 1997.
56. Weitzman, M. L., 1974: "Prices vs. quantities," *Review of Economic Studies,* 41: 477.
57. Pizer, W. A., 1997: "Prices vs. quantities revisited: the case of climate change," Resources for the Future, Discussion Paper No. 98-02.
58. See Goulder, L. H., 1995: "Environmental taxation and the 'double dividend': a reader's guide," *International Tax and Public Finance,* 2: 157.
59. Hahn, R. W., 1984: "Market power and transferable property rights," *Quarterly Journal of Economics,* 99: 753. Alternatively, a few large buyers could try to exercise monopsony power, depressing allowance prices.
60. Dudek, D. J. and J. B. Wiener, 1996: "Joint implementation, transaction costs, and climate change," OECD/GD (96) 173, pp. 20–21.
61. In addition, recent research suggests that under more realistic models with imperfect enforcement policies, quantity instruments are preferable to price instruments. See Montero, J. P., 1999: "Prices versus quantities under imperfect enforcement," MIT CEEPR Working Paper.
62. Cf. Merrill, T. W., 1997: "Golden rules for transboundary protection," *Duke Law Journal,* 46: 981.
63. Wiener, "Global environmental regulations," 1999, pp. 748 & n.266, 761 & n.311.

64. This concept is advanced and discussed in Wiener, "Global environmental regulation," 1999, pp. 742–770.
65. Baumol and Oates, 1988, pp. 279–281.
66. Ibid., p. 281. Negotiators might limit the perverse incentive effect by constraining the availability of the subsidy to certain countries or circumstances, but that might undercut the ability of the subsidy to attract participation.
67. Hoel and Schneider, 1997, p. 165.
68. Chang, H. F., 1995: "An economic analysis of trade measures to protect the global environment," *Georgetown Law Journal,* 83: 2131.
69. Barrett, S., 1996: "Building property rights for transboundary resources," in S. S. Hanna, C. Folke, and K. G. Mäler (eds.), *Rights to Nature* (Washington, DC: Island Press), pp. 265, 280–282.
70. Haass, R. N., 1997: "Sanctioning madness," *Foreign Affairs,* November–December: 74, 77–80.
71. Jacobson, H. K. and E. B. Weiss, 1997: "Compliance with international environmental accords: achievements and strategies," in M. Rolen, H. Sjöberg, and U. Svedin (eds.), *International Governance on Environmental Issues* (Dordrecht: Kluwer Academic Publishers), pp. 78, 109.
72. Hurrell, A. and B. Kingsbury, 1992: "The international politics of the environment: an introduction," in A. Hurrell and B. Kingsbury (eds.), *The International Politics of the Environment* (Oxford: Clarendon Press), pp. 7–8.
73. And by delivering the side payments through market trades rather than official government aid, this system would yield abatement investments that are less impeded by bureaucratic costs, more cost-effective, and more innovation-enhancing. See Burtraw, D. and M. A. Toman, 1992: "Equity and international agreements for CO_2 containment," *Journal of Energy Engineering,* 118: 122, 131–132. Ironically, the reduced role of the bureaucracy in allowance trading (as compared with official government aid) could lead government representatives in both industrialized and developing countries to oppose allowance trading. This domestic power struggle is one possible reason for the opposition to allowance trading by government officials whose national economies would benefit greatly from trading. See Wiener, "On the political economy of global environmental regulation," 1999, pp. 780–781.
74. Victor, D., 2001: *The Collapse of the Kyoto Protocol* (Princeton: Princeton University Press); Cooper, R., 1998: "Toward a real global warming treaty," *Foreign Affairs,* March–April: 66, 70–72, 74, 78.
75. Hahn, R. W., 1998: *The Economics & Politics of Climate Change* (Washington, DC: American Enterprise Institute), p. 43.
76. Coase, R., 1960: "The problem of social cost," *Journal of Law & Economics,* 3:1. The Coase theorem holds that in a world of zero transaction costs, the initial assignment is irrelevant to efficiency and to the ultimate allocation after bargaining. To be sure, distributional impacts would still matter. The point here is that the more a regime is designed to reduce the transaction costs of postassignment contractual reallocations, the easier it will be to negotiate that regime initially.
77. Barrett, S., 1998: "A theory of international cooperation," Fondazione Eni Enrico Mattei Working Paper No. 43-98, p. 7.

78. Stewart, R. B., J. B. Wiener, and P. Sands, 1996: *Legal Issues Presented by a Pilot International Greenhouse Gas Trading System* (Geneva: UNCTAD), p. 45.
79. Wiener, "Global environmental regulation," 1999, pp. 785–787.
80. In any case, the higher transaction costs of project-based CDM credits may ensure that they trade at a lower price than formal allowances would. See Dudek and Wiener, 1996. And rules for buyer liability under Article 12 (but not under Article 17, where emission limits are enforced through national emission inventories) could also make CDM credits less attractive to buyers than formal allowances. These steps would help distinguish CDM credits from the more environmentally dependable commodity of formal allowances and also encourage developing countries to join the formal cap-and-trade system (with headroom allowances).
81. This suggestion is advanced and elaborated in Stewart, R. B. and J. B. Wiener, 2002: *Reconstructing Climate Policy* (Washington, DC: American Enterprise Institute).
82. Wigley, T., R. Richels, and J. Edmonds, 1996: "Economic and environmental choices in the stabilization of atmospheric CO_2 concentrations," *Nature,* 379: 240–243. Cf. Schneider, S. H. and L. H. Goulder, 1997: "Achieving low-cost emissions targets," *Nature,* 189: 13–14.
83. Hammitt, J. K., 1999: "Evaluation endpoints and climate policy: atmospheric stabilization, benefit–cost analysis, and near-term greenhouse-gas emissions," *Climatic Change,* 41: 447–468.
84. The atmospheric CO_2 concentration in the year 2000 was about 375 ppm.
85. In Hammitt's model, the marginal abatement costs associated with the optimal path are $10/ton of carbon in 2000, $40 in 2050, $110 in 2100, and $190 in 2150. The marginal abatement costs associated with the least-cost stabilization paths are close to zero through 2070, 2050, and 2010 for stabilization at 750, 650, and 550 ppm, respectively, and then rise steeply from zero to more than $300 per ton of carbon within about 50 years after those dates to accomplish stabilization.
86. Hammitt finds that several alternative assumptions—a higher climate sensitivity (4.5°C increase in temperature caused by a doubling in GHG concentrations, rather than 2.5°C), a higher damage function (15 percent of world GDP rather than 2 percent loss caused by a warming of 2.5°C), a damage function related to the rate of climate change rather than the level of climate change, and earlier technological innovations that significantly reduce abatement costs—each call for even greater near-term emission reductions to achieve the optimal path. With high climate sensitivity, Hammitt's optimal path requires an 8 percent reduction in emissions below BAU in 2010 yet remains 40 percent above 1990 levels through 2100 before declining. Marginal abatement costs in the high-sensitivity case are $25 per ton of carbon in 2000, $90 in 2050, and $230 in 2100. With high damages, Hammitt's optimal path requires emissions equal to 1990 levels through 2050 and then declining. Marginal abatement costs in the high-damages case are $70 per ton of carbon in 2000, $220 in 2050, and $500 in 2100.
87. Hahn, 1998; Schmalensee, 1998; Council of Economic Advisers, Economic Report of the President 2002 (Washington, DC: Government Printing Office).

88. See Oppenheimer, M. and R. H. Boyle, 1990: *Dead Heat* (New York: Basic Books). This is also the implication drawn by advocates of the "precautionary principle."
89. Hammitt, J. K., R. J. Lempert, and M. E. Schlesinger, 1992: "A sequential decision strategy for abating climate change," *Nature,* 357: 315–318.
90. Lempert, R. J. and M. E. Schlesinger, 2001: "Adaptive strategies for climate change," in *Innovative Energy Strategies for CO_2 Stabilization* (Cambridge, England: Cambridge University Press).
91. The Kyoto target of 5 percent below the 1990 level for industrialized countries (roughly 15 percent below BAU) by 2012 appears to lie somewhere between Hammitt's high-damages case and his case of both high climate sensitivity and high damages (which requires about a 20 percent reduction below 1990 levels through 2020). In Hammitt's model, marginal abatement costs in the case of both high sensitivity and high damages are $170 per ton of carbon in 2000, $400 in 2050, and $500 in 2100.
92. Schneider, S. H., 1997: "Integrated assessment modeling of global climate change: transparent rational tool for policymaking or opaque screen hiding value-laden assumptions?" *Environmental Modeling and Assessment,* 2: 229–248.

CHAPTER 6

Carbon Abatement with Economic Growth: A National Strategy

Stephen Bernow, Alison Bailie, William Dougherty, Sivan Kartha, and Michael Lazarus

The risk of catastrophic global climate disruption from human activities could be mitigated if atmospheric CO_2 concentrations are stabilized at approximately 450 ppm, about 60 percent above preindustrial concentrations. This entails keeping total global carbon emissions within 500 billion tons over the twenty-first century rather than the 1,400 billion tons toward which the world is headed. To achieve this goal, annual global carbon emissions from fossil fuels must be at least halved by the end of the century and deforestation halted. Global annual per capita carbon emissions must decrease from today's 1 ton to less than 0.3 tons, notwithstanding growing populations and economies. For the United States, which currently emits about one-fourth of the global total at almost 6 tons per capita, this implies a twenty-fold decrease in carbon intensity and more than ten-fold decrease in emissions over the century, if national emissions converged during the century to equal per-capita limits. Whatever burden-sharing approach is adopted, it is clear that the United States must radically reduce its carbon emissions over the next several decades.

This chapter presents the results of a study showing that the United States could dramatically reduce its greenhouse gas (GHG) emissions over the next two decades while the economy continues to grow.[1] It examines a set of policies to increase energy efficiency, accelerate adoption of renewable energy, reduce air pollution, and shift to less carbon-intensive fuels. The policies are targeted within and across sectors—residential and commercial buildings, industrial facilities, transportation, and power generation. They include incentives, standards, codes, market mechanisms, regulatory reforms, research and development, public outreach, technical assistance, and infrastructure investment.

Together with steps to reduce emissions of non-CO_2 GHGs and land-based CO_2 emissions and the acquisition of a limited amount of allowances internationally, this portfolio of policies would allow the United States to meet its obligations under the Kyoto Protocol, reducing its GHG emissions to 7 percent below 1990 levels by 2010, with far greater reductions by 2020. It would bring overall economic benefits to the United States because lower fuel and electricity bills would more than pay the costs of technology innovation and program implementation. In 2010, the annual savings would exceed costs by $50 billion and by 2020 by approximately $135 billion. At the same time, jobs, gross domestic product (GDP), and incomes would increase, and pollutant emissions would decrease.

Energy use in buildings, industries, transportation, and electricity generation was modeled for this study using the U.S. Department of Energy's (DOE's) Energy Information Administration (EIA) National Energy Modeling System (NEMS). The NEMS model version, data, and assumptions used in this study were those of the EIA's *Annual Energy Outlook,*[2] which also formed the basis for the base case. We refined the NEMS model with advice from the EIA, based on their ongoing model improvements and drawing on expertise from colleagues at Union of Concerned Scientists, the National Laboratories, and elsewhere.

Table 6.1 summarizes overall energy and GHG impacts and economic impacts of the policy set for the base case and climate protection case for 2010 and 2020. The policies cause reductions in primary energy consumption that reach 11 percent by 2010 and 30 percent in 2020 (Fig. 6.1), relative to the base case in those years, through increased efficiency and greater adoption of cogenerated heat and power (CHP).

Relative to today's levels, use of nonhydro renewable energy roughly triples by 2010 in the climate protection case because of a renewable portfolio standard (RPS, described later in this chapter), whereas in the base case it increases by less than 50 percent. Given the entire set of policies, nonhydro renewable energy doubles relative to the base case in 2010, accounting for about 10 percent of total primary energy supplies. The absolute amount of renewables does not increase substantially between 2010 and 2020 because the 10 percent RPS electric sector targets in 2010 give the same absolute amount as the 20 percent in 2020 because demand declines sharply as a result of the efficiency policies. A more aggressive renewable policy for the 2010–2020 period could be considered.[3]

The reductions in energy-related carbon emissions are even more dramatic than the reductions in energy consumption because of the shift toward lower-carbon fuels and renewable energy. Carbon emissions have already risen by more than 15 percent since 1990, and in the base case they will rise a total of 35 per-

TABLE 6.1. Summary of Results

	*1990**	*2010 Base Case*	*2010 Climate Protection*	*2020 Base Case*	*2020 Climate Protection*
End-use energy (quads)	63.9	86.0	76.4	97.2	72.6
Primary energy (quads)	84.6	114.1	101.2	127.0	89.4
Renewable energy (quads)					
Nonhydro	3.5	5.0	10.4	5.5	11.0
Hydro	3.0	3.1	3.1	3.1	3.1
Net GHG emissions (Mt Ce/year)	1,648	2,205	1,533	—	—
Energy carbon	1,338	1,808	1,372	2,042	1,087
Land-based carbon	—	—	–58	—	—
Non-CO_2 gases	310	397	279	—	—
International trade	—	—	–60	—	—
Net savings†					
Cumulative present value (billion)	—	—	$105	—	$576
Levelized annual (billion)	—	—	$13	—	$49
Levelized annual per household	—	—	$113	—	$375
Macroeconomic impacts (changes in year)‡					
GDP (billion)	—	—	$23.2	—	$43.9
Jobs (billion)	—	—	0.7	—	1.3
Wages (per household)	—	—	$220	—	$400

*Under Kyoto, the base year for three of the non-CO_2 GHGs (HFCs, PFCs, SF_6) is 1995, and the 1995 levels for these emissions are reported here.

†Savings are in 1999 dollars. The 2010 savings include $2.3 billion per year ($9 billion cumulative through 2010) of non–energy-related costs needed to meet the Kyoto target. Costs are not included in 2020 because these measures policies do not extend past 2010.

‡Impacts were calculated using an Input-Output model, taking account of productivity trends and assuming that there is otherwise less than full employment in these job areas that would be required by the shifts from energy to other demands caused by the policies.

cent by 2010, in stark contrast to the 7 percent emission reduction that the United States negotiated at Kyoto. In the climate protection case, the United States promptly begins to reduce energy-related carbon emissions; by 2010 emissions are only 2.5 percent above 1990 levels, and by 2020, emissions are well below 1990 levels. Relative to the base case, the 2010 reductions[4] amount to 436 megatons of carbon (Mt C) per year.

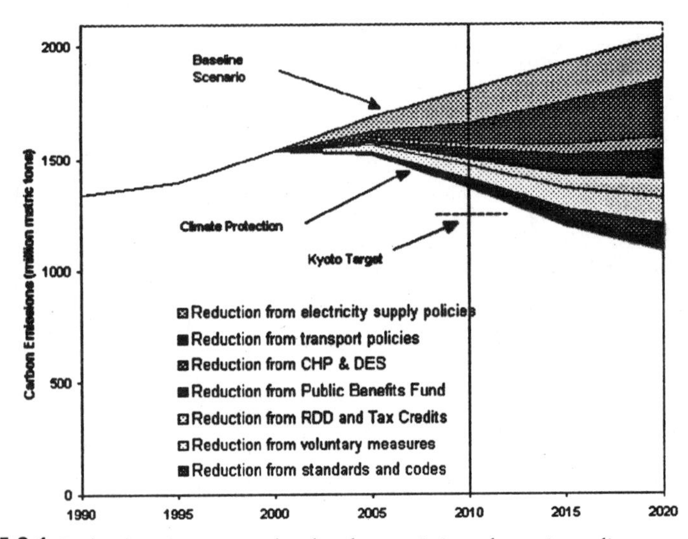

FIGURE 6.1. Reductions in energy-related carbon emissions, by major policy group.

Land-based activities, such as forestry, land use, and agriculture, yield another 58 Mt C/year of reductions. Methane emissions are also reduced through measures aimed at landfills, natural gas production and distribution systems, mines, and livestock husbandry. The potent fluorine-containing GHGs are reduced by substituting non-GHGs, implementing alternative cleaning processes in the semiconductor industry, reducing leaks, and investing in more efficient gas-using equipment. In total, the climate protection case adopts reductions of these other GHGs equivalent to 118 Mt C/year by 2010.

Together the reduction measures for energy-related carbon (436 Mt C/year), land-based carbon (58 Mt C/year), and noncarbon gases (118 Mt carbon equivalents [Ce]/year) amount to 612 Mt Ce/year of reductions in 2010. Through these measures, the United States is able to accomplish most of its emission reduction obligation under the Kyoto Protocol through domestic actions. This leaves the United States slightly shy of its Kyoto target, with only 60 Mt C/year worth of emission allowances to procure from other countries through the "flexibility mechanisms" of the Kyoto Protocol (emissions trading, joint implementation, and the Clean Development Mechanism). The climate protection case assumes that the United States will take steps to ensure that allowances procured through these flexibility mechanisms reflect legitimate mitigation activity. In particular, we assume that the United States does not use so-called hot air allowances (i.e., allowances sold by countries that negotiated excessively high Kyoto targets).

The set of policies in the climate protection case also reduces air pollutants

TABLE 6.2. Impact of Policies on Air Pollutant Emissions (Million Tons)

	*1990**	*2010 Base Case*	*2010 Climate Protection*	*2020 Base Case*	*2020 Climate Protection*
CO	65.1	69.8	63.8	71.8	59.8
NO_x	21.9	16.5	13.9	16.9	12.0
SO_2	19.3	12.8	6.2	12.7	3.3
VOC	7.7	5.5	5.1	5.9	4.9

that cause or aggravate human health problems and adversely affect agriculture, forests, water resources, and buildings. The policies would significantly reduce energy-related emissions, as summarized in Table 6.2. Sulfur oxide emissions would decrease the most: by half in 2010 and by nearly 75 percent in 2020. The other pollutants are reduced between 7 and 16 percent by 2010 and between 17 and 29 percent by 2020, relative to base case levels in those years.

The complete climate protection package provides net economic benefits to the United States while improving public health and the environment. In dramatically reducing energy consumption, the climate protection strategy reduces our dependence on insecure energy supplies and positions the United States as a supplier of innovative and environmentally superior technologies and practices.

Far from being the economically crippling burden that some allege and others fear, ratifying the Kyoto Protocol and ambitiously reducing GHG emissions could initiate a national technological and economic renaissance with cleaner energy, industrial processes, and products in the coming decades. In the United States, we therefore face an important challenge. We can be followers, leaving more forward-looking countries to assume the global leadership in charting a sustainable path. Or we can embrace the opportunity to usher in a technological and environmental transition, providing world markets with the advanced and clean energy technologies needed to sustain the new century's economic growth.

Energy Policies

Analyses of the investment costs and energy savings of policies to promote energy efficiency and cogeneration in the residential, commercial, and industrial sectors and efficiency for light-duty vehicles were taken primarily from the American Council for an Energy-Efficient Economy (ACEEE).[5] Analyses of avoided energy, costs and emissions, pollutant emission caps, renewable energy, and other transportation modes followed the approaches taken in Bernow et al.[6] Later in this chapter we group these policies into the particular sector where they take effect and describe the key assumptions made concerning the technological impacts of the individual policies. Unless otherwise indicated, each of the policies is assumed to start in 2003.

In evaluating the avoided energy, costs, and emissions of these policies we relied primarily on the DOE's NEMS model, data, and assumptions. We adapted the Energy Information Administration's 2001 Reference Case Forecast[7] to create a slightly revised base case. Our policies build on those included in this base case forecast (i.e., we avoid taking credit for emission reductions, costs, or savings already included in the EIA 2001 Reference Case).

Policies in the Building and Industrial Sectors

Carbon emissions from fuel combustion in residential and commercial buildings account for about 10 percent of U.S. GHG emissions, and emissions from the industrial sector account for another 20 percent. When emissions associated with the electricity consumed are counted, these levels reach more than 35 percent for buildings and 30 percent for industry. We analyzed a set of policies that include new building codes, new appliance standards, tax incentives for the purchase of high-efficiency products, a national public benefits fund, expanded research and development, voluntary agreements, and support for combined heat and power.

Building Codes

Building energy codes require all new residential and commercial buildings to be built to a minimum level of energy efficiency that is cost-effective and technically feasible. Good practice residential energy codes, defined as the 1992 (or a more recent) version of the Model Energy Code (now known as the International Energy Conservation Code), have been adopted by 32 states.[8] Good practice commercial energy codes, defined as the American Society of Heating, Refrigeration and Air-Conditioning Engineers (ASHRAE) 90.1 model standard, have been adopted by 29 states.[9] However, the Energy Policy Act of 1992 (EPAct) requires all states to adopt a commercial building code that meets or exceeds ASHRAE 90.1 and requires all states to consider upgrading their residential codes to meet or exceed the 1992 Model Energy Code.

This policy assumes that the DOE enforces the commercial building code requirement in EPAct and that states comply. We also assume that relevant states upgrade their residential energy codes to the 1995 or 1998 Model Energy Code either voluntarily or through the adoption of a new federal requirement. Furthermore, we assume that the model energy codes are significantly improved during the next decade and that all states adopt mandatory codes that go beyond current good practice by 2010. To quantify the impact of these changes, we assume a 20 percent energy savings in heating and cooling in buildings in half of new homes and commercial buildings.

New Appliance and Equipment Efficiency Standards

The track record for electricity efficiency standards is impressive, starting with the National Appliance Energy Conservation Act of 1987 and continuing through the various updates that were enacted in early 2001 for washers, water heaters, and central air conditioners. These standards have removed the most inefficient models from the market while still leaving consumers with a variety of products. An analysis of DOE figures by the American Council for an Energy Efficient Economy estimates that nearly 8 percent of annual electricity consumption will be saved in 2020 because of standards already enacted.[10] However, many appliance efficiency standards haven't kept pace with legal update requirements or technological advances. The DOE is many years behind its legal obligation to regularly upgrade standards for certain appliances to the "maximum level of energy efficiency that is technically feasible and economically justified."

In this study, we assume that the government upgrades existing standards or introduces new standards for several key appliances and equipment types: distribution transformers, commercial air conditioning systems, residential heating systems, commercial refrigerators, exit signs, traffic lights, torchière lighting fixtures, ice makers, and standby power consumption for consumer electronics. We also assume higher energy efficiency standards for residential central air conditioning and heat pumps than were allowed by the Bush administration. These are all measures that can be taken in the near term, based on technologies that are available and cost-effective.

Tax Incentives

A wide range of advanced energy-efficient products have been proven and commercialized but have not yet become firmly established in the marketplace. A major reason for this is that conventional technologies get "locked in"; they benefit from economies of scale, consumer awareness and familiarity, and existing infrastructure that make them more attractive to consumers, while alternatives are overlooked although they could be financially viable once mass produced and widely demonstrated.

In this study, we include initial tax incentives for a number of products. For consumer appliances, we considered a tax incentive of $50 to $100 per unit. For new homes that are at least 30 percent more efficient that the Model Energy Code, we considered an incentive of up to $2,000 per home; for commercial buildings with at least 50 percent reduction in heating and cooling costs relative to applicable building codes, we applied an incentive equal to $2.25 per square foot. Regarding building equipment such as efficient furnaces, fuel cell power systems, gas-fired heat pumps, and electric heat pump water heaters, we considered a 20 percent investment tax credit. Each of these incentives would be introduced with a sunset clause, terminating them or phasing them out in approximately 5 years to avoid their becoming permanent subsidies. Versions of all the tax incentives considered here have already been introduced into bills before the Senate or House.[11]

National Public Benefits Fund

Electric utilities historically have funded programs to encourage more efficient energy-using equipment, assist low-income families with home weatherization, commercialize renewables, and undertake research and development (R&D). Such programs typically have achieved electricity bill savings for households and businesses that are roughly twice the program costs.[12] Despite the proven effectiveness of such technologies and programs, increasing price competition and restructuring have caused utilities to reduce these "public benefit" expenditures over the past several years. To preserve such programs, 15 states have instituted public benefit funds that are financed by a small surcharge on all power delivered to consumers.

This study's policy package includes a national public benefits fund (PBF) fashioned after the proposal introduced by Senator Jeffords (S. 1369) and Representative Pallone (H. 2569) in the 106th Congress. The PBF would levy a surcharge of 0.2 cents per kilowatt-hour on all electricity sold, costing the typical residential consumer about $1 per month. This federal fund would provide matching funds for states for approved public benefit expenditures. In this study, the PBF is allocated to several different programs directed at improvements in lighting, air conditioning, motors, and other cost-effective energy efficiency improvements in electricity-using equipment.

Expand Federal Funding for Research and Development in Energy-Efficient Technologies

Federal R&D funding for energy efficiency has been a spectacularly cost-effective investment. The DOE has estimated that the energy savings from 20 of its energy efficiency R&D programs has been roughly $30 billion so far—more than three times the federal appropriation for the entire energy efficiency and renewables R&D budget throughout the 1990s.[13]

Tremendous opportunities exist for further progress in material-processing technologies, manufacturing processing, electric motors, windows, building shells, lighting, heating and cooling systems, and superinsulation, for example. The Environmental Protection Agency's (EPA's) Energy Star programs have also saved large amounts of energy, building on the achievements of R&D efforts and ushering efficient products into the marketplace. By certifying and labeling efficient lighting, office equipment, homes, and offices, Energy Star has helped foster a market transformation toward much more efficient products and buildings. Currently, roughly 80 percent of personal computers, 95 percent of monitors, 99 percent of printers, and 65 percent of copiers sold are Energy Star certified.[14] In light of these successes, the EPA should be allocated the funds to broaden the scope of its Energy Star program, expanding to other products (refrigerators, motors) and building sectors (hotels, retailers), and the vast market of existing buildings that could be retrofitted. In this study, we assume that increased funding to expand research and development efforts in industry (e.g., motors), buildings (e.g., advanced heating and cooling), and transport (e.g., more fuel efficient cars and trucks) will lead to more energy-savings products becoming commercially available.

Support for Cogeneration

Cogeneration (or CHP) is a superefficient means of coproducing two energy-intensive products that are usually produced separately: heat and power. The technical and economical value of CHP has been widely demonstrated, and some European countries rely heavily on CHP for producing power and providing heat to industries, businesses, and households. The thermal energy produced in cogeneration can also be used for building and process cooling or to provide mechanical power.

CHP already provides about 9 percent of all electricity in the United States, but there are barriers to its wider cost-effective implementation.[15] Environmental standards should be refined to recognize the greater overall efficiency of CHP systems by assessing facility emissions on the basis of fuel input rather than useful energy output, for example. Nonuniform tax standards discourage CHP implementation in certain facilities. Moreover, utility practices generally are highly hostile to prospective CHP operators, imposing discriminatory pricing and burdensome technical requirements and costs for connecting to the grid.

In this study, we include policies that would establish a standard permitting process, uniform tax treatment, accurate environmental standards, and fair access to electricity consumers through the grid. Such measures would help to unleash a significant portion of the enormous potential for CHP. In this study we assumed 50 gigawatts (GW) of new CHP capacity by 2010 and an additional 95 GW between 2011 and 2020. With electricity demand reduced by the various energy efficiency policies adopted in this study, cogenerated electricity reaches 8 percent of total remaining electricity requirements in 2010 and 36 percent in 2020.

Policies in the Electric Sector

A major goal for U.S. energy and climate policy is to dramatically reduce carbon and other pollutant emissions from the electric sector, which is responsible for more than one-third of all U.S. GHG emissions. We analyzed a set of policies in the electric sector that include standards and mechanisms to help overcome existing market barriers to investments in technologies that can reduce emissions. The three policies—a renewable portfolio standard, a cap on pollutant emissions, and a carbon cap and trade system—are described here.

Renewable Portfolio Standard

A renewable portfolio standard (RPS) is a flexible, market-oriented policy for accelerating the introduction of renewable resources and technologies into the electric sector. An RPS sets a schedule for establishing a minimum amount of renewable electricity as a fraction of total generation and requires each generator that sells electricity to meet the minimum by producing that amount of renewable electricity in its mix or acquiring credits from generators that exceed the minimum. The market determines the portfolio of technologies and geographic distribution of facilities that meet the target at least cost. This is achieved by a trading system that awards credits to generators for producing renewable electricity and allows them to sell or purchase these credits. Thirteen states—Arizona, Connecticut, Hawaii, Iowa, Maine, Massachusetts, Minnesota, Nevada, New Jersey, New Mexico, Pennsylvania, Texas, and Wisconsin—already have RPSs, and Senator Jeffords introduced a bill in the 106th Congress (S. 1369) to establish a national RPS.

In this study, we have applied an RPS that starts at a 2 percent requirement in 2002, grows to 10 percent in 2010, and grows to 20 percent in 2020, after all efficiency policies are included. Wind, solar, geothermal, biomass, and landfill gas are eligible renewable sources of electricity, but environmental concerns exclude municipal solid waste (because of concerns about toxic emissions from waste-burning plants) and large-scale hydro (which, in any event, need not be treated as an emerging energy technology because it already supplies nearly 10 percent of the nation's electricity).

We also tighten the existing SO_2 cap to reduce sulfur emissions to roughly 40 percent of current levels by 2010 and one-third of current levels by 2020. We also impose a cap-and-trade system on NO_x emissions in the summer, when NO_x contributes more severely to photochemical smog. This system expands the current cap-and-trade program, which calls on 19 states to meet a target in 2003 that then remains constant, to include all states with a cap that is set first in 2003 but decreases in 2010, relative to 1999 levels. The cap results in a 25 percent reduction of annual NO_x emissions by 2003 and a 50 percent reduction by 2010.

Carbon Cap-and-Trade Permit System

This study introduces a cap-and-trade system for carbon in the electric sector, with the cap set to achieve progressively more stringent targets over time, starting in 2003 at 2 percent below current levels, increasing to 12 percent below current by 2010 and 30 percent below by 2020. Restricting carbon emissions from electricity generation has important benefits, including reduced emissions of SO_2 and NO_x, fine particulate matter (which is a known cause of respiratory ailments), and mercury (which is a powerful nervous system toxin and already contaminates more than 50,000 U.S. lakes and streams). A progressively more stringent target also reduces demand for coal and hence mining-related pollution of streams and degradation of landscapes and terrestrial habitats.

In the SO_2, NO_x, and CO_2 trading systems, permits are distributed through an open auction, and the resulting revenues can be returned to households (e.g., through a tax reduction or as a rebate). Recent analyses suggest that an auction is the most economically efficient way to distribute permits, meeting emission caps at lower cost than allocations based on grandfather allowances or equal per–kilowatt-hour allowances.[16] Implementing such auctions for the electric sector will also clear the way for an economy-wide approach in future years based on auctioning. In this study, the price of auctioned carbon permits reaches $100 per metric ton carbon.

Though not specifically targeted by the trading programs, the operators of the 850 grandfathered coal plants built before the Clean Air Act of 1970, which emit three to five times as much pollution per unit of power generated as newer coal power plants, probably will retire these plants rather than buying the credits necessary to keep them running. When the Clean Air Act was adopted, it was expected that these dirty power plants would eventually be retired. However, utilities are continuing to operate these plants beyond their design life and have increased their output over the last decade. If these old plants are subjected to the same requirements as newer facilities, as has been done or is being considered in several states including Massachusetts and Texas, operators would be obliged to modernize or retire them in favor of cleaner electric generation alternatives.

Policies in the Transport Sector

Another goal for U.S. energy and climate policy is to reduce carbon emissions from the transport sector, which is responsible for about one-third of all U.S. GHG emissions. We analyzed a set of policies in the transportation sector that include improved efficiency (light-duty vehicles, heavy-duty trucks and aircraft), a full fuel-cycle GHG standard for motor fuels, measures to reduce road travel, and high-speed rail (HSR).

Strengthened CAFE Standards

Today's cars are governed by fuel economy standards that were set in the mid-1970s. The efficiency gains made in meeting those standards have been entirely wiped out by increases in population and driving and the trend toward gas-guzzling SUVs. When the fuel economy standards were implemented, light-duty trucks accounted for only about 20 percent of vehicle sales. Light trucks now account for nearly 50 percent of new vehicle sales; this has brought down the overall fuel economy of the light-duty vehicle fleet, which now stands at its lowest average fuel economy since 1981. If the fuel economy of new vehicles had held at 1981 levels rather than tipping downward, American vehicle owners would be importing half a million fewer barrels of oil each day.

We introduce in this study a strengthened Corporate Average Fuel Economy (CAFE) standard for cars and light trucks, along with complementary market incentive programs. Specifically, fuel economy standards for new cars and light trucks rise from EIA's projected 25.2 mpg for 2001 to 36.5 mpg in 2010, increasing to 50.5 mpg by 2020. This increase in vehicle fuel economy would save by 2020 approximately twice as much oil as could be pumped from Arctic National Wildlife Refuge oil field over its entire 50-year lifespan.[17] Based on assessments of near-term technologies for conventional vehicles and advanced vehicle technologies for the longer term, we estimate that the 2010 CAFE target can be met with an incremental vehicle cost of approximately $855 and the 2020 CAFE target with an incremental cost of $1,900. To put these incremental costs in perspective, they are 1/3 to 1/2 the cost of gasoline saved at the pump over the vehicle's lifetime.[18]

Improving Efficiency of Freight Transport

We also consider policies to improve fuel economy for heavy-duty truck freight transport, which accounts for approximately 16 percent of all transport energy consumption. A variety of improvements such as advanced diesel engines, drag reduction, rolling resistance, load reduction strategies, and low friction drive-trains offer opportunities to increase the fuel economy of freight trucks.

To accelerate the improvement in heavy-duty truck efficiency, we have considered measures that expand R&D for heavy-duty diesel technology, vehicle labeling and promotion, financial incentives to stimulate the introduction of new technologies, efficiency standards for medium- and heavy-duty trucks, and fuel taxes and user fees calibrated to eliminate the existing subsidies for freight trucking. Together, it is estimated that these policies could bring about a fuel economy improvement of 6 percent by 2010 and 23 percent by 2020.

Improving Efficiency of Air Travel

Air travel is the fastest-growing mode of travel, far more energy intensive than vehicle travel. One passenger mile of air travel today uses about 1.7 times as much fuel as vehicle travel.[19] We consider policies for improving the efficiency of air travel, including R&D in efficient aircraft technologies, fuel consumption standards, and a revamping of policies that subsidize air travel through public investments.

We assume that air travel efficiency improves by 23 percent by 2010 and 53 percent by 2020. This is in contrast to the base case, where efficiency increases by 9 percent by 2010 and 15 percent by 2020 through a combination of aircraft efficiency improvements (advanced engine types, lightweight composite materials, and advanced aerodynamics), increased load factors, and acceleration of air traffic management improvements.[20] We assume that air travel can increase to 82 seat-miles per gallon by 2020 from its current 51.

Greenhouse Gas Standards for Motor Fuels

Transportation in the United States relies overwhelmingly on petroleum-based fuels, making it a major source of GHG emissions. We introduce here a full fuel-cycle GHG standard for motor fuels, similar in concept to the RPS for the electric sector. The standard is a cap on the average GHG emissions from gasoline and would be made progressively more stringent over time. Fuel suppliers would have the flexibility to meet the standard on their own or by buying tradable credits from other producers of renewable or low-GHG fuel.

The policy adopted in this study requires a 3 percent reduction in the average national GHG emission factor of fuels used in light-duty vehicles in 2010, increasing to a 7 percent reduction by 2020. The policy would be complemented by expanded R&D, market creation programs, and financial incentives. Such a program would stimulate the production of low-GHG fuels such as cellulosic ethanol and biomass- or solar-based hydrogen.

For this modeling study, we assume that most of the low-GHG fuel is provided as cellulosic ethanol, which can be produced from agricultural residues, forest and mill wastes, urban wood wastes, and short-rotation woody crops.[21] Because cellulosic ethanol can be coproduced with electricity, in this study we assume that electricity output reaches 10 percent of ethanol output by 2010 and 40 percent by 2020.[22] Because of the accelerated development of the production technology for cellulosic ethanol, we estimate that the price falls to $1.4 per gallon of gasoline equivalent by 2010 and remains at that price thereafter.[23]

Improving Alternative Modes to Reduce Vehicle Miles Traveled

The amount of travel in cars and light-duty trucks continues to grow because of increasing population and low vehicle occupancy. Between 1999 and 2020, the rate of growth in vehicle miles traveled is projected to increase in the base case by about 2 percent per year. The overall efficiency of the passenger transportation system can be significantly improved through measures that contain the growth in vehicle miles traveled through land use and infrastructure investments and pricing reforms to remove implicit subsidies for cars, which are very energy intensive.

We assume that these measures will affect primarily urban passenger transportation and result in a shift to higher-occupancy vehicles, including carpooling, vanpooling, public transportation, and telecommuting. We consider that these measures can achieve reductions in vehicle miles traveled of 8 percent by 2010 and 11 percent by 2020 relative to the Base Case in those years.

High-Speed Rail

HSR is an attractive alternative to intercity vehicle travel and short-distance air travel. In both energy cost and travel time, high-speed rail may be competitive with air travel for trips of roughly 600 miles or less, which account for about one-third of domestic air passenger miles traveled. Investments in rail facilities for key intercity routes (such as the northeast corridor between Washington and Boston, the east coast of Florida between Miami and Tampa, and the route linking Los Angeles and San Francisco) could provide an acceptable alternative and reduce air travel in some of the busiest flight corridors.[24]

In this analysis we have taken the U.S. Department of Transportation's (DOT's) recent estimates of the potential HSR ridership, which, based on projected mode shifts from air and automobile travel in several major corridors of the United States, reaches about 2 billion passenger miles by 2020.[25] Although this level of HSR ridership provides only small energy and carbon benefits by 2020, it can be viewed as the first phase of a longer-term transition to far greater ridership and more advanced, faster, and more efficient electric and magnetic levitation (MAGLEV) systems in the ensuing decades.

Summary Results

Table 6.3 summarizes the carbon reductions and the net costs (generally net benefits) of each energy policy through 2010 and 2020. Carbon reductions reach 436 Mt C in 2010 (about 24 percent below the base case in that year) and 955 Mt C in 2020 (about 47 percent below the base case in that year). The costs were computed by discounting and summing the incremental annualized capital costs, administrative costs, incremental operations and maintenance (O&M)

TABLE 6.3. Carbon Reductions, Net Costs, and Costs of Saved Carbon in 2010 and 2020

	2010			*2020*		
	Carbon Savings (Mt C/yr)	*Cumulative Net Cost**	*Cost of Saved Carbon/ Ton (1999 $)*	*Carbon Savings (Mt C/yr)*	*Cumulative Net Cost*	*Cost of Saved Carbon/Ton (1999 $)*
Buildings and industry sectors						
Appliance standards	29	–$24	–$315	86	–$84	–$256
Building codes	7	–$5	–$353	30	–$23	–$244
Voluntary measures	61	–$50	–$229	118	–$112	–$179
Research and design	21	–$18	–$257	71	–$53	–$186
Public benefits fund	50	–$29	–$224	134	–$101	–$187
Tax credits	4	–$4	–$292	11	–$8	–$152
CHP and DES[1]	21	–$53	–$611	59	–$151	–$554
Subtotal	193	–$183	–$301	509	–$532	–$242
Electric sector						
RPS; NO_x/SO_x cap and trade; carbon cap and trade						
Subtotal	147	$140	$258	190	$258	$188

(*continues*)

TABLE 6.3. *Continued*

	2010			2020		
	Carbon Savings (Mt C/yr)	*Cumulative Net Cost**	*Cost of Saved Carbon/ Ton (1999 $)*	*Carbon Savings (Mt C/yr)*	*Cumulative Net Cost*	*Cost of Saved Carbon/Ton (1999 $)*
Transport sector						
Vehicle travel reductions	29	–$50	–$496	37	–$126	–$495
Light-duty vehicle efficiency improvements	38	–$19	–$270	136	–$149	–$296
Heavy-duty vehicle efficiency improvements	8	–$3	–$179	33	–$22	–$214
Aircraft efficiency improvements	10	–$3	–$106	28	–$14	–$129
Greenhouse gas standards	11	$7	$227	22	$25	$237
Subtotal	96	–$68	–$272	256	–$286	–$265
TOTAL	436	–$111	–$80	955	–$561	–$121[2]

[1]District energy systems using cogeneration.
[2]"Subtotals" and "Totals" are sums for "Carbon Savings" and are weighted averages for "Net Cost" and "Cost of Saved Carbon per Ton."
*Cumulative Net Cost = present value in $ billion (1999 dollars)

and fuel costs, and subtracting the discounted O&M and fuel cost savings, using a 5 percent real discount rate. Overall the net savings achieved by the demand policies more than offset the net costs for the electric supply policies. The climate protection policy package as a whole results in cumulative net savings of $114/t C through 2010, and $574/t C through 2010.[26]

It is important to note that the large net savings achieved by the energy efficiency policies create the "economic space" into which policies for fuel shifting to low emissions and renewable energy resources and technologies can step while retaining overall net economic benefits. Rather than limiting policies to those with net benefits at the margin, this approach takes the longer view by bringing cutting-edge options into early use, thereby inducing technology learning and setting the stage for the deeper carbon reductions for which they will be needed in the future while getting deeper carbon and emission reductions in the near term.

Achieving Kyoto

Energy-related CO_2 emissions are the predominant source of U.S. GHG emissions for the foreseeable future, and their reduction is the central and ultimate challenge for protecting the climate. Yet with its delayed and weak emission mitigation policies, the United States may not be able to rely solely on energy sector policies and technologies to meet its Kyoto obligation of emissions 7 percent below 1990 levels with no net economic cost. As our analysis has shown, such efforts, if aggressively pursued, would slow our growth in energy sector CO_2 emissions from a projected 35 percent to 2.5 percent above 1990 levels by 2010 and still achieve a small net economic benefit. This would be a major accomplishment but would still leave us 128 Mt C/year short of achieving a target of 1,244 Mt C/year by 2010 if the Kyoto target were confined only to the domestic energy sector. A tighter carbon cap for the electric sector could increase domestic energy-related emission reductions to meet the Kyoto requirement, but this would incur incremental costs that could eliminate the net benefit and lead to a modest overall net cost.

Of course, there is more to the Kyoto agreement. The Kyoto targets cover six gases—methane (CH_4), nitrous oxide (N_2O), perfluorocarbons (PFCs), hydrofluorocarbons (HFCs), sulfur hexafluoride (SF_6), and carbon dioxide. The use of these gases is growing because of the ongoing substitution of ozone-depleting substances (ODS) with HFCs and, to a lesser extent, growth in CH_4 emissions from livestock and coal and natural gas systems, in N_2O from fertilizer use, and growth in PFC emissions from semiconductor manufacture.[27]

The U.S. commitment requires emissions of all six gases, in aggregate, to be reduced to 7 percent below their baseline levels.[28] When all of the six "Kyoto gases" are considered, base year emissions amount to 1,680 Mt Ce/year, making the –7 percent Kyoto reduction target equal to 1,533 Mt Ce/year, as shown in the third column of Fig. 6.2. The projected 2010 emissions for all six gases is 2,205 Mt Ce/year (first column), so the total required reduction is expected to be 672 Mt Ce/year. The energy-CO_2 policies described in the previous sections yield 436 Mt Ce/year in reductions by 2010 (second column), leaving the United States with 236 Mt Ce/year additional reductions to achieve from other policies and measures.

The Kyoto agreement gives us several options for obtaining the additional 236 Mt Ce/year of reductions. Two of these options involve domestic reductions: the control of non-CO_2 gases (multigas control) and the use of sinks or biotic sequestration through the land use, land use change, and forestry options

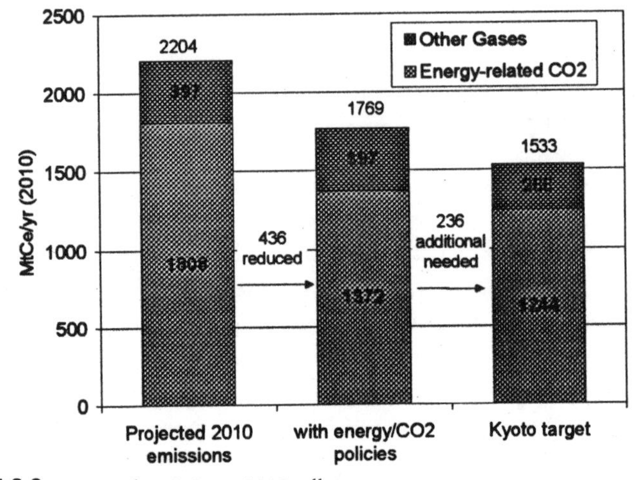

FIGURE 6.2. Projected emissions, 2010, all gases.

allowed under the protocol. The other options involve obtaining credits and allowances from international sources. Under the Kyoto Protocol, countries can purchase credits and allowances through the Clean Development Mechanism (CDM), joint implementation, or emission trading (ET) to offset domestic emissions exceeding our 7 percent reduction target. This section examines how we might meet the Kyoto target through the use of these options and what the costs and other implications might be.

Domestic Options

Articles 3.3 and 3.4 and Sinks

GHG emissions and removals from land use and land use change and forestry (LULUCF) are a subject of great controversy and scientific uncertainty. The Kyoto Protocol treats LULUCF activities in two principal categories: afforestation, reforestation, and deforestation under Article 3.3; and "additional human-induced activities" such as forest and cropland management under Article 3.4. Different interpretations of these two articles can have widely varying impacts on the U.S. reduction commitment.[29] For instance, the U.S. estimate of business-as-usual forest uptake during the first commitment period is 288 Mt Ce/year. If fully credited as an Article 3.4 activity, this uptake could provide credit equal to more than 40 percent of the U.S. reduction requirement with no actual mitigation effort. However, most countries do not interpret the protocol as allowing credit for business-as-usual offsets and therefore believe they should be excluded.

Because our analysis was conducted before the July 2001 COP6bis meetings in Bonn, we based our LULUCF analysis on the "consolidated negotiating text" issued by Jan Pronk, president of COP6, in the weeks before the meeting.[30] The Pronk text reflected an attempted compromise between various parties on a number of contentious issues and was the basis for the final COP6bis outcome on LULUCF issues.[31] The Pronk text capped total U.S. crediting from Article 3.4 activities and afforestation and reforestation projects in the CDM and JI at roughly 58 Mt Ce/year.[32] Domestic forest management activities would be subject to an 85 percent discount. Thus, if one assumes the U.S. estimate mentioned earlier, the Pronk rules would result in 42 Mt Ce/year of essentially zero-cost credit for forest management activities that are expected to occur anyway.[33] In addition, agricultural management (e.g., no-till agriculture, grazing land management, and revegetation) would be allowed under a net–net accounting approach that would allow the United States to count another expected 10 Mt Ce/year of business-as-usual (i.e., zero-cost) credit toward the cap. In sum, the Pronk proposal translates to 52 Mt Ce/year of free carbon removals and another 6 Mt Ce/year that could be accrued through new domestic forest or agricultural management activities.[34]

Based on a recent summary of LULUCF cost estimates, we assumed that the 6 Mt Ce/year of "new" offsets allowable under the Pronk text would be purchased for $10/t Ce.[35] A total of 58 Mt Ce/year of LULUCF credit therefore would be available to help meet the reduction requirement of 236 Mt Ce/year remaining after adoption of the energy-related CO_2 policies described earlier.

The net result of our analysis is slightly different from the implications of the COP6bis agreement. The agreement would allow approximately 28 Mt Ce/year of existing forest management, up to 16 Mt Ce/year of reforestation and afforestation through the CDM, and an unlimited amount of new Article 3.4 forest and agricultural management activities.[36] The difference is that the United States would receive fewer free credits from business-as-usual activity and would need to pay a bit for domestic and CDM projects to reach the 58 Mt Ce/year of assumed LULUCF activity modeled here. However, the United States would no longer be capped with respect to further Article 3.4 offsets, potentially offering an expanded pool of lower-cost reduction opportunities than modeled here.

Multigas Control

Multigas control is a fundamental aspect of the protocol, and its potential for lowering the overall cost of achieving Kyoto targets has been the subject of several prominent studies (Reilly et al., 1999 and 2000). Table 6.4 shows baseline and projected emission levels for the non-CO_2 gases.[37]

TABLE 6.4. Baseline and Projected Emissions for the Non-CO_2 Kyoto Gases (Mt Ce/yr)

Gas	*Base Year (1990/1995)*[1]	*7% Below Base Year*	*Projected 2010*	*Reductions Required**	*Sources*
Methane	170	158	186	28	U.S. EPA, 1999
Nitrous oxide	111	103	121	18	Reilly et al., 1999b; U.S. EPA, 2001a
High-GWP gases (HFC, PFC, SF_6)	29	27	90	63	U.S. EPA, 2000
TOTAL	310	288	397	109	

*These are the reductions that would be needed if each gas were independently required to be 7 percent below its base year level.

[1]Within the Kyoto Protocol, the base year for CH_4 and N_2O is 1990, whereas for the High-GWP gases it is 1995.

Methane emissions are expected to grow by only 10 percent from 1990 to 2010, largely because of increased natural gas leakage and venting (resulting from increased consumption), enteric fermentation, and anaerobic decomposition of manure (caused by increased livestock and dairy production). Methane from landfills, which accounted for 37 percent of total methane emissions in 1990, is expected to decline slightly as a consequence of the Landfill Rule of the Clean Air Act,[38] which requires all large landfills to collect and burn landfill gases.

Several measures could reduce methane emissions well below projected levels. The EPA estimates that capturing the methane from landfills not covered by the Landfill Rule and using it to generate electricity is economically attractive at enough sites to reduce projected landfill emissions by 21 percent.[39] At a cost of $30/t Ce, the number of economically attractive sites increases sufficiently that 41 percent of landfill emissions can be reduced. Similarly, the EPA has constructed methane reduction cost curves for reducing leaks and venting in natural gas systems, recovering methane from underground mines, using anaerobic digesters to capture methane from manure, and reducing enteric fermentation by changing how livestock are fed and managed.

We have used a similar EPA study to estimate the emission reductions available for the high–global warming potential (GWP) gases.[40] Table 6.1 shows that the high-GWP gases, though only a small fraction of baseline emissions (first column), are expected to rise so rapidly that they will account for majority of net growth in non-CO_2 emissions relative to the 7 percent reduction target (last column). In many applications, other gases can be substituted for HFCs and PFCs, new industrial process can be implemented, leaks can be reduced, and more efficient gas-using equipment can be installed. For instance, minor repairs of air-conditioning and refrigeration equipment could save an estimated 6.5 Mt Ce/year in HFC emissions by 2010 at cost of about $2/t Ce. New cleaning processes for semiconductor manufacture could reduce PFC emissions by 8.6 Mt Ce/year by 2010 at an estimated cost of about $17/t Ce. In all, the EPA identified 37 measures for reducing high-GWP gases, a list that is likely to be far from exhaustive given the limited experience with and data on abatement methods for these gases.

The major source of nitrous oxide in the United States is the application of nitrogen fertilizers, which results in about 70 percent of current emissions. Given the tendency of farmers to apply excess fertilizer to ensure good yields, effective strategies for N_2O abatement from cropping practices have been elusive. Therefore, aside from measures to reduce N_2O from adipic and nitric acid production (amounting to less than 1 Mt Ce/year) and from mobile sources as a result of transportation policies, we have not included a full analysis of N_2O reduction opportunities.[41]

Relying largely on recent EPA abatement studies,[42] we developed the cost curve for reducing non-CO_2 gases depicted in Fig. 6.3.[43] In addition to what is covered in the EPA studies, we assumed that:

Only 75 percent of the 2010 technical potential found in the EPA studies would actually be achieved and that policies and programs needed to promote these measures would add a transaction cost of $5/t Ce.

The savings in 2010 fossil fuel use resulting from the policies and measures implemented in the energy sector will yield corresponding benefits for several categories of non-CO_2 emissions. In particular, we assumed that reduced oil use in the transport sector (down 14 percent) will lead to a proportional decrease in N_2O emissions from mobile sources;[44] reduced natural gas demand (down 13 percent) will result in proportionately fewer methane emissions from leaks and venting; and reduced coal production (down 49 percent) will lead to decreased underground mining and its associated emissions.[45]

Figure 6.3 shows that domestic options, taken together, are insufficient to reaching the Kyoto target. The line on the left is the "supply curve" of non-CO_2 abatement options, and the line on the right is the reduction requirement after both energy-related and Article 3.3 and 3.4 sinks are accounted for. Under cur-

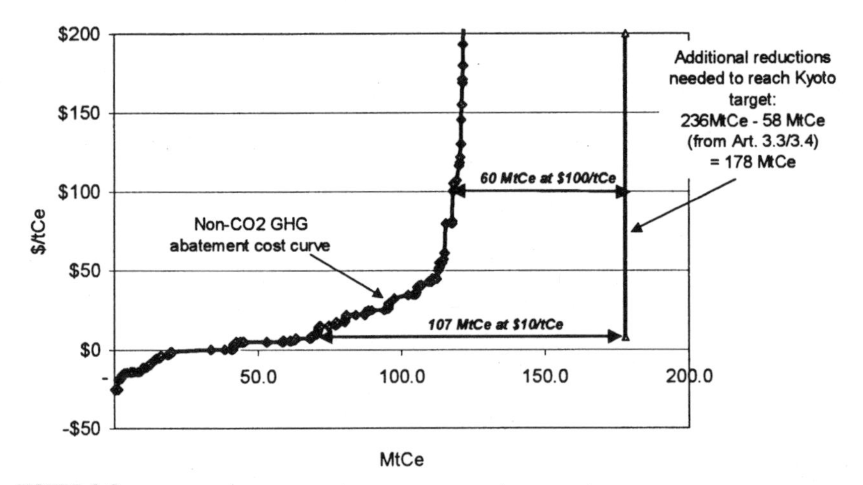

FIGURE 6.3. Non-CO_2 emission reductions, costs and potential, 2010.

rent conditions (only 9 years left until 2010), the supply of remaining domestic options appears insufficient to satisfy demand. This gap ranges from 107 Mt Ce/year at $10/t Ce to 60 Mt Ce/year at $100/t Ce. Therefore, to meet our Kyoto obligations, we are looking to the international market to fill this gap.

International Options

The Kyoto Protocol creates two principal types of GHG offsets in the international market: the purchase of surplus allowances from countries that are below their Kyoto targets and the creation of carbon credits through project-based mechanisms, CDM, and JI.

Emissions Allowance Trading and Hot Air

The combination of emission targets based on circa 1990 emissions and the subsequent restructuring and decline of many economies in transition (EITs) means that these countries could have a large pool of excess emission allowances, typically called "hot air" (see Appendix B). We assume that hot air will constitute no more than 50 percent of all international trading, and we assume a maximum availability of 200 Mt Ce/year, based on a recent analysis.[46]

CDM and JI

CDM and JI projects can be an important part of a comprehensive climate policy, providing they truly contribute to sustainable development in the host countries and create genuine additional GHG benefits. It is reasonable to expect that the U.S. government and other stakeholders will want to develop the CDM and JI market to involve developing countries, engage in technology transfer, develop competitive advantages, and prepare for future commitment periods.

Similarly, the possibility of limited crediting lifetimes, or discounting of carbon reductions in future project years, as proposed by some, could increase the effective cost per t Ce. In a recent analysis, Bernow et al. (2000) illustrate how different approaches to standardizing baselines could lead to differences in additional power sector activity (t Ce) of a factor of 4. These considerations are rarely included in CDM and JI analyses, either bottom-up or top-down.

Given the small differences between the two different approaches, we adopt the top-down model results[47] because they provide a fuller CDM curve, include multiple gases, and provide a cost curve for JI investments as well.

Combining the Options

There are two ways to combine the available options to meet our Kyoto target. We can prioritize which options to rely on more heavily, based on their strategic advantages and benefits, as we have done for energy and CO_2 policies. Or we can simply seek lowest-cost solution for the near term. A long-term climate policy perspective argues for the former approach. For example, rules and criteria for JI, and especially CDM, should be designed so that additionality, sustainability, and technology transfer are maximized. Ideally, our cost curves for CDM and JI would reflect only investments that are consistent with those criteria. However, our current ability to reflect such criteria in quantitative estimates of CDM and JI potential is limited.[48]

It is possible to model priority investment in the domestic reductions of non-CO_2 gases by implementing some measures that are higher cost than the global market clearing carbon price. Just as energy and CO_2 measures such as RPSs can be justified by the technological progress, long-term cost reductions, and other benefits that they induce, so can some non-CO_2 measures. Although we have not attempted to evaluate specific policies for non-CO_2 gases as we have for CO_2, we have picked a point on the non-CO_2 cost curve, \$100/t Ce, to reflect an emphasis on domestic action. At \$100/t Ce, domestic non-CO_2 measures can deliver 118 Mt Ce/year of reductions, still about 60 Mt Ce/year short of the Kyoto goal, to which we must turn to the international market.

To model the global emission trading market, we used the CDM and JI cost curves and hot air assumptions described earlier, together with assumptions about the demand for credits and allowances from all Annex B parties.[49] This model yields market-clearing prices and quantities for each of the three princi-

TABLE 6.5. Reductions Available in 2010 from Various Sources (in Mt Ce)

	Domestic Options		*International Trade*			
	Non-CO_2 Gases	*Sinks*	*CDM*	*JI*	*Hot Air (ET)*	*Total*
Amount available at or below $0/t Ce	41	52				93
Amount available at $0–$100	77	6				83
Amount available at $8			30	6	25	61
Annual costs ($ million)	$1,783	$60	$235	$48	$196	$2,322

pal flexible mechanisms: CDM, JI, and ET and hot air.[50] The results are shown in Table 6.5.

The first row of the table shows that 93 Mt Ce/year is available at net savings or no net cost, more than half from the nonadditional forest management and other Article 3.4 sinks implicit in the Pronk text. Another 77 Mt Ce/year of non-CO_2 gas savings is available as we climb the cost curve from $0 to $100/t C (second row). The net result is that nearly $1.8 billion per year is invested in technologies and practices to reduce non-CO_2 GHG emissions by 118 Mt Ce/year in 2010. Another $60 million per year is directed toward the 6 Mt Ce/year of expected additional sinks allowed under the Pronk proposal. The third row shows that of the 60 Mt Ce/year of international trading, half comes from CDM projects, and much of the rest comes from hot air. The model we use estimates a market-clearing price of about $8/t Ce for this 60 Mt C/year of purchased credits and allowance, amounting to a total annual cost of less than $500 million.[51]

In summary, of the 672 Mt Ce/year in total reductions needed to reach Kyoto by 2010, nearly 65 percent comes from energy sector CO_2 reduction policies, 18 percent from domestic non-CO_2 gas abatement, 9 percent from domestic sinks, and 9 percent from the international market. The net economic benefits deriving from the energy-related carbon reductions reach nearly $50 billion/year in 2010. The total annual cost for the 35 percent of 2010 reductions coming from those last three options—non-CO_2 control, sinks, and international trading—is estimated at approximately $2.3 billion, making the total package a positive economic portfolio by a large margin. Had we taken the other approach noted at the beginning of the section—aiming for the lowest near-term compliance cost—we would rely more heavily on international trading.

We modeled this scenario and found that it would nearly double the amount of international trading, lower the overall annual cost to $0.9 billion, and reduce the amount of non-CO_2 control by more than 40 percent. This additional benefit is minor in comparison to the economic and environmental benefits of the entire policy portfolio.

Conclusions

This study shows that the United States can achieve its carbon reduction target under the Kyoto Protocol: 7 percent below 1990 levels for the first budget period of the protocol. Relying on national policies and measures for GHG reductions and accessing the flexibility mechanisms of the Kyoto Protocol for a small portion of its total reductions, the United States would enjoy net economic savings as a result of this climate protection package. Such action would lead to carbon emission reductions of about 24 percent by 2010 relative to the base case, bringing emissions to about 2.5 percent above 1990 levels. Furthermore, emissions of other pollutants would also be reduced, thus improving local air quality and public health.

Adopting these policies at the national level through legislation not only will help America meet its Kyoto targets but also will lead to economic savings for consumers because households and businesses would enjoy annual energy bill reductions greater than their investments. These net annual savings would increase over time, reaching nearly $113 per household in 2010 and $375 in 2020. The cumulative net savings would be about $114 billion (present value 1999$) through 2010 and $576 through 2020.

Although implementing this set of policies and additional nonenergy related measures is an ambitious undertaking, it is an important transitional strategy to meet the long-term requirements of climate protection. It builds the technological and institutional foundation for much deeper long-term emission reductions needed for climate protection. Such actions would stimulate innovation and invention in the United States while positioning the United States as a responsible international leader in meeting the global challenge of climate change.

Notes

1. This study—*The American Way to the Kyoto Protocol* (2001)—was undertaken by Tellus Institute for World Wildlife Fund for Nature (WWF). Important input to that study was provided on energy efficiency (by colleagues at American Council for an Energy-Efficient Economy) and renewable energy (by colleagues at Union of Concerned Scientists and other experts). Bailie, A., S. Bernow, W. Dougherty, M.

Lazarus, S. Kartha (Tellus), and M. Goldberg (MRG and Associates), 2001. Employment income and GDP impacts were developed in *Clean Energy: Jobs for America's Future* (Washington, DC: World Wildlife Fund).

2. U.S. EIA, 2001a: *Annual Energy Outlook 2001 with Projections to 2020* (Washington, DC: U.S. Department of Energy); U.S. EIA, 2001b: *U.S. Carbon Dioxide Emissions from Energy Sources, 2000 Flash Estimate* (Washington, DC: U.S. Department of Energy), http://www.eia.doe.gov/oiaf/1605/flash/sld001.htm.
3. ACEEE, 1999: *Meeting America's Kyoto Protocol Targets,* H. Geller, S. Bernow, and W. Dougherty (eds.) (Washington, DC: American Council for an Energy-Efficient Economy).
4. Throughout this report the U.S. emission target for the year 2010 is the average of the 5-year period from 2008 to 2012.
5. ACEEE, 1999.
6. Bernow, S., K. Cory, W. Dougherty, M. Duckworth, S. Kartha, and M. Ruth, 1999: *America's Global Warming Solutions* (Washington, DC: World Wildlife Fund).
7. U.S. EIA, 2001a, 2001b.
8. BCAP, 1999: *Status of State Energy Codes* (Washington, DC: Building Codes Assistance Project), Sept./Oct.
9. Ibid.
10. Nadel, S. and H. Geller, 2001: *Smart Energy Policies: Saving Money and Reducing Pollutant Emissions through Greater Energy Efficiency,* Report No. E012 (Washington, DC: American Council for an Energy-Efficient Economy, with Tellus Institute).
11. The bills include those introduced by Senators Murkowski and Lott (S.389), Bingaman and Daschle (S.596), Smith (S.207), and Hatch (S.760), and Representative Nussle (H.R. 1316).
12. Nadel, S. and M. Kushler, 2000: "Public benefit funds: a key strategy for advancing energy efficiency," *The Electricity Journal,* October: 74–84.
13. EERE, 2000: *Scenarios for a Clean Energy Future,* Prepared by the Interlaboratory Working Group on Energy-Efficient and Clean-Energy Technologies (Washington, DC: U.S. Department of Energy, Office of Energy Efficiency and Renewable Energy).
14. U.S. EPA, 2001a. *The Power of Partnerships, Climate Protection Partnerships Division, Achievements for 2000: In Brief* (Washington, DC: U.S. Environmental Protection Agency); Brown, R., C. Webber, and J. Koomey, 2000: "Status and future directions of the Energy Star program," in *Proceedings of the 2000 ACEEE Summer Study on Energy Efficiency in Buildings,* 6: 33–43 (Washington, DC: American Council for an Energy-Efficient Economy).
15. Elliott, R. N. and M. Spurr, 1999: *Combining Heat and Power: Capturing Wasted Energy* (Washington, DC: American Council for an Energy-Efficient Economy).
16. Burtraw, D., K. Palmer, R. Barvikar, and A. Paul, 2001: *The Effect of Allowance Allocation on the Costs of Carbon Emissions Trading,* Discussion Paper 01-30 (Washington, DC: Resources for the Future).
17. Assuming a mean value at a market price of oil of $20/barrel. U.S. Geological Survey, 2001: *Arctic National Wildlife Refuge, 1002 Area, Petroleum Assessment, 1998, Including Economic Analysis,* Fact Sheet FS-028-01, April. (See also U.S. Geological

Survey, 1999: *The Oil and Gas Resource Potential of the Arctic National Wildlife Refuge 1002 Area, Alaska,* USGS Open File Report 98-34.)

18. Assuming a retail price of gasoline of $1.50/gallon, a 10-year life of the vehicle, and 12,000 miles per year.
19. Assuming typical load factors of 0.33 for autos and 0.6 for air.
20. Lee, J. J., S. P. Lukachko, L. A. Waitz, and A. Schaefer, 2001: "Historical and future trends in aircraft performance, cost and emissions," *Annual Review of Energy and the Environment,* 26 (November); Office of Technology Assessment, 1994: *Saving Energy in U.S. Transportation,* OTA-ETI-589, Washington, DC; Interlaboratory Working Group, 2000: *Scenarios for a Clean Energy Future,* Argonne National Laboratory, the National Renewable Energy Laboratory, Lawrence Berkeley National Laboratory, Oak Ridge National Laboratory, and Pacific Northwest National Laboratory. Commissioned by DOE Office of Energy Efficiency and Renewable Energy, http://www.ornl.gov/ORNL/Energy_EFF/CEF.htm.
21. Walsh, M., B. Perlack, D. Becker, A. Turhollow, and R. Graham, 1997: *Evolution of the Fuel Ethanol Industry: Feedstock Availability and Price,* Biofuels Feedstock Development Program (Oak Ridge, TN: Oak Ridge National Laboratory); Walsh, M., R. Perlack, A. Turhollow, D. Ugarte, D. Becker, R. Graham, S. Slinsky, and D. Ray, 1999: *Biomass Feedstock Availability in the United States: Draft* (Oak Ridge, TN: Oak Ridge National Laboratory), April.
22. Lynd, L. 1997: "Cellulosic ethanol technology in relation to environmental goals and policy formulation," in J. DeCicco and M. DeLucchi (eds.), *Transportation, Energy and Environment: How Far Can Technology Take Us?* (Washington, DC: American Council for an Energy-Efficient Economy).
23. Interlaboratory Working Group, 2000.
24. U.S. DOT, 1997: *High-Speed Ground Transportation for America* (Washington, DC: Federal Railroad Administration).
25. Ibid.
26. A 5 percent discount rate was used for carbon as well as costs in the cost of saved carbon computations, based on the presumption that they will have a commodity value within some form of tradable permits regime.
27. U.S. EPA, 2001a.
28. These gases can be controlled interchangeably, using 100-year GWPs, as long as the total carbon equivalents (Ce) are reduced to 93 percent of their baseline levels. In contrast to the main three gases (CO_2, CH_4, and N_2O), which have a 1990 base year, the high-GWP gases have a base year of 1995.
29. For instance, different accounting methods and rules have been considered regarding what constitutes a forest, which biotic pools and lands are counted, which activities are considered eligible for crediting under Article 3.4, and uncertainties in measuring above- and below-ground carbon stocks.
30. See "Consolidated negotiating text proposed by the president," as revised June 18, 2001, FCCC/CP/2001/2/Rev.1, http://www.unfccc.int/resource/docs/cop6secpart/02r01.pdf. COP6bis refers to the Sixth Session of the Conference of the Parties to the United Nations Framework Convention on Climate Change, part two (Bonn, 16–27 July 2001).

31. See FCCC/CP/2001/L.7: "Review of the implementation of commitments and of other provisions of the convention. Preparations for the first session of the Conference of the Parties serving as the meeting of the parties to the Kyoto Protocol (Decision 8/CP.4). Available online: http:/unfccc.int/resource/docs.html.
32. The Pronk text, along with the COP6bis agreement, prohibits first commitment period crediting of CDM projects that avoid deforestation.
33. This figure is drawn from the Annex Table 1 of the April 9 draft of the Pronk text, which adopts the accounting approach for Article 3.3. activities suggested by the IPCC Special Report of LULUCF. This approach yields an Article 3.3 debit of 7 Mt Ce/year from net afforestation, reforestation, and deforestation activity, which under the Pronk approach could be offset fully by undiscounted forest management activities. Thus, the 42 Mt Ce/year estimate is based on 85 percent × (288 – 7) Mt Ce/year.
34. The Pronk proposal also allowed this cap to be filled through afforestation and deforestation activities in the CDM.
35. Missfeldt and Haites (2001) use a central estimate of 50 Mt Ce/year at $7.50/t Ce for CDM afforestation and reforestation projects. They also assume the availability of 150 Mt Ce/year at $15/t Ce for Article 3.4 sinks in Annex B countries. However, the Pronk 85 percent discount on forest management projects would, in principle, increase their cost accordingly (by 1/.15 or 6.7 times). However, given the small quantity (6 Mt Ce) that could be purchased, lower-cost opportunities in cropland management or the CDM should more than suffice. Missfield, F. and E. Haites, "The potential contribution of sinks to meeting Kyoto Protocol commitments" *Environmental Science Policy* (2001), 4: 269–292.
36. This figure is listed in a footnote to the agreement because the United States was not a party to it.
37. U.S. EPA (1999, 2000) expects voluntary Climate Change Action Plan (CCAP) activities to reduce 2010 methane and high-GWP gas emissions by about 10 percent and 15 percent, respectively, reductions that are not included in their 2010 projections shown in Table 6.1. Instead, these reductions are embodied in both their and our cost curves. See U.S. EPA, 1999: *U.S. Methane Emissions 1990–2020: Inventories, Projections, and Opportunities for Reductions* (Washington, DC: U.S. Environmental Protection Agency, Office of Air and Radiation), September, http://www.epa.gov/ghginfo; U.S. EPA, 2000: *Estimates of U.S. Emissions of High–Global Warming Potential Gases and the Costs of Reductions,* Review Draft, Reid Harvey (Washington, DC: U.S. Environmental Protection Agency, Office of Air and Radiation), March, http://www.epa.gov/ghginfo.
38. U.S. EPA, 1999.
39. Ibid.
40. U.S. EPA, 2000.
41. U.S. EPA, 2001b: *Draft U.S. Nitrous Oxide Emissions 1990–2020: Inventories, Projections, and Opportunities for Reductions* (Washington, DC: EPA), September 2001.
42. U.S. EPA, 1999, 2000, 2001b.
43. The result is a cost curve that is similar and more up-to-date than that used in widely cited multiple gas studies. Reilly, J. et al., (1999a): "Multi-Gas Assessment

of the Kyoto Protocol," *Nature* 401: 549–555 (October 7, 1999). Reilly, J., R. G. Prinn, J. Harnisch, J. Fitzmaurice, H. D. Jacoby, D. Kicklighter, P. H. Stone, A. P. Sokolov, and C. Wang, 1999b: "Multi-Gas Assessment of the Kyoto Protocol," Report No. 45, MIT Joint Program on the Science and Policy of Global Change, Boston, MA, January 1999. Available online: http://web.mit.edu/globalchange/www/rpt45.html. EERE, 2000.

44. A similar assumption is used by the European Commission (1998). European Commission, 1998: *Options to Reduce Nitrous Oxide Emissions* (Final Report): A report produced for DGXI by AEA Technology Environment, November. Available online: http://www.europa.eu.int/comm/environment/enveco/climate_change/nitrous_oxide_emissions.pdf. Approximately 15 percent of N_2O emissions are a byproduct of fuel combustion, largely by vehicles equipped with catalytic converters (U.S. EPA, 2001a).
45. We assume that coal production is a proportional to coal use (i.e., we ignore net imports and exports). U.S. EPA expects that the marginal methane emission rate will increase with production because an increasing fraction is expected to come from deeper underground mines (U.S. EPA, 1999).
46. Victor, D. G., N. Nakicenovic, and N. Victor, 2001: "The Kyoto Protocol emission allocations: windfall surpluses for Russia and Ukraine," *Climatic Change* 49 (3): 263–277, May 2001.
47. Grütter, J. 2001: "World market for GHG emission reductions: an analysis of the world market for GHG abatement, factors and trends that influence it based on the CERT model," prepared for the World Bank's National AIJ/JI/CDM Strategy Studies Program, March 2001.
48. We did briefly examine the potential contribution of a CDM fast track for renewables and efficiency, as embodied in the Pronk text. Applying the power sector CDM model developed by Bernow et al. (2001), we found that a carbon price of $20/t Ce would induce only 3 Mt Ce/year of new renewable energy project activity by 2010. At a price of $100/t Ce, this amount rises to 18 Mt Ce/year. Given that a large technical potential for energy efficiency projects exists at low or negative cost per t Ce, fast-track efficiency projects (less than 5 MW useful energy equivalents according to Pronk text) could significantly increase the amount available at lower costs. Bernow, S., S. Kartha, M. Lazarus, and T. Page. 2001: "Cleaner generation, free riders, and environmental integrity: Clean development mechanism and the power sector," *Climate Policy 23,* (2001) 1–21.
49. For the estimated demand for CDM, JI, and ET and hot air from other Annex 1 parties, we used a combination of cost curves (Reilly et al., 1999b; Ellerman, A. D., and A. Decaux, 1998, Analysis of Post-Kyoto Emissions Trading Using Marginal Abatement Curves, MIT Joint Program on the Science and Policy of Global Change *Report No. 40,* October, Cambridge, MA.; Vrolijk, C. and M. Grubb, 2000: "Quantifying Kyoto: How will COP-6 decisions affect the market?" Report of a workshop organized by the Royal Institute of International Affairs, UK, in association with: Institute for Global Environmental Strategies, Japan: World Bank National Strategies Studies Program on JI and CDM; National Institute of Public Health and the Environment, Netherlands, Erik Haites, Canada; and Mike Toman,

U.S. on 30–31 August 2000, Chatham House, London.) embodied in a spreadsheet model (Grütter, J. 2001. World Market for GHG Emission Reductions: An analysis of the World Market for GHG abatement, factors and trends that influence it based on the CERT model. Prepared for the World Bank's National AIJ/JI/CDM Strategy Studies Program, March, 2001.).

50. Our approach is similar to that used in a few other recent studies, Grütter, 2001; Haites, E., 2000; Missfeldt and Haites, 2001; Krause, F., P. Baer, and S. DeCanio, 2001: "Cutting carbon emissions at a profit: Opportunities for the U.S.," International Project FOR Sustainable Energy Paths, El Cerrito, California, April. Available online: http://www/ipsep.org.; Vrolijk and Grubb, 2000).
51. The market clearing price is lower here than in other similar studies, largely because of a much lower U.S. demand for international trade, which results from our aggressive pursuit of domestic abatement options and the fact that we assume that domestic policies and investments should be done as a matter of sound energy and environmental policy (i.e., they are price-inelastic).

PART III

POLICY CONTEXT

CHAPTER 7

U.S. Climate Change Policy

Armin Rosencranz

During the first 3 months of 2001, there were two startling developments in climate change policy. In January, the Intergovernmental Panel on Climate Change (IPCC) reported unequivocally that the world's climate is warming and that anthropogenic sources—mostly burning coal, oil, and gas to produce electricity—are at least partially responsible.[1] In March, his second month as U.S. president, George W. Bush both reversed his earlier position on regulating domestic emissions of CO_2 and repudiated the Kyoto Protocol of 1997. The Kyoto Protocol had been a signal accomplishment of the Clinton administration. President Clinton signed it in 1998.[2]

There is no way to reconcile these two developments. The IPCC 2001 report offered the considered assessment of the overwhelming majority of the world's climate scientists, and President George W. Bush's reversal and repudiation seemed a head-in-the-sand response driven by ignorance, short-sightedness, and the interests of elements of the American business community.

U.S. Climate Policy Since 1988

The Bush (I) Presidency

On June 13, 1988, in the midst of some of the hottest summer temperatures ever recorded, U.S. Senator Tim Wirth (D-CO) launched Senate hearings on climate change. National Aeronautics and Space Administration scientist James Hansen announced at these hearings that he was almost certain that hot weather was part of a pattern of human-induced climate change.[3] Later that summer, presidential candidate George Bush (I) observed, "Those who think we are powerless to do anything about the greenhouse effect are forgetting about the White House effect. As President, I intend to do something about it."[4]

Indeed, on January 30, 1989, President Bush's tenth day in office, Secretary of State James Baker endorsed no-regrets policies—measures to combat climate change that were already justified on other economic grounds.[5] But that was about as far as the Bush (I) administration was prepared to go. Over the next two years, the United States called for "national strategies" and resisted the European Community's call for binding commitments to carbon reduction targets and timetables.[6]

At the World Environmental Summit in Rio de Janeiro in 1992, which President Bush attended, the U.S. delegation successfully opposed targets and timetables, called for a comprehensive climate agreement that included sources and sinks of all greenhouse gases, and refused to make any explicit grant of financial support to developing countries.[7] The United States promptly signed the watered-down UN Framework Convention on Climate Change, and the Senate ratified it in October 1992.[8]

The First Clinton Administration (1993–1997)

In February 1993, the new Clinton administration sought both to raise new revenue to pay down the federal deficit and reduce domestic carbon emissions with a BTU tax based on the heat content of all fossil fuels. This tax was expected to reduce carbon emissions through energy conservation and more efficient energy consumption.[9] In his Earth Day speech on April 21, 1993, President Clinton announced "our nation's commitment to reducing our emissions of greenhouse gases to their 1990 levels by the year 2000."[10] As climate policy analysts Shardul Agrawala and Steinar Andresen have observed, "This marked a significant departure in U.S. climate policy—from *whether* to reduce greenhouse emissions to *how* and *by when*."[11]

The Clinton BTU tax passed the House but was defeated in the Senate after aggressive lobbying by industry groups. After this rebuff, the Clinton–Gore Climate Change Action Plan (CCAP) advocated a much more modest approach based on tax incentives and support for research rather than revenue measures.[12] The CCAP has enlisted more than 5,000 companies and organizations but has not reduced carbon emissions. The program was poorly funded and seemed a low White House priority. Economic growth, coupled with low fuel prices, actually caused emissions to increase by 13 percent during the Clinton years (15 percent over 1990 levels).[13]

In its international negotiations, the Clinton team made some progress toward international action on climate change. At the Geneva Ministerial Meeting (Second Session of the Conference of the Parties [COP-2]) in 1996, Tim Wirth, former senator and then Undersecretary of State for Global Affairs, agreed to targets and timetables on behalf of the United States, although he opposed legally binding commitments before 2010.[14] The U.S. delegation argued for flexibility in implementation through multiyear targets and national discretion in pacing emission reductions. The United States also sought emission trading between developed countries and argued that industrialized countries should get credits for reductions they brought about in developing countries. The U.S. delegation also echoed the Bush (I) inclusion of all carbon sources and sinks. Finally, to stave off domestic criticism, the U.S. delegation drew away from its COP-1 agreement in 1995 that no new commitments would be required from developing countries. Instead, at COP-2, U.S. negotiators urged all countries to limit carbon emissions.[15] Just before Kyoto, President Clinton announced that "participation" by developing countries would be required before the U.S. signed a climate agreement ("participation" was left undefined).[16]

On July 25, 1997, anticipating the COP-3 negotiations at the end of the year in Kyoto, Japan, the U.S. Senate passed a resolution by a unanimous vote (95–0) declaring its opposition to any international climate agreement that either harmed the U.S. economy or did not require the participation of developing countries.[17] The resolution was cosponsored by the Senate's longest-serving member, Robert C. Byrd, Democrat of West Virginia (a major coal-producing state), and climate change skeptic Senator Chuck Hagel, Republican of Nebraska. A Senate resolution has no legal effect, but this resolution presumably was designed to send a signal to the president that the U.S. Senate was unlikely to ratify the kind of climate agreement that seemed likely to emerge from the next conference of the parties to the Framework Convention on Climate Change.

Supporting the Byrd–Hagel Resolution, the Global Climate Coalition, an industry lobby, reportedly spent $13 million in an advertising campaign before the Kyoto meeting, in which it raised the specter of lost jobs and doubled gasoline prices among major voting groups (labor, farmers, minorities, and retirees).[18]

President Clinton and His Party

Despite President Clinton's and Vice President Gore's consistent rhetoric advocating programs to moderate climate change, reduce carbon emissions, develop non–carbon-based energy sources, and promote energy efficiency, the Clinton administration shied away from significant public investment on climate-related programs. In fact, as shown in Chapter 18, public spending on alternative energy research and development continued the decline that had begun in the Reagan and Bush (I) years (1981–1993).

Environmentalists certainly can be counted on to support carbon-reduction programs, but other significant Democratic Party core constituencies, such as labor, minorities,[19] and low-income people, including retirees, are ambivalent about programs that may affect jobs and the cost of living.[20] This could happen through job losses in coal, oil, and gas production, steel and automobile manufacture, increases in home heating and gasoline prices, and price rises in energy-intensive goods.[21] The ambivalence toward carbon reduction within major Democratic constituencies may help explain the defeat of President Clinton's 1993 BTU tax and his wariness ever after.[22] It may also explain why West Virginia, with twice as many registered Democrats as Republicans, preferred oil man George W. Bush over environmentalist Al Gore in the 2000 presidential election.[23]

The Clinton—Gore Record

In his final State of the Union Address on January 27, 2000, President Clinton observed,

> The greatest environmental challenge of the new century is global warming. . . . If we fail to reduce the emission of greenhouse gases, deadly heat waves and droughts will become more frequent, coastal areas will flood and economies will be disrupted. That is going to happen, unless we act. . . . New technologies make it possible to cut harmful emissions and provide even more growth.[24]

President Clinton proposed $2.4 billion in the Fiscal Year (FY) 2001 budget to combat global climate change. This is 40 percent more than what Congress enacted in FY 2000 but still a very modest amount of funding: about the cost of one B-1 bomber or 0.13 percent of the federal budget for 2001. The new initiatives included increased efforts to develop clean energy sources and a new Clean Air Partnership Fund to spur state and local efforts to reduce air pollution, with collateral benefits in reduced greenhouse gases and a 5-year package of tax incentives to promote clean energy and energy-efficient technologies and renewable energy, with a special emphasis on biomass energy (biofuels) to support farm incomes and strengthen the rural economy.[25] In addition to these programs, the president's budget called for $1.7 billion for scientific research on the earth's climate system and adaptation to climate change through the U.S. Global Change Research Program.[26]

In FY 2000, Congress underfunded the president's climate change related budgetary requests by 40 percent.[27] For example, President Clinton asked for $100 million for the Clean Air Partnership Fund, and Congress appropriated only $41 million. Another showcase program, the Partnership (between the federal government and automakers) for a New Generation of Vehicles was underfunded by $58 million below the president's budget allocation of $264 million.[28] This program seeks to develop cars driven by fuel cells that would deliver the same performance as current models with three times greater economy.

Interviewed by actor Leonardo DiCaprio for Earth Day 2000, President Clinton returned to the basic theme of proposing everyday, common-sense programs such as tax breaks to encourage manufacturers to develop and consumers to buy energy-efficient products and greater spending on research into energy-efficient vehicles and alternative (noncarbon) fuels.[29] The natural response is that no one can fault tax incentives or research, but neither President Clinton nor Vice President Gore mentioned that a total of $4 billion for everything, including basic scientific research, is a negligible amount of money in a large economy with large public spending. Their Climate Change Technology Initiative is an umbrella for Department of Energy programs aimed at energy efficiency and renewable energy, Environmental Protection Agency (EPA) voluntary programs such as Energy Star and Green Lights, and Agriculture Department work on sequestering carbon in soils. It also includes proposals for tax credits on energy-efficient vehicles and homes and for producing electricity from wind and biomass.

All these programs are very modestly funded because the president asked for so little and the Congress appropriated even less. The overall thrust of the Clinton–Gore climate change policies was that we could address the threat of climate change and even meet our Kyoto target of a 7 percent reduction in greenhouse emissions from 1990 levels by 2008–2012[30] for little cost in federal spending and no cost to the economy.

The government-funded five-lab study,[31] discussed in Chapter 16, contains much the same message: The U.S. can meet Kyoto targets with no-regrets measures such as energy efficiency standards and clean energy technologies.[32] The Clinton–Gore record is mixed: modest progress on energy efficiency standards[33] and no restrictions on air pollution beyond those provided in the Clean Air Act of 1990. But Clinton did send Gore to rescue COP-3 at Kyoto in 1997 and did sign the Kyoto Protocol a year later at COP-4.

Apparently, President Clinton trusted Vice President Gore's insistence that global climate change was an important issue—important enough to rescue the Kyoto protocol from its stalled negotiations and commit to a 7 percent reduction in carbon emissions by 2008–2012, using 1990 as the base year. But the economists among President Clinton's senior policy advisors continued to worry about the costs to the economy of carbon reduction. Their advice, the opposition of industry and labor toward carbon reduction, and congressional resistance to funding even modest climate change research and education programs seems to account for the Clinton administration's modest efforts to reduce domestic carbon emissions.

At Kyoto, the United States advocated emission trading, multiyear targets, credit for carbon sinks (agriculture and trees), and a "basket" of six gases—CO_2, methane, nitrous oxide, sulfur hexafluoride, hydrofluorocarbons, and perfluorocarbons.[34] Vice President Gore flew in for the meeting's final 16 hours and called for increased flexibility for all negotiations. The chief U.S. negotiator, Stuart Eizenstat, thereupon agreed to a 7 percent reduction from, rather than a mere rollback to, 1990 carbon levels by 2008–2012. This concession led to a last-minute agreement at Kyoto. Other countries, especially those of the European Union (EU), receded in their opposition to the basket of six gases, the inclusion of sinks, and the use of multiyear targets.[35] Emission trading was recognized as a future mechanism, and the Clean Development Mechanism (CDM) was invented to incorporate and supersede joint implementation. The Group of 77 and China prevailed on the exemption of developing countries from specific emission reductions.[36]

At the 1998 COP-4 in Buenos Aires, the United States continued to try to flesh out the two market-based approaches that emerged in general terms at Kyoto: emission trading and the CDM. It also continued to call for voluntary participation by developing countries, presumably to appease the U.S. Senate.[37] Argentina, the host country for COP-4, made a gesture of voluntary participation that played well in the United States. Finally, the United States signed the Kyoto Protocol in Buenos Aires—subject, of course, to eventual Senate ratification by a two-thirds majority. But President Clinton made it clear that the Kyoto Protocol would not be placed before the Senate during his tenure as president.

COP-6 at The Hague

Between COP-4 in November 1998 and COP-6 in November 2000, significant domestic opposition to emission cuts developed in corporate circles and in Congress. Because it became clear that an emission trading system would take many years to be realized, the only alternative to lifestyle-changing cuts in carbon emissions was for U.S. negotiators at The Hague to seek credits for carbon sinks in U.S. agriculture and forestry practices.[38] The EU insisted that at least half of industrialized countries' reduction targets be achieved through domestic cuts in fossil fuel emissions.[39] It seemed at one point in the negotiations that the gap between the EU and U.S. positions could be bridged. But the EU delegates, perhaps under pressure from their Green Party members and environmental non-government organizations (NGOs), eventually refused to bridge this gap, and COP-6 ended without agreement. Because George W. Bush was declared president-elect in the three weeks after the end of COP-6, it seemed to many observers that the gap between the EU and the United States probably would widen in the future.[40]

The George Bush (II) First Hundred Days

That likelihood was confirmed on March 13, 2001, when President Bush wrote to four Republican senators that he was not willing to regulate CO_2 emissions in light of the ongoing California energy shortage and "the incomplete state of scientific knowledge of the causes of, and solutions to, global (climate) change, and the lack of commercially available technologies for removing and storing carbon dioxide."[41]

Sixteen days later, the second shoe dropped when President Bush repudiated the Kyoto Protocol by stating, "I will not accept a plan that will harm our economy and hurt American workers."[42]

Reaction to President Bush s March 2001 Pronouncements

Most environmental NGOs deplored President Bush's March 2001 pronouncements on global climate change. One typical response came from the Union of Concerned Scientists (UCS):

> The president cited two reasons for his decision, both of which are ill founded, and without merit.
>
> The first is that he does not believe the evidence of global warming is clear. Nothing could be further from the truth. A panel of the world's leading scientists recently released the most comprehensive study ever on global warming, and found that it is well underway, will have devastating impacts if emissions go unchecked, and can be limited at little or no net economic cost.
>
> The second is that including caps on carbon dioxide emissions will significantly increase electricity costs for the nation's consumers. His claim is based on a fatally flawed study commissioned by former Representative David McIntosh, a hard-line opponent of action on global warming. Other recent analyses by the Department of Energy, Environmental Protection Agency and private groups

> demonstrate that major reductions in power plant pollutants including carbon dioxide can be achieved at modest cost.[43]

The UCS e-mail to its supporters goes on to explore "solutions to deflate soaring electricity prices," including presidential support of clean energy sources and energy efficiency measures to reduce demand. UCS noted that wind energy is the fastest growing energy supply in the world.[44] It is hard to comprehend President Bush's political strategy in these March 2001 reversals. Whereas conservatives and the business community make up perhaps 20 percent of the electorate,[45] polls indicate that a large majority of Americans of both major parties consistently favor protecting the environment and conserving open spaces.[46] Conservatives and businesspeople may be more deeply committed to their beliefs than environmentally minded people are to theirs, but the political calculations of the Bush team still seem to risk a backlash from voters in 2002 and 2004.

In the tradition of President Nixon going to China, the historical moment seemed ripe for Republican presidential leadership Vice President Gore, who championed the global warming issue in his 1992 bestseller, *Earth in the Balance,* and who had been the lightning rod for congressional opposition to carbon abatement since 1993, was off the public stage. Republicans seemed poised to co-opt this issue and divide his supporters. Several industry leaders, including CEOs of oil companies, had announced that global climate change was here to stay and was to be taken seriously. Several large corporations, reluctant to appear regressive on the issue of climate change, had left the Global Climate Coalition, the industry lobby that had been so vocal in opposing to an international climate agreement for most of the 1990s. The IPCC had just predicted a global temperature rise as high as 6°C in the twenty-first century unless greenhouse gas emissions are reduced. Seizing the moment to announce progressive, market-based carbon dioxide policies would seem to have been the safest political course, with the possibility of reducing the usual swing to the opposition party in the 2002 midterm elections and nailing home President Bush's reelection in 2004.

The February, 2002 Bush Climate Change Strategy

On February 14, 2002, President Bush announced his long awaited strategy to address climate change. His target is to cut the rate of annual domestic carbon emissions through voluntary corporate action from 183 metric tons per million dollars of GDP to 151 metric tones by 2012.[47] His aim is to slow the *growth* of emissions rather than reducing them—thereby avoiding harm to the U.S. economy. He talked of cutting greenhouse gas "intensity" by 18 percent over the next decade. The 18 percent cut is not a cut in emissions but rather a cut in the level of emissions per unit of economic output.[48] Growth in economic output between 2002 and 2012 make it likely that U.S. carbon emissions would be significantly *higher* in 2012 than they are today.

The Bush climate change strategy included a proposed $4.6 billion in tax credits over five years, averaging $900 million per year, to stimulate investments in clean energy sources, hybrid and fuel cell vehicles, and emissions reducing technologies.[49] Notwithstanding much talk by Bush administration officials and the Council of Economic Advisers about market based initiatives, there is nothing in the new strategy about carbon emissions trading—one of the flexibility mechanisms of the Kyoto Protocol contributed by the U.S. delegation to COP-3 in 1997.

The President said his 2003 budget commits $4.5 billion to climatic change, "more than any other nation's commitment in the entire world."[50] This includes $588 million toward energy conservation research and development (R & D), $408 million towards renewable energy R & D, and $510 million for the new Department of Energy "Freedom Car Initiative."[51]

President Bush observed that, under the Kyoto Protocol, the U.S. would have had to "make deep and immediate cuts in our economy to meet an arbitrary target. It would have cost our economy up to $400 billion and we would have lost 4.9 million jobs."[52] He also noted that "developing countries such as China and India already account for a majority of the world's greenhouse gas emissions,"[53] but failed to acknowledge that China and India *together* contain 2.3 billion people and produce fewer carbon emissions than the U.S. with 280 million people.

President Bush's budget allocations to address climate change are on the same order and roughly the same tiny percentage of GDP that President Clinton allocated. Like the Bush-Cheney energy policy, there is scant emphasis in the new climate change strategy on energy conservation, renewable energy, or fuel efficiency standards. An editorial in the *New York Times* concluded that President Bush does not regard global warming as a problem: "There seems no other way to interpret a policy that would actually increase the gases responsible for heating the earth's atmosphere . . . By his own figures, actual emissions . . . could rise by 14 percent, which is exactly the rate at which they have been rising for the last 10 years."[54]

Senator James Jeffords (Ind., VT), chair of the Senate Environment and Public Works Committee, described the Bush climate change strategy as "divorced from the reality of global warming."[55] Environmentalists must now appeal to Congress, where they seek legislation to require corporations to publicly disclose their carbon emissions—in contrast to the voluntary disclosure advocated by President Bush.

Prognosis

Stanford Senior Fellow David Victor has argued that the Kyoto targets were "symbolically high but hopelessly unrealistic, and should be abandoned."[56] *The Economist* believes that Bush's formal repudiation of Kyoto contains some good news and some bad news.[57] The bad news is that President Bush has retrogressed by alleging continued uncertainties about the science of climate change (in the face of an overwhelming consensus among climate scientists) that global warming and its damaging effects are real and are caused by human activity. Additionally, President Bush's team has argued that developing countries get a "free ride" by not being required to cut carbon emissions, while developed countries suffer economic loss.[58] But developing countries' per capita carbon emissions are now a small fraction of per capita emissions in the United States and other developed countries; the developed countries' emissions account for the bulk of the greenhouse effect, so it is fair that they act first and all climate negotiations envision a carbon emission role for developing countries at a later stage.[59]

The good news, according to *The Economist,* is that the Bush administration is focusing on the costs of complying with Kyoto targets. An international climate change treaty could be implemented in a flexible way that gives broad play to market forces and encourages innovation and development of clean technologies. Europeans seem to be unreasonably skeptical of market approaches, and the Bush repudiation may get them to rethink their position. *The Economist* cites with approval Victor's argument that the cause of the Kyoto Protocol's collapse is its cap-and-trade system, which allows ambitious targets but puts no limits on compliance costs.[60]

Michael Grubb, a scholar at London's Imperial College, believes that the EU must take the lead and act boldly in international climate negotiations. Contrary to conventional wisdom among economists, he believes that technical change driven by corporate research and development in response to market conditions will tend to accelerate carbon abatement to induce cost reductions, rather than deferring abatement to await cost reductions. Grubb's model encourages early action to accelerate the development of cost-effective technologies.[61]

Moreover, Grubb seems to have accurately predicted that with the EU, eager to accommodate its environmentalists, taking the lead, and with flexible policies and mechanisms, the Kyoto Protocol would come into force without the United States. Russia would be motivated to ratify because it could sell its unused emission credits to other industrialized countries, and Japan would not scuttle a treaty bearing the name *Kyoto.*[62] With the EU, Russia, Australia, Japan, and a scattering of other signatory states, the protocol would come into force during this decade with the required ratification of 55 signatory states representing 55 percent of global carbon emissions. With such a demonstration of serious purpose and without an "elephant" (the United States) in the room, developing countries might voluntarily commit to reduction targets. This commitment, together with flexible market-based mechanisms developed in tandem with Bush policy advisors and with growing pressure from Democrats, scientists, environmentalists, and progressive U.S. industrialists, could eventually bring the United States back into global climate change negotiations.

The declaration that the Kyoto Protocol is dead seems premature in light of its adoption by 178 nations in Bonn in July 2001. In fact, the EU agreed to compromises in Bonn that well exceed those they were unwilling to consider only eight months earlier in The Hague.

Conclusions

Benefiting from the international shock over President Bush's withdrawal from the Kyoto negotiating process, the EU went along with sweeping compromises in July 2001 (COP-6bis) that they had not considered in November 2000 at COP-6. In the wake of Bush's repudiation and unilateralism, the EU was willing to accept a partial deal rather than a continued stalemate. Thus, U.S. policy has had a major, though obviously unintended, influence on the entire climate change negotiation.

The major challenge for U.S. climate change advocates seems to be to persuade the Bush administration to act now to reduce carbon dioxide domestically and to collaborate in shaping carbon reduction policies, strategies, and mechanisms with fellow member states of the Framework Convention on Climate Change, all sharing the same carbon-loaded atmosphere.

Notes

1. *IPCC Third Assessment Report (TAR) Synthesis Report,* April 2001.
2. Agrawala, S. and S. Andresen, 2001: "U.S. climate policy: evolution and future prospects," *Energy and Environment,* Summer: 9.
3. Hansen, J., 1988: *Testimony Before Senate Energy and Natural Resources Committee,* June 23.
4. Agrawala and Andresen, 2001, n.2, p. 3.
5. Bodansky, D., 1993: "The United Nations Framework Convention on Climate Change: a commentary," *Yale Journal of International Law,* 451.
6. Agrawala and Andresen, 2001, n.2, p. 4.
7. Ibid., p. 4. The Rio Conference did acknowledge that all countries had "common but differentiated responsibilities" for carbon abatement and recognized the need

for industrial countries to support developing countries in these efforts to protect the global commons.

8. *Congressional Record,* 1992: S17, 156, Daily Edition, October 7, 1992.
9. Agrawala and Andresen, 2001, n.2, p. 4.
10. Ibid., p. 5.
11. Ibid., p. 6 (emphasis added).
12. *United States: Taking Action on Climate Change,* October 1999. The White House: Climate Change Task Force, available at http://www.epa.gov/globalwarming/publications/actions/cop5/usaction_99.html.
13. "U.S. pledges to seek ways to protect world climate," *International Herald Tribune,* April 4, 2001, p. 3.
14. Undersecretary of State Timothy Wirth at COP-2 in Geneva in July 1996, cited in Grubb, M., C. Vrolijk, and D. Brack, 2000: *The Kyoto Protocol: A Guide and Assessment* (London: Royal Institute of International Affairs), p. 54.
15. Agrawala and Andresen, 2001, n.2, p. 6.
16. Ibid.
17. *Congressional Record,* July 27, 1997, S8113–8138.
18. Agrawala and Andresen, 2001, n.2, p. 12.
19. "Refusing to Repeat Past Mistakes: How the Kyoto Climate Change Protocol would disproportionately threaten the economic well-being of blacks and Hispanics in the United States," a study conducted by Management Information Services, Inc., Washington, DC, June, 2000. Available at http://www.ceednet.org.
20. Agrawala and Andresen, 2001, n.2, p. 10.
21. Ibid., p. 10–11.
22. Ibid., p. 11.
23. Ibid. Agrawala and Andresen note that West Virginia's five electoral votes would have secured Gore's election in 2000, even without Florida.
24. President Bill Clinton, State of the Union Address, January 27, 2000. Four months later, the U.S. Global Change Research Program released its draft report analyzing the potential impacts of global climate change on the United States over the next 100 years. The report predicts a 5–10°F temperature increase; heavier rainfall and increased drought, varying region by region; the disappearance of some natural ecosystems (alpine meadows in the Rockies and some barrier islands); reduced availability of freshwater in some regions; and increased crop productivity. See http://www.gcrib.org/NationalAssessment and http://sedac.ciesin.org/NationalAssessment.
25. The White House, Office of the Press Secretary, Press Release, February 3, 2000.
26. Ibid. This money was to be divided among 10 federal agencies.
27. Ibid.
28. Ibid.
29. "Interview of the President by Leonardo DiCaprio for Earth Day," The White House, Office of the Press Secretary, April 23, 2000, available at http://www.pub.whitehouse.gov/rui- . . . /oma.eop.vs/2000/4/2/5.text.1.
30. As noted earlier, greenhouse gas emissions increased by 15 percent during 1990–2000

and are expected to increase by another 15 percent in the decade 2001–2010 in the absence of significant domestic programs to reduce these emissions.

31. "Scenarios for a clean energy future," November 15, 2000, available at http://www.ornl.gov/ORNL/Energy_EFF/CEF.htm.
32. Ibid.
33. In a letter to the *New York Times* dated January 8, 2001, White House Chief of Staff John Podesta claimed that the Clinton administration's "new efficiency standards for home appliances will save consumers $50 billion and avoid 224 million metric tons of greenhouse gases over the next ten years." *New York Times,* January 10, 2001, p. A22.
34. Agrawala and Andresen, 2001, n.2, p. 6.
35. Ibid., p. 7.
36. Ibid.
37. Ibid.
38. Ibid. David Sandalow, Assistant Secretary of State for Oceans, Environment, and Science and head of the U.S. delegation at COP-6 in The Hague, said in his opening statement, "The United States will work with all Parties to craft sound decisions that include:
    - Strong, market-based rules for the flexible mechanisms;
    - Binding legal consequences for failure to meet targets;
    - Rules that recognize the role of forest and farmlands in fighting climate change;
    - A prompt start to the Clean Development Mechanism, with rules to ensure its workable operation and environmental integrity;
    - Help to provide the technology and capacity [that] developing countries need to combat climatic change and adapt to its impacts."

    See also Sandalow, D. B. and I. A. Bowles, 2001: "Climate change: fundamentals of treaty-making on climate change," *Science,* 292 (5523): 1839.
39. "Hotting up in The Hague," *The Economist,* November 18, 2000, p. 83.
40. Beyond these EU/U.S. divisions, Agrawala and Andresen argue that American "national culture" influences the shape of climate change policy: "There is unlikely to be domestic support for measures that seemingly enhance government control over citizen behavior. This sentiment has played a major role in the U.S. insistence on flexible, market-based approaches, as opposed to more top-down measures . . . supported by most European countries." Agrawala and Andresen, 2001, n.2, p. 16.
41. "Letter from the president to Senators Hagel, Helms, Craig, and Roberts," The White House, Office of the Press Secretary, March 13, 2001.
42. Press briefing at the White House, March 29, 2001. See also "Rage over global warming," *The Economist,* April 7, 2001, p. 18.
43. E-mail from Lloyd Ritter to the UCS list, March 20, 2001.
44. Ibid.
45. To be sure, this 20 percent of the electorate may represent 80 percent or more of the country's wealth.
46. In an April 2001 CBS poll, 61 percent of respondents said that protecting the envi-

ronment was more important to them than producing energy. Only 29 percent chose energy over the environment. See *Wall Street Journal,* April 20, 2001, p. 16.

47. See "Bush to Unveil Plan Linking Economy and Environment," *Wall Street Journal,* February 14, 2002, A22.
48. See "Hot Air," Economist.com Global Agenda, February 14, 2002.
49. Remarks by the President on Climate Change and Clean Air, The White House, Office of the Press Secretary, February 14, 2002. Available at http://usa.or.th/news/wf/epf416.htm.
50. Ibid.
51. Ibid.
52. Ibid.
53. Ibid.
54. "Backward on Global Warming," Editorial, *New York Times,* February 16, 2002.
55. "Bush Plan Deepens Divide Over Kyoto Protocol," *Nature,* 21 February, 2002, p. 821.
56. Victor, D. G., 2001: *The Collapse of the Kyoto Protocol and the Struggle to Slow Global Warming* (Princeton: Princeton University Press), 160 pp.
57. "Oh no, Kyoto," *The Economist,* April 7, 2001, pp. 73–75.
58. Ibid.
59. Ibid.
60. Ibid., citing Victor, 2001, n.46.
61. Grubb, M. and J. Koehler, 2001: "Induced technical change; Evidence and Implications for Energy-Environmental Modeling and Policy." Report to the Deputy Secretary-General, Organisation for Economic Co-operation and Development, Paris, 2001, p. 4.
62. "Oh no, Kyoto," p. 47.

CHAPTER 8

The Climate Policy Debate in the U.S. Congress

Kai S. Anderson

The Clinton administration and the U.S. Congress professed strong but dissimilar preferences regarding how the domestic and international climate change debates should evolve. Throughout the 106th Congress (1999–2000), the Clinton administration kept a low profile on the issue, and Congress remained deeply divided regarding how the United States might best meet the challenge or whether it should address the issue at all.

Context of the Climate Policy Debate

The international climate change negotiations pursuant to the UN Framework Convention on Climate Change (FCCC) serve as a lightning rod for the criticism of congressional and corporate climate change skeptics. These skeptics have skillfully shifted the domestic debate from the science of climate change to the contentious provisions of the Kyoto Protocol and effectively linked the two issues in the minds of many legislators. The global warming debate in the U.S. Congress has focused largely on the pros and cons of the protocol, which requires that developed countries reduce their greenhouse gas (GHG) emissions below specified levels over a 5-year period beginning in 2008 but imposes no similar targets on developing countries. The lack of binding commitments by developing countries is a divisive issue for Congress.

A comprehensive review of climate change negotiations to date is beyond the scope of this chapter. However, a brief review of some of the major scientific, legislative, and diplomatic events that contributed to the current policy atmosphere provides a basis for considering how Congress might cut through the Kyoto Protocol smokescreen and move the climate change debate forward.

Framing the Framework Convention

The Intergovernmental Panel on Climate Change (IPCC), which convened in 1988 at the request of the UN Environment Program and the World Meteorological Organization, serves as the primary source of objective scientific and technical information to the FCCC. The IPCC issued its first scientific assessment of climate change in 1990.[1] This report found that increases in the concentrations of GHGs in the earth's atmosphere will cause global warming.[2] The 15 months of negotiations that led to the FCCC largely reflected this assessment.

By signing the FCCC at the Earth Summit, the United States joined 160 other nations in a global effort to stabilize atmospheric concentrations of GHGs at levels that will "prevent dangerous anthropogenic interference" with the climate system.[3] The use of the subjective term *dangerous* left open to negotiation the level of GHG emission control that the international community needs to achieve.

In October 1992, the U.S. Senate consented to ratification of the FCCC,[4] and President Bush signed and submitted the ratification document to the UN secretary general. The entry into force of the FCCC sanctified the nonbinding pledge of the United States and the other signatory Annex I parties (i.e., developed countries and countries transitioning toward market economies such as Russia) to return their GHG emissions to 1990 levels by the year 2000. By ratifying the FCCC, Annex I countries agreed to adopt national plans to mitigate climate change by enhancing carbon sinks and reducing GHG emissions.[5]

In an effort to meet the American commitment under the FCCC, the Bush administration adopted a no-regrets policy on climate change with the expectation that the United States could achieve its FCCC emission control obligations by encouraging voluntary actions included in the Energy Policy Act (EPAct) of 1992.[6] The voluntary nature of the no-regrets policy and the Senate approval of the FCCC reflected both the nonbinding nature of the FCCC and the expectation, fostered by a 1991 National Academy of Sciences (NAS) report, that the United States could "reduce or offset greenhouse gas emissions by between 10 and 40 percent of 1990 levels at low cost or some net savings."[7] Pursuant to the provisions of the FCCC, the Bush and Clinton administrations have published three climate change action plans.

The first of these climate action plans, released by the Bush administration in December 1992, focused primarily on estimating U.S. emissions and assessing how existing initiatives, such as the Environmental Protection Agency's (EPA's) "green programs" and initiatives authorized by EPAct, could help control emissions. The assumption that climate change uncertainty precluded taking any steps that would not be done for other beneficial reasons, such as reducing air pollution or cutting energy expenses, underpinned the no-regrets policy.

The Clinton administration submitted climate change action plans to the FCCC in 1994 and 1997.[8] These plans, like the Bush plan, focused largely on catalyzing voluntary action to increase American energy efficiency, promote energy conservation, and encourage the use of low-carbon and carbon-free fuels. Whereas the first Clinton climate plan suggested that U.S. emissions would meet the goal of the FCCC, the second plan estimated that emissions in 2000 would exceed 1990 levels by 13 percent.[9]

Prelude to the Kyoto Protocol

Following the precedent of the 1985 Vienna Convention, which led to the highly successful 1987 Montreal Protocol on Substances that Deplete the Ozone Layer, the FCCC provided for a "conference of parties" framework to organize the ongoing climate change deliberations.[10] To establish a scientific basis for the convention process, the IPCC published a second supplementary report in 1994.[11] In part because it became increasingly clear that major developed countries, including the United States and Japan, would not meet the voluntary emission goals of the FCCC, the first Conference of Parties (COP-1) convened in Berlin in the spring of 1995. The goal of COP-1 was to establish a common approach to protecting the earth's climate. The ministerial declaration of COP-1, the Berlin Mandate, outlined the process that eventually led to the Kyoto Protocol.

Despite resistance from U.S. negotiators, the Berlin Mandate exempted developing countries from near-term, legally binding GHG emission reductions. Because the Berlin Mandate did not include developing countries in the first round of emission limits under the FCCC, some U.S. legislators claimed that the treaty would disadvantage U.S. companies. Concerns regarding American competitiveness with developing countries also raised questions about the accuracy of the 1991 NAS, which projected modest cost of GHG mitigation.

The development and release of the IPCC's three-volume Second Assessment Report (SAR)[12] in 1995 met with hostility from congressional climate change skeptics. A series of emotional congressional oversight hearings focused on the science reported in the SAR.[13] These hearings reflected a widening rift between the Clinton administration and congressional skeptics.

After U.S. Undersecretary of State for Global Affairs Timothy Wirth advocated midterm legally binding emission targets at the second Conference of the Parties (COP-2) in Geneva, both the cost of controlling GHG emissions and the consequences of noncompliance with the FCCC invited much more attention. The COP-2 ministerial declaration bore a stronger resemblance to the U.S. negotiating position than did the Berlin Mandate, but the U.S. position attracted sharp congressional criticism nonetheless.

Several months before the third Conference of the Parties (COP-3) in July 1997, Senate Resolution 98, the so-called Byrd–Hagel Resolution, passed by a vote of 95–0.[14] The Byrd–Hagel Resolution advised the Clinton administration that the "U.S. should not be a signatory to any protocol" that would either "mandate new commitments to limit or reduce greenhouse gas emissions for [developed countries] unless the protocol . . . also mandates new specific scheduled commitments to limit or reduce greenhouse gas emissions for developing country parties within the same compliance period" or "result in serious harm to the economy of the United States."[15]

The Byrd–Hagel Resolution further stipulated that in the event that any Kyoto agreement was submitted for the advice and consent of the Senate, it "should be accompanied by a detailed explanation of any legislation or regulatory actions that may be required to implement" such an agreement, including "an analysis of the detailed financial costs and other impacts on the economy of the United States" that would result from implementation of the agreement.[16] More than 3 years later, the Byrd–Hagel Resolution remains the most influential expression of the Senate's intent regarding climate change.

The Kyoto Protocol: Congressional Debate Heats Up

The Kyoto negotiations proved difficult because parties found it difficult to achieve consensus on whether developing countries as well as developed countries should assume binding emissions limitations, which GHGs to cover, what level of emission reductions to require of each country, and whether to include flexible compliance mechanisms. Negotiators finalized the text of the Kyoto Protocol only after an eleventh-hour intervention by Vice President Gore. The protocol requires the United States to reduce its annual GHG emissions to 7 percent below 1990 emissions during the first commitment period scheduled for 2008–2012.[17]

In light of the inherent conflict between the Berlin Mandate exemption for developing countries and the Byrd–Hagel Resolution provision requiring commitments from all countries, it came as no surprise that the protocol did not satisfy the Senate. However, the protocol does require that all parties to the FCCC undertake programs to improve national emissions and sequestration monitoring, publish and update their climate mitigation and adaptation efforts, and participate in the promotion and dissemination of climate change science and climate-friendly technologies.[18] These capacity-building and information-gathering efforts are vital to the long-term climate change mitigation efforts of developing countries.

American negotiators successfully lobbied for the inclusion of the flexibility mechanisms—joint implementation (JI), emission trading, and the Clean Development Mechanism (CDM)—in the protocol. JI, which is popular with many U.S. companies and some conservation groups, allows countries to earn credit for emission reductions that result from conservation projects funded abroad. Emission trading allows countries bound by emission limits (i.e., Annex I countries) to sell emission allowances to other countries so that emission reductions can be achieved at lower net cost. The CDM serves as a clearinghouse for developed country–financed projects in the developing countries, without direct bilateral linkage that characterizes JI projects. The inclusion of the flexibility mechanisms in the protocol promises to lower compliance costs for developed countries.

Partly because of the complexity of the flexibility mechanisms and disagreement over the consequences of noncompliance, negotiators deferred finalizing timelines and implementation specifics until the fourth Conference of the Parties (COP-4) to the FCCC in Buenos Aires, Argentina, in November 1998.

The Buenos Aires Workplan and Subsequent Negotiations

The unremarkable result of the Buenos Aires negotiations was a 2-year workplan intended to guide the development of protocol implementation rules, such as how the flexibility mechanisms and the GHG emission accounting systems will work. Despite the Byrd–Hagel Resolution, the United States signed the protocol during the COP-4. And although Argentina and Kazakhstan indicated their intentions to adopt emission targets, some members of Congress viewed the administration's decision to sign the protocol before securing binding commitments from critical developing countries, notably China and India, as an affront to the Senate. For its part, the Clinton administration hoped that by demonstrating American commitment to the process, signing the treaty would enhance the United States' credibility and leverage in the COP-4 negotiations.

Subsequent high-level negotiations in Bonn, Germany (COP-5), and The Hague, Netherlands (COP-6), failed to produce agreement on the Buenos Aires workplan. After the election of President George W. Bush and confirmation of General Colin Powell as secretary of state, the United States called for a delay in further climate change discussions.

The American Political Dynamic

The Fate of the Kyoto Protocol

Although the United States signed the Kyoto Protocol in Buenos Aires, the Senate must ratify the treaty before it becomes binding. Given the current political climate in Washington, a Senate vote on ratification would result in rejection of the protocol. Understanding this political reality and recognizing that there is no mandatory timeline for submitting the treaty to the Senate, the Clinton administration avoided a ratification vote. The Clinton administration's decision to postpone ratification angered many climate change skeptics. This frustration manifested itself as a vigorous effort, including highly partisan oversight hearings, designed to prevent "back-door implementation" of the protocol. The unfortunate result of these antiprotocol efforts was that the congressional debate, which should focus on the science, economics, and politics of climate change and options to mitigate against it, deteriorated into a stand-off between the Clinton administration and some members of Congress.

Despite deep disagreement in the United States about its necessity, efficacy, attainability, and economic impacts, the protocol remains the focus of ongoing international negotiations, which began after the entry into force of the FCCC. Before the protocol becomes binding, more than 55 countries, together accounting for more than 55 percent of 1990 global GHG emissions, must ratify it.[19] Because the United States contributes more than 20 percent of global anthropogenic GHGs, American ratification is almost prerequisite for the protocol to enter into force. From an environmental standpoint, the engagement of developing countries, some of which have extremely high emission potential, is equally critical for long-term GHG emission reductions.

Together, these considerations constitute a catch-22: The U.S. Senate claims that it will defeat the treaty and prevent any agency from implementing programs aimed "solely" at achieving Kyoto emission goals unless developing countries commit to limiting their emissions; however, developing countries claim that they will not accept any emission restrictions or emission growth targets until the developed countries, which account for the majority of current and historical anthropogenic GHG emissions, make real reductions and provide technical and financial assistance to developing countries.

Searching for Traction

Many stakeholders agree that the crux of the congressional climate change policy debate is what the United States should do in light of the uncertainties of climate science, the uncertainties inherent in policies designed to mitigate GHG emissions, and the potential consequences of global warming. Congressional views vary widely regarding the importance and reliability of climate science, the economic impacts of climate policy, the potential for technology to mitigate GHG emissions, and the environmental implications of global climate change.

Despite the quantum advances in climate science during the 1990s,[20] on balance, congressional opinions regarding climate change have evolved only modestly since the United States signed and ratified the FCCC in 1992. Moreover, although polling data suggest that most Americans believe climate change poses a threat, the issue lacks the immediacy that tends to galvanize broad-based action on environmental issues. This political stasis also reflects the fact that legislators stand to gain little or nothing politically from championing climate change legislation. In light of the highly polarized political atmosphere, any meaningful climate change legislation acts as a magnet for criticism.

The lack of political traction on the issue of climate change ensures that although a wide range of policy options to mitigate GHG emissions exist—including energy conservation and energy efficiency improvement programs, carbon sequestration initiatives, fuel switching and substitution incentives, carbon taxes, new source performance standards, and emission trading schemes—the range of politically viable options is narrow.

Crediting Voluntary Early Action

In the waning days of the 105th Congress, momentum began building behind the idea that the scientific, economic, and political uncertainties associated with climate change justify taking proactive steps to mitigate GHG emissions rather than further inaction.

Public statements made by fossil fuel giants British Petroleum and Shell, as well as energy-intensive companies including American Electric Power, Boeing, 3M, Sun Microsystems, United Technologies, Toyota, and Weyerhaeuser, acknowledging the potential threat of climate change, catalyzed this change in attitude. The change in rhetoric reached a head when Senators John Chafee (R-RI), Connie Mack (R-FL), and Joseph I. Lieberman (D-CT) introduced the Credit for Voluntary Early Action Act (S. 2617) in October 1998. The Credit for Voluntary Early Action Act was designed to encourage the proactive involvement of businesses to reduce U.S. GHG emissions and to refocus the debate on what Congress should do about climate change rather than the advisability of the protocol.[21]

The sponsors of S. 2617 viewed the financial and political uncertainty of how corporate good deeds might be treated in the event of future GHG regulation as a potent deterrent to corporate activism and innovation that might otherwise help limit GHG emissions. Some businesses fear that if they voluntarily improve their efficiency now, they will encounter more difficulty achieving mandatory emission reductions in the future than their competitors who continue business-as-usual emissions. Therefore, one of the most compelling arguments for credit for early action legislation is that it would reduce the uncertainty that might otherwise deter some companies from taking voluntary, cost-effective actions to mitigate their GHG emissions.[22]

At the beginning of the 106th Congress, six Republicans and six Democrats introduced the Credit for Voluntary Early Reductions Act (S. 547), a bill nearly identical to S. 2617. The spectrum of criticism attracted by S. 547 underscores the difficulty of advancing any climate change legislation. At one extreme, many conservative climate change skeptics voiced their fear that the bill—though voluntary—would go too far too fast; at the other extreme, many environmental advocates complained that the bill would do too little too slowly to control emissions.

Critics on the right included former vice presidential candidate Jack Kemp, who alleged that S. 547 would pave the way for "back-door implementation" of the protocol.[23] Kemp asserted,

> Awarding early credits to companies that reduce emissions now might encourage a trend toward cutting back on GHG production. This would win points with the global warming crowd. And the plan also looks attractive to industries, which would find it advantageous to reduce the long-term costs of the Kyoto treaty, or domestic version of it, and secure some tangible gain in return.[24]

Many of those who hope that the climate change issue can be killed, including industry-funded lobbying groups such as the Global Climate Coalition and the Cooler Heads Society, benefit from the widely held perception that the protocol includes unreachable targets that would compromise the American economy. Any early credit bill that reduced compliance cost uncertainty represents a threat to their efforts defeat any form of climate change legislation.

Critics on the left complained that S. 547 fell well short of the effort necessary to adequately address climate change. Somewhat paradoxically, environmentalists contended that the United States would jump the gun on international climate change negotiations by passing S. 547. However, stakeholders interested in protecting earth's climate should favor the United States establishing itself as a leader in environmental stewardship by reducing its aggregate contribution to atmospheric GHG concentrations without waiting for an international consensus.

Some in the environmental community objected to the notion, embraced by

S. 547, of creating fungible emission credits as a means of minimizing compliance costs. Their criticism reflects the fact that under emission trading scenarios some individual sources could continue to pollute at high rates by purchasing emission credits from other sources for whom compliance is less expensive. This criticism is appropriate for emissions that concentrate adverse air pollution effects locally or regionally (e.g., sulfur dioxide contributing to acid rain); however, because carbon is a global pollutant, it makes no difference to the atmosphere where those emission reductions occur.

Bashing Carbon Regulation

Whereas S. 547 represented a moderate, albeit controversial effort to address the challenge of mitigating GHG emissions, the Small Business, Family Farms, and Constitutional Protection Act (H.R. 2221), introduced in June 1999 by Representative David McIntosh (R-IN), represented an attempt to simultaneously kill early credit legislation and the protocol. It provided that

> Federal funds may not be used to propose or issue rules, regulations, decrees, or orders or for programs designed to implement, or in preparation for implementing, the Kyoto Protocol to the United Nations Framework Convention on Climate Change before the date on which the Senate gives its advice and consent to ratification of the Kyoto Protocol.

In addition, H.R. 2221 would have explicitly prohibited the use of federal funds to support advocacy, development, or implementation of any early credit system before ratification of the protocol and prevented federal agencies from regulating carbon emissions. The implications of H.R. 2221 with regard to controlling air pollution or conserving energy were grave even for nonregulatory initiatives because any action construed as potentially paving the way for implementation of the protocol was at risk. Although H.R. 2221 was too extreme to pass, the vocal antiprotocol rhetoric of climate change skeptics thwarted early credit legislation in the 106th Congress.

Climate change skeptics in the House and Senate attempted to attach numerous antiprotocol riders to various spending bills in 1999 and 2000. One such amendment, offered by Senator Kit Bond (R-MO) on the Senate Interior Appropriations Bill, would have prevented implementation of President Clinton's federal energy efficiency executive order, which the administration estimated would save taxpayers about $750 million annually. However, because the amendment violated the recently reinstated Senate Rule XVI prohibition on amending appropriation bills with legislative language, the amendment failed.

The Knollenberg Language

Perhaps the most effective tactic used against the Clinton Administration's climate change policies came not as free-standing bill but as a legislative rider in the appropriation process. Representative Joseph Knollenberg (R-MI) inserted language designed to prevent "back-door implementation" of the protocol into numerous FY 2000 and 2001 appropriation bills. The Knollenberg language prohibited federal agencies from pursuing any effort that would contribute to meeting the goals of the protocol before ratification by stipulating that "none of the funds appropriated by this Act shall be used to propose or issue rules, regulations, decrees, or orders for the purpose of implementation or in preparation for implementation, of the Kyoto Protocol."

Opponents of the Knollenberg language, including the Clinton administration, contended that such language stifled many effective initiatives by expanding restrictions on Kyoto-related activities to include nonregulatory policies, programs, and initiatives. The breadth of activities covered by the Knollenberg language served to further confuse the climate change debate and to drain the administration's political capital and concentration on the issue of GHG emissions. The Knollenberg language affected the climate debate by shifting discussion from funding for climate protection initiatives proposed by the administration to the perceived shortcomings of the Kyoto Protocol. This rhetorical distraction, combined with the substantive prohibitions of the Knollenberg language, left the administration's climate change budget substantially weakened.

Calling for More Research

The corporate momentum that developed behind early crediting proposals early in 1999 served as a wake-up call for some climate change skeptics. In response to S. 547, Senator Frank Murkowski (R-AK) and nine other senators introduced the Energy and Climate Policy Act of 1999 (S. 882). Although the findings of S. 882 reiterated that the protocol "fails to meet the minimum conditions" of the Byrd–Hagel Resolution, they marked an important shift in skeptic rhetoric. Importantly, S. 882 found that "although there are significant uncertainties surrounding the science of climate change, human activities may contribute to increasing global concentrations of GHGs in the atmosphere, which in turn may ultimately contribute to global climate change beyond that resulting from natural variability."[25]

Although they acknowledged the threat of climate change, S. 882 and the similarly research-oriented Climate Change Energy Policy Response Act (S. 1776), introduced by Senator Larry Craig (R-ID), did not delineate a new vision for mitigating GHG emissions. For example, S. 882 would have provided $200 million annually between 2001 and 2010 for a "climate technology research, development, and demonstration program," replaced the Department of Energy's director of climate protection position with an Office of Global Climate Change, and required reevaluation and strengthening of the climate change provisions of EPAct. Under S. 1776 the secretary of energy would have to review myriad scientific issues and coordinate a new National Resource Center on Climate Change.

Critics of these bills agreed that long-term research and development are crucial to addressing climate change but contended that the bills did not include enough focus on deploying climate-friendly technologies now and therefore postponed an inevitable investment. In addition, naysayers pointed out that S. 882 provides no incentive, beyond public recognition already provided by the voluntary reporting provisions of EPAct, for companies to undertake GHG emission reductions. Critics also argued that S. 882 did nothing to provide baseline protection for companies that address GHG emissions or sequestration now and therefore would not stimulate significant climate benefits.

The Clinton administration and Congress battled to a draw over antiprotocol riders, and although hearings were held on various legislative proposals, no major climate change bills advanced in the 106th Congress. However, the opportunity cost of the 2 years of gridlock were a major victory for climate change skeptics and a frustrating setback for the administration.

Breaking American Ice on Climate Change

Fossil fuels are the energy lifeblood of industrialized society, and the inevitable transition to non–carbon-based energy systems is a profound challenge. But the magnitude of the challenge does not diminish the importance—some would argue the necessity—of minimizing anthropogenic impacts on the earth's climate. Meeting the climate change challenge is imperative for continued economic prosperity and a crucial component of our obligation to leave future generations a planet more healthy and productive, and a global economy more equitable and vibrant, than we enjoy today.

If American ingenuity and entrepreneurial spirit are quickly brought to bear on the climate challenge, the United States will enjoy a head start in the international race to develop the technologies that will fuel the sustainable economies of the future. In terms of economic competitiveness in the multi–billion-dollar global marketplace, it is a race that America can ill afford to lose. From an environmental perspective, it is a race in which the United States, by virtue of its disproportionate aggregate and per capita contributions to GHG pollution, must participate.

The necessity of U.S. participation in addressing climate change and the seemingly intractable state of the domestic policy debate require further discussion. The remainder of this chapter focuses on three keys to breaking the U.S. climate change policy logjam: overcoming the inertia inherent in the current uncertainty paradigm, encouraging and capitalizing on the trend toward increased corporate environmental commitments, and educating and cultivating climate-savvy decision makers.

The first challenge requires legislators to resist the use of scientific uncertainty as an excuse for postponing difficult decisions on climate change. The second challenge is to further catalyze and promote the greening of corporate America. The final, perhaps most difficult challenge is to enhance policymakers' understanding of the causes and consequences of climate change, a goal that relies heavily on overcoming the inertia engendered by uncertainty.

Overcoming the Inertia of Uncertainty

A common argument against mitigating climate change is that because climate science is uncertain, the extent to which efforts to protect the climate will prove beneficial remains unclear. Some climate skeptics contend that the United States should postpone climate change initiatives until research provides a better foundation for understanding climate change and an improved basis for developing a national strategy to mitigate the consequences of climate change.

The problems with this uncertainty argument are threefold. First, the idea that someday we will have a perfect understanding of the climate system is a fallacy. Second, the substantial opportunity cost incurred by deferring climate change mitigation efforts is compounded by the fact that GHGs remain in the atmosphere for years to millennia. Finally, the idea that decisions should not be made with imperfect information runs contrary to the reality of the public policymaking process and the precautionary principle.

Scientists will never completely understand complex natural systems. For example, there are human diseases, such as some cancers, that medical professionals do not entirely understand. Even when they know the causes and symptoms of a given disease, doctors often cannot cure it. However, even when no cure exists, physicians work to minimize incidence of the disease and treat the associated symptoms.

The earth's climate is a complex system, and scientists will never completely understand how it works. However, climate scientists do understand many of the key factors that affect the earth's climate and strive to expand that knowledge daily. Policymakers should not use imperfect knowledge of earth's climate as an excuse to abdicate their responsibility to oversee climate change mitigation efforts. On the contrary, scientific uncertainty coupled with socioeconomic uncertainty regarding the potential impacts of climate change represent a strong argument in favor of precautionary action.

Once policymakers recognize that a wait-and-see approach to climate change is a risky strategy, the obvious question is what to do in light of the substantial uncertainties. Policymakers must dismiss the notion that they will find a silver bullet that solves the problem of climate change; moreover, they must accept that the first policy steps certainly will prove either imperfect or incomplete. Therefore, policymakers must design initial policy actions to lead the United States in the right direction but allow enough flexibility to change course when climate protection goals become more clear and scientific understanding of the earth's climate improves. In other words, policymakers need to take steps that begin mitigating GHG emissions now but remain mindful that prescribing a long-term climate agenda probably would prove counterproductive.

Creating a Climate Change Constituency

Jack Kemp's observation that early credit legislation would sow the seeds of a climate change constituency is on target. Early crediting proposals are designed to remove barriers to action by providing credits valid under any future regulatory regime. However, any GHG mitigation credit is valueless until a regulatory regime takes effect. Therefore, the more credits a company amasses by mitigating its GHG emissions now, the stronger that company's incentive for supporting a future regulatory regime.

The advance planning needed by major corporations makes it imperative that they plan for current and future regulatory regimes. Some businesses view the regulation of GHGs as an inevitable constraint on their future activities. As a result, regardless of whether they welcome such regulation, they hope to help shape or at least anticipate the parameters of that regulatory regime. By doing so, businesses can anticipate and manage their GHG control costs.

When a critical mass of business interests steps forward asking to undertake proactive climate change mitigation in trade for a greater degree of regulatory certainty provided by early credit legislation (or some other baseline protection program), policymakers will respond by sponsoring such legislation. At that time, some environmentalists undoubtedly will criticize any program as too weak on polluters, and some companies probably will demand a more lenient credit scheme. However, all sides will be forced to reach a compromise that provides at least modest climate protection and regulatory certainty.

Looking for Leadership

The legislative machinery of Capitol Hill is not well tuned for resolving complex, contentious environmental issues in an expedient fashion. On the contrary, although the legislative system usually prevents bad legislation from passing, the process does not promote the passage of constructive legislation. In the absence of an obvious environmental crisis, most complex environmental laws result from many years of methodical deliberation. This protracted legislative deliberation helps prevent policymakers from forcing poorly conceived bills into law, but it also guarantees that passing any climate change legislation in a timely fashion will be a difficult challenge.

Passage of proactive climate change legislation would be greatly aided by either a major climate change–related catastrophe or a consensus by a critical mass of legislators. Successful consensus-building efforts require a combination of significant staff time and genuine commitment and willingness of legislators to compromise. If climate change legislation results from consensus building, it will take much longer than if it stems from an environmental emergency. However, developing and reaching agreement on a sophisticated climate change bill will take time. In light of the notoriously high rate of Hill staff turnover, unless a group of legislators takes personal interest in climate change legislation, the odds of reaching such a deal seem remote.

Alternatively, if a coalition of progressive companies and moderate environmentalists steps forward with a reasonable plan for addressing climate change, viable legislation could be developed and enacted within a given Congress. The involvement of businesses in such an initiative will be critical because without business support, many legislators will dismiss any climate change legislation as either an environmentalist ploy to reduce American resource consumption or an unsubstantiated, and therefore unimportant, scientific notion.

Conclusions

Despite the growing body of scientific evidence that indicates that climate change poses a very real threat to the global environment, the 106th Congress failed to move any meaningful climate change legislation. For its part, the Clinton administration was unable to force Congress to engage constructively on the issue. These failures, combined with the election of President George W. Bush, represent a tremendous source of concern for environmentalists interested in climate change.

Momentum in Congress has shifted away from credit for early action proposals to a four-pollutant regulatory approach endorsed by President Bush during his campaign (a commitment he abandoned early in his term). In the Congress, bipartisan legislation to simultaneously control emissions of mercury, nitrogen, sulfur, and carbon such as bills introduced by Representative Thomas Allen and 19 cosponsors (H.R. 1335) and Senator James Jeffords and 14 cosponsors (S. 556) has already been introduced in the 107th Congress. In all likelihood, however, some version of credit for early action will resurface as any bill to regulate carbon emissions moves forward.

Whether policymakers disavow the safety blanket of scientific uncertainty, businesses recognize their potentially tremendous climate change liability. Whether Congress and the Bush administration decide to advance environmental issues will determine whether the United States assumes a leadership role in the international effort to protect the earth's climate.

If Congress and the White House agree to tackle the climate change challenge together, the Kyoto Protocol or something like it will follow the Montreal Protocol into history as a case study in international environmental stewardship and global cooperation. If no such accord is reached, the Kyoto Protocol will join the Comprehensive Test Ban Treaty as a failed ratification vote by which the United States abdicates its international leadership responsibility to the detriment of the global environment, the global community, and future generations.

Notes

1. Intergovernmental Panel on Climate Change (IPCC), 1990: *Climate Change: The IPCC Scientific Assessment,* J. T. Houghton, G. J. Jenkins, and J. J. Ephraums (eds.) (Cambridge, England: Cambridge University Press).
2. Ibid.
3. United Nations Framework Convention on Climate Change, 1992.
4. *Congressional Record,* 1992: S. 17, 156, *Daily Edition,* October 7, 1992.
5. Ibid.
6. The Energy Policy Act of 1992 was considered by Congress at the same time the FCCC was being developed in international negotiations.
7. National Academy of Sciences, 1991: *Policy Implications of Greenhouse Warming* (Washington, DC: National Academy Press).
8. Climate Change Action Plans. Bureau of Oceans and International Environmental and Scientific Affairs Office of Global Change, 1992: National Action Plan for Global Climate Change, Departments of State Publication 10026.
   Clinton, William, J., and Albert Gore, Jr., 1993: The Climate Change Action Plan.
   Clinton, William, J., and Albert Gore, Jr., 1997: Climate Action Report, 1997 Submission of the United States of America Under the United Nations Framework Convention on Climate Change.
9. Ibid.
10. http://www.unfccc.de/resource/conv/conv_002.html.
11. Intergovernmental Panel on Climate Change, 1994: *Climate Change 1994: Radia-*

tive Forcing of Climate Change and an Evaluation of the IPCC IS92 Emissions Scenarios, J. T. Houghton, L. G. Meira Filho, J. Bruce, H. Lee, B. A. Callander, E. Haites, N. Harris, and K. Maskell (eds.) (Cambridge, England: Cambridge University Press).

12. Intergovernmental Panel on Climate Change, 1996: *Climate Change 1995: The Science of Climate Change: Contribution of WGI to the Second Assessment Report of the Intergovernmental Panel on Climate Change,* J. T. Houghton, L. G. Meira Filho, B. A. Callander, N. Harris, A. Kattenberg, and K. Maskell (eds.) (Cambridge, England: Cambridge University Press).
13. Gelbspan, R., 1998: *The Heat Is On* (Reading, MA: Perseus Books).
14. *Congressional Record,* July 27, 1997, pp. S8113–8138.
15. *Congressional Record,* July 27, 1997, pp. S8113–8138 and June 12, 1997, pp. S5622–S5626.
16. *Congressional Record,* July 27, 1997, pp. S8113–8138.
17. http://www.unfccc.de/resource/docs/convkp/kpeng.html.
18. Ibid.
19. Ibid.
20. IPCC, 1996; National Academy of Sciences, 2001: *Climate Change Science: An Analysis of Some Key Questions* (Washington, DC: National Academy Press).
21. *Congressional Record,* October 12, 1998, pp. S12309–12312.
22. For an analysis of several early credit proposals, see Nordhaus, R. R. and S. C. Fotis, 1998: *Early Action and Global Climate Change: An Analysis of Early Action Crediting Proposals* (Washington, DC: Pew Center on Global Climate Change).
23. Kemp, J. and F. L. Smith, Jr., 1999: "Beware the Kyoto Compromise," *New York Times,* January 13. Section A, p. 19.
24. Ibid.
25. *Congressional Record,* April 27, 1999, pp. S4267–4270.

CHAPTER 9

Population and Climate Change Policy

Frederick A. B. Meyerson

The rapid rise of atmospheric carbon dioxide levels and average global surface temperatures in the twentieth century was accompanied by the most dramatic human population increase in history—almost 300 percent between 1900 and 2000. Although the causal links between these rising trends are complex, the underlying facts are straightforward. As the global population increased from 1.6 billion in 1900 to 2.5 billion in 1950 and 6.1 billion in 2000, people progressively consumed greater quantities of fossil fuel. At the same time, we have expanded agriculture, deforestation, the production of certain chemicals, and other activities that produce carbon dioxide and other greenhouse gases (GHGs).

Through the 1960s, global emission increases were the product of both rising population and per capita increases. Since 1970, average per capita emissions have been stable, so that on a global scale the rise in industrial carbon dioxide emissions over the last three decades correlates closely with population growth.[1] Population trends and policy therefore have played a major part in the trajectory of past emissions, and they could have an even greater role in the future.[2] The size of the human population and its activities in the twenty-first century will be a key factor affecting the extent of emissions, biotic carbon sinks, and climate change. Similarly, the impact of warming on humanity will be greatly affected by population size: Larger numbers of humans will effectively reduce the options for mitigating or adapting to sea level rise, changes in precipitation patterns, and other projected byproducts of warming.[3]

The Effect of Climate Change on Population

The projected impacts of climate change on the human species are a serious concern. Rising global surface temperatures and changes in precipitation magnitude, intensity, and geographic distribution may well redraw the world renewable resource map. The Intergovernmental Panel on Climate Change (IPCC) best-estimate scenario also projects a sea level rise of about half a meter by 2100 (with a range of 9–88 cm), substantially greater than the increase over the last century.[4] The human and ecological impacts of rising oceans would be substantial, including increased flooding, coastal erosion, aquifer salinization, and loss of coastal cropland, wetlands, and living space. The intensity and frequency of hurricanes and other hazardous weather may also increase, endangering the growing human population in coastal areas, although there is still scientific uncertainty in this area.[5]

Whether or not these climatic changes affect net global agricultural production, they are almost certain to shift productivity between regions and countries and within nations.[6] For example, projections developed in one study indicate that although net U.S. agriculture production may not be diminished by global warming, certain regions of the country are likely to suffer as a result of changes in precipitation and temperature.[7] This suggests that climate change policy must address the demographic issues related to changing regional and national fortunes as well as the aggregate global economic and biological impact.

A warming climate also poses a significant public health threat. Higher average temperatures mean longer and more intense heat waves, with a corresponding potential for more cases of severe heat stress. The redistribution of precipitation patterns would markedly increase the number of people living in regions under extreme water stress, a problem that would be compounded by increasing population.[8] The geographic range of temperature-sensitive tropical diseases, such as malaria and dengue fever, would also expand.[9]

The combined effects of population growth and climate change can produce regional resource shortages, which in turn could result in the exploitation of environmentally sensitive areas such as hillsides, floodplains, coastal areas, and wetlands.[10] These conditions may also increase the numbers of environmental refugees, international economic migration, and associated sociopolitical challenges.[11] In general, the geographic distribution and movement of people in the twenty-first century, as well as their absolute numbers, are likely to be significant climate policy and environmental issues.

Historical Population and Emission Trends

Human population grew from about 1 billion in 1800 to more than 6 billion in 2000, and in that period growth rates reached levels unprecedented in the 2-million-year history of our species (Fig. 9.1). The global population growth rate peaked around 1970 at 2.1 percent per year and declined slowly to 1.3 percent by the end of century, but the 1980s and 1990s saw the greatest numbers of added people, almost 800 million in each decade.[12]

During the last two centuries, the human species has evolved from a global culture dominated by agriculture and the use of domestic animals for work and travel to a society propelled by and dependent on fossil fuels. These two phenomena, the population explosion and the energy/industrial revolution, are inextricably intertwined. It is unlikely that the human population would have reached its present levels in the absence of the discovery and exploitation of fossil fuels. Likewise, the needs of the growing population have provided an ever-expanding market for oil, gas, and coal exploration and production.

In this exceptional era, anthropogenic dominance of the biological assets of the planet has grown to the point where humans now use or dominate an estimated 39–50 percent or more of terrestrial biological production through agriculture, forestry, and other activities.[13] In achieving this preeminence, we have also decreased the ability of the global ecosystem to absorb and store carbon by replacing complex natural ecosystems such as tropical forests and tallgrass prairies with much simpler, less diverse, lower-biomass agricultural systems such as cornfields. This process of ecosystem simplifcation, though it supports more

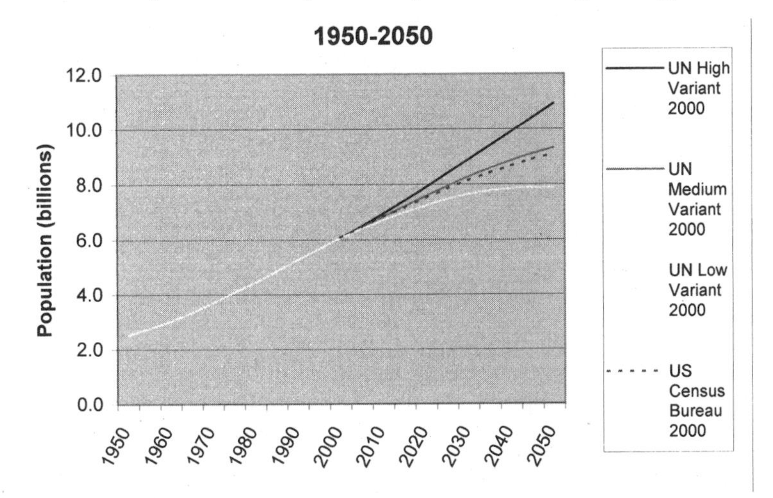

FIGURE 9.1. World population history and projections, 1950–2050.

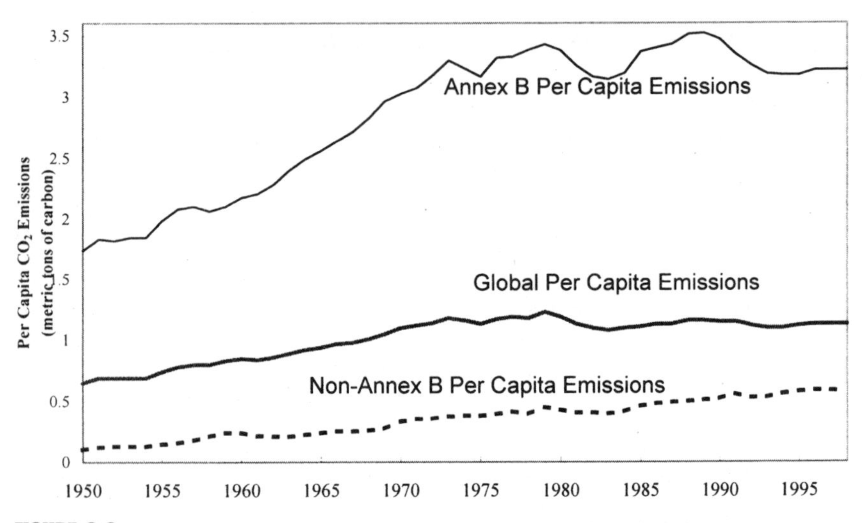

FIGURE 9.2. Global, Annex B, and non–Annex B per capita carbon dioxide emissions 1950–1998.

people and greater per capita consumption, has also raised the species extinction rate to 100 to 1,000 times the historical rate.[14]

Global emissions of carbon dioxide (the dominant GHG) from fossil fuel combustion and related activities grew from 8 million metric tons (Mmt) in 1800 to 534 Mmt in 1900 and 6,608 Mmt in 1998.[15] Over the same period, global per capita emissions have risen from less than 0.01 metric ton (mt) in 1800 to 0.3 mt in 1900 and 1.13 mt in 1998.[16]

The trends in population and emissions since 1950 are particularly relevant with respect to present and future climate policy. Whereas population has more than doubled during this period (from 2.5 to 6 billion), global per capita carbon dioxide emissions have risen by only 74 percent. More importantly, all of this per capita rise took place between 1950 and 1970. Since then, global per capita emissions have fluctuated in a narrow range between 1.1 and 1.2 mt per capita and averaged 1.14 mt from 1970 to 1998 (Fig. 9.2).[17] The obvious result is that the increase in global carbon dioxide emissions (62 percent) between 1970 and 1998 correlates closely with population growth over that period (57 percent).[18] At the global scale, as well as within the United States by itself, greater and more widespread affluence (consumption) has effectively canceled out gains in energy efficiency.[19] Should this phenomenon continue (stable per capita carbon dioxide emissions), population growth would continue to be a key determinant of future emission increases.[20]

Population Projections to 2050

Long-term human population projections are extrapolations of past fertility and mortality trends, combined with educated guesses about future family size preference; population policy; availability of reproductive health services and abortion; and general human health conditions. Each of these factors will in turn be affected by future economic and political conditions, which are difficult to predict. Finally, future environmental changes, including the effects of climate change itself, may alter both fertility and mortality.

These areas of demographic uncertainty are balanced to some degree by biological factors. With very few exceptions, women's primary childbearing years fall between 15 and 44 years of age, which means that the entire potentially reproductive human population is known 15 years in advance. Also useful but somewhat less reliable has been the historical tendency for the global fertility rate (the number of children per woman) to change slowly, from five children in the early 1950s, to four in the late 1970s, and less than three in the late 1990s. Similarly, global average human life expectancy has followed a fairly smooth slow upward trend, from 46 years in the early 1950s to 65 in the late 1990s.[21] These fertility and mortality trends are projected to continue, but of course there is no guarantee.

As a result of these consistent past patterns in fertility and life expectancy, as well as population momentum,[22] demographers can provide a credible (although not infallible) range of population projections for a reasonable period into the future.[23] Among the facets of measurable human behavior, near- and mid-term population projections are likely to be far more reliable than, for instance, projections of economic trends, for which not only the rate but also the direction (positive vs. negative) can change precipitously from year to year.

The most recent UN demographic projections suggest that by 2050, the global population could be as low as 7.9 billion and declining or as high as 10.9 billion and continuing to increase rapidly, although both of these extremes are unlikely. The medium variant scenario, often incorrectly called a prediction, projects a population of 9.3 billion by 2050, which would be an almost 50 percent increase over the current level (Fig. 9.1).[24] All three variants assume a continuation of the global fertility rate decline, from 2.82 children in the late 1990s to 2.62 (high variant), 2.03 (medium variant), and 1.68 (low variant) in the 2045–2050 period. Note that the small difference in the fertility assumption between the high and low variants, of less than one child per woman, results in a net difference of 3.4 billion people by 2050, as much as the entire global human population in the mid-1960s.

One important implication relevant to climate change is that the size of the global population a few decades from now could potentially be greatly affected by population policy, which in turn could have a major impact on both GHG emissions and our ability to adapt to global warming.[25] This conclusion is supported by an analysis of global reproductive health and contraceptive prevalence data. As a general rule, the fertility rate tends to reach or decline below replacement level (2.1 children per woman) in countries in which contraceptive prevalence rises to about 70 percent. Even today, however, only about 50 percent of the world's women use modern contraception, and the rate is below 10 percent in a significant number of the least-developed countries.[26]

For this reason, one of the most cost-effective long-term GHG emission reduction strategies may be to provide reproductive health services to the hundreds of millions of people who still do not have adequate access.[27] The climate-related costs associated with each future birth have been estimated to be in the range of several hundred to several thousand dollars. This amount is in the same general range as the estimated costs for providing universal access to contraception, calculated on a per-birth-averted basis.[28]

One complicating demographic factor is that even in countries with high levels of contraceptive use, such as the United States (where 67 percent of women use modern contraceptive methods), almost half of pregnancies are unplanned or unintended.[29] The principle cause of this counterintuitive phenomenon is that many methods of modern contraception are not perfect even under ideal conditions; moreover, women and men do not use contraception consistently or carefully. The composite actual (as opposed to theoretical) annual effectiveness for contraception in the United States is estimated to be 91 percent, which means that on average, about 1 out of every 11 women will get pregnant each year while using contraception.[30] American women who use contraception continuously during their reproductive years are estimated to have, on average, 1.8 contraceptive failures (unintended pregnancies) during their lives.[31] In the United States, approximately 50 percent of all unintended pregnancies are carried to term, and the other half result in abortions.[32] Reported rates of unplanned pregnancy vary from country to country, but throughout the world a substantial proportion of pregnancies are unintended.

The demographic implications for both the United States and world are substantial. The high failure rate of contraception in actual use means that women continue to have more pregnancies than they would otherwise choose, and unless they have access to and then choose abortion, they are having unintended children. This is a global phenomenon; the data for developing countries also suggest that women on average have about one more child than their stated family size preference.[33]

One consequence is that abortion has been and continues to be a demographic factor of enormous worldwide significance. There are an estimated 46–60 million abortions per year globally (both legal and illegal), a number close to the current net world population increase of 78–84 million per year.[34] Therefore, for the foreseeable future and until the actual effectiveness of contraception can be improved, continued movement toward population stability is likely to depend in substantial part on access to abortion.

The underlying message is that future trends in the availability and effectiveness of contraception and access to abortion services will have a major impact on whether the population in 2050 most resembles the high, low, or medium UN scenario.[35] Fortunately, reproductive health care and family planning are areas in which improvements are possible in many parts of the world with a small economic investment compared with other social expenditures.[36]

The 1994 UN Cairo Population Agreement

Every 10 years, the UN holds an international population conference. Most recently, in 1994, representatives of more than 180 countries met in Cairo at the International Conference on Population and Development (ICPD). The Programme of Action signed at that conference reaffirmed broad principles of human sexual and reproductive rights and also set a number of specific goals. One ICPD target is that by 2015 all countries should provide universal access to a broad range of safe and reliable family planning methods.[37] As of 2000, the populations of many developing countries still lack sufficient access to reproductive health services, although the situation in most nations has been improving slowly.[38]

Based on the range of historical experience in both developed and developing countries, it is reasonable to suggest that universal access to family planning, if achieved by 2015, could result in a global fertility rate at or below replacement level by the mid–twenty-first century. Universal access in turn would be likely to result in a global population by 2050 closer to the low or medium than the high UN projection.

The estimated cost of achieving the reproductive health goals of the Cairo (ICPD) agreement, including universal access to family planning, is estimated to be $17 billion per year in 2000 and $22 billion annually by 2015.[39] Several analyses have estimated that these modest investments in family planning would be cost-effective in climate policy terms.[40] Cost estimates per unwanted birth averted are in the range of hundreds to a few thousand dollars, which would place the cost of fertility reductions in approximately the same range as that of other GHG emissions reduction strategies.[41] An advantage of this approach is that most policymakers consider that providing universal access to family planning to be a no-regrets approach that would provide other environmental, economic, social, and human rights benefits.

Progress toward the ICPD goals has been mixed. The 180 countries that met in Cairo agreed that the developing countries should provide two-thirds of the projected $17 billion annual cost by 2000, with developed countries donating the remaining third. Unfortunately, as of 2000, less than $10 billion was being spent annually on family planning and other Cairo-related goals in the developing world. A major reason for this shortfall is that developed countries have lived up to only about one-third of the commitment they made in Cairo.[42] There has been a general downward trend in international development aid by industrialized nations in the 1990s. The United States, historically the largest single donor country, has cut its international family planning assistance by about 35 percent since fiscal year (FY) 1995, despite a period of unparalleled domestic prosperity.[43] As a result, the United States provided only about $392.5 million in international family planning funding in FY 2000, rather than the approximately $1 billion per year U.S. share envisioned by the Cairo agreement.[44]

Although there is an international consensus that improving reproductive health and family planning has positive economic, social, and environmental effects, a few fundamentalist countries, along with the Vatican, have been able to slow down progress toward many of the Cairo goals. Some of the Cairo opponents are against family planning in general for religious reasons, others are opposed to abortion, others want to limit sexual education, some are opposed to free speech about reproductive health topics, and others fear coercion by family planning providers. Although these countries constitute a very small minority in terms of their number and relative population size, the bureaucratic structure and political realities within the UN system give them disproportionate power. A parallel situation exists in the U.S. Congress, where a handful of House members opposed to family planning assistance, domestic and international, have been able to block or weaken population-related legislation in the late 1990s and early twenty-first century.

Population and the Kyoto Protocol

The 1997 Kyoto Protocol to the Framework Convention on Climate Change, if ratified, would commit 38 developed (Annex B) countries to cut their national GHG emissions by an average of 5.2 percent between 1990 and 2008–2012 (herein after referred to as 2010).[45] Developing (non–Annex B) nations face no specific emission limitation obligations in the protocol, on the principle that industrialized nations have contributed the most to the problem and thus have an obligation to take the first steps.[46]

Population is not specifically referenced in the Kyoto Protocol, but it will play a major role in terms of ratification of and compliance with the protocol, future climate policy negotiations, and the viability of both. The protocol is based on national caps; these will not be adjusted for increases or decreases in population caused by either fertility or migration between 1990 and the end of the first commitment period (2008–2012). Because population increases result in more houses, cars, and other consumption that produces GHG emissions, countries with rising populations are at a comparative disadvantage under the national cap formula used in Kyoto.[47]

The Kyoto national caps were negotiated on the basis of 1990 emission levels of the Annex B countries, at which point per capita emissions among these countries averaged 3.24 mt, approximately three times the global average and six times the per capita average among non–Annex B countries (Fig. 9.2).[48] The Kyoto Protocol also preserved as benchmarks the wide array of national per capita carbon dioxide emission levels among the Annex B countries for 1990, from Portugal (1.17 mt per capita) to Luxembourg (6.88 mt per capita).[49] It is unclear from an equity perspective why such dissimilar countries should be treated equally, but in fact the Kyoto Protocol requires both Portugal and Luxembourg to reduce their national emissions by 8 percent from 1990 levels.[50] The Kyoto national cap formula thereby effectively creates an entitlement for countries with the highest per capita emission levels, such as Luxembourg and the United States (5.18 mt per capita in 1990), and an effective ceiling for low-emission Annex B countries.[51]

Demographic trends further complicate the per capita inequities established by the Kyoto national cap formula. Within the Kyoto Protocol Annex B group, some countries' populations are growing rapidly, others are stable, and some have declining populations (Table 9.1). The net effect over the Kyoto first commitment period (1990–2012) is that some countries with fast-growing populations, including the United States, must dramatically reduce their per capita emissions, and others with declining population can actually become less GHG-efficient.[52]

These demographically driven windfalls and penalties appear not to have been adequately anticipated or appreciated by the Kyoto Protocol negotiators.[53] However, if the protocol enters into force and its terms are met, the demographic effect on allowable per capita emissions among Annex B countries will be substantial. For instance, the United States, with 1990 per capita emissions of 5.17 mt, would be effectively allowed only 3.89 mt by 2010, primarily

TABLE 9.1. Selected Annex B and non–Annex B Countries. Population Changes 1990–2010, 1990–2050, Kyoto Per Capita Effects

Country	*%/1990 Em. Required by 2010*	*1990 Pop. (millions)*	*2010 Pop. proj. (U.N. medium)*	*% Change (population) 1990–2010*	*%/per cap. CO_2* reduction *required 1990–2010*	*2050 Pop. proj. (U.N. medium)*	*Change (population) % 1990–2050*
Annex B:							
Germany	92	79.4	81.4	2.5%	11%	70.8	–10.8%
Italy	92	57.0	56.4	–1.1%	7%	43.0	–24.7%
Japan	94	123.5	128.2	3.8%	10%	109.2	–11.6%
New Zealand	100	3.4	4.0	20.3%	20%	4.4	32.1%
Russian Federation	100	148.3	137.0	–7.6%	–8%	104.3	–29.7%
United States	93	249.4	308.6	23.7%	33%	397.1	59.2%
Non–Annex B:							
Brazil	n/a	147.9	191.4	29.4%	n/a	247.2	67.1%
China	n/a	1155.3	1366.2	18.3%	n/a	1462.1	26.6%
India	n/a	850.8	1164.0	36.8%	n/a	1572.1	84.8%
Indonesia	n/a	182.8	237.7	30.0%	n/a	311.3	70.3%
Mexico	n/a	83.2	112.9	35.6%	n/a	146.7	76.2%
Nigeria	n/a	87.0	146.9	68.8%	n/a	278.8	220.3%
Pakistan	n/a	119.2	181.4	52.2%	n/a	344.2	188.8%
Republic of Korea	n/a	42.9	49.6	15.8%	n/a	51.6	20.3%

Source: U.N. Population Division 2000 (medium variant).

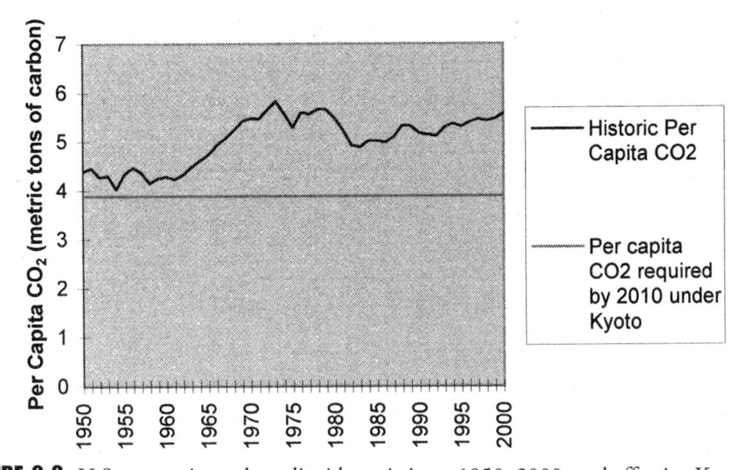

FIGURE 9.3. U.S. per capita carbon dioxide emissions, 1950–2000, and effective Kyoto target for 2010.

because its population its projected to increase by 59 million between 1990 and 2010 (Table 9.1).[54] This would require a 33 percent reduction of per capita emissions by the United States by the end of that period (Fig. 9.3).

The Russian Federation will be allowed to increase its per capita emissions by 8 percent from 1990 to 2010 because of the projected decline of Russia's population by 7.6 million during that period. New Zealand, at the lower end of the Annex B per capita emission spectrum (2.1 mt per capita in 1990), will have to reduce its per capita emissions by 20 percent from 1990 to 2010 because of its projected 20 percent population increase (Table 9.1).[55] Overall, the effective Annex B per capita emission allotments by 2008–2012 will be significantly different than they were in 1990, but they will make no more sense from an equity or environmental standpoint than they did in that year.

These two demographic factors—the initial per capita inequity established by the Kyoto Protocol and the greatly different population growth trajectories of the Annex B countries—put additional strain on an already problematic and politically besieged international environmental agreement. Changing demographics may well force countries with growing populations to seek nondomestic emission reduction solutions, which are controversial or unpopular for a variety of reasons. For instance, emission trading may well result in unforeseen and seemingly accidental financial windfalls for certain countries, and the credits under the Clean Development Mechanism, if allowed, may be difficult to measure and verify. In this regard, demographic distortions are at best distracting and unfair; at worst they could damage the protocol beyond repair.

The United States, in particular, seems almost certainly incapable of reducing its domestic emissions within the remaining years before the 2008–2012 period. From 1990 to 2000, U.S. carbon dioxide emissions rose somewhat more quickly than the country's population (17 percent and 13 percent, respectively).[56] Per capita emissions in the United States have therefore actually risen since 1990 (5.17 mt per capita) to 5.58 mt in 2000, although this spread will be reduced due to the greater than expected population increase recorded in the 2000 Census (Fig. 9.3).[57] All five of the projections prepared by the U.S. Department of Energy in 1999 suggest that the close correlation between emissions and population growth is likely to continue for the next two decades.[58]

The drop in U.S. emissions effectively required by the Kyoto Protocol, to 3.89 mt per capita by 2010, would be a more than 30 percent decrease in 10 years, to a per capita level not seen in the United States since before 1950. The largest single drop in per capita emissions in recent U.S. history occurred in the 11-year period between 1973 and 1983, when per capita emissions dropped by 18 percent as the result of the economic hardships caused by the OPEC oil embargo of the 1970s and the severe recession of the early 1980s.[59] It is unlikely that those conditions will be repeated, at least not voluntarily, so the reality of the protocol seems to drive the United States towards nondomestic solutions or, worse yet, nonratification. It can be reasonably argued that the likely U.S. failure to meet the Kyoto Protocol requirements has primarily demographic roots (its 59 million [24 percent] population increase from 1990 to 2010) and that this alone may be enough to sink the protocol.

Twenty-First-Century Projections: Climate Policy in a Demographically Diverging World

The sharply diverging demographic trends among the major Kyoto Protocol Annex B countries are projected to continue after 2012 and, if anything, increase. This is likely to render the 1990 benchmark emission formula and approach even less viable in the future. The U.S. population, for instance, is projected to rise from 249 million in 1990 to 397 million in 2050 (middle scenario), a 59 percent increase.[60] Meanwhile, Germany is projected to experience a population decline from 79 to 71 million over the same period, an 11 percent decrease, and the Russian Federation is projected to fall from 148 to 104 million people, an 30 percent decrease.[61]

Population projections for the developing world (non–Annex B countries) vary even more dramatically. For example, Pakistan's population is projected to rise from 119 to 344 million between 1990 to 2050 (a 189 percent rise), whereas South Korea's population is projected to grow only from 43 to 52 million (a 20 percent increase) over the same period (Table 9.1).[62] How to equitably and practically incorporate and adjust for these substantial projected differences in population is an unresolved piece of the climate policy puzzle. A simple extension of the Kyoto approach is not likely to work.

Recent trends in per capita emissions are also relevant. For the developed (Annex B) countries as a whole, per capita emissions have been relatively flat since 1970, fluctuating above 3 mt per person. In 1950, the developing (non–Annex B) country per capita average emission was only 0.1 mt, but it increased to 0.6 mt by 1997 and continues to rise (Fig. 9.2).[63] Developing country emissions are still far lower than those of developed countries on a per capita basis, but the gap is narrowing, from 17:1 to 5:1 from 1950 to 1997,[64] and this trend is expected to continue.

One factor driving this closing gap is the projected decrease in average household size as family size drops in developing countries. Significant economies of scale in energy use are lost as household size decreases (and therefore the number of households per capita increases), as it already has in the United States and other developed countries. In 1990, average household size in developed and developing countries was 2.7 and 4.8 people, respectively. By 2050, one analysis projects that the ratio may be only 2.6 to 3.4 people per household.[65] An important related factor is population aging, an inevitable byproduct of demographic transition and increasing life expectancy, which has significant implications for household and per capita GHG emissions.[66]

At present, the current emission spectrum between rich and poor countries is as wide as the differences in per capita income. In 1995, the 20 percent of the world's population living in countries with the highest per capita emissions contributed 63 percent of the world's fossil fuel CO_2 emissions. The low emitters—the 20 percent of the world's population in countries at the opposite end of the spectrum—contributed just 2 percent of global fossil-fuel CO_2 emissions.[67] Despite the narrowing trend, substantial variations between regions and countries are likely to persist, not only because of economic reasons but also because of differences in climate, geography, and natural resources.

Further complicating the equity issues raised by Kyoto, some non–Annex B countries, such as South Korea and South Africa, already produce per capita emissions that exceed those of some Annex B countries, such as Portugal, Switzerland, and Romania. China, with its enormous population and rising per capita emissions, is projected to surpass the United States as the largest contributor to GHGs within the next few decades and may well pass some Annex B countries in per capita emissions in that period.

The aggregate emissions of the developing countries are rising rapidly and are expected to surpass those of the developed countries within the first few decades of the twenty-first century as a result of both rising population and per

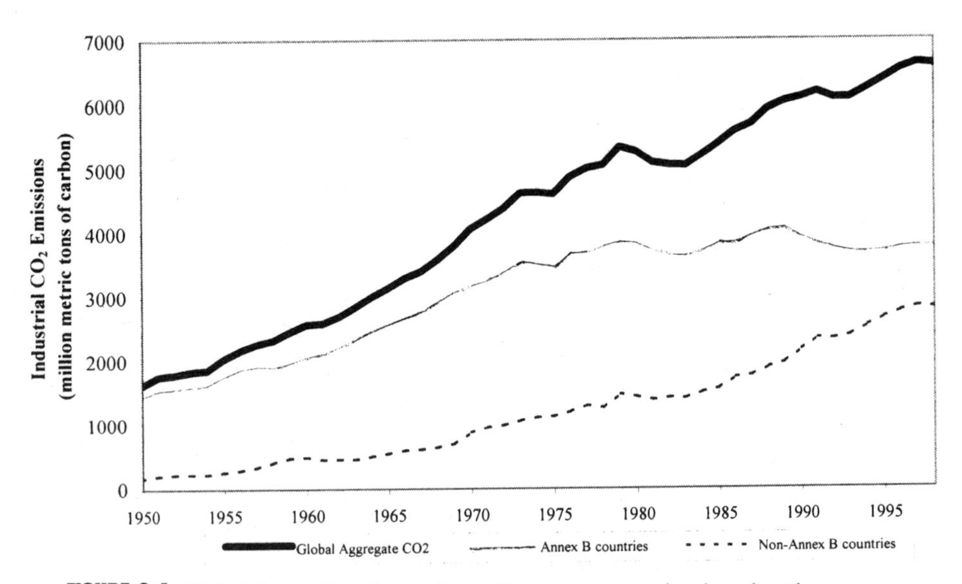

FIGURE 9.4. Global, Annex B, and non–Annex B aggregate annual carbon dioxide emissions, 1950–1998.

capita emissions in the developing world (Fig. 9.4). Clearly, no long-term climate policy can be effective without incorporating both the developed and developing world under an equitable GHG plan and global emissions cap or trajectory of caps.

Beyond the Kyoto Protocol: Incorporating Demographic Change into Climate Policy

Beyond the important first step that the Kyoto Protocol represents, it is evident that a future global climate change agreement will need to incorporate the physical and political reality of human population growth and decline, international migration, and changing relative levels of per capita emissions.[68] Almost all additional significant population growth is projected to occur in developing countries (the notable exception being the United States).[69] Although developed countries have been the dominant source of GHG emissions in the past, developing country emissions will become the major factor early in the twenty-first century, and a future treaty will need to respond to this coming demographic reality.

A few general principles and concepts come to mind. First, the next agreement should have an explicit, scientifically informed environmental goal. This necessarily means a global GHG emission cap, which will also effectively create a global per capita standard. One of the fundamental flaws of the Kyoto Protocol is that it caps the emissions of only certain countries, so that even if ratified and fully complied with, it would make no appreciable difference in 2012 atmospheric greenhouse gas concentrations or the long-term emission trend.[70] It has been difficult to market the Kyoto plan even to environmentalists for this reason.

Realistically, an explicit environmental goal that leads to actual reductions in emissions and atmospheric GHG concentrations can be accomplished only over many decades.[71] This longer-term approach should naturally put demographic change into focus as a factor that will affect both overall per capita emission levels and the relative potential rights and responsibilities of individual countries.

Second, future population growth matters a great deal in terms of emissions and climate change policy. Additional population growth is almost inevitable for several decades, barring an unforeseen catastrophe. However, the extent of future population growth can almost certainly be substantially affected by national and international population policy and family planning assistance. And global population growth and size will determine the allowable global per capita emission level under any GHG stabilization or reduction plan. For instance, a return to 1990 global emission levels by 2050 would require a 0.78 mt per capita level under the UN (2000) low variant population scenario but a much more difficult 0.56 mt per capita under the high variant scenario (Fig. 9.5).

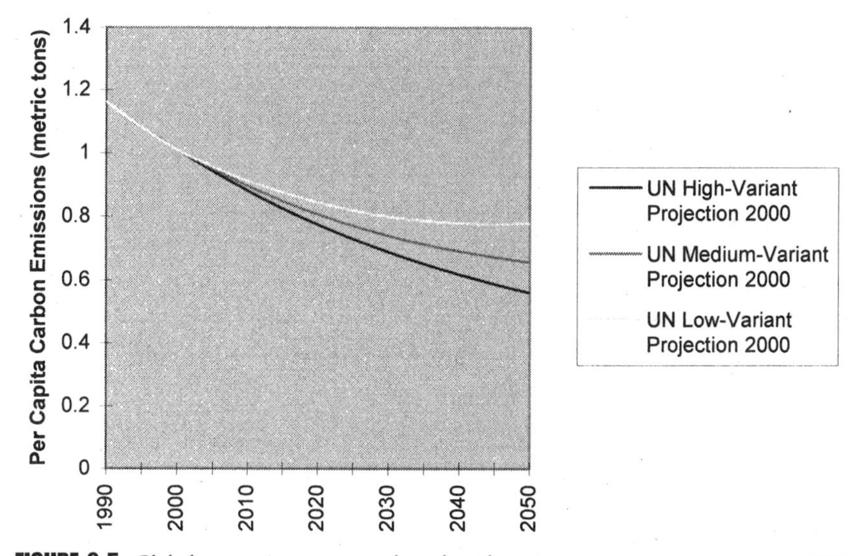

FIGURE 9.5. Global per capita average carbon dioxide emissions necessary to maintain 1990 emission level.

Universal access to family planning, and other health and education goals committed to in Cairo in 1994 but far from fully implemented, could be accomplished at a reasonable cost and with essentially no regrets. The estimated impact of those actions, in terms of GHG emission reductions alone, makes them a viable, cost-effective complement (or alternative) to other climate policy strategies. A world that has 7 rather than 10 billion people in 2050 is likely to result in conditions in which policymakers and individuals can better mitigate and adapt to the effects of climate change such as sea level rise and changes in agriculture, storm frequency, and precipitation. It is unlikely that efforts to offset fossil fuel GHG emissions by changes in land use can be achieved in the face of continued rapid population growth and the resultant pressures on forests and other biotic resources.

Third, population and demographic change should be an explicit part of the next phase of climate policy negotiations. Given the demographically driven distortions of the Kyoto approach and the even larger population issues associated with a truly global agreement, the issue cannot be avoided. In the twenty-first century, the demographic map of the world will be rewritten. Eastern and western Europe and the Russian Federation are likely to experience significant population declines. The population of the United States, on the other hand, is projected to double, primarily as a result of immigration. India will almost certainly become the most populous nation on Earth, and China's population is projected to stabilize and begin to decline. Pakistan, Nigeria, and other developing countries may well triple or quadruple.

Migration and other demographic change caused by climate change itself may further alter global population distribution in ways that cannot be foreseen. All of this argues for climate policy that is demographically flexible. Because GHG emissions ultimately are attributable to people, it makes sense to set emission targets or entitlements on the basis of where population is located. Although a global per capita emission standard may be politically impossible to attain in the near term, it is worthwhile to consider a plan that converges toward such a standard over the course of many decades.[72] It is hard to imagine that developing countries will ultimately agree to any plan that does not provide equitable access to the global atmospheric commons.

For the past three decades, global per capita emissions have been stable, and there is no indication that this trend will change substantially in the near future. Yet a decline in per capita emissions must occur, not just among the developed countries but also in major developing nations such as China and Mexico, assuming continuation of present population and emission trajectories. One of the keys to progress will be to break through the traditional political divisions between developed and developing countries, groupings that make less and less sense given the reality of the wide, overlapping spectrum of demographic, socioeconomic, and emission characteristics of the world's nations.

In the twentieth century, the near quadrupling of human population and more than tripling of per capita carbon dioxide emissions produced a situation in which the human species has had a significant impact on the earth's climate. Future demographic trends will likewise affect the climate. The outcome depends on fertility and mortality trends and access to reproductive health services and education, particularly in the developing world, where most of the growth will occur. Which path the population takes over the next 50 years will have a major impact on the extent of global warming and its economic, social, and environmental consequences.

Acknowledgments

The author thanks Armin Rosencranz, Laura Meyerson, John O. Niles, Brian O'Neill, Holly Pearson, Marc Feldman, Stephen Schneider, and Daniel Esty for their helpful comments on drafts of the manuscript. During the writing, the author was supported in part by a research fellowship from the National Science Foundation.

Notes

1. Meyerson, F. A. B., 1998: "Population, carbon emissions, and global warming: the forgotten relationship at Kyoto," *Population and Development Review,* 24: 115–130. The global per capita average carbon dioxide emission was 1.10 mt in 1970 and 1.13 mt in 1998, and the average for the period from 1970 to 1998 was 1.14 mt per capita. Carbon Dioxide Information Analysis Center, 2002: "Global, regional and national carbon dioxide emissions from fossil-fuel burning, cement production, and gas flaring: 1751–1998, available at http://cdiac.esd.ornl.gov.
2. See Dietz, T. and E. A. Rosa, 1997: "Effects of population and affluence on CO_2 emissions," *Proceedings of the National Academy of Sciences USA,* 94: 175–179. See also Chapter 3 of O'Neill, B. C., F. L. MacKellar, and W. Lutz, 2000: *Population and Climate Change* (Cambridge, England: Cambridge University Press).
3. Meyerson, F. A. B., 1999: *Population Dynamics and Global Climate Change* (Washington, DC: Population Resource Center).
4. See Chapter 11 of Intergovernmental Panel on Climate Change (IPCC), 2001: *IPCC Third Assessment Report, Working Group I.* Available online: http://www.ipcc.ch. See also Chapter 7 of Houghton, J. T., L. G. Meira Filho, B. A. Callander, N. Harris, A. Kattenberg, and K. Maskell (eds.), 1996: *Climate Change 1995: The Science of Climate Change* (Cambridge, England: Cambridge University Press).
5. Henderson-Sellers, A., 1998: "Tropical cyclones and global climate change: a post-

IPCC assessment," *Bulletin of the American Meteorological Society,* 79: 19–38; Mahlman, I. D., 1997: "Uncertainties in projections of human-caused climate warming," *Science,* 278: 1416–1417.

6. Rosenzweig, C. and D. Hillel, 1998: *Climate Change and the Global Harvest: The Potential Impacts of the Greenhouse Effect on Agriculture* (New York: Oxford University Press).
7. Mendelsohn, R. and J. R. Neumann (eds.), 1999: *The Impact of Climate Change on the United States Economy* (Cambridge, England: Cambridge University Press).
8. Hadley Centre for Climate Prediction and Research, 1998: *Climate Change and Its Impacts* (London: The U.K. Meteorological Office and DETR).
9. Epstein, P. R., H. F. Diaz, S. Elias, G. Gradherr, N. E. Graham, W. J. M. Martens, E. Mosley-Thompson, and J. Susskind, 1998: "Biological and physical signs of climate change: focus on mosquito-borne diseases," *Bulletin of the American Meteorological Society,* 79: 409–417.
10. In the view of many ecologists, current global population and consumption patterns are already unsustainable in terms of maintaining biodiversity and the habitat that supports it. See Meffe, G. K., A. H. Ehrlich, and D. Ehrenfeld, 1993: "Human population control: the missing agenda," *Conservation Biology,* 7: 1–3; Wilson, E. O., 1992: *The Diversity of Life* (New York: W.W. Norton). See also Root, T. L. and S. H. Schneider, 1995: "Ecology and climate: research strategies and implications," *Science* 269: 331–341.
11. For discussion of the history and potential for environmental refugees, see Ramlogan, R., 1996: "Environmental refugees: a review," *Environmental Conservation,"* 23: 81–88; Myers, N., 1994: "Environmental refugees in a globally warmed world," *Bioscience* 43: 752–761.
12. Although the rate of population growth was lower in the 1980s and 1990s than in the 1970s and 1960s, the base population was higher in the latter period, resulting in greater net population gain. United Nations Department for Economic and Social Information and Policy Analysis Population Division, 2000: *World Population Prospects: The 2000 Revision* (New York: United Nations); United Nations Population Fund (UNFPA), 1999: *The State of World Population 1999* (New York: United Nations).
13. Vitousek, P. M., H. A. Mooney, J. Lubchenko, and J. M. Melillo, 1997: "Human domination of Earth's ecosystems," *Science* 277: 494–499.
14. Wilson, 1992.
15. Carbon Dioxide Information Analysis Center, 2002.
16. Ibid.
17. Ibid.
18. Ibid.; Meyerson, "Population, carbon emissions, and global warming," 1998.
19. For instance, in the United States, although the emission intensity per unit of gross domestic product has improved (become more efficient), consumption increased at an equal or greater rate, so that per capita carbon dioxide (and other GHG) emissions have remained level or slowly rising since 1990. Energy Information Administration, 2000: "Emissions of greenhouse gases in the United States 1998," U.S. Department of Energy, available online: http://www.eia.doe.gov/.

20. While there is agreement that population growth leads to increased GHG emissions, the exact percentage contribution depends on the decomposition methodology used. See Bongaarts, J., 1992: "Population and global warming," *Population and Development Review,* 18: 299–319; O'Neill et al., 2000.
21. United Nations Department for Economic and Social Information and Policy Analysis Population Division, 1998.
22. Population momentum is the potential for future increase in population size that is inherent in any present age and sex structure even if fertility levels were to drop immediately to replacement levels. See Weeks, J. R., 1999: *Population: An Introduction to Concepts and Issues,* 7th ed. (Belmont, CA: Wadsworth Publishing Company).
23. The global population projections prepared by the United Nations and U.S. Census Bureau are created by aggregating individual country projections.
24. United Nations Department for Economic and Social Information and Policy Analysis Population Division, 2000: *World Population Prospects: The 2000 Revision* (New York: United Nations, February 28, 2001 draft). Note that the U.S. Census Bureau, which also released its latest projections in 2000 but produces only a single (medium variant) global projection, has a somewhat different projection for 2050 (9.1 billion). U.S. Census Bureau, 2000: "International data base," U.S. Census Bureau, www.census.gov. The current and historical differences between the U.S. Census Bureau and UN medium variant scenarios underscore the high sensitivity of long-range demographic projections and trends to small changes in fertility and mortality assumptions. For two other approaches to population projections, see Lutz, W., W. Sanderson, and S. Scherbov, 1996: "Probabilistic population projections based on expert opinion," in W. Lutz (ed.), *The Future Population of the World* (London: Earthscan), pp. 397–428; and Lee, R. D. and S. Tuljapurkar, 1994: "Stochastic population forecasts for the United States: beyond high, medium and low," *Journal of the American Statistical Association,* 89: 1175–1189.
25. For an analysis of the sensitivity of carbon dioxide emissions to population size, affluence, energy intensity, and carbon intensity, see Yang, C. and S. H. Schneider, 1997–98: "Global carbon dioxide emissions scenarios: sensitivity to social and technological factors in three regions," *Mitigation and Adaptation Strategies for Global Change,* 2: 373–404.
26. UNFPA, 1999. Globally, an estimated 58 percent of women use some form of contraception, including nonmodern methods such as the rhythm method, and 50 percent use modern contraception.
27. Ibid.
28. For a discussion of the projected sensitivity of GHG levels to childbearing, see O'Neill, B. C. and L. Wexler, 2000: "The greenhouse externality to childbearing: a sensitivity analysis," *Climatic Change.*
29. United Nations Department for Economic and Social Information and Policy Analysis Population Division, 2000; UNFPA, 1999; Henshaw, S. K., 1998: "Unintended pregnancy in the United States," *Family Planning Perspectives,* 30: 24–29, 46.
30. Trussell, J. and B. Vaughan, 1999: "Contraceptive failure, method-related discon-

tinuation and resumption of use: results from 1995 National Survey of Family Growth," *Family Planning Perspectives,* 31: 64–72, 93.

31. Ibid. Another study, which corrected the same data for underreporting of induced abortion, estimated contraceptive failure rates in the United States to be even higher. See Fu, H., J. E. Darroch, T. Haas, and N. Ranjit, 1999: "Contraceptive failure rates: new estimates from the National Survey of Family Growth," *Family Planning Perspectives,* 31: 56–63.
32. Henshaw, 1998.
33. Pritchett, L. H., 1994: "Desired fertility and the impact of population policies," *Population and Development Review,* 20: 1–55.
34. The United Nations estimates annual population growth as of 1999 to be 78 million per year; UNFPA, 1999. The Population Reference Bureau estimates annual population growth to be 84 million per year; Population Reference Bureau, 2000: *Population Today* (Washington, DC: Population Reference Bureau). The lowest recent estimate for the global number of abortions is 46 million (1995), of which 20 million (estimated) were illegal. Approximately 25 percent of the 180 million global annual pregnancies are ended by abortion. Henshaw, S. K., S. Singh, and T. Haas, 1999: "The incidence of abortion worldwide," *International Family Planning Perspectives,* 25 (Supplement): S30–38.
35. See also Bongaarts, J., 1994: "Population policy options in the developing world," *Science,* 263 (5148): 771–776.
36. It is important to note that the extent to which population policies and programs affect fertility continues to be a topic of debate. See Jain, A. (ed.), 1998: *Do Population Policies Matter? Fertility and Politics in Egypt, India, Kenya and Mexico* (New York: The Population Council); Bongaarts, J., 1999: "The role of family planning programs in contemporary fertility transitions," in G. W. Jones and J. Caldwell (eds.), *The Continuing Fertility Transition* (London: Oxford University Press); Pritchett, L. H., 1994: "Desired fertility and the impact of population policies," *Population and Development Review,* 20: 1–55; Bongaarts, J., 1994: "The impact of population policies: comment," *Population and Development Review,* 20: 616–620.
37. United Nations, 1994: *International Conference on Population and Development Programme of Action* (New York: United Nations).
38. UNFPA, 1999.
39. United Nations Population Fund (UNFPA), 1994: *Background Note on the Resource Requirements for Population Programmes in the Years 2000–2015* (New York: United Nations).
40. Cline, W. R., 1992: *Global Warming: The Economic Stakes* (Washington, DC: Institute for International Economics); Birdsall, N., 1994: *Another Look at Population and Global Warming. Population, Environment and Development* (New York: United Nations); O'Neill and Wexler, 2000.
41. See Chapter 6 of O'Neill et al., 2000. See also Nordhaus, W. D. and J. Boyer, 1998: "What are the external costs of more rapid population growth? Theoretical issues and empirical estimates," 150th Anniversary Meeting of the American Association for the Advancement of Science, Philadelphia.
42. Developed (donor) countries are providing only $1.9–2.0 billion of the $5.7 billion

annual expenditure they committed to in Cairo (for the year 2000). Developing countries are providing about $7.7 billion of the $11.3 billion they committed to in Cairo. United Nations, 1999: *Report of the International Forum for the Operational Review and Appraisal of the Programme of Action of the International Conference on Population and Development (ICPD)* (New York: United Nations).

43. The U.S. budget for international family planning assistance in FY 1995 was $541.6 million (A.I.D.), with an additional $35 million in funding for the United Nations Population Fund (UNFPA). United States Agency for International Development, 2000: www.info.usaid.gov.
44. In FY 2000, the U.S. budget for international family planning assistance was $372.5 million (A.I.D.), with a probable additional $20 million for UNFPA, depending on the interpretation and execution of withholding language. United States Agency for International Development, 2000: www.info.usaid.gov.
45. The formula used in the Kyoto Protocol is a measurement of the average national emission for the years 2008–2012. United Nations, 1997: *Conference of the Parties, Third Session. Kyoto Protocol to the United Nations Framework Convention on Climate Change* (Kyoto, Japan: United Nations). The year 2010 will be used as the reference year here, to facilitate analysis of demographically related issues.
46. Ibid.
47. Meyerson, F., November 10, 1997: "Pollution and our people problem," *The Washington Post,* p. A21.
48. Carbon Dioxide Information Analysis Center, 2000.
49. Ibid.
50. United Nations, 1997.
51. Carbon Dioxide Information Analysis Center, 2000.
52. Meyerson, 1997.
53. In response to Meyerson "Population, carbon emissions, and global warming," 1998, one author has argued that population projections were an implicit part of the Kyoto negotiations, though not explicitly discussed. See de Sa, P., 1998: "Population, carbon emissions and global warming: comment," *Population and Development Review,* 24: 797–803. However, there is little evidence that the negotiators considered the magnitude of the differences in population growth or its effect on emissions. See Meyerson, F. A. B., 1998: "Toward a per capita–based climate treaty: reply," *Population and Development Review,* 24: 804–810; Benedick, R. E., 2001: "Contrasting approaches: the ozone layer, climate change, and resolving the Kyoto dilemma," in E. D. Schulze, S. P. Harrison, M. Heimann, E. A. Holland, J. Lloyd, I. C. Prentice, and D. Schimel (eds.), *Global Biogeochemical Cycles in the Climate System* (San Diego, CA: Academic Press/Max Planck Institute for Biogeochemistry).
54. Meyerson "Population, carbon emissions, and global warming," 1998; U.S. Census Bureau, 2000: "International data base," U.S. Census Bureau, www.census.gov. The U.S. population increase (and therefore necessary per capita emission reduction) is likely to be even greater than projected for the 1990–2010 period. The 2000 Census revealed that, as of that year, U.S. population was 6.7 million (2.4 percent greater than previous medium variant projection). U.S. Census Bureau, 2000: "First Census 2000 results: resident population and apportionment counts," U.S. Census

Bureau, http://www.census.gov/main/www/cen2000.html.

55. United Nations Department for Economic and Social Information and Policy Analysis Population Division, 2000; Carbon Dioxide Information Analysis Center, 2000.
56. Energy Information Administration, 2002.
57. The 2000 census indicates that U.S. population is 6.7 million higher than projected, primarily due to greater than anticipated immigration during the 1990s. However, the adjusted U.S. per capita emission for 2000 would be approximately 5.35 mt, roughly the same as it was in 1980. See U.S. Census Bureau, 2000: "First Census 2000 results."
58. Energy Information Administration, 2000: "Annual energy outlook 2000 with projections to 2020," U.S. Department of Energy, http://www.eia.doe.gov/.
59. Meyerson, "Population, carbon emissions, and global warming," 1998.
60. U.S. Census Bureau, 2002: "International data base," U.S. Census Bureau, www.census.gov. Note that the UN Population Division medium variant projection for the United States in 2050 (349 million) is substantially lower than that of the Census Bureau (404 million) because of differences in fertility and net migration assumptions.
61. United Nations Department for Economic and Social Information and Policy Analysis Population Division, 2000.
62. Ibid.
63. Meyerson, "Population, carbon emissions, and global warming," 1998.
64. Carbon Dioxide Information Analysis Center, 2002.
65. See Chapter 2 of O'Neill et al., 2000.
66. See MacKellar, L. and D. R. Vining, Jr., 1995: "Population, households, and CO_2 emissions," *Population and Development Review,* 21: 849–865; Meyerson, F. A. B. 2001: "Replacement migration: a questionable tactic for delaying the inevitable effects of fertility transition," *Population and Environment,* 22: 401–409. Note also that urbanization is an additional factor related to both household size and aging that affects emissions. The urban proportion of the world's population increased from 30 percent in 1950 to about 50 percent in 2000 and is projected to exceed 60 percent by 2030; United Nations Department for Economic and Social Information and Policy Analysis Population Division, 1998. The effect on emissions is complex because urbanization tends to increase per capita income and creates both economies and diseconomies of scale as city size increases. For a brief discussion, see Chapter 2 of O'Neill et al., 2000.
67. Engelman, R., 1998: *Profiles in Carbon: An Update on Population, Consumption and Carbon Dioxide Emissions* (Washington, DC: Population Action International). Disparities between individual countries are even greater. For instance, the average person in the United States contributed 5.3 tons of fossil fuel carbon emissions to the atmosphere in 1995, more than 16,000 times as much as the average Somali and almost five times as much as the average Mexican. Within individual countries, unequal wealth distribution may also mean that a small percentage of the population may be responsible for a large share of GHG emissions.
68. Meyerson, "Toward a per capita–based climate treaty," 1998.

69. Between 1990 and 2000, the U.S. population increased by 32.7 million, the greatest addition in any decade in U.S. history. U.S. Census Bureau, 2000: "First Census 2000 results."
70. Malakoff, D., 1997: "Thirty Kyotos needed to control warming," *Science,* 278:2048; Bolin, B., 1998: "The Kyoto negotiations on climate change: a science perspective," *Science,* 279: 330–331.
71. Yang and Schneider, 1997–98.
72. See Meyerson, "Toward a per capita–based climate treaty," 1998. For a different approach to developing country emissions commitments, based on GHG intensity per unit of domestic product, see Baumert, K. A., R. Bhandari, and N. Kete, 2000: *What Might a Developing Country Climate Commitment Look Like?* (Washington, DC: World Resources Institute).

CHAPTER 10

Global Climate Change: A Business Perspective

Thomas G. Burns

Global climate change is one of the most critical issues facing society at the start of the new millennium. It has the potential to cause serious disruptions in the world's economic system in at least two ways:

- Unfavorable changes in the climate could have an impact on the ability of the earth to sustain life as we know it today.
- Alternatively, a drastic effort to prevent such changes, for example by sharply reducing fossil fuel consumption, could fundamentally alter the way modern society operates.

For these reasons, it is important to improve our understanding of the workings of the global climate and to recognize the full impact of proposed solutions to potential problems. Failing to respond appropriately to a real problem, trying to solve the wrong problem, or solving the correct problem in the wrong way will result in additional costs that must be borne by society as a whole. Regardless of the reason, incurring such unnecessary costs will slow economic growth, reduce standards of living, and limit the resources available to cope with other emerging but perhaps unforeseen problems.

As a starting point, the business community strongly supports ongoing climate science research to improve the understanding of both the extent of any potential change in the climate and the underlying causes. Then, if further research confirms that climate change outside the range of natural climate variability can be both detected and attributed to human causes, the basic approach of business would be to choose solutions that address the problem in ways that optimize benefits and minimize costs.

In addressing problems of this type, business generally believes that, on a continuum of policies and measures, it is far more efficient to maximize the use of the democratic free market system and to minimize the use of command-and-control regulatory frameworks to achieve environmental goals.

Climate change is a long-term problem that is amenable only to long-term solutions. Precipitous action is likely to increase the costs and to result in an inferior outcome for society. Policies responding to any long-term change in climate should include all the approaches at our disposal, including prevention, mitigation, and adaptation.

Wealth creation is the area in which the private sector has been most successful and has the most to offer. This is not the process that some derisively call trickle-down economics. The whole point of any business endeavor is to create wealth by producing products of more value than the costs of the inputs used by the business.

It is true that industrialization has created some of the environmental problems we face today, including the potential threat from greenhouse gas (GHG) emissions. However, it is equally true that the tools developed in the process of industrialization permit us to grapple with these problems far more effectively than less wealthy societies. The reality is that a well-developed environmental ethic is associated primarily with wealthy societies that have long since satisfied the basic human needs of their citizens.

Throughout history, wealthy societies have always been able to educate their populaces more effectively. The wealth created in the private sector ultimately funds not only government and government-supported education but also private educational institutions, foundations, and even environmental nongovernment organizations (NGOs). So it is in everyone's best interests to ensure that the private sector can continue doing what it does best: creating wealth.

The Business Approach to Global Climate Change

Businesses deal with economic and technical uncertainties in the course of their normal activities. What business does not want to happen is for society to move so quickly toward a particular solution that better, less costly alternatives are precluded. Because global climate change is a long-term, slowly evolving issue, it is important to move incrementally toward solutions. This does not mean that there are not economically attractive strategies that can be (and already are being) used today. By starting now but moving cautiously, the world will always be in a position to adapt to changing circumstances while using the best, most current technologies.

In a period of rapid growth and transition, old-style businesses that fail to adapt will lose markets to more nimble competitors and go out of business. Such business failures occur far more often in a market economy than in one that is centrally directed. For example, one may ask how many obsolete factories were ever shut down in the Soviet Union and which economy—ours or theirs—was more productive, innovative, and dynamic?

In general, the experience of business suggests that smooth transitions are preferable to abrupt ones. In the past, radically new technologies typically have taken a long time to enter the market. Even such major technological innovations as the steam engine, the electric light, the telephone, the automobile, the airplane, and the personal computer took several decades or longer to achieve significant market penetration. There is some evidence that modern information and communication technologies are shrinking the time it takes for a new technology to reach maturity, but it remains true that a new technology is rarely an instantaneous success.

Business also wants to be sure that its customers understand the significance of new regulations purporting to reduce anthropogenic GHG emissions. Often, although the benefits of environmental protection appear attractive, the costs are not always apparent. When product prices rise, business often is blamed and must remind the public that at least part of the reason is to be found in the increased costs associated with today's more environmentally friendly products and processes. These products now incorporate many of the costs that were previously considered to be external to the production process.

Another example might be found in the area of urban planning. Everyone likes the idea of reducing urban sprawl, at least until it becomes apparent that the logical consequence is an increase in population density. If urban growth boundaries mean that needed housing cannot be built on the periphery of today's urban areas, it must be built within current city limits or shortages will occur and real estate prices and rents will rise. Similarly, everyone likes the idea of an alternative fuel vehicle that would release them from their dependence on gasoline purchases until they learn what they might have to give up in terms of flexibility, comfort, power, range, cost, or convenience.

Unresolved Issues in Climate Change Negotiations

Based on experience in the rapidly changing, highly competitive world of business, the private sector generally would like to see the negotiators take a practical, flexible approach to resolving the remaining open issues in the Kyoto Protocol. This means that it is important to set rules that encourage, rather than penalize, behaviors and projects that can lead to reduced GHG emissions. It is also important to develop a framework that is flexible enough to cope with the innovations and unforeseen opportunities that will undoubtedly occur as experience in this area grows.

Business recognizes that there is a definite role for government intervention in our complex economy, particularly to respond to clear-cut market failures. Creating conditions that lead to internalizing the externalities of environmental burdens is another area in which government has successfully demonstrated leadership. One example is the U.S. sulfur dioxide emission trading system that efficiently encourages electric utilities to reduce acid rain precursors.

Unfortunately, many of the participants in the negotiating process oppose this kind of flexibility because it relies on markets, not on regulators, as the primary control mechanism. Proponents of increased government intervention in the economy generally don't trust the judgment of the market, in the belief that regulators can do a better job.

For a vision of a future based on the regulatory model, we need only to observe the experience of the former eastern bloc socialist countries that carried the command-and-control approach to its logical conclusion. It will probably take several generations before they build the wealth that will let them recover from their centrally planned economic and environmental disasters and begin to approach the levels of health, safety, and environmental ethics already prevailing in the developed countries enjoying a democratic free market.

Many developing countries, as well as many environmental NGOs, also have reservations about the use of flexible mechanisms to achieve GHG reductions. They "fear that the developed countries will use their great financial power to buy their way out of emissions restrictions and transfer those limits to poorer countries, where they will interfere with industrial development."[1]

However, no one can compel an economy in transition or a developing country to sell its excess credits. They will elect to do this only at a fair market price. After all, the hallmark of a fair market transaction is one that makes both buyer and seller better off. Otherwise, the transaction won't take place.

One tactic of the opponents of flexible mechanisms is to try to make the procedures so restrictive and complex that the mechanisms will be impossible to use. Business, on the other hand, would like to see these tools defined in such a way that they can be easily understood and applied. Complex regulations always add a component of additional risk to any project because it is often impossible to know in advance how the regulations will be interpreted. Unnecessary complexity makes any project more expensive and often leads to costly delays that only postpone the time when the project can start to provide benefits to the business, its customers, and the global environment.

Compliance mechanisms are also a source of continuing controversy. As long as the focus is on penalizing failure rather than rewarding success, there will be resistance to any agreement by many of the participants in the negotiations.

At least part of the problem in achieving agreement on practical, flexible ways to meet the overarching goals of the UN Framework Convention on Climate Change (UN FCCC) can be traced to the fact that issues unrelated to the climate cloud the negotiating process. Issues that were not explicitly on the table in the climate negotiations but were important in formulating national positions include trade and tariff relationships, protectionism, development aid, economic competitiveness, internal politics, regional considerations, and many others.

What a Typical Business Wants from a Climate Change Policy

Policies designed to prevent, mitigate, or adapt to a changing climate should be similar to other environmental regulations.

Because business has put capital at risk to produce a product or service, business would like some assurance from society that the ground rules will not change abruptly during the economic life of the investment. This is part of the level playing field sought by business. Establishing the rules that specify conduct and apply to all competitors is an important function of government.

Business prospers most when it is able to grow, either by creating new markets or by taking market share for an existing product from a competitor. This becomes more difficult when artificial restrictions are placed on the demand for its products or services or on its ability to meet consumer demand. Consumers are also made worse off when their choices are artificially restricted. Although it sometimes appears self-serving, the business community often has to speak not only for itself but also for its customers. In general, the broad mass of consumers is not well organized. Many special interest groups that purport to represent consumers do not always have the best interests of the consumer in mind. Business interacts continuously with its customers and, by observing their behavior, gains insights into their interests that may not be obvious to others.

For example, consumer surveys generally report that 60–70 percent of all consumers say they would be willing to pay more for "green" products, but only about 10 percent of them "buy green" when faced with a real-life buying situation. Consumer advocates use the higher number to claim that there is an unfilled demand for such products; business faces the reality of the marketplace.

To summarize, there are a number of elements of a climate change policy that many businesses would subscribe to:

Good science: Business believes that it is important to have some assurance that we understand the size, source, and scope of the climate problem well enough to be reasonably certain that proposed solutions are appropriate and effective.[2]

Market mechanisms: It is important to use the power of the marketplace in support of the environment.

Customer focus: Ultimately, important decisions balancing current costs and future benefits must be made by a society that has a sound understanding of the consequences.

Smooth transitions: Efficient use of previous capital investments suggests that incremental progress will be less costly than abrupt lifestyle changes.

New technologies: Keep an open mind with respect to all current and future opportunities.

Avoid subsidies: Guessing exactly which of several competing new technologies will develop into a market-dominant position has little likelihood of success.

How We Arrived at the Current Situation

Before 1973, energy consumption tended to rise at about the same rate as economic growth. Oil consumption was increasing much faster than economic growth as consumers and industry switched from coal to oil. After 1973, energy consumption growth dropped to only about half of the rate of gross domestic product (GDP) growth as efficiencies were discovered and implemented. Oil consumption remained flat until the mid-1990s as nuclear power plants were commissioned, natural gas achieved significant growth, and coal enjoyed a resurgence. Even today, both oil and energy consumption are growing more slowly than GDP, and only natural gas, mainly because of increased use in electricity generation, is growing more rapidly.

Although the economic system is never perfectly efficient, competition tends to drive businesses to improve productivity by making more efficient use of resources. For example, when the first oil price shocks hit the United States in the early 1970s, our energy-consuming industrial and commercial equipment was designed for our historically low-energy-cost, high-wage-cost environment. This meant that industrial machinery and commercial buildings were designed for maintenance-free operation and not primarily for energy efficiency. The same held true for buildings, appliances, and vehicles.

Improved energy efficiency continues to have a big payoff in terms of both economic efficiency and reduced GHG emissions. As a result of the sheer size of our economy, we still have a large stock of older automobiles, appliances, buildings, and industrial equipment that were designed based on materials and standards that may need updating. Clever companies see the opportunities for competitive advantage in life-cycle-costing and total-cost-of-ownership approaches to design, purchasing, and operations.

However, these opportunities should not be taken to mean that reduction in energy consumption is to be pursued at all costs. Doing something more efficiently is not the same as doing less of that particular activity in its impact on economic output. Pursuing energy efficiency means doing more with less. Simply doing less of some desirable activity means lower economic growth and less wealth creation.

Making Better Decisions on Global Climate Change

Decision analysis provides a set of tools that can help reduce the risks associated with investment decisions that are largely irreversible. For example, if you decide to build a steel mill, you are stuck with equipment designed for a specific purpose. It is costly, if not impossible, to change your mind after the investment is made and decide that you really would prefer to make computer chips in that facility.

"Most investment decisions share three important characteristics in varying degrees. First, the investment is partially or completely *irreversible.* Second, there is *uncertainty* over the future rewards from the investment. Third, you have some leeway about the *timing* of your investment."[3] These three characteristics of almost all investment decisions create value associated with obtaining additional information about the future, including market demand, price, technology, and economic conditions. After all, if you make the wrong decision, scarce capital is wasted. Every investment decision offers at least three choices: "Yes," "No," and "Wait" (to gather more information before making the decision). To the extent that an irreversible decision can be deferred, risks are reduced and an option value is created.

Environmental investments are no different from other opportunities to invest a society's scarce capital resources. Just as in business, wrong choices that waste resources can be made. Current technologies that may prove to be inappropriate in the long run may be the only ones available now, and a delay that would not have a significant effect on the long-term outcome might allow better technologies to be developed and applied.

Business Reaction to the Kyoto Protocol

For the moment, the Kyoto Protocol[4] to the UN FCCC[5] is still the approach being followed by the international community to deal with climate change.

There are two basic parts to the protocol: the targets and timetables, coupled with the compliance measures (and associated enforcement mechanisms and bureaucracies) and the flexible mechanisms.

There are varying reactions to the Kyoto Protocol in the business community, depending on the particulars of the businesses involved. Businesses that are strongly supportive of the protocol's targets and timetables often see market opportunities that will be created as a result of the government support of climate-friendly technologies promised by the protocol. A second group recognizes the inevitability of some effort to reduce human impact on the climate and, in this context, regards the Kyoto targets as a reasonable first step. Others believe that the targets and timetables are unrealistic and that the negotiators have vastly underestimated the economic consequences of trying to achieve them. This third group tends to oppose these, but generally not all, aspects of the Kyoto Protocol. A further reality is that so far most businesses remain untouched by the negotiations or unconvinced by the science and thus indifferent to the issues.

Nonetheless, most of the involved businesses do support some parts of the protocol—specifically the flexible mechanisms, which include three market-based methods to ensure that any required GHG reductions are made in the most efficient way possible. These market mechanisms include joint implementation (JI), the Clean Development Mechanism (CDM), and emission trading. Use of these three techniques will ensure that any emissions reduction targets will be reached efficiently and that development and use of new and improved technologies will be stimulated.

The CDM, an innovative approach conceived in Kyoto, permits developed countries to work with developing countries to support sustainable development in ways that result in reduced GHG emissions. The protocol says that CDM projects should be host country driven, support sustainable development, help build capacity and institutions, transfer cutting-edge technology, and generate GHG emission credits to be shared on a negotiated basis among the parties.

Emission trading calls for development of a systematic market that will allow holders of excess emission reduction credits to sell them to those who need additional credits to meet their treaty obligations. Countries, companies, and other entities that can develop emissions reduction credits at low cost will be encouraged to do so, knowing that they can sell any that are in excess of their needs. The funds generated by such sales can be reinvested in productive, wealth-creating enterprises. On the other side of the transaction, countries, companies, and other entities that face costly (or impossible) emission reduction targets can meet their needs by going to the market and paying a going market rate for such credits. The interaction of suppliers and purchasers in the market will set the price of the credits.

How Market Mechanisms Can Support Environmental Values

It may come as a surprise to some that the market system that has created our prosperity has come to play a major role in supporting society's environmental values.

One of the key problems facing society is how to correctly account for the external or nonmarket costs of any particular activity. It is difficult to know just where to draw the boundaries and exactly which costs are relevant to include. In principle, the end user or consumer of any particular good or service should be expected to pay all of the costs associated with the production and use of the product. Anything less amounts to a subsidy to the consumer. Putting this simple statement into practice is extremely complex and difficult, however.

There have been many attempts to do this exercise for transportation, which ranks right after food, clothing, and shelter in the purchasing hierarchy of most consumers. Unrecovered external costs often are described as a subsidy for the automobile, but the subsidy actually is for personal transportation. If government provides roads and parking places as part of society's infrastructure, motorists may only pay for part of the costs via motor fuel taxes. Because almost all taxpayers use personal transportation, the same people ultimately pay these costs, even if not in direct proportion to their use of the facilities. In this case, some of the fixed costs of transportation remain hidden in the tax code. What is certain is that anyone using this infrastructure benefits from the subsidy, which is not limited to any particular fuel or vehicle type.

Another common practice that is often identified as a subsidy for transportation is the provision of large parking lots by suburban shopping malls. In this case, the costs are incorporated into the cost structure of the merchants in the mall, and are ultimately paid by the very same customers who use the lots to park their cars. In contrast, downtown parking garages, which are unable to identify the merchants benefiting from the parking spaces, generally charge motorists for the service. Although the motorist/shopper pays a market rate for either parking space, the motorist views mall parking as a fixed cost and the downtown garage as a variable cost.

Environmental impacts are often regarded as an external cost of doing business that should be incorporated more directly into purchasing decisions. Traditional environmental regulations often were based on a command-and-control approach. The regulatory agency in effect says "do this, do it this way, and do it on this time schedule." This approach often overspecifies the solution, limiting flexibility and stifling creativity.

Beginning with Project 88, sponsored by then Senators Tim Wirth and John Heinz, there has been an ongoing effort to shift this traditional regulatory paradigm to one that works with, rather than against, market forces. In general, the market-based approach sets environmental performance goals and then lets the private sector find the most efficient way to reach them. This engenders creativity and technological development, which in turn create a competitive advantage for the innovator and generate greater environmental benefits at lower cost. Our experience with previous applications of market-based approaches to environmental regulations can be applied to the climate change issue.

One of the best examples of this new market-based regulatory approach with direct application to a proposed system of GHG emission trading is the often-cited SO_2 emission trading system in the United States. The efficiencies associated with this system have lowered the costs of SO_2 reductions far below even the most optimistic estimates made before the system was launched.

Another example of a very successful market-based trading system was the one that permitted oil refiners to trade lead additive credits. This approach not only reduced the cost but also helped speed up the nation's transition to lead-free gasoline.

Use of markets to determine a price for the environmental impacts of a particular product helps ensure that the best, most cost-effective solutions are developed. To be effective, any GHG emission reduction program must incorporate market mechanisms in some fashion. Greater use of voluntary, market-based mechanisms will result in more GHG emission reductions at lower cost and less economic disruption than other approaches.

The Role of New Technologies

Most businesses are convinced that technology is the only viable solution to the problems of a changing climate because the drastic changes in lifestyle that would result from reducing the world's energy consumption using today's technologies would be unacceptable to most people. In the twentieth century, the rate of technological change was extremely high, accelerating throughout the century. There is no reason to believe that the pace of change will not continue to accelerate in the coming century. We can expect to see at least as much change in the next 100 years as we saw in the last 100 years.

A number of industry-led programs have highlighted both existing and potential technologies that are already reducing GHG emissions and will contribute to even lower GHG emissions in the future.[6] These and other efforts strongly suggest that development of innovative new technologies offers the lowest-cost, most acceptable way to deal with the problems associated with a changing climate.

A wide range of alternative energy technologies show promise, particularly in specific locations and applications. Geothermal electricity generation is economical where there is a large source of geothermal energy close to a major market. Windpower works where there is a source of steady wind and a backup system to provide electricity when the wind subsides. Solar photovoltaic electricity is appropriate in locations with small power demands and that are unlikely to be connected to a grid in the near future.

Although technological change is inevitable, it is important to note that there are good reasons why existing, dominant technologies enjoy the significant markets that they do. They supply consumers with products or services that they want at prices they are willing to pay. Some argue that new products and new technologies to reduce GHG emissions must be subsidized to overcome the market inertia associated with doing things the old way.

Subsidies are needed only when the new technology is neither more efficient nor cheaper than the existing technology it seeks to replace. Historically, successful new technologies have offered the consumer clear improvements at lower costs. For example, it didn't take a government-subsidized program to encourage the rapid transition from long-playing records to cassette tapes and then to compact discs.

Outside the business community, many observers believe that the introduction of new technologies can be jump-started by subsidies or other nonmarket incentives. However, as economics professor Lester Thurow of the Massachusetts Institute of Technology has said, "Never has an industry been restored to economic health after introduction of a subsidy." Furthermore, "picking winners" is seldom a successful technology strategy—there are simply too many unknowns.

In the past 10 years, California has led the effort to develop a zero-emission vehicle (ZEV). A recent review of the status of the program and the state of the art of battery development has resulted in a gradual refocusing of the program. Instead of requiring manufacturers to make only pure electric vehicles, California will allow them to offer a variety of hybrid, fuel cell, and super-ultra-low-emission vehicles using improved gasoline engine technology. This flexibility has resulted in true innovations in technology, even though the original goal of a battery-powered electric vehicle proved once again to be beyond reach.[7]

Although these alternative vehicle technologies continue to require significant subsidies (a new state law offers up to $9,000 per vehicle) to make them cost-competitive with conventional automobiles, there is at least some prospect that they can provide a satisfactory option for part of the personal transportation market.

Undoubtedly, new developments in personal transportation systems will continue to be made. However, to be successful they must offer a better, more reliable product at lower cost to the consumer. This is not an easy task.

Major Policy Issues in the Climate Change Debate

The policy debate on climate change often conflates a number of issues, including some that are only loosely connected to the issue of climate change. The fundamental issues include the following:

- How serious is the problem of climate change? (detection)
- To what extent is it caused by human activity? (attribution)
- How soon will it affect us? (timing)
- What can we do about it? (options)
- Who should take action? (responsibility)
- How much will it cost? (efficiency)
- Who will pay for it? (burden sharing)

The first three on the list are issues that can be resolved by improved understanding of the science of the global climate. Although by far the largest influences on the climate are the result of natural processes, as the number of people on earth has grown, human influence on the climate system has also grown. Humans today account for only about 3–4 percent of the total carbon flux; the rest is the result of natural processes.[8] If the climate system is in a stable range, it is difficult to see how such a small increment could have a major influence. However, if the climate system is at or near a point of instability or transition, even a small increment could tip the climate balance into another state, either warmer or colder than at present.

Climate science is difficult because it is impossible to run controlled experiments. Therefore, scientists turn to models of the climate to analyze the past relationships between the key variables in a way that provides insights into the future under various emission scenarios. Over time, modelers have improved their understanding of the climate to the extent that the models can replicate most of the historical climate record, improving the credibility of their efforts to extrapolate current conditions into the future under changing conditions.

However, a model is only a representation of the extremely complex combination of variables that creates the earth's climate. Most modelers believe that it will be at least a decade before the questions of detection, attribution, and timing of temperature changes are satisfactorily resolved. "For other variables like precipitation, and for changes at smaller spatial scales, the human signal will emerge from the background noise much more slowly. In some cases, it may be many decades before we can clearly see these signals."[9]

One of the key uncertainties arising from climate science is which of the many GHGs is the most important in causing temperature change.[10] It is well known that various GHGs have different global warming potentials, and the ease and cost of reducing emissions of these gases vary widely. The most effective strategy may be to start with reducing emissions of some of the high GHG potential gases, such as methane, and move to controlling carbon dioxide emissions only as energy supply and consumption technologies improve.

In trying to manage the problems caused by a changing climate, there are several fundamental strategies, which fall into the broad categories of prevention, mitigation, and adaptation. A prevention strategy attempts to get people to act in ways that do not lead to human-caused changes in climate, such as sharply curtailing automobile use to reduce the output of GHGs.

Mitigation strategies focus on offsetting the effects of the emissions. For example, the CO_2 emitted as a combustion byproduct might be offset by a forestry project or by injection of the produced CO_2 into an underground reservoir.

An adaptation strategy attempts to reduce the impact of climate change on society by, for example, moving cities or factories away from floodplains to reduce the risk of flooding. An intensified public health effort to respond to changing disease vectors is also a form of adaptation.

Industry generally focuses its efforts on the first two strategies because it has some degree of control over its own emissions. Waste reduction and energy efficiency programs are effective ways to reduce the impact of activities needed to meet consumer demand.

Emerging technologies, such as gas-to-liquids (GTL), offer the opportunity to convert unmarketable byproduct natural gas into valuable liquid fuel, achieving a significant overall reduction in GHGs emitted. Unfortunately, much of the known resources of natural gas lie in areas that are too remote from markets to justify the expense of pipelines and other infrastructure. GTL offers a way to bring these resources to market.

The energy industry also is increasingly exploring mitigation opportunities, including reforestation and gas reinjection, to ensure that overall GHG emission reductions are made in the most cost-effective way possible.

At almost all major energy-producing companies, many of these strategies have already been used in an effort to reduce the impact of company operations on the global environment. Flare reduction has long been a major industrywide effort. A flare gas capture and international transportation project in West Africa could eliminate 100 million tons of carbon dioxide emissions over a 20-year project life.

A major refiner's ongoing efforts to conserve energy enable it to produce the same amount of output using 18 percent less fuel than would have been needed in 1991. Cogeneration of electricity in large refineries and oil producing operations with a significant need for both power and steam is an important part of these energy conservation efforts.

Several oil companies are working to develop and commercialize gas-to-liquids technology that will convert remote sources of natural gas to a sulfur- and nitrogen-free liquid feedstock with clean burning qualities. Widespread availability of this product would reduce the need for conventional crude oil.

The oil industry is also exploring the possibility of reinjecting carbon dioxide into depleted underground oil and gas reservoirs and investigating a variety of forestry projects to sequester carbon. Other technologies used to produce energy with little or no climate impact include geothermal electricity production.

Many other energy companies have made substantial investments in alternative energy technologies in the expectation that they could be used to supply energy to the market. These efforts included solar water heaters, solar photovoltaic arrays, wind farms, and alcohol fuels. After much effort and investment with no prospects of commercial success, most of these projects were abandoned by about 1990. As the issue of climate change came to the fore in the 1990s, energy companies renewed their interest in these technologies, especially in view of the GHG emission reduction credits they might generate.

An Opportunity or a Threat?

The world of business is inherently dynamic. All businesses are trying to grow, both by creating new products to serve new markets and by replacing the products of their competitors in existing markets. This dynamism creates value for the consumer and wealth for society. But it also results in winners and losers.

When new, better, more efficient, and cheaper products edge out older, less desirable options, investors and workers in the obsolete industries suffer, but society benefits. The costs of the new products (which use less raw material, labor, and capital) are less than costs of the old, resulting in a net benefit to society.

If society wants to require substitution of newer but more expensive, less efficient, and less productive technologies,[11] it must be willing to accept the associated reduction in wealth and standard of living that necessarily follows. From society's point of view, any new jobs associated with a replacement technology are part of the cost of its introduction. Conversely, the jobs lost by workers in obsolete industries (painful as they may be for those involved) are a benefit to society as a whole because their talents can be more productively employed elsewhere.

Today, jobs are sometimes claimed to be a benefit of introducing a new technology. It is often said that alternative energy production will result in the creation of many new jobs as compared with the low-employment, capital-intensive oil industry. However, these jobs are a cost of the new technology. Moving from high-productivity employment to a job that is less productive must be regarded as detrimental to society's total output.

Conclusions

Global climate change is a major strategic issue for businesses of all types. Either a change in future global climate or the effects of our responses to avert such a change will alter our existing social and economic system. All sectors of the economy are likely to be affected in some way by the climate issue, either directly as a result of regulations or technological obsolescence or indirectly through impacts on economic growth and development.

Because of uncertainties about the size, timing, and impact of climate change, businesses that have studied this issue generally believe that an incremental approach is warranted. This will give markets a chance to work, inventors the time to develop new technologies, and existing facilities the ability to live out their productive lives.

There are two main parts to any climate policy. One aspect, primarily value based, is the degree to which society decides to devote its scarce resources to solving the identified climate problem. The second aspect is the way in which these resources are deployed.

The private sector has valuable input to the former and strong interests in the latter. Deciding the relative significance of any particular problem and how much of society's scarce resources should be allocated to address it is fundamentally a political decision that must be made based on good information and the values of the society. The private sector, as an agent of society, will pursue whatever goals the society and its elected representatives agree upon.

With its focus on creating wealth by producing the most output using the least input, business has a strong interest in ensuring that implementation rules are written in ways that lead to efficient and effective use of the resources under its control. Once a political decision is made to achieve a particular goal, it becomes essential to craft policies that encourage efficient use of resources to achieve that goal. In the end, the private sector will have to implement whatever policies society agrees are appropriate.

Notes

1. Anderson, J. W., 1999: "A 'crash course' in climate change," in W. E. Oates (ed.), *The RFF Reader in Environmental and Resource Management* (Washington, DC: Resources for the Future).
2. Although some in the business community continue to harbor strong doubts about the science of global climate change, particularly the conclusions drawn from interpreting the climate models, most businesses recognize that there is at least a significant chance that human activity is having some impact on the global climate. Discussions have therefore moved from the fundamentals to the specifics, including the best way to reverse, mitigate, or offset human impacts.
3. Dixit, A. K. and R. S. Pindyck, 1994: *Investment Under Uncertainty* (Princeton: Princeton University Press), p. 3.
4. Negotiated in Kyoto, Japan, in December 1997, subsequently signed by more than 85 countries and ratified by 16 developing nations but no developed nations. This treaty must be ratified by 55 parties to the UN FCCC, including parties representing at least 55 percent of the CO_2 emissions generated by the countries with reduction targets specified in the protocol, before it enters into force.
5. The UN FCCC, a treaty negotiated in Rio de Janeiro in 1992, has been ratified (including by the United States) and has entered into force.
6. "The role of technology in responding to concerns about global climate change," *The Business Roundtable,* July 1999, www.brtable.org; "Voluntary Actions by the Oil and Gas Industry, a conference on industry best practices," American Petroleum Institute, December 1999, www.api.org; "Technology assessment in climate change mitigation: a workshop summary," International Petroleum Industry Environmental Conservation Association, May 1999, www.ipieca.org.
7. Burns, T. G., 2000: "The California Air Resources Board . . . and the 'elsewhere emissions vehicle,'" http://www.PIRINC.org, December.
8. Intergovernmental Panel on Climate Change (IPCC): *Climate Change 2001: The Scientific Basis,* contribution of Working Group I to the Third Assessment Report of the IPCC, Fig. 3.1, Houghton, J. T., Y. Ding, D. J. Griggs, M. Noguer, P. J. van der Linden, and D. Xiaous (eds.) (Cambridge, England: Cambridge University Press), 944 pp.
9. Wigley, T. M. L., 1999: *The Science of Climate Change, Global and U.S. Perspectives* (Washington, DC: National Center for Atmospheric Research, Pew Center for Global Climate Change), p. 15.
10. Hansen, J., M. Sato, R. Ruedy, A. Lacis, and V. Oinas, 2000: "Global Warming in the Twenty-First Century: An Alternative Scenario," *Proceedings of the National Academy of Sciences of the United States of America,* 97 (18): 9875–9880, August 29. Veizer, J., Y. Godderis, and L. M. François, 2000: "Evidence for Decoupling of Atmospheric CO_2 and Global Climate During the Phanerozoic Eon," *Nature,* December, 408 (7): 698–701.

11. This is akin to the situation in a country that imposes highly protectionist tariff barriers to stifle competition from abroad. Although this may benefit current manufacturers and their employees, it raises costs and reduces choices for consumers, leading to a less productive economy.

CHAPTER 11

Activities Implemented Jointly[1]

Reimund Schwarze

The first Conference of the Parties (COP-1) to the UN Framework Convention on Climate Change (FCCC) in Berlin in 1995 established a program of activities implemented jointly (AIJs). Under this program, greenhouse gas (GHG) reduction and sequestration projects can be carried out through partnerships between an investor from a developed country and a host from a developing country or a country with an economy in transition (EIT).[2] The purpose of this program is to enhance technology transfer from developed to developing countries and to gather experience on the opportunities and obstacles for the joint implementation of climate protection measures. The AIJ experience will help elaborate the design of project-based mechanisms outlined in Articles 6 and 12 of the Kyoto Protocol, known as joint implementation (JI) and the Clean Development Mechanism (CDM), respectively.

A pattern emerges of region-specific investment portfolios that differ significantly between the United States, Japan, and Europe. These investment patterns can be traced to differing objectives and criteria for government project approval in the national AIJ programs of investor and host countries.[3] Established national links of trade and general development aid also influence these investment patterns.[4]

This study consists of 103 activity reports on 143 projects[5] from 36 countries (27 host and 9 investor countries) that reduced emissions by approximately 170 million tons (Mt).[6] Except for the Japanese projects, the data were found in the standardized reporting formats (Uniform Reporting Format [URF] or United States Initiative on Joint Implementation-Uniform Reporting Document [USIJI-URD]).

The results presented in this chapter proved to be robust in two sampling experiments, one in which I omitted 25 projects in the planning stage and another based on a smaller sample of 96 projects.[7]

Regional Distribution of AIJs

The UN FCCC's report on AIJs points out that "the geographical distribution of activities . . . shows a marked imbalance."[8] Indeed, as indicated in Fig. 11.1, 68 percent of the total projects were located in EITs, with those remaining hosted mostly in Latin American nations. A more balanced picture emerges when one considers the amount of GHG reduced because the greatest effect of GHG reduction will be achieved in Latin America.

This imbalance can be traced back to a general heterogeneity of AIJ projects: The average amount of GHGs reduced by projects is five times larger in Latin America than in central and eastern Europe because some European investor countries with heavy EIT investment have stipulated in their national programs that AIJs shall be "small to allow for quick implementation."[9]

Figure 11.1 also confirms the UN FCCC's view that "the bulk of current AIJ is between Annex I Parties."[10] Currently at 68 percent of the total, this predominance likely links with the projects' timing. Most of the projects were initiated before the Kyoto Conference in 1997, implying a reasonable expectation that AIJ projects would be incorporated after 2000 into a program of joint implementation between Annex I countries only. Before Kyoto's introduction of

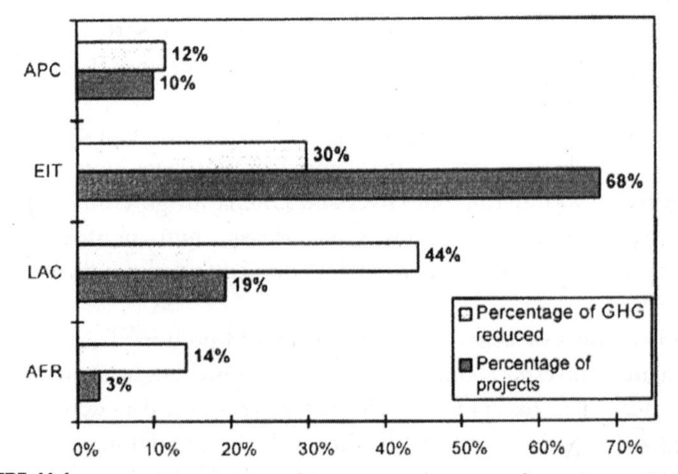

FIGURE 11.1. Regional distribution of AIJs. APC = Asian-Pacific countries, EIT = economies in transition, LAC = Latin American countries, AFR = African countries.

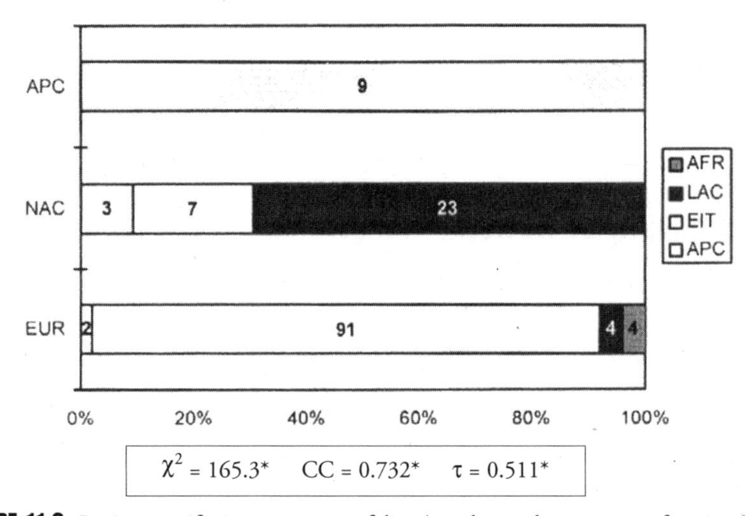

FIGURE 11.2. Region-specific investment portfolios (number and percentage of projects). APC = Asian-Pacific countries, AFR = African countries, LAC = Latin American countries, EIT = economies in transition.

the CDM, it was unclear whether AIJs with non–Annex I countries would fulfill Annex I countries' emission reduction commitments, spurring their investment in the Annex I countries of central and eastern Europe rather than in non–Annex I countries elsewhere. However, this is only one explanation for the observed pattern of cooperation in AIJs. In view of Fig. 11.2's regional investment trading portfolios, a "neighborhood trading" hypothesis emerges.

Reviewing the investment portfolio of Asian-Pacific investor countries (APC, i.e., Australia and Japan), we find that all nine projects take place in neighboring Asian-Pacific economies. Projects in EITs make up almost 90 percent of the European AIJ portfolio, with a similar percentage of regional focus marking U.S. investment. This pattern of intra-Asian, intra-European, and intra-American cooperation ("neighborhood trading") is significant according to contingency analysis indicators—which test the existence (Pearson's χ^2), the strength (contingency coefficient) and the direction (Goodman and Kruskal's τ) of claimed contingencies—and can be explained by reference to the established institutional links of development cooperation. The hemisphere partnership for development of the 1994 Summit of the Americas[11] and the Baltic Sea Region Initiative of the Nordic States[12] and the European Union (EU) exemplify pertinent institutions of development cooperation.[13]

Australia and Japan's focus on neighboring Asia–Pacific economies is a stated trade and development policy objective of the national AIJ programs. Australia's International Greenhouse Partnerships Office has as a main objective "to enhance Australian trade and investment links in environmental technology and services . . . particularly in the Asia–Pacific region."[14] The government of Japan states that "the role of Japan is to share technologies with other countries, especially with Asian economies."[15]

Distribution of Activity Types

Climate change mitigation activities can be classified as related to energy efficiency (EEF), renewables (REN), fuel switching (FUE), fugitive gas capture (FGC), land use change and forestry (LUCF),[16] agriculture, industrial processes, solvents, waste disposal, and bunker fuels.

Only the first five of these activity types have been implemented in AIJs (Fig. 11.3).[17] Most projects (83 percent) have been in energy-related activities (EEF, FUE, REN, or FGC). Only 17 percent of the projects have been related to LUCF activities (forest preservation, reforestation, and afforestation). However, LUCF projects account for more than 38 percent of the GHG reductions, largely because LUCF projects have a much longer lifetime (39 years on average) than AIJs (21 years).

This difference also can be attributed to project heterogeneity. As Fig. 11.4 indicates, LUCF-related projects reduce GHGs more than fossil fuel–related activities, on average. A notable exception is fugitive gas capture, which shows the single largest GHG reduction per project.

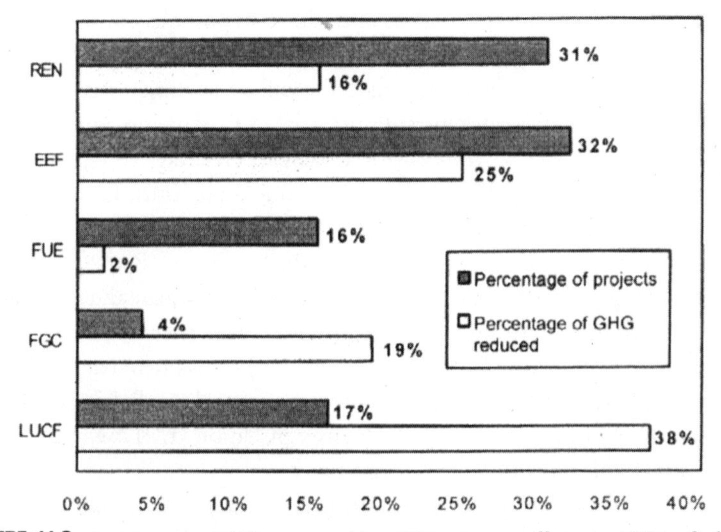

FIGURE 11.3. Activity type. REN = renewables, EEF = energy efficiency, FUE = fuel substitution, FGC = fugitive gas capture, LUCF = land use change and forestry.

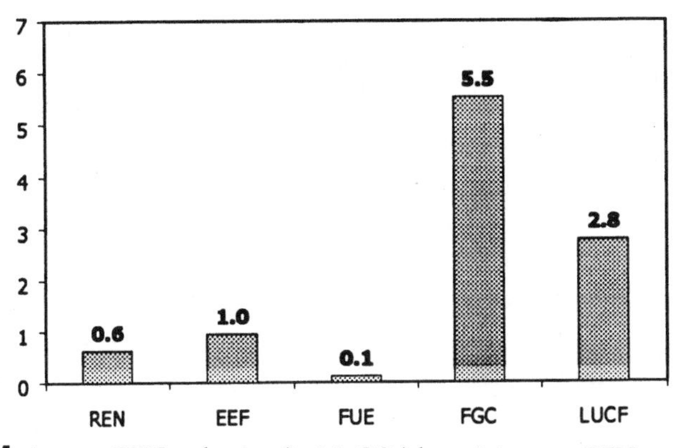

FIGURE 11.4. Average GHG reduction (in Mt CO_2) by activity type. REN = renewables, EEF = energy efficiency, FUE = fuel substitution, FGC = fugitive gas capture, LUCF = land use change and forestry.

It is of some interest to look at the regional distribution of investment on different activities (Fig. 11.5). Europe, Japan, and Australia focus on technological mitigation activities, whereas the United States has a much greater involvement in forestry projects. Several European national AIJ programs exhibit preferences for fuel switching and energy efficiency–related projects (e.g., in Germany and Switzerland) or for quickly implementable small investments, whereas the

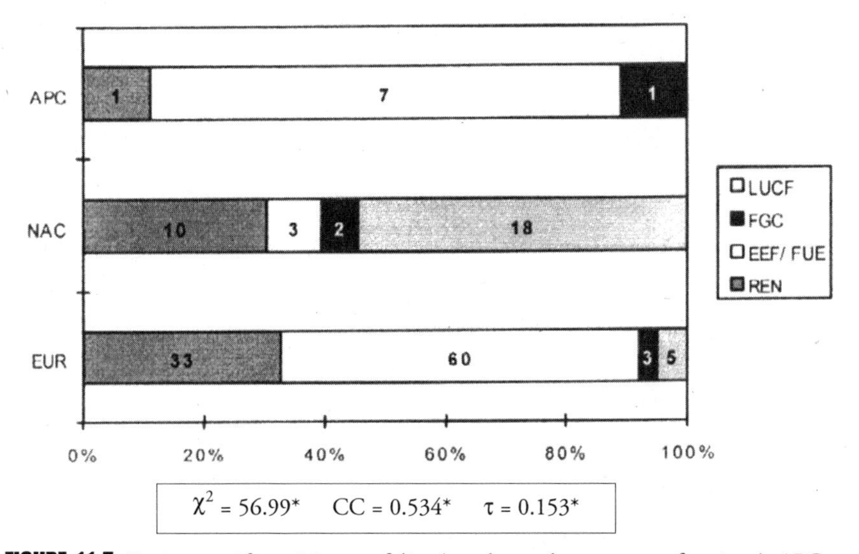

FIGURE 11.5. Region-specific activity portfolios (number and percentage of projects). APC = Asian-Pacific countries, REN = renewables, EEF = energy efficiency, FUE = fuel substitution, FGC = fugitive gas capture, LUCF = land use change and forestry.

American AIJ program exhibits no historical focus on project type or size. National AIJ priorities of host countries also influence this observed distribution of activity types. For example, the government of Poland, host to numerous coal-to-gas conversion and energy efficiency enhancement projects, selected this portfolio with the aim "to achieve technological development and upgrade equipment in activities that directly reduce the generation of GHG in production of goods and services."[18]

On the other hand, the government of Costa Rica promoted projects in the forestry sector "to claim the cost of environmental services executed by private forest owners at international level."[18a] Costa Rica hosts 15 AIJ projects, 11 of which are LUCF-related (i.e., in sustainable forestry, reforestation, and forest preservation).

There is a suggestive link between the observed regional pattern of AIJs (neighborhood trading) and the sectoral pattern of AIJs. Private U.S. investors, looking for AIJs in their neighborhood, find hosts who favor forestry projects, which confirm to the USIJI criteria of project diversity and cost-effectiveness. Public or publicly cofinanced private investors in Europe and Japan find hosts in their neighborhood (EIT and China), who share the goals of energy efficiency and fuel substitution. Of course, the actual reasoning for each AIJ project may have been quite different, but this is the typical reasoning that the aggregate data suggest.

Private Sector Participation in AIJs

Figure 11.6 indicates a large amount of private investment in AIJ funds in addition to a large share of public non–AIJ-related funding, as from the World Bank's Global Environmental Facility or bilateral direct aid.

As shown in Fig. 11.7, U.S. investors account for most of the private AIJ support, funding 31 of the total 38 projects, in comparison to six private Euro-

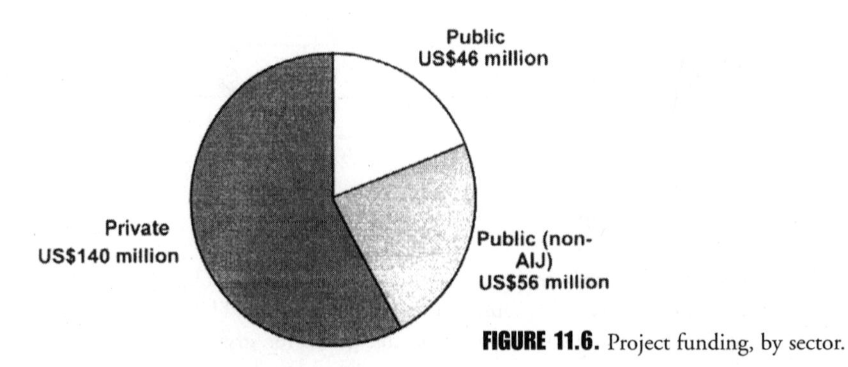

FIGURE 11.6. Project funding, by sector.

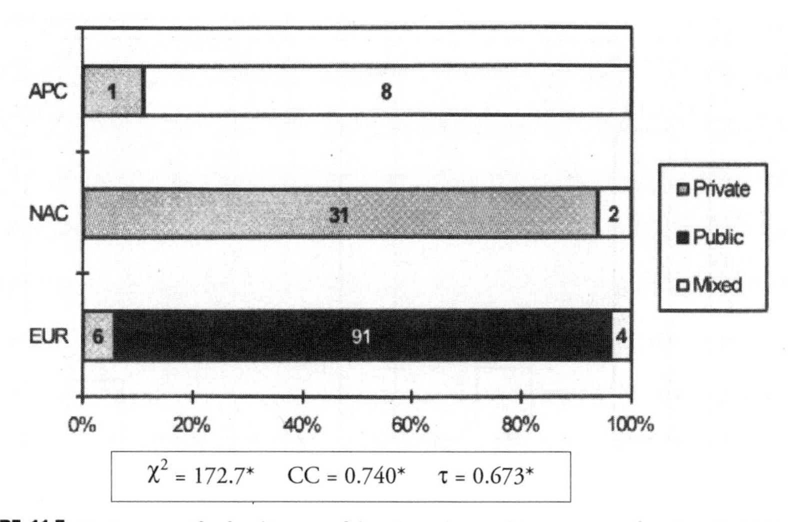

FIGURE 11.7. Region-specific funding portfolios (number and percentage of projects). APC = Asian-Pacific countries.

pean projects and one Australian project. This result is a likely consequence of the U.S. Initiative on Joint Implementation's explicit encouragement of privately initiated projects.[19]

Interestingly, AIJ investment exhibits regional specificity. U.S. projects typically are large in cost and effects, not particularly focused on technologies, and privately initiated. They are overwhelmingly implemented in Latin American countries. European projects, on the other hand, typically are small, publicly funded, and related to energy efficiency and fuel substitution. They are located predominantly in EITs. Asian projects are somewhat similar to European projects (i.e., they are small and publicly cofinanced) but are focused on the Asia–Pacific region. The public funding of AIJ projects is much smaller in Japan than in Europe.

Cost of AIJs

Considering aggregate data, one can establish a credible and consistent picture of AIJ cost. Figure 11.8 expresses the average total cost per ton of CO_2 equivalent reduced by activity type as a sum of investment cost, operating cost (if available), project development cost, and transaction cost for monitoring, verification, and general project administration. These figures indicate fugitive gas capture as the cheapest way to reduce GHG emissions, followed by LUCF activities and those related to energy efficiency. Each of these options costs less than US$4 per ton of CO_2.

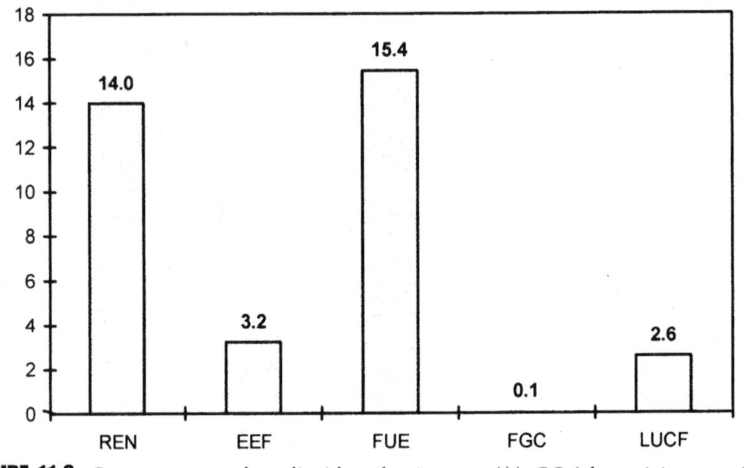

FIGURE 11.8. Gross average carbon dioxide reduction cost ($/t CO_2) by activity type. REN = renewables, EEF = energy efficiency, FUE = fuel substitution, FGC = fugitive gas capture, LUCF = land use change and forestry.

Given our gross total cost approach, US$4 per ton of CO_2 is a rock-bottom figure, especially in comparison with a UNEP estimate of US$14 based on a non–Annex I multicountry study.[20]

Furthermore, if one considers that most AIJs generate revenues from fuel savings or selling of forest products and services (the amount of which cannot easily be extracted from available data), these AIJ options are clearly no-regrets options. We find that all AIJs, except for coal-to-gas fuel substitution projects, provide revenues higher than costs (Fig. 11.9).[21]

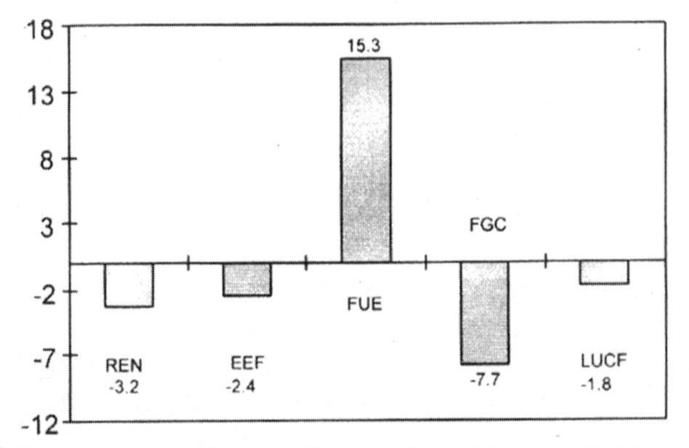

FIGURE 11.9. Net average reduction cost ($/t CO_2) by activity type. REN = renewables, EEF = energy efficiency, FUE = fuel substitution, FGC = fugitive gas capture, LUCF = land use change and forestry.

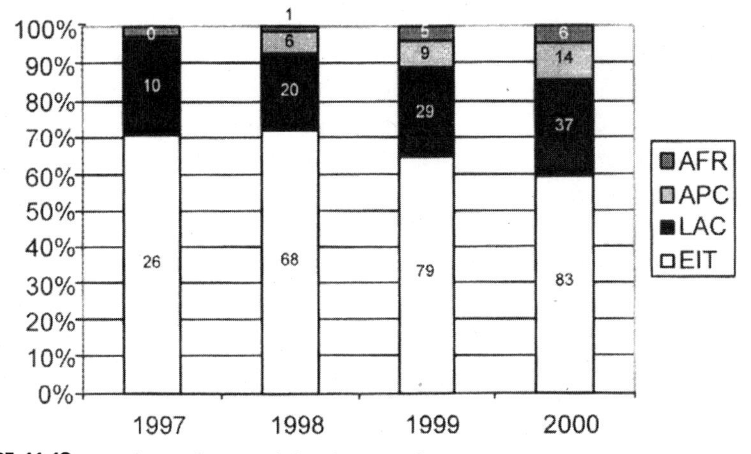

FIGURE 11.10. Number and regional distribution of AIJ activities, 1997–2000. AFR = African countries, APC = Asian-Pacific countries, LAC = Latin American countries, EIT = economies in transition. (Adapted from Fig. 1 of FCCC/SB2000/6, p. 7).

Recent Developments

In its latest UN FCCC synthesis report,[22] the climate secretariat describes recent AIJ developments, largely confirming the results presented here and noting new developments. First, the number of AIJ projects and parties involved in AIJs increased from the year 1999 to the year 2000. The emission reductions increased even more dramatically, largely because of a few additional fugitive gas capture projects. Second, the regional distribution of projects is gradually changing in favor of non–Annex I host countries, with the EIT countries' share of AIJ hosting decreasing by more than 5 percent of the total projects between 1999 and 2000 (Fig. 11.10).

Conclusions

From the original data, I find that AIJs are very much influenced by regional factors, particularly by national AIJ programs, with a marked divergence between the regional investment portfolios of the United States and of Europe and Japan. U.S. projects typically are large in cost and environmental effects, privately initiated, and located in the Latin American countries. European and Japanese projects typically are small, publicly funded, and located in EITs and China. This regional pattern of AIJs can be related to proximity (neighborhood trading) and the established institutions of trade and development aid (lower transaction cost).

Another general result of this analysis is the cost-minimizing feature of AIJs.

AIJs are overwhelmingly no-regret measures with zero or negative cost. How does this finding of national preferences affect the efficiency of project-based Kyoto mechanisms such as CDM and JI? Because these Kyoto mechanisms clearly resemble AIJs in the need for government approval, we expect that future JI and CDM national programs will influence the amount and structure of trading, as in AIJs. Basic economic reasoning indicates that regulatory preferences diminish the efficiency of trade compared with a free market because marginal costs are not equal between participants. However, this result must be qualified. As long as we are talking about no-regrets strategies (i.e., as long as mitigation projects come at zero or negative cost), national preferences only diminish the amount of cost savings that will be exploited. This may be the price worth paying to achieve sustainable development in non–Annex I countries and meaningful participation of these countries in the early phase of international climate change policy.

Notes

1. This is a revised and updated version of a paper originally published in *Ecological Economics* Schwarze, R., 2000: "Activities Implemented Jointly: Another Look at the Facts," *Ecological Economics,* 32: 255–267.
2. EITs are central and eastern European countries and states of the former Soviet Union that are in transition to a market economy.
3. UN Framework Convention on Climate Change/Subsidiary Body for Scientific and Technological Advice (UN FCCC/SBSTA), 1997b: "Activities implemented jointly under the pilot phase. Synthesis report on activities implemented jointly." Addendum, FCCC/SBSTA/1997/12/Add.1. Available online: http://www.unfccc.de.
4. Government approval is a legal prerequisite for AIJs under the Berlin Mandate. AIJs must be backed by a mutual agreement between the host and the investor country known as letter of intent.
5. Several UN FCCC reports are on more than one project; for example, the AIJ between Norway and Poland (norpol01a) is on 21 coal-to-gas boiler conversion projects. To make this information compatible with other, more disaggregated reports (e.g., on boiler conversion between Sweden and Estonia: estswe01–estswe16), I have changed the UN FCCC definition of projects ("activities") in this study to "projects as declared in the DNA [Designated National Authority] reports." Following this new definition of projects, my set of data contains 143 projects, as compared to 96 projects considered by the UN FCCC.
6. These figures compare to an expected 5-year (2008–2012) Kyoto market of approximately 10,000 Mt CO_2, which is derived from the International Institute for Applied System Analysis' "most likely" estimate of emission surpluses for western Europe, North America, and the Pacific, Organisation for Economic Co-operation and Development (OECD) in 2008–2012 (2,697 Mt C). See Victor, D.G., N. Nakicenovic, and Victor, N., 2001: "The Kyoto Protocol Emission Allocations:

Windfall Surpluses for Russia and Ukraine," *Climatic Change*: 49 (3): 263–277, Table 2.

7. Schwarze, R., 1998: *Activities Implemented Jointly: Another Look at the Facts [Diskussionspapier der Wirtschaftswissenschaftlichen Dokumentation der TU Berlin]*, Report number: 98-18 (unpublished).
8. UN Framework Convention on Climate Change/Conference of the Parties: *Activities Implemented Jointly: Review of Progress under the Pilot Phase* (Decision 5/CP.1). "Second synthesis report on activities implemented jointly." Note by the secretariat (FCCC/CP/1998/2).
9. A detailed list of national AIJ programs can be found in UN FCCC (1997b). An example of a national AIJ program is the USIJI.
10. UN FCCC, 1997a: United Nations FCCC/Subsidiary Body for Scientific and Technological Advice, *Activities Implemented jointly under the Pilot Phase.* "Synthesis report on activities implemented jointly" (FCCC/SBSTA/1997/12), p. 4. Available online: http://www.unfccc.de. Reaffirmed in UN FCCC, 1998, UN Framework Convention on Climate Change/Conference of the Parties, *Activities Implemented Jointly: Review of Program under the Pilot Phase* (Decision 5/CP1). "Second synthesis report on activities implemented jointly," p. 4. Notes by the secretariat (FCCC/CP/1998/2) from www.unfccc.de. Annex I countries include the 24 original OECD member countries and 11 former states of the Soviet bloc as well as the European Community as a regional economic integration organization.
11. U.S. Department of State, 1995: "Summit of the Americas action plan. Implementation of the summit action plan," americas.fiu.edu/state.
12. Baltic Sea States Summit, 1996: "Presidency Declaration," Visby, Sweden, May 3–4, 1996, available online: www.baltinfo.org/Docs/headsofgov/2/summit.htm.
13. Commission of the European Communities, 1996: "Final communication from the commission's Baltic Sea Region Initiative," Brussels, October 4, 1996, SEC (96) 608, www.baltinfo.org/Docs/eu/communi.htm.
14. International Greenhouse Partnerships Office, 1998: "International greenhouse partnerships. Project guidelines and priorities," available at www.dist.gov.au/resources/energy_greenhouse/2nd_round.pdf.
15. Matsuo, N., 1996: "AIJ Japan program update," www.northsea.nl/jiq/japan.htm.
16. The Kyoto Protocol (Report 3) distinguishes between three different LUCF activities: afforestation, reforestation, and deforestation. Experience in the pilot phase shows that this systematically important distinction is not very useful in classifying AIJ projects in this field because most of these projects combine measures to protect existing forest and activities to reforest nonforest lands (e.g., abandoned pasture). Following this, USIJI has classified its latest projects in this sector as LUCF (see www.unfccc.de/program/aij/aijact99/criusa04-99.html).
17. Arguably, two projects of fugitive gas capture (nldrus01) could also be classified as related to waste disposal because they are on sanitary landfilling with energy recovery. In this study, I follow the classification of these projects as FGC by the UN FCCC.
18. The mentioned host country priorities in the AIJ pilot phase are drawn from www.unfccc.de/program/aij/aijprog/aij_ppol.html (Poland); and

18a. www.unfccc.de/program/aij/aijprog/aij_pcri.html (Costa Rica).

19. USIJI, 1998: "Unique aspects of U.S. Initiative on Joint Implementation Projects," www.ji.org.usiji/unique.html.

20. UNEP, 1994: "UNEP greenhouse gas abatement costing studies," Phase Two Report, Part 1: "Main report," UNEP Collaborating Centre on Energy and Environment, Riso National Laboratory, Denmark.

21. Negative cost does not necessarily imply that these projects are economically viable because most AIJs do not factor the opportunity cost of capital (interest on investment). Because of this general flaw in the costing of AIJs, we may assume that the rate of return for most AIJs is below the market rate of return. I am grateful to Mike Toman of Resources for the Future for this important qualification.

22. United Nations Framework Convention on Climate Change/Subsidiary Body for Scientific and Technological Advice/Subsidiary Body for Implementation, 2000: *Activities Implemented Jointly under the Pilot Phase*. Fourth synthesis report and draft revised uniform reporting format. Note by the secretariat (FCCC/SBSTA/2000/6). Available online: http://www.unfccc.de.

PART IV

FORESTS AND AGRICULTURE

CHAPTER 12

Climate Change and Agriculture: Mitigation Options and Potential

Holly L. Pearson

Growing crops and raising livestock are responsible for approximately 20 percent of the anthropogenic greenhouse effect.[1] Worldwide, these activities contribute approximately 5, 50, and 70 percent of anthropogenic carbon dioxide (CO_2), methane (CH_4), and nitrous oxide (N_2O) emissions, respectively.[2] In addition, tropical deforestation and land degradation account for an estimated one-fifth of the annual atmospheric increase in CO_2, and more than half of this newly cleared land is used for agriculture.[3]

Because agriculture is an important source of greenhouse gases (GHGs), countries could choose to meet part of their commitments under the Kyoto Protocol to the UN Framework Convention on Climate Change (FCCC) through activities in the agricultural sector. If ratified, the protocol will require industrialized (Annex I) countries to reduce their aggregate GHG emissions in 2008–2012 by an average of 5.2 percent below 1990 emission levels.[4] On July 23, 2001, the Conference of the Parties (excluding the United States) agreed on an important series of measures for implementation of the Kyoto Protocol: the Bonn Agreement.[5] The specific policies that individual countries will develop to meet their protocol commitments are still unclear.

Agricultural mitigation activities fall into two broad categories: reducing GHG emissions from agricultural activities and increasing the amount of carbon that is stored in vegetation and soils on agricultural lands.[6] In addition to domestic options, the protocol includes the Clean Development Mechanism (CDM), defined in Article 12, which would allow Annex I countries to earn credits to meet domestic emission reduction commitments by investing in projects in non–Annex I countries that reduce emissions.[7] Under the Bonn Agreement, only certain activities are eligible for credit in the first commitment period. A variety of economic, political, and scientific issues will affect the extent to which countries use their domestic agricultural sectors to meet emissions-reduction commitments. In the longer term, factors such as increasing human population size, growing and changing food demands, and climate change itself come into play, with profound effects on mitigation through agricultural activities.

Emission Reductions in the Agricultural Sector

Treatment in the Kyoto Protocol

A key feature of the Kyoto Protocol is that Annex I countries can combine reductions in any of the listed GHGs as long as the total reduction meets the commitment specified in Annex I.[8] The gases can be compared by their global warming potentials (GWPs), which reflect the ability of gases to trap heat in the atmosphere and the atmospheric lifetime of the gases relative to CO_2.[9] Thus, the more Annex I countries can reduce emissions of non-CO_2 gases, the less they will have to reduce CO_2.[10]

The GHGs emitted by the agricultural sector are CO_2, CH_4, and N_2O, and the sources listed in Annex A of the Kyoto Protocol include enteric fermentation, manure management, rice cultivation, agricultural soils, prescribed burning of savannas, field burning of agricultural residues, and other emissions. Annex I countries currently estimate sources of these gases in their national inventories following guidelines developed by the IPCC[11] and submit data to the UN through National Communications.[12]

Although it is not yet known which sources countries will regulate, some countries may want to pursue reductions in the agricultural sector because the practices that reduce emissions offer economic or other environmental benefits. In assessing mitigation opportunities, however, uncertainties in measurement and inventory methods must be considered. Emission reductions for some agricultural activities—for instance, those involving many small sources with much site-to-site variation—may be more difficult to estimate than other emission sources,[13] and sufficient measurement protocols could be prohibitively costly.

Opportunities to Reduce Emissions in Annex I Countries

Carbon Dioxide

Direct CO_2 emissions result from the use of farm machinery and energy, and indirect emissions arise in the manufacture of synthetic fertilizers, pesticides, and other agrochemicals.[14] Overall, the agricultural sector is a small source of

CO_2 in Annex I countries because it is responsible for only an estimated 3–4.5 percent of fossil fuel use.[15] Field burning of agricultural residues and prescribed burning of savannas also result in CO_2 emissions, but these are small in Annex I countries. Land use activities and land use change associated with agricultural production can also produce CO_2 (see below).

Regulation of CO_2 emissions through market approaches, such as tradable emission permits or taxes, or nonmarket mechanisms, such as emission quotas, could increase fuel and energy prices. In response, farmers might switch to farming practices that use less fuel or energy, thereby reducing their costs as well as their CO_2 emissions. For instance, reducing the number of field operations through conservation tillage can decrease fuel use by up to 55 percent.[16] The net benefit to the atmosphere depends on the extent to which increases in use of herbicides or other agrochemicals accompany conservation tillage. Reducing use of synthetic nitrogen fertilizers can decrease CO_2 emissions because production of nitrogen fertilizer uses large amounts of fossil fuel—the equivalent of 1.2 kg of fossil carbon for each kilogram of fixed nitrogen.[17] The IPCC has identified additional agricultural CO_2 mitigation options.[18]

Methane

By far the largest source of agricultural CH_4 in Annex I countries is livestock. Methane is a natural byproduct of digestion in livestock, a source known as enteric fermentation.[19] Methane is also produced when manure decomposes in the absence of oxygen. Though important on a global basis, CH_4 emissions from paddy rice cultivation are a minor CH_4 source in most Annex I countries, with Japan a notable exception.[20] Field burning of agricultural residues and prescribed burning of savannas also result in minor CH_4 emissions in some Annex I countries.

The data available to date indicate that agriculture is a major source of CH_4 in most Annex I countries, averaging 44 percent of total CH_4 emissions (Table 12.1). It is a more modest contributor to overall GHG emissions, ranging from 1.3 percent in Japan to nearly 40 percent in New Zealand, and averaging 6 percent. As Table 12.1 indicates, there are substantial differences between countries in the sources and magnitude of agricultural CH_4 emissions.

Agricultural CH_4 emissions in Annex I countries as a group declined between 1990 and 1998, mainly as a consequence of decreases from countries with economies in transition (EITs).[21] In the early 1990s in EIT countries, there was a major restructuring of the agricultural sector that included the breakup of many large farms, development of the private sector, a rise in fuel and fertilizer prices, and a partial loss of traditional markets for agricultural products.[22] Herd

TABLE 12.1. Agricultural Emissions of CH_4 in 1998 (or latest year available, as indicated) from Annex I Countries

Country	*Enteric Fermentation (10^3 tons)*	*Manure Management (10^3 tons)*	*Other Agriculture (10^3 tons)**	*Total Agriculture (10^3 tons)*	*Percentage Agriculture of Total CH_4 Emissions*	*Percentage Agricultural CH_4 of Total GHG Emissions†*	*Percentage Change in Agricultural CH_4 Emissions 1990 to 1998 (latest year available)*
Australia	2,887	83	376	3,346	59.8	14.5	−2.1
Austria	131	26	35	192	41.8	5.1	−11.5
Belgium	206	130	14	351	60.4	5.1	−7.9
Bulgaria	82	30	2	114	17.4	2.9	−55.6
Canada	855	242	0	1,097	25.7	3.4	+11.9
Czech Republic	86	35	0	121	22.9	1.7	−68.6
Denmark	138	45	0	184	64.1	5.1	−4.9
Estonia	23	7	0	30	29.7	2.9	−100.0
Finland	71	9	0	81	40.9	2.2	−13.6
France	1,329	175	31	1,535	59.4	5.8	−6.3
Germany	1,016	514	26	1,556	44.7	3.2	−22.2
Greece	143	27	109	279	54.8	4.9	+2.9
Hungary	81	34	3	118	17.2	3.0	−47.9
Iceland (1995)	10	1	0	11	78.6	8.7	−9.1
Ireland	494	70	0	564	86.9	18.6	+8.9
Italy	628	183	82	893	45.3	3.5	−1.9
Japan (1997)	NA	NA	NA	789	56.8	1.3	−6.7
Latvia	32	4	0	36	37.1	6.5	−208.3

Lithuania	73	10	0	83	46.9	7.3	–118.1
Luxembourg (1995)	16	1	0	17	77.3	3.5	–5.9
Netherlands	342	94	0	435	40.8	4.0	–16.1
New Zealand	1,389	17	0	1,406	88.3	39.7	–6.1
Norway	94	16	0	110	31.8	4.3	+9.1
Poland	534	46	1	582	24.9	3.0	–46.0
Portugal	116	147	9	272	39.8	7.6	–10.7
Romania (1994)	NA	NA	NA	357	24.4	4.6	–56.9
Russian Federation (1996)	NA	NA	NA	3,362	18.1	3.7	–50.5
Slovakia	55	11	0	66	24.5	2.6	–104.5
Slovenia (1990)	38	6	0	44	25.0	4.8	NA
Spain	620	361	19	999	48.1	5.8	+5.9
Sweden	143	16	0	159	62.1	4.8	–0.6
Switzerland	118	19	0	137	62.3	5.4	–10.2
Ukraine	NA	NA	NA	1,196	18.5	5.5	–88.5
United Kingdom	883	112	0	995	37.7	3.2	–4.2
United States	5,885	3,990	511	10,386	32.9	3.3	+15.6
	TOTAL				AVERAGE		
All EU	6,276	1,910	325	8,512	54.8	5.7	–9.1
All Annex I	18,518	6,461	1,218	31,902	44.2	6.0	–30.3

Source: Greenhouse Gas Inventory Database, available at http://unfccc.int/resource/ghg/tempemis2.html (last visited May 15, 2002). Most data are from Greenhouse Gas Inventory submissions by Annex I parties from 2000. Data for Croatia, Liechtenstein, and Monaco were not available. Totals may not add up due to rounding.

*Includes source categories of rice cultivation, prescribed burning of savannas, and field burning of agricultural residues.

†Total GHG emissions include CO_2, CH_4, and N_2O. The other greenhouse gases—HFCs, PFCs, and SF_6—were not included because the data set is not complete. A 100-year time horizon was assumed, with a global warming potential of 21 for CH_4.

sizes of all livestock tended to decline sharply, with consequent effects on CH_4 emissions. How long these lowered emissions will be sustained depends on the economic transition process as well as agricultural policies. In addition, to the extent that agricultural activities were merely displaced to non–Annex I countries (a phenomenon known as leakage), overall emissions to the atmosphere did not decrease.

Declines also occurred in several non-EIT countries. For instance, in New Zealand, removal of government support for livestock production contributed to a decline in livestock-associated CH_4 between 1984 and 1994.[23] Product quotas and controls on livestock density in the European Union have decreased herd sizes and CH_4 emissions.[24] However, CH_4 emissions increased between 1990 and 1998 in six countries, notably in Canada (12 percent) and the U.S. (16 percent)[25] (Table 12.1).

Technologically feasible and cost-effective practices that reduce the amount of CH_4 emitted from enteric fermentation or manure management are available in some places. Improving diet quality, nutrient balance, and digestibility of feed allows more carbon to be directed toward milk or meat production, with less released as CH_4.[26] In many Annex I countries, livestock have been moved off the land and into large livestock facilities, which increases CH_4 emissions because adding water to manure to aid control and storage increases anaerobic CH_4 production. Technologies such as digesters and covered lagoons can capture the CH_4 produced in liquid-based manure management systems and make it available for on-farm use as fuel,[27] reducing emissions of both CO_2 and CH_4 because farms have to purchase less energy. A guaranteed way to reduce livestock CH_4 emissions is to reduce herd size through decreases in meat consumption, but per capita consumption is high in Annex I countries and increasing worldwide.[28]

Nitrous Oxide

N_2O is produced by natural microbial processes in soils (nitrification and denitrification), but applying synthetic and manure fertilizers to soils adds nitrogen and, consequently, may increase N_2O emission rates.[29] Livestock manure is the second most important source of agricultural N_2O emissions, followed by prescribed burning of savannas, field burning of agricultural residues, and rice cultivation.[30]

The GHG emission estimates available to date indicate that agriculture produces an average of 62 percent of total N_2O emissions in Annex I countries, but country-to-country variation is large (Table 12.2). Agricultural emissions of N_2O in Annex I countries as a group declined between 1990 and 1998. As with CH_4, much of this decline can be attributed to reductions in EIT countries. Despite a decline at the aggregate level, agricultural N_2O emissions increased between 1990 and 1998 in 9 countries (Table 12.2).

TABLE 12.2. Agricultural Emissions of N_2O in 1998 (or latest year available, as indicated) from Annex I Countries

Country	*Agricultural Soils (10^3 tons)*	*Manure Management (10^3 tons)*	*Other Agriculture (10^3 tons)**	*Total Agriculture (10^3 tons)*	*Percentage Agriculture of Total N_2O Emissions*	*Percentage Agricultural N_2O of Total GHG Emissions†*	*Percentage Change in Agricultural N_2O Emissions 1990 to 1998 (latest year available)*
Australia	51.8	1.7	17.3	70.7	78.9	4.5	+16.4
Austria	3.3	NA	NA	3.3	44.1	1.3	–1.2
Belgium	9.3	1.5	0.0	10.8	32.0	2.3	–0.9
Bulgaria	33.7	1.5	0.0	35.2	73.7	13.0	–38.7
Canada	130.8	16.3	0.0	147.0	70.2	6.7	+13.4
Czech Republic	16.0	1.4	0.0	17.4	64.2	3.6	NA
Denmark	26.1	1.5	0.0	27.6	90.6	11.3	–16.5
Estonia	1.2	0.0	0.0	1.2	90.8	1.7	+31.1
Finland	11.5	1.4	0.0	12.9	50.3	5.3	–16.2
France	166.9	10.1	0.2	177.2	65.2	10.0	–2.5
Germany	76.0	8.0	0.0	84.0	52.6	2.6	–12.5
Greece	19.0	0.6	0.1	19.7	65.1	5.1	–7.2
Hungary	32.9	1.6	0.0	34.5	98.5	13.0	NA
Iceland (1995)	0.2	0.0	0.0	0.2	47.5	2.2	–13.6
Ireland	23.0	2.4	0.0	25.3	78.0	12.3	+10.8
Italy	69.0	12.3	0.0	81.3	65.4	4.7	+4.8
Japan (1997)	NA	NA	NA	6.5	9.9	0.2	–29.6
Latvia	3.2	0.0	0.0	3.2	84.0	8.6	–85.5
Lithuania	1.7	0.0	0.0	1.7	14.9	2.2	–84.7
Luxembourg (1995)	0.5	0.0	0.0	0.5	69.6	1.4	0.0
Netherlands	25.2	0.7	0.0	25.9	36.2	3.6	+16.7
New Zealand	37.3	0.4	0.0	37.6	96.7	15.7	+1.3

(*continued*)

TABLE 12.2. *Continued*

Country	*Agricultural Soils (10^3 tons)*	*Manure Management (10^3 tons)*	*Other Agriculture (10^3 tons)**	*Total Agriculture (10^3 tons)*	*Percentage Agriculture of Total N_2O Emissions*	*Percentage Agricultural N_2O of Total GHG Emissions†*	*Percentage Change in Agricultural N_2O Emissions 1990 to 1998 (latest year available)*
Norway	8.5	0.0	0.0	8.5	51.5	4.8	–1.9
Poland	31.2	0.0	0.1	31.2	60.6	2.4	–23.8
Portugal	7.5	0.1	0.1	7.7	35.8	3.2	–9.2
Romania (1994)	NA	NA	NA	6.8	27.2	1.3	–67.2
Russian Federation (1995)	0.0	111.0	0.0	111.0	49.2	1.7	–44.5
Slovakia	7.3	1.8	0.0	9.1	85.2	5.4	–44.9
Slovenia (1990)	4.6	0.0	0.0	4.6	90.0	7.4	NA
Spain	61.1	51.2	1.2	113.5	80.2	9.8	+5.0
Sweden	13.8	1.9	0.0	15.7	61.2	6.9	–6.2
Switzerland	7.0	1.4	0.0	8.3	71.8	4.9	–9.8
Ukraine	NA	NA	NA	6.4	40.5	0.4	–77.1
United Kingdom	90.7	5.0	0.0	95.7	53.0	4.5	–4.7
United States	991.9	47.3	1.4	1,040.6	73.7	4.9	+11.7
	TOTAL				AVERAGE		
All EU	602.8	96.66	1.6	701.08	58.6	5.6	–2.7
All Annex I	2,073.2	170.1	20.4	2,282.9	61.7	5.4	–14.8

Source: Greenhouse Gas Inventory Database, available at http://unfccc.int/resource/ghg/tempemis2.html (last visited May 15, 2002). Most data are from Greenhouse Gas Inventory submissions by Annex I parties from 2000. Data for Croatia, Liechtenstein, and Monaco were not available. Totals may not add up due to rounding.

*Includes source categories of rice cultivation, prescribed burning of savannas, and field burning of agricultural residues.

†Total GHG emissions include CO_2, CH_4, and N_2O. The other greenhouse gases—HFCs, PFCs, and SF_6—were not included because the data set is not complete. A 100-year time horizon was assumed, with a global warming potential of 310 for N_2O.

If estimation and measurement methods improve to the extent that N_2O emissions reductions can be credited, some countries may choose these reductions as part of their GHG emission reduction strategy. One way to reduce N_2O emissions is to improve the management of nitrogen fertilizers. An estimated half to two-thirds of applied nitrogen fertilizers leave farmers' fields through leaching or gaseous losses, instead of being incorporated into crop plants.[31] These gaseous losses contribute to climate change, and the leaching losses pollute waterways and nonagricultural ecosystems. Thus, increasing the amount of fertilizer that is actually used by crops (i.e., fertilizer use efficiency) could yield a variety of environmental benefits. Manipulating fertilizer type, the method, rate, and timing of application, water management, and use of nitrogen transformation inhibitors may decrease N_2O emissions while maintaining crop productivity.[32]

The methodological challenges for N_2O are substantial. Under the IPCC inventory methods, the amount of nitrogen fertilizer applied to fields is multiplied by emission factors to estimate the amount of nitrogen lost as N_2O.[33] Countries are encouraged to develop region-specific emission factors, but doing so will entail substantial field research and data analysis. Most countries rely on the IPCC default values, potentially leading to greater inaccuracy in emission estimates. To deal with these issues, the IPCC has recently developed good practice guidelines that complement inventory methods and improve uncertainty management.[34]

Opportunities to Reduce Emissions in Non–Annex I Countries

Annex I countries contribute a smaller proportion (though not amount) of worldwide GHG emissions each year. For instance, the developed: developing country ratio of GHG emissions was 89:11 percent in 1950 and 56:44 percent in 1995; developing country emissions are projected to exceed developed country emissions early in this century.[35] Like total GHG emissions, emissions from the agricultural sector will increase dramatically in non–Annex I countries. Numbers of livestock and their associated CH_4 emissions are steadily growing. For instance, the quantity of grain used for feed in China, the largest non–Annex I country, increased more than 500 percent in the past two decades and continues to increase.[36] Nitrogen fertilizer use is predicted to more than double from 1990 to 2020 in developing countries.[37]

Projects undertaken by Annex I countries that reduce agricultural CO_2, CH_4, or N_2O emissions in non–Annex I countries could result in emission reductions credits. Agricultural projects are not specifically mentioned in the Bonn Agreement.[38] Along with renewable energy and energy efficiency projects however, the agreement allows credit for small-scale projects that "both reduce anthropogenic emissions by sources and directly emit less than 15 kilotonnes of carbon dioxide equivalent annually."[39] Livestock feed and forage quality often are low in non–Annex I countries, and diet improvements may reduce CH_4 emissions from enteric fermentation. Small manure digesters, already in widespread use in China, could be used in other non–Annex I countries.[40] With nitrogen fertilizer use increasing rapidly in the tropics, improving application methods, timing, and rates could simultaneously reduce N_2O emissions, increase farmer profits, and protect nonagricultural ecosystems from nitrogen pollution.[41] Recent work on intensive wheat production in Mexico suggests that lower applications of nitrogen fertilizer, better timed to crop needs, could maintain crop yields and quality while increasing farm profits and decreasing N_2O emissions.[45]

All projects would require accurate establishment of baseline emission levels and measurement of changes relative to that baseline in a verifiable manner, which may be difficult for certain types of projects. For example, although the processes that produce N_2O in soils have been studied extensively, it is still not possible to predict reliably N_2O emissions after application of nitrogen fertilizer in a specific agricultural field.[43] In addition, the N_2O emission factors that are used in the IPCC Guidelines for National Greenhouse Gas Emissions may not be appropriate for tropical agricultural systems,[44] but developing site- or even region-specific factors may be prohibitively expensive and time-consuming. Difficulties such as these may limit crediting of N_2O emission reductions in the agricultural sector.

Future Considerations

The most recent UN scenario (middle variant) projects a human population of 8.9 billion by 2050 (see Chapter 9). Given the close link between population size and food production, this near 50 percent increase over the current population will have profound implications for agriculture worldwide. Humans rely on three main systems to supply food: oceanic fisheries, rangelands, and croplands.[45] Production in the first two systems generally has leveled off, implying that much of the necessary increase in food production in the future will have to come from croplands.[46]

The option of expanding food production by cultivating more land (known as extensification) is constrained by several factors. Conversion of farmland to non-agricultural uses and abandonment of land because of soil erosion already are reducing the land area currently cultivated.[47] In addition, a substantial area of land that was formerly used to grow grain has been converted to soybean production, with soy meal used as a protein supplement for livestock and poultry feeds.[48] Although some new land will certainly be brought into cultivation, much of it may be inherently low in fertility, requiring inputs of fertilizer and agrochemicals, the production of which results in GHG emissions.

Food production will have to increase through intensification, that is, through increases in production per unit land area (Table 12.3). The great productivity increases that the world has seen since the mid-1900s have resulted largely from plant breeding, irrigation, and increased use of synthetic fertilizers.[49] Although the capacity of these techniques, as well as biotechnology, to increase future food production is under debate, the important point here is that both irrigation and use of fertilizers result in GHG emissions; irrigation is energy intensive, and nitrogen fertilizers require large energy inputs to produce and increase N_2O emissions from fields. These factors may limit future mitigation potential.

The problem of feeding a growing human population is compounded by the fact that consumption patterns are also changing; as wealth increases, meat consumption generally increases, which requires more grain production per person.[50] Eating higher on the food chain results in more GHG emissions. Analyses of climate change mitigation options beyond the near term must incorporate these substantial upcoming changes in the agricultural sector.

TABLE 12.3. Expected Contributions of Changes in the Area of Land Cultivated (Extensification) and Crop Production per Unit Land Area (Intensification) to Increasing Food Production Between 1988 and 2010

		Intensification	
	Extensification	*Increased Number of Crops*	*Increased Yield per Crop*
Africa (sub-Saharan)	30%	17%	53%
Near East & North Africa	9%	20%	71%
South Asia	7%	13%	80%
East Asia	27%	9%	64%
Latin America	28%	19%	53%

Source: Table 9.4 in P. J. Gregory, J. S. I. Ingram, B. Campbell, J. Goudriaan, L. A. Hunt, J. J. Landsberg, S. Linder, M. Stafford Smith, R. W. Sutherst, and C. Valentin, 1999: "Managed production systems," in B. Walker, W. Steffen, J. Canadell, and J. Ingram (eds.), *The Terrestrial Biosphere and Global Change, Implications for Natural and Managed Ecosystems* (Cambridge, England: Cambridge University Press).

Another important factor for mitigation is the effect of climate change on agriculture. Although a review of the potential impacts of climate change on agriculture is beyond the scope of this chapter, it is important to note that almost every aspect of agriculture could be affected: crop yields, pest populations, availability of water for irrigation, crop and livestock diseases, and weed distributions and abundance.[51] Although higher yields are expected to provide most of the necessary increase in food production (Table 12.3), food production scenarios do not yet incorporate the effects of climate change. As climate changes, different quantities of fertilizer, agrochemicals, and irrigation or more land may be needed. These changes could affect both national emissions and mitigation options.

Increasing Carbon Storage on Agricultural Lands

Treatment of Carbon Sinks in the Kyoto Protocol and the Bonn Agreement

The Kyoto Protocol is not comprehensive in terms of the carbon sources and sinks that will be creditable toward emissions-reduction commitments. Article 3.3 states that sinks must result from "direct human-induced land-use change and forestry activities, limited to afforestation, reforestation, and deforestation since 1990."[52] Thus, in terms of the agricultural sector, only carbon storage associated with forestation of former agricultural lands since 1990 would be covered under Article 3.3.

Article 3.4 of the protocol raised the possibility of crediting additional sinks. Under the protocol, the Conference of the Parties (COP) will develop "modalities, rules and guidelines as to how, and which, additional human-induced activities related to changes in greenhouse gas emissions by sources and removals by sinks in the agricultural soils and the land-use change and forestry categories shall be added to, or subtracted from, the assigned amounts for Parties included in Annex I."[53]

In the Bonn Agreement, the COP (excluding the United States) agreed on a number of principles, definitions, and accounting rules pertaining to land use, land-use change, and forestry.[54] Regarding agricultural activities, "cropland management," "grazing land management," and "revegetation" were deemed "eligible land-use, land-use change, and forestry activities under Article 3, paragraph 4, of the Kyoto Protocol" for the first commitment period.[55] To claim credit, a party must demonstrate that the activities occurred since 1990 and were human-induced.[56] The parties agreed to account for agricultural activities on a net–net basis: net emissions or removals over the commitment period minus net removals in the base year, times five.[57]

A variety of agricultural practices can increase carbon storage in agricultural soils. Soil carbon content is a function of both inputs (which depend on the quantity, quality, and location of plant material deposited on or in the soil) and outputs (i.e., the quantity of carbon leaving the soil through soil erosion, decomposition of organic matter, fire, or other processes). On currently cropped land, practices such as conservation tillage, crop rotation, use of winter cover crops, elimination of summer fallow, and judicious application of manure as fertilizer may lead to soil carbon increases.[58] On grazing lands, managing the density and movement of animals, fertilization, fire, irrigation, and grassland species can affect soil carbon content.[59] Soil carbon levels in land taken out of crop production and planted in perennial grasses or trees generally increase relative to cropped land.[60] If these lands remain out of production, they may continue to slowly accumulate carbon for decades.[61] Carbon would also accumulate for years in the aboveground vegetation.[62]

Intense debate over which sinks should be included beyond those listed in Article 3.3 preceded the Bonn Agreement. The alternatives ranged from only the sinks defined in Article 3.3 to a scheme based on full-carbon accounting in which carbon changes in all ecosystems would be estimated and the net source or sink applied to meeting Kyoto commitments. The Special Report on Land Use, Land-Use Change, and Forestry (LULUCF) published by the IPCC in 2000 examined the scientific and technical implications of carbon sequestration for climate change mitigation and aided implementation of the Kyoto Protocol.[63]

Several arguments can be made for allowing credit for carbon storage on agricultural lands. First, if accounting is done accurately, carbon storage represents a real reduction in atmospheric CO_2, which may be more feasible for countries to pursue than emission reductions alone in the short term. Allowing some mitigation through sinks could motivate mitigation efforts. Second, picking and choosing which sinks to count—in effect creating a large number of small areas of land that must be monitored—could result in loopholes that are detrimental to the atmosphere.[64] If all sinks are counted, there is less potential for double-counting and leakage, i.e., an increase in carbon sinks in one area that leads to a decrease in an area outside the accounting system.[65] Finally, sink projects could provide myriad environmental benefits in addition to climate change mitigation.[66] Practices that increase soil carbon tend to reduce erosion and increase water- and nutrient-holding capacity of soil, thus improving overall soil quality and long-term productivity of the land.[67]

However, there are a number of reasons to be cautious about sink crediting. Biotic carbon sinks may be vulnerable to weather events, political instability, pests, fire, and climate change itself.[68] In addition, as the IPCC Special Report on LULUCF makes clear, many scientific and technical issues must be resolved before the integrity of credits is ensured, and this resolution must occur rapidly. There is also an active debate in the scientific community over the net effect of carbon-sequestering activities, given that the activities themselves may use extra energy and may result in increased emissions of other GHG.[69] Finally, a concern that has received less attention is the connection between the conservation tillage, one of the main practices suggested for increasing soil carbon, and genetically modified (GM) crops.[70]

In a recent essay on the potential to increase carbon storage in agricultural soils, Rosenberg and Izaurralde identified critical outstanding questions involving spatial and temporal limitations to carbon storage, methodological issues for monitoring and verification, and policy and economic problems in implementing soil carbon sequestration programs worldwide.[71] Below I discuss these issues in the context of Annex I and non–Annex I countries.

Agricultural Lands in Annex I Countries: A Source or a Sink of CO_2?

Until the mid-twentieth century, land use changes in North America and Europe were a large source of atmospheric CO_2, exceeding emissions from fossil fuel use.[72] Conversion of forests or grasslands to agricultural production results in CO_2 emissions: Vegetation, an important stock of containing carbon, is often removed from the site, and soil carbon is lost as CO_2 to the atmosphere through enhanced decomposition of soil organic matter and soil erosion. In addition, land conversion often is accompanied by biomass burning, which releases CH_4 and N_2O as well as CO_2.

Today there is little conversion of forests, grasslands, or other native ecosystems to agricultural use in temperate regions. Most agricultural soils have been cultivated for decades to centuries and may have reached a new equilibrium carbon content.[73] At least some of the carbon that was lost from temperate agricultural soils could be regained through carbon-sequestering practices.[74]

Annex I countries, with the exceptions of the United Kingdom and Australia, reported that their land use change and forestry sectors were a net sink of carbon in 1990 and 1995.[75] As with CH_4 and N_2O emissions, however, the inventory data must be used cautiously. Current estimates focus more on forests than agricultural lands and do not include changes in carbon stocks caused by most agricultural activities. Although the land use change and forestry sector detailed in the IPCC Guidelines for National Greenhouse Gas Inventories is inclusive in theory,[76] countries may have difficulty estimating carbon changes associated with all land uses and land use changes. For instance, in the U.S. inventory of GHG emissions and sinks, only two categories of agricultural land use and land management activities are included: agricultural use of organic soils and liming.[77] Activities on mineral soils (the majority of arable land), which could either increase or decrease soil carbon, are not yet included because of insufficient data.

Many Annex I countries are likely to have an agricultural carbon sink in the short term if current land set-aside and diversion policies are maintained and the same land that is taken out of production remains uncultivated. Historically, land set-aside programs have been used for regulation of commodity supply and environmental conservation. For example, as part of the European Union's Common Agricultural Policy (CAP) reforms in 1992, 6.3 million hectares of previously cultivated land were left idle.[78] In the United States, the Conservation Reserve Program (CRP) and other federal conservation programs result in a set aside of more than 12 million hectares[79] (compared with a total cropland area of approximately 134 million hectares).[80]

Measuring, verifying, and monitoring soil carbon storage in a manner sufficiently accurate and inexpensive to support crediting under the Kyoto Protocol remains a challenge. It is possible to monitor changes in soil carbon content, but methods are expensive and not sufficiently sensitive to detect year-to-year changes.[81] Recently, Post et al. reviewed the various methods of soil carbon measurement—both direct (field sampling, laboratory analysis, eddy covariance) and indirect (geographic information, remote sensing, process modeling)—and suggested a monitoring plan.[82] There is a need for technical development at all spatial scales. Methodological issues are discussed extensively in the IPCC's Special Report on LULUCF.[83]

Crediting for Agricultural Carbon Sequestration in the United States

Several efforts have been under way in the United States to improve estimates of soil carbon content and carbon sequestration resulting from various agricultural activities. (Although the United States did not participate in the Bonn Agreement, these activities pre-dated the Bonn negotiations, and interest in carbon sequestration continues.) The U.S. Department of Agriculture's (USDA's) Natural Resource Conservation Service and Agricultural Research Service are developing field carbon equations that will incorporate many cropping practices,[84] and researchers for the Iowa Carbon Storage Project are collecting data from around the state and using them to calibrate a well-established soil carbon model

(Century).[85] Despite the fact that measurement, estimation, and verification methods are still under development, a private market for carbon has already emerged, and deals that involve agricultural carbon are already under way.[86]

Agricultural carbon credit programs could help offset some of the negative publicity that the Kyoto Protocol has gotten in the U.S. agricultural community. The possibility of price increases for energy and fuel stirred up anti-Kyoto sentiment. For instance, the American Farm Bureau cited studies predicting that the average farmer could lose one-quarter to one-half of his or her annual income if Kyoto is adopted[87] and launched Farmers Against the Climate Treaty (FACT), a group of agricultural organizations working against U.S. ratification of the Kyoto Protocol.[88] However, the USDA predicted a much more modest impact, with net cash returns to farmers falling by 0.5 percent.[89] The potential for farmer income through carbon credits was not factored into these analyses.

Opportunities in Non–Annex I Countries

The extent to which activities involving land use, land-use change, and forestry (LULUCF) should be included in the CDM has been the subject of heated discussion within the policy community.[90] Article 12 of the Kyoto Protocol is silent on the issue of carbon sink projects. Under the Bonn Agreement, afforestation and reforestation are the only eligible LULUCF projects allowed under the CDM in the first commitment period.[91] Thus, it is only in later commitment periods that carbon sequestration in agricultural soils could become eligible for credit through the CDM. In addition, the agreement set a limit on the extent to which the eligible LULUCF activities could be used to meet protocol commitments.[92]

It can be argued that the focus of climate change mitigation should be domestic emission reductions. On the other hand, CDM projects may provide a way for non–Annex I countries to participate more fully in climate change mitigation and for other environmental benefits to be gained (see Chapter 13). In many non–Annex I countries, conversion of native ecosystems to agricultural land is proceeding rapidly.[93] Whereas soil organic matter losses are faster during the first 25 years of cultivation in the temperate zone, losses may be more rapid in the tropics, in part because of warmer temperatures.[94] Only half of the land that is converted from tropical forest to agriculture actually increases the productive agricultural area; the other half replaces previously cultivated land that has been degraded and taken out of production.[95] Thus, if agriculture can be made more efficient, less forest and savanna land will need to be converted to agricultural production, reducing GHG emissions and potentially increasing soil carbon.[96] Marginal agricultural land offers the opportunity for reforestation (plantation or natural regeneration), agroforestry (i.e., interspersing woody species with crops), and improvement of agricultural practices, such as tillage, fertilizer, and water management.[97] Some of the same carbon-sequestering practices that are useful in the temperate zone could be used on productive land.[98]

If these types of activities are to be eligible for credit in the future, there must be safeguards to ensure that projects produce real climate benefits as well as positive environmental and societal results. The technical issues that face carbon accounting in Annex I countries may be more difficult in non–Annex I countries because soil inventories, land use surveys, remote sensing, and other methods are less likely to be available; greater technical and institutional capacity would be needed. In addition, because host nations would get credit per unit of avoided GHG emissions and investor countries would get to add credits to their emission budget, there would be incentive to inflate the accounting. Ideally, climate change mitigation activities will facilitate ways of supplying food, fuel, and fiber that benefit farmers and reduce net GHG emissions while paving the way for broader participation from developing countries in binding emission reduction commitments.

Future Considerations

Carbon storage in agricultural soils is limited both temporally and spatially. It is generally assumed that soils, with the possible exception of those in wetlands, will eventually reach an equilibrium carbon content and not sequester above that amount.[99] When that amount is reached, carbon-storing practices must be continued to prevent release of the carbon back to the atmosphere.[100] Thus, unlike some reductions in GHG emissions, activities that increase carbon storage will not yield benefits indefinitely, and it is generally acknowledged that soil carbon sequestration would play a strategic role in mitigation only in the short term, complementing other mitigation activities.[101]

Beyond these limitations, changes in climate, increases in atmospheric CO_2 concentration, and nutrient deposition may influence the storage potential of soils. Changes in temperature and moisture regimes are likely to affect rates of accumulation and decomposition of carbon in soils.[102] In addition, climate change could affect soil carbon storage through changes in plant growth (i.e., through the amount of plant carbon that is deposited in and on soils).[103] Increases in atmospheric CO_2 concentration may influence plant growth, plant tissue allocation patterns, and activity of soil microbes.[104] All of these factors could affect soil carbon storage even if farmer practice remains constant or shifts to carbon-storing activities.

Perhaps most significantly, because of increases in human population size and food demand, land use will continue to change in ways that affect soil carbon storage. As more land is converted from forest or grassland to cropland or pasture, especially in developing countries (Table 12.3), soil carbon storage will decline. Even in developed countries, taking land out of production, as is currently done in the European Union and the United States, is possible only if adequate supplies of food, fiber, and energy can be produced from the remaining area, often to feed people in both Annex I and non–Annex I countries. This is currently possible in many Annex I countries through intensive farming on existing agricultural land, but increases in human population size and political or environmental factors could preclude this possibility in the future.

Total Mitigation Potential of the Agricultural Sector

In 1995, the IPCC estimated annual worldwide mitigation potential of the agricultural sector to be 1,669–3,417 million metric tons of CO_2, 24–92 million metric tons of CH_4, and 0.4–1.1 million metric tons of N_2O (Table 12.4). Combining these estimates, total mitigation potential would be 620–1,546 million metric tons of carbon equivalent. This represents 13–33 percent of 1990 emissions of CO_2, CH_4, and N_2O from all sectors in Annex I countries,[105] a sizable amount given protocol commitments ranging from –8 to +10 percent of 1990 levels.

The carbon sink value in Table 12.4 is similar to other recent estimates.[106] The IPCC Special Report on LULUCF has further divided carbon sinks geographically (Annex I and non–Annex I) and by agricultural practice (Table 12.5). This latest estimate is based on peer-reviewed studies in the literature and assumptions about the economic, social, and technical constraints that limit land available for these practices.[107]

Estimates of mitigation potential such as these must be used cautiously. The scientific literature available to support analyses is still limited. Inaccuracies in measurement and estimation of emissions and carbon storage are inherent in the estimates. Even in the United States, where data sets are large, the estimates of carbon storage vary more than twofold.[108] These analyses also do not reflect the recent decisions of the COP in the Bonn Agreement (e.g., projects involving agricultural soils are not eligible for credit under the CDM in the first commitment period), and current estimates cannot anticipate policies beyond the first commitment period. According to the IPCC Special Report on LULUCF, the largest uncertainty in carbon sequestration estimates is where and to what

TABLE 12.4. Estimated Worldwide Mitigation Potential of the Agricultural Sector (Excluding Biofuel Production)

Category	*Estimated Decrease Through Practice (Mt gas/yr)*	*Estimated Decrease Through Practice (MMTCE/yr)**
CO_2		
Emission reductions†	37–183	10–50
Increasing carbon sinks‡	1,632–3,234	445–882
TOTAL	1,669–3,417	455–932
CH_4		
Ruminant animals	12–45	69–258
Animal waste	2–7	11–40
Rice paddies	8–35	46–200
Biomass burning	1.5–4.5	9–26
TOTAL	24–92	135–524
N_2O		
Mineral and organic fertilizers	0.3–0.9	25–76
Tropical biomass burning and land conversion§	0.06–0.17	5–14
TOTAL	0.36–1.1	30–90

Source: Tables 23-5 and 23-11 in V. Cole, C. Cerri, K. Minami, A. Mosier, N. Rosenberg, and D. Sauerbeck, 1996: "Agricultural options for mitigation of greenhouse gas emissions," in R. T. Watson, M. C. Zinyowera, and R. H. Moss (eds.), *Climate Change 1995: Impacts, Adaptations and Mitigation of Climate Change: Scientific-Technical Analyses* (Cambridge, England: Cambridge University Press).

*In converting Mt of gas to million metric tons of carbon equivalent (MMTCE), a 100-year time horizon was assumed, with global warming potentials of 21 and 310 for CH_4 and N_2O, respectively.
†Assuming a reduction of 10–50% in developed countries only.
‡Based on carbon sequestration over a 100-year period. Category includes better management of existing agricultural soils globally, permanent set-aside of 15% of surplus agricultural land in temperate regions, and restoration of soil carbon on degraded lands globally.
§Includes biomass burning, management of soils after burning, and forest conversion.

extent LULUCF activities may occur.[109] Finally, the IPCC estimates do not yet take into account the effects of climate change and elevated CO_2 on agricultural production or soil carbon storage.[110] Instead, they reflect current agricultural conditions. Thus, although estimates can be useful for conducting near-term policy analyses and stimulating additional climate change research, assessment of mitigation options for coming decades must consider additional factors.

TABLE 12.5. Estimated Carbon Storage Potential of Agricultural Activities in Two Time Periods

Practices	*Site*	*Area (10^6 ha)*	*Adoption or Conversion (% of area)*		*Rate of Storage (t C ha^{-1} yr^{-1})*	*Storage Potential (Mt C/yr)**	
			2010	*2040*		*2010*	*2040*
Improved management							
Cropland (includes conservation tillage, crop rotation, cover crops, erosion control, fertility and irrigation management)	Annex I	589	40	70	0.32	75	132
	Non–Annex I	700	20	50	0.36	50	126
Rice paddies (irrigation, fertilizer, and plant residue management)	Annex I	4	80	100	0.10	<1	<1
	Non–Annex I	149	50	80	0.10	7	12
Agroforestry (improved management of trees on cropland)	Annex I	83	30	40	0.50	12	17
	Non–Annex I	317	20	40	0.22	14	28
Grazing land (improved herd, plant, and fire management)	Annex I	1,297	10	20	0.53	69	137
	Non–Annex I	2,104	10	20	0.80	168	337
Land-use change							
Agroforestry (conversion from unproductive cropland and grasslands)	Annex I	~0	~0	~0	~0	0	0
	Non–Annex I	630	20	30	3.1	391	586
Grassland (conversion from cropland)	Annex I	602	5	10	0.8	24	48
	Non–Annex I	855	2	5	0.8	14	34

Source: Table 4-1 in Intergovernmental Panel on Climate Change, 2000: *Land Use, Land-Use Change, and Forestry* (Cambridge, England: Cambridge University Press), pp. 198–204. Note that all listed activities will not be eligible for credit in Annex I and non–Annex I countries under the Kyoto Protocol and Bonn Agreement.

*Rates of carbon gain are averages for approximately 20–40 years; after this period of accumulation, rates typically will approach zero. Uncertainty in estimates may be as high as +50%.

Moving Ahead with Mitigation

Growing crops and raising livestock for human consumption necessarily result in GHG emissions. Agriculture will always produce substantial quantities of CO_2, CH_4, and N_2O and influence soil carbon storage. The need for increased agricultural production in the future is also certain: human population size will increase by more than a billion, potentially by nearly 5 billion (see Chapter 9), by the mid–twenty-first century. However, the foods we choose to produce and the agricultural practices used in production differ in their relative impacts on the atmosphere.

Through a variety of practices, such as use of cover crops and judicious application of manure as fertilizer, crops can be grown in ways that increase soil carbon and overall soil quality. Increasing the use of these practices may be most feasible in the temperate zone, where soils have lost carbon over years of cultivation and where agricultural extension and research services often are available to assist with adoption of new practices. Pilot field projects in different agricultural regions would be invaluable in both developed and developing countries. Without more field research, there is danger of both sham credit awards and the belief that crediting soil carbon simply cannot be done. Although multiple forestry projects are up and running (see Chapter 13), progress on the agricultural front has been slower.

Another mitigation opportunity is the improvement of nitrogen fertilizer management. In many areas, the majority of fertilizer applied to crops is lost through leaching and gaseous emissions. Changing fertilizer application techniques, without increasing the amount of fertilizer applied, holds promise for increasing the amount of fertilizer used by crop plants. If credit is to be awarded for reductions in N_2O emissions, however, estimation techniques must improve to prevent uncertainty from overwhelming reductions. With its Good Practice Guidance and Uncertainty Management in National Greenhouse Gas Inventories, the IPCC is moving in the right direction, but techniques for measuring and estimating N_2O emissions must be improved. As with soil carbon storage, pilot field projects would be very useful.

For livestock manure management, greater use of CH_4 recovery systems, such as digesters, could reduce CH_4 emissions. These systems may also protect surface water and groundwater quality by eliminating the danger of overflow and leakage from uncovered manure lagoons. In the absence of climate policies or regulations, such systems may already be profitable as well as desirable environmentally. Stronger government programs for outreach and financial assistance, combined with stricter water pollution legislation for farms, could help increase use of these technologies. Although small manure digesters are already used in many developing countries, there is much room for improvement, especially as livestock numbers grow in response to population increases.

This last point illustrates the most important agricultural mitigation activities: controlling human population size and changing consumption patterns. Population size directly affects how much crop production is needed each year, how big livestock herds must be, and how much land can be set aside or left forested. While not disparaging the importance of agricultural mitigation activities and the development of methods to assess their effects, I believe that international family planning, education of women and girls worldwide, and national population policies deserve much more attention in this context. In addition, the environmental effects of a diet rich in livestock products—often to an unhealthy degree—especially in industrialized countries should be highlighted in climate change assessments and nutritional education programs. It is through family planning and consumption choices that every person can play a role in mitigating agriculture's contribution to climate change.

Acknowledgments

I thank Janine Bloomfield, Elsa Cleland, Frederick Meyerson, John O. Niles, Tracey Osborne, and Vern Ruttan for helpful comments on the manuscript and Helene Poulshock and Nick Thompson for research assistance. I acknowledge the support of a Lokey Research Fellowship at Environmental Defense.

Notes

1. Cole, V., C. Cerri, K. Minami, A. Mosier, N. Rosenberg, and D. Sauerbeck, 1996: "Agricultural options for mitigation of greenhouse gas emissions," in R. T. Watson, M. C. Zinyowera, and R. H. Moss (eds.), *Climate Change 1995: Impacts, Adaptations and Mitigation of Climate Change: Scientific–Technical Analyses* (New York: Cambridge University Press).
2. Watson, R. T., M. C. Zinyowera, and R. H. Moss (eds.), 1996: "Technologies, policies and measures for mitigating climate," Intergovernmental Panel on Climate Change Technical Paper I, available online: http://www.ipcc.ch/pub/techrep.htm.
3. Houghton, R. A., 1994: "The worldwide extent of land-use change," *BioScience*, 44: 305–313.
4. Kyoto Protocol to the UN FCCC, 1997: UN Doc. FCCC/CP/1997/7/Add.1 (hereinafter Kyoto Protocol), available online: http://cop3.unfccc.int.
5. Decision 5/CP.6, Implementation of the Buenos Aires Plan of Action, 2001: UN doc. FCCC/CP/2001.L.7 (hereinafter Bonn Agreement), available online: http://unfccc.int/resource/docs/cop6secpart/l07.pdf.
6. When carbon is held in soil, it is prevented from acting as a GHG in the atmosphere; thus, agricultural soils can sequester carbon, or act as a carbon sink. An addi-

tional agricultural mitigation option is increased use of agricultural crops and crop residues to produce fuels and energy (biofuels), as discussed in Chapter 16.

7. Dudek, D., J. Goffman, M. Oppenheimer, A. Petsonk, and S. Wade, 1998: *Cooperative Mechanisms Under the Kyoto Protocol* (New York: Environmental Defense Fund).
8. Kyoto Protocol, Article 3.1.
9. Working Group I, 1996: "Technical summary," in J. T. Houghton, L. G. Meira Filho, B. A. Callander, N. Harris, A. Kattenberg, and K. Maskell (eds.), *Climate Change 1995: The Science of Climate Change* (Cambridge, England: Cambridge University Press). The GWPs used by the IPCC for CH_4 and N_2O are 21 and 310, respectively. However, there is still scientific debate over the use of GWPs. O'Neill, B. C., 2000: "The jury is still out on GWPs," *Climatic Change,* 44: 427–443.
10. Wigley, T. M. L., 1998: "The Kyoto Protocol: CO_2, CH_4 and climate implications," *Geophysical Research Letters,* 25: 2285–2288.
11. Intergovernmental Panel on Climate Change (IPCC), 1997: *Revised 1996 IPCC Guidelines for National Greenhouse Gas Inventories,* Vol. 1 [hereinafter IPCC Guidelines]. Houghton, J. T., L. G. Meira Filho, B. Lim, K. Treanton, I. Mamaty, Y. Bonduki, D. J. Griggs, and B. A. Callender (eds.). UK Meterological Office, Bracknell. Available online: http://www.ipcc-nggip.iges.or.jp/public/gl/inrs4.htm.
12. National Communications for Annex I countries and for some non–Annex I countries are available at http://www.olis.oecd.org/olis/1997doc.nsfl.
13. OECD, 1997: "Agriculture and forestry: Identification of options for net greenhouse gas reduction," Annex I Expert Group on the UN FCCC Working Paper No. 7, available online: http://www.olis.oecd.org/olis/1997doc.nsfl.
14. In national inventories, these emissions are counted in the energy and industrial, rather than the agricultural, sectors. IPCC Guidelines, 1997, pp. 1.2–1.8.
15. Cole et al., 1996, p. 754. Fuel use by the food sector as a whole, including processing, storage, and distribution, is much higher, however, accounting for 10–20 percent of total CO_2 emissions.
16. Frye, 1984, as cited in Cole et al., 1996, p. 754.
17. Cole et al., 1996, p. 754.
18. Ibid.
19. Ruminant animals, such as cattle and sheep, produce more CH_4 per unit of food digested than nonruminant animals, such as pigs and horses, but emissions from all major livestock are estimated in national inventories.
20. In 1995 Japan's total agricultural CH_4 emissions were 831 Mt, of which 379 Mt, or 46 percent, resulted from rice cultivation. Emission data are available at http://unfccc.int/resource.ghg.tempemis2.html. See also Japan's Second National Communication Under the UN FCCC, http://unfccc.int/resource/natcom/index.html.
21. According to Annex B of the Kyoto Protocol, EIT parties include Bulgaria, Croatia, Czech Republic, Estonia, Hungary, Latvia, Lithuania, Poland, Romania, Russian Federation, Slovakia, Slovenia, and Ukraine.
22. For example, see Estonian Second National Report to the UN FCCC, available online: http://unfccc.int/resource/natcom/index.html.

23. Storey, M., 1997: "The climate implications of agricultural policy reform," Working Paper 16, Policies and Measures for Common Action, Annex I Experts Group on the FCCC, supported by the OECD and the International Energy Agency, available at http://www.oecd.org/env/docs/cc/wpaper16.pdf.
24. Ibid., p. 18.
25. As the largest emitter of livestock CH_4, the United States has several voluntary programs that address livestock CH_4 emissions but no mandatory programs. Voluntary programs include the AgStar program, launched by the Clinton administration's Climate Change Action Plan and the Ruminant Livestock Efficiency Program, a joint U.S. Environmental Protection Agency (EPA) and Department of Agriculture program. For more information, see http://www.epa.gov/outreach/rlep/index.htm.
26. Hogan, K. B. (ed.), 1993: *Options for Reducing Methane Emissions Internationally,* Vol. 1: *Technological Options for Reducing Methane Emissions,* EPA 430-R-93-006 A and Vol. 2: *International Opportunities for Reducing Methane Emissions,* EPA 430-R-93-006 B (Washington, DC: U.S. EPA). Feed additives, such as antibiotics and hormones, increase animal growth rates and efficiency of feed use, resulting in less CH_4 per unit of product, but concerns about their safety for both livestock and people may outweigh their climate benefits. For example, see Jacobs, Paul, 1999: "U.S., Europe lock horns in beef hormone debate," *Los Angeles Times,* April 9, A1; Belsie, Laurent, 1999: "Signs of the food fight to come: Gene-spliced plants and hormone-treated beef raise ethical questions about how much to fool with nature," *Christian Science Monitor,* March 10, 1; Grady, D., 1999, "A move to limit antibiotic use in animal feed," *New York Times,* March 8.
27. See Lusk, P., 1998: *Methane Recovery from Animal Manures, The Current Opportunities Casebook* (Golden, CO: National Renewable Energy Laboratory).
28. See "United States leads world meat stampede," *Worldwatch Press Briefing,* July 2, 1998, available online: http://www.worldwatch.org/alerts/pr980704.html.
29. N_2O emissions can occur through a direct pathway from soils where nitrogen was added or through two indirect pathways: from volatilization as ammonia and nitric oxide and subsequent redeposition on unfertilized soils or from leaching and runoff. See IPCC Guidelines, 1997, p. 4.53.
30. For GHG emission data, see http://unfccc.int/resource/ghg/tempemis2.html.
31. Tilman, D., 1998: "The greening of the Green Revolution," *Nature,* 396: 211–212; Matson, P. A., W. J. Parton, A. G. Power, and M. J. Swift, 1997: "Agricultural intensification and ecosystem properties," *Science,* 277: 504–509.
32. Mosier, A. R., J. M. Duxbury, J. R. Freney, O. Heinemeyer, and K. Minami, 1998: "Assessing and mitigating N_2O emissions from agricultural soils," *Climatic Change,* 40: 25–28.
33. IPCC Guidelines, 1997, Vol. 3.
34. At the 16th IPCC plenary session, May 1–8, 2000, guidelines for four emission categories were accepted, one of which was agriculture. IPCC, 2000: "Good practice guidance and uncertainty management in national greenhouse gas inventories," available online: http://www.ipcc-nggip.iges.or.jp/public/gp/gpgaum.htm.
35. Meyerson, F. A. B., 1998: "Population, carbon emissions, and global warming: the forgotten relationship at Kyoto," *Population and Development Review,* 24: 115–130.

36. "United States leads world meat stampede," 1998.
37. Galloway, J. N., W. H. Schlesinger, H. Levy II, A. Michaels, and J. L. Schnoor, 1995: "Nitrogen fixation: anthropogenic enhancement–environmental response," *Global Biogeochemical Cycles,* 9: 235–252.
38. See Bonn Agreement, 2001.
39. Ibid.
40. Cole et al., 1996, p. 758.
41. For a review of off-site consequences of nitrogen fertilizer use in tropical ecosystems, see Matson, P. A., W. H. McDowell, A. R. Townsend, and P. M. Vitousek, 1999: "The globalization of N deposition: ecosystem consequences in tropical environments," *Biogeochemistry,* 46: 67–83.
42. Matson, P. A., R. Naylor, and I. Ortiz-Monasterio, 1998: "Integration of environmental, agronomic, and economic aspects of fertilizer management," *Science,* 280: 112–115.
43. Mosier et al., 1998, p. 10.
44. Recent work suggests that N_2O losses may be proportionally greater in some tropical systems. Bouwman, A.F., 1998: "Nitrogen oxides and tropical agriculture," *Nature,* 392: 866–867.
45. Brown, L. R., 1999: "Feeding nine billion," in L. R. Brown, C. Flavin, and H. F. French (eds.), *State of the World 1999* (New York: W.W. Norton).
46. Ibid.
47. Ibid.
48. Ibid.
49. Ibid.
50. Gregory, P. J., J. S. I. Ingram, B. Campbell, J. Goudriaan, L. A. Hunt, J. J. Landsberg, S. Linder, M. Stafford Smith, R. W. Sutherst, and C. Valentin, 1999: "Managed production systems," in B. Walker, W. Steffen, J. Canadell, and J. Ingram (eds.), *The Terrestrial Biosphere and Global Change, Implications for Natural and Managed Ecosystems* (Cambridge, England: Cambridge University Press), pp. 231–270.
51. The potential impacts of climate change on agriculture globally and regionally have been the subject of a number of excellent reviews. For example, see Adams, R. M., B. H. Hurd, and J. Reilly, 1999: *A Review of Impacts to U.S. Agricultural Resources* (Washington, DC: Pew Center on Global Climate Change); Rosenzweig, C. and D. Hillel, 1998: *Climate Change and the Global Harvest, Potential Impacts of the Greenhouse Effect on Agriculture* (New York: Oxford University Press); Rosenzweig, C. and F. N. Tubiello, 1997: "Impacts of global climate change on Mediterranean agriculture: current methodologies and future directions, an introductory essay," *Mitigation and Adaptation Strategies for Global Change,* 1: 219–232; Reilly, J., et al., 1996: "Agriculture in a changing climate: impacts and adaptation," in R. T. Watson, M. C. Zinyowera, and R. H. Moss (eds.), *Climate Change 1995: Impacts, Adaptations and Mitigation of Climate Change: Scientific–Technical Analyses* (New York: Cambridge University Press). The U.S. National Assessment of the potential consequences of climate variability and change is available online: http://www.usgcrp.gov/usgcrp/nacc/default.htm.
52. Kyoto Protocol, 1997, Article 3.3.

53. Ibid., Article 3.4.
54. Bonn Agreement, 2001, Part VII.
55. Ibid.
56. Ibid.
57. Ibid.
58. See IPCC, 2000: *Land Use, Land-Use Change, and Forestry* [IPCC LULUCF] (Cambridge, England: Cambridge University Press), pp. 198–204; Lal, R., J. M. Kimble, R. F. Follett, and C. V. Cole, 1998: *The Potential of U.S. Cropland to Sequester Carbon and Mitigate the Greenhouse Effect* (Chelsea, MI: Sleeping Bear Press).
59. IPCC LULUCF, 2000, pp. 204–208; Schuman, G. E., J. D. Reeder, J. T. Manley, R. H. Hart, and W. A. Manley, 1999: "Impact of grazing management on the carbon and nitrogen balance of a mixed-grass rangeland," *Ecological Applications,* 9: 65–71.
60. See Gebhart, D. L., H. B. Johnson, H. S. Mayeux, and H. W. Polley, 1994: "The CRP increases soil organic carbon," *Journal of Soil and Water Conservation,* 49: 488–492.
61. See Burke, I. C., W. K. Laurenroth, and D. P. Coffin, 1995: "Soil organic matter recovery in semiarid grasslands: implications for the Conservation Reserve Program," *Ecological Applications,* 5: 793–801.
62. IPCC LULUCF, 2000, pp. 208–213.
63. Ibid., pp. 183–236. The Special Report discusses definitional and carbon accounting issues (e.g., land-based vs. activity-based approaches), the boundary between natural phenomena and human-induced activities, uncertainty management, leakage, and baselines.
64. Ibid., pp. 185–186, 195–198; IGBP Terrestrial Carbon Working Group, 1998: "The terrestrial carbon cycle: implications for the Kyoto Protocol," *Science,* 280: 1393–1394.
65. For example, if one farmer took land out of corn production and planted trees to store carbon, it is possible that another farmer in the region would bring idle land into corn production or intensify production on existing lands in a way leading to soil carbon losses that do not get measured.
66. See Lal et al., 1998, p. 61.
67. For instance, recent work has shown that judicious application of manure and use of legumes, instead of conventional chemical fertilizers, can sustain crop yields and reduce nitrogen leaching while increasing soil carbon content and overall quality. Drinkwater, L. E., P. Wagoner, and M. Sarrantonio, 1998: "Legume-based cropping systems have reduced carbon and nitrogen losses," *Nature,* 396: 262–265.
68. Lashof, D. and B. Hare, 1999: "The role of biotic carbon stocks in stabilizing greenhouse gas concentrations at safe levels," *Environmental Science and Policy,* 2: 101–109.
69. Schlesinger, W. H., 1999: "Carbon sequestration in soils," *Science,* 284: 2095; Smith, P. and D. S. Powlson, 2000" "Considering manure and carbon sequestration," *Science,* 287: 427; IPCC LULUCF, 2000, pp. 183, 192, 199–200.
70. When tillage is reduced, weeds generally must be controlled by herbicides because

they are no longer plowed under the soil. GM techniques have enabled companies to make crops that are resistant to specific herbicides, such as Monsanto's Round-Up Ready soybeans and other crops, so applications of the herbicide kill the weeds but leave the crop unaffected. These GM crops may make it easier for farmers to use conservation tillage practices. Therefore, efforts to increase the area under conservation tillage may increase the area planted in GM crops (at least in countries that permit their planting). However, the question of whether these crops are safe for human health and for the environment has inspired heated controversy, and some of the safety issues will take additional years of study. See Ferber, D., 1999: "GM crops in the cross hairs," *Science,* 286: 1662–1666; Enserink, M., 1999: "Ag biotech moves to mollify its critics," *Science,* 286: 1666–1668.

71. Rosenberg, N. J. and R. C. Izaurralde, 2001: "Storing carbon in agricultural soils to help head-off a global warming: an editorial essay," *Climatic Change,* 51 (1), October 2001, p. 1–10.
72. Flach, K. W., T. O. Barnwell, Jr., and P. Crosson, 1997: "Impacts of agriculture on atmospheric carbon dioxide," in E. A. Paul, K. Paustian, E. T. Elliott, and C. V. Cole (eds.), *Soil Organic Matter in Temperate Agroecosystems, Long-Term Experiments in North America* (Boca Raton, FL: CRC Press).
73. Cole et al., 1996, p. 752.
74. For a discussion of technologies that could enhance soil carbon sequestration, including precision agriculture, plant biotechnology, microbial biotechnology, and chemical technology, see Metting, F. B., J. L. Smith, J. S. Amthor, and R. C. Izaurralde, 2001: "Science needs and new technology for increasing soil carbon sequestration," *Climatic Change* 51 (1): 11–34.
75. UN FCCC Conference of the Parties, 1998: "Review of the implementation of commitments and of other provisions of the Convention," at p. 15, UN Doc. FCCC/CP/1998/11/Add.1, available online: http://www.cop4.org. In the UK, CO_2 emissions from wetland drainage and peat extraction outweighed LULUCF sinks; see "Climate change: the United Kingdom programme," the United Kingdom's second report to the FCCC, available online: http://unfccc.int/resource/natcom/index.html. In Australia, clearing of forest and grasslands led to large CO_2 emissions; see "Climate change," Australia's second national report under the UN FCCC, available online: http://unfccc.int/resource/natcom/index.html.
76. IPCC Guidelines, 1997, pp. 5.3–5.4.
77. U.S. EPA, 1999: *Inventory of U.S. Greenhouse Gas Emissions and Sinks: 1990–1997* (3/2/99 draft) (Washington, DC: U.S. EPA).
78. Storey, 1997, p. 13.
79. USDA Farm Service Agency, 1999: "The Conservation Reserve Program 18th sign-up," available online: http://www.fsa.usda.gov/DAFP/cepd/18thcrp/18thBooklet-Final.PDF.
80. Lal et al., 1998, p. 27.
81. Rosenberg and Izaurralde, 2001.
82. Post, W. M., R. C. Izaurralde, L. K. Mann, and N. Bliss, 2001: "Monitoring and verifying changes in organic carbon in soil," *Climatic Change,* 51 (1): 73–99.
83. IPCC LULUCF, 2000, pp. 75–104.

84. An informational brochure from the USDA on climate change and agriculture and the development of a carbon market is available online: http://www.swcs.org/docs/carbon_brochure.pdf.
85. Iowa Natural Resources Conservation Service, "Carbon Sequestration and Trading," Iowa NRCS Issue Paper, 1998.
86. The Greenhouse Emissions Management Consortium (GEMCo), 1999: "GEMCo members agree to buy emission reduction credits from Iowa farmers," available online: http://www.gemco.org/Iowa_Farm_Project.htm; Rosenberg and Izaurralde, 2001.
87. Lipton, D. and M. Thornton, 1999: "Study finds climate treaty to cut farm income 50 percent," American Farm Bureau Federation News Release, February 16; Francl, T., R. Nadler, and J. Bast, 1998: *The Kyoto Protocol and U.S. Agriculture* (Chicago: The Heartland Institute); Sparks Companies, Inc., 1999: *The Kyoto Protocol: Potential Impacts on U.S. Agriculture* (McLean, VA: Sparks Companies, Inc.).
88. See the American Farm Bureau's Web site at http://www.fb.com/climate/index.html.
89. USDA, 1999: *Economic Analysis of U.S. Agriculture and the Kyoto Protocol* (Washington, DC: USDA).
90. For a discussion of some of the scientific, political, and economic issues involved in crediting soil carbon storage under the CDM, see Marland, G., B. A. McCarl, and U. Schneider, 2001: "Soil carbon: policy and economics," *Climatic Change,* 51 (1): 101–117.
91. Bonn Agreement, 2001.
92. Ibid. For the first commitment period, the total of additions to and subtractions from the assigned amount of a party resulting from eligible LULUCF activities under the CDM cannot exceed 1 percent of base year emissions of that party, times five.
93. Houghton, 1994.
94. Matson et al., 1997, p. 506.
95. Houghton, 1994.
96. Lal, R. and J. P. Bruce, 1999: "The potential of world cropland soils to sequester C and mitigate the greenhouse effect," *Environmental Science and Policy,* 2: 177–185.
97. Bloomfield, J. and H. L. Pearson, 2000: "Land use, land-use change, forestry, and agricultural activities in the Clean Development Mechanism: estimates of greenhouse gas offset potential," *Mitigation and Adaptation Strategies for Global Change,* 5: 9–24.
98. Ibid.
99. Lal et al., 1998.
100. On the issue of longevity of agricultural carbon storage, see Marland et al., 2001.
101. Rosenberg and Izaurralde, 2001; IPCC LULUCF, 2000, p. 183 (considering carbon sequestration estimates to be applicable for a maximum of 50 years).
102. Gregory et al., 1999.
103. Ibid..
104. Ibid.
105. Calculated from data in Table B.15 in UN FCCC Conference of the Parties, 1998:

"Review of the implementation of commitments and of other provisions of the convention," FCCC/CP/1998/11/Add.2, http://www.cop4.org.

106. Lal and Bruce, 1999; Paustian, K., C. V. Cole, D. Sauerbeck, and N. Sampson, 1998: "CO_2 mitigation by agriculture: an overview," *Climatic Change,* 40: 135–162.

107. IPCC LULUCF, 2000, p. 183.

108. Metting et al., 2001.

109. IPCC LULUCF, 2000, p. 183.

110. Recent studies have begun to estimate these effects. Ibid., pp. 200–201.

CHAPTER 13

Tropical Forests and Climate Change

John O. Niles

Most major international environment and sustainable development issues touch on tropical deforestation. At the 1992 UN Conference on Environment and Development (UNCED), five major agreements were signed concerning tropical forests. Tropical deforestation bears directly on the subject of two of these accords: the Biodiversity Convention and the Framework Convention on Climate Change (FCCC). Tropical deforestation is the leading cause of biodiversity loss. Behind fossil fuel combustion, tropical deforestation is the second leading cause of greenhouse gas (GHG) emissions. The other three agreements signed at UNCED—Agenda 21, The Rio Declaration, and the Statement on Forest Principles—demonstrate the linkage between tropical forest conservation and sustainable development.

In addition to jeopardizing ecosystems rich in species and carbon pools, the continued dismantling of tropical forests affects tens of millions of the world's poorest people. Tropical deforestation endangers the lives, health, and future sustainability of people who rely on tropical forests for their survival.

Despite the UNCED agreements and more than 100 multilateral tropical forest accords, deforestation in the tropics continues at a rapid pace. The most recent estimates suggest that every year approximately 40 million acres of tropical forest are destroyed. Many of the consequences of deforestation are irreversible, at least on human time scales. Tropical deforestation is arguably one of the planet's most urgent issues in need of new solutions.

There is still substantial scientific uncertainty in most aspects of the global carbon cycle with respect to deforestation. In addition to acting as stores, sinks, and potential sources of carbon, tropical forests also are a green blanket over large equatorial areas, the region of the world where incoming solar radiation is most intense. Tropical forests provide a historically stable land surface for key biophysical processes that influence climate and weather. These biophysical processes, such as the strength of large-scale circulation cells, regional rainfall patterns, and energy balances, are even less understood than carbon dynamics. Outside of their role in carbon dynamics, tropical forests help stabilize our planet's climate. Safeguarding tropical forests, when and where appropriate, will help maintain hydrologic and other conditions that humans take for granted.

In July 2001, political decisions at the sixth Conference of the Parties, part b (COP6b), excluded tropical forest conservation from the first commitment period of the Kyoto Protocol. This means that projects that conserve tropical forests and reduce GHG emissions will not be eligible for carbon trading. Therefore, these types of projects will not attract financial resources that would have otherwise been enabled if forest conservation were a sanctioned type of mitigation. Specific wording at COP6b restricted forest activities in developing countries to reforestation and afforestation activities.

The decision to exclude tropical forest conservation was highly contentious and politically charged. The issue pitted Europeans against Americans, Brazilians against other Latin American countries, and environmental groups against other environmental groups. The treatment of tropical forests in the protocol reached the highest levels of government and environmental activism. In the final negotiating days of COP6a in November 2000, before the George Bush administration abandoned the Kyoto process, President Clinton is reported to have called President Cardoso of Brazil to discuss the matter. Also at COP6, forest activist Julia Butterfly Hill announced her support for keeping rain forest conservation in the protocol. Indigenous groups issued press releases for and against including forest conservation in the protocol. Key differences remained between many groups and nations on this matter after the COP6a talks collapsed. At COP6b, with the United States on the sidelines, Brazil, Europe, and other countries and groups opposed to tropical forest conservation in the treaty prevailed. Projects that reduce emissions by stopping tropical deforestation would not be credited under the terms of the agreement. The World Wide Fund for Nature (WWF) announced victory in their press statement of July 22, 2001: "Avoided deforestation—which would have provided a particularly destructive loophole—has been excluded from the [Kyoto Protocol]." However, the Nature Conservancy demurred and said, "Today's agreement represents the worst possible outcome for those interested in international biodiversity conservation. It excludes projects that reduce emission from tropical deforestation, the source of over 20 percent of annual emissions."

Despite the interrelated nature of tropical deforestation and climate change, these two issues will not be integrated in the near term under the Kyoto Protocol. Yet timing is a critical issue for tropical forests. Forests in the tropics are being lost at a rate of roughly 1 percent per year,[1] causing roughly 20 percent of the climate change problem. Large tracts of forests will disappear in the next few decades, even if new powerful political actions are implemented rapidly. If we continue to lose tropical forests at current rates, by the end of this century there will be few primary tropical forests left to protect.[2] This will have untold consequences for the people and species that rely on these ecosystems and could compromise global environmental security.

The Biological Context

The biogeochemical role of forests in the cycles of GHGs is not a trivial exercise in careful accounting. This confusion can be compounded by natural variability between forests, year-to-year fluctuations, and uncertainties associated with GHG accounting within forests. A few general observations about GHG accounting in forests may help inform the CDM debate.[3]

Before delving into the science of tropical forests and climate change, it may be helpful to distinguish between sources of GHGs (emissions) and the sequestration of GHGs (also called sinks or uptake). The sink debate in the Kyoto Protocol has been one of the largest stumbling blocks in negotiations. For clarity, definitions used by the IPCC are provided in Box 13.1.[4]

Sources and sinks are distinct biogeochemical processes that deserve different policy treatments. A sink is a process or mechanism that draws carbon out of the atmosphere, whereas a source is a process or mechanism that puts carbon into the atmosphere. Stopping a transfer of carbon into the atmosphere is different from initiating measures to draw carbon down, even if on a ton-by-ton basis the net impact for the atmosphere is equal. For example, stopping the destruction of forests[5] is distinctly different from creating new forests, even if the two measures result in the same net balance of carbon in the atmosphere.

Tropical deforestation creates a net flow of carbon into the atmosphere.[6] Therefore, preventing a tropical forest from being destroyed is an emission reduction. Calling this effort a sink confuses the fundamentals of carbon cycling. In forestry terms, a sink consists of trees planted or forests manipulated to increase the flow of carbon from the atmosphere to vegetation on or in the ground. Preventing deforestation prevents or reduces emissions. This is legally and biologically distinct from creating a sink by planting trees or other measures.

BOX 13.1

Definitions of a Sink and Source

Sink: Any process or mechanism that removes a GHG, an aerosol, or a precursor of a GHG from the atmosphere. A given pool (reservoir) can be a sink for atmospheric carbon if, during a given time interval, more carbon is flowing into it than is flowing out of it.

Source: Opposite of a sink. A carbon pool (reservoir) can be a source of carbon to the atmosphere if less carbon is flowing into it than is flowing out of it.

Tropical Forests as Sources of GHGs

Once degraded, tropical forests rarely regain their original carbon storage capacity, at least not in a politically meaningful time span. Some carbon accumulates in the vegetation and associated soils after deforestation.[7] Additionally, some carbon may be bound in charcoal created by burning. However, most carbon is transferred from the biosphere to the atmosphere after deforestation and will not be recovered on that site.

Two primary types of carbon contribute to rising atmospheric GHG levels: fossilized carbon (such as coal, oil, and gas) and biogenic (biologically active) carbon. The Intergovernmental Panel on Climate Change (IPCC) estimates that about 6.3 ± 0.6 gigatons (Gt) of fossilized carbon are released annually through the combustion of fuels and during cement production. Meanwhile, roughly 1.6 ± 0.8 Gt of biogenic carbon are emitted from changes in land use practices.[8] Deforestation in the tropics accounts for most of the carbon released from land use changes. This suggests that although carbon fluxes vary from year to year, tropical deforestation probably causes around 20 percent of net GHG emissions (Fig. 13.1). It is possible that because of forest degradation that does not appear on satellite images, this figure is an underestimate.[9] A recent study that suggests tropical deforestation may release 3 Gt C.[10]

After fossil fuel burning, tropical deforestation is the second leading cause of GHG emissions to the atmosphere. There are two broad approaches to reducing human-caused GHG emissions: lessen emissions from tropical deforestation or lessen emissions from fossil fuels. Because CO_2 and other GHGs are assumed to be well mixed and exert their warming globally, it shouldn't matter whether an emission reduction comes from a forest or from a fossil fuel source. Or does it?

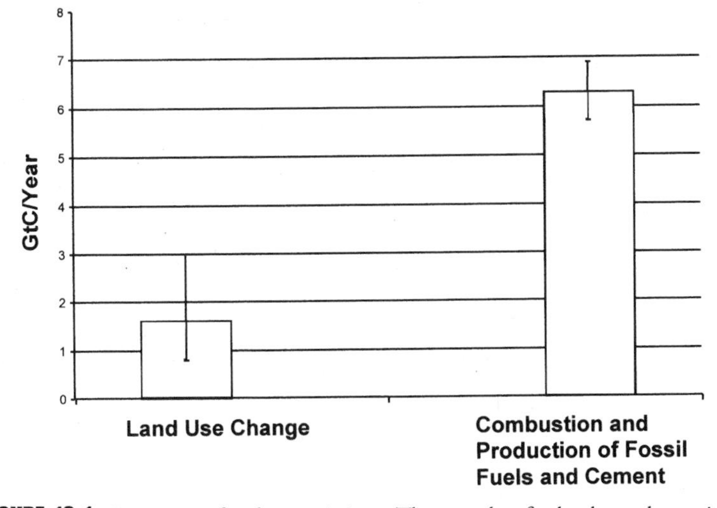

FIGURE 13.1. Categories of carbon emissions. The error bar for land use change includes a higher estimate for carbon emissions from tropical deforestation in Malhi & Grace. (From IPCC, 2000; Malhi & Grace, 2000).

Tropical Forests as Sinks for GHGs

Complicating matters slightly, tropical forests are not only potential sources of carbon, they also can act as a sink of carbon already in the atmosphere.[11,12] How much carbon is still open to debate, but some preliminary estimates are possible. But before estimating the size of the carbon sink in mature tropical forests, a brief foray into the two types of carbon sinks is necessary.

The first kind of sink occurs when an ecosystem is recovering from a prior disturbance that released carbon, such as a previously logged forest that is regrowing. Any accumulation of carbon in an ecosystem recovering from a previous disturbance implies that there was a loss of carbon at an earlier time.[13,14] After tropical deforestation, it takes centuries for carbon to reach predisturbance levels. In some cases, this may never occur.[15] The inability of mature ecosystems to fully recover lost carbon (biomass) may result from myriad human-caused environmental disturbances such as acid rain, nitrogen deposition, or invasive species. Also, ecosystems may not regain their original carbon content because direct human pressures (e.g., firewood needs, continued deforestation, or fires) prevent a stasis period long enough for these systems to recover. As an ecosystem recovers previously lost carbon, no real carbon benefit occurs since the ecosystem is merely "catching up" for carbon previously emitted.

The second kind of sink, CO_2 fertilization, occurs when undisturbed ecosystems accumulate more carbon than they release over a period of time.[16] This is the case in some mature forest ecosystems and in other ecosystems. A useful way to think about CO_2 fertilization is to consider plants as machines that combine carbon in the air (in the form of CO_2) with solar radiation to create stored carbon and energy. The more CO_2 in the air, the more efficient the plants are at converting airborne carbon into biomass carbon. Rising CO_2 levels fertilize plants and help them grow bigger or faster.[17] This type of sink is an important ecosystem service that helps slow the buildup of atmospheric GHGs.

Some studies suggest that tropical forests will be some of the most resilient CO_2 sinks in coming decades.[18] Tropical forests may have an advantage in absorbing excess atmospheric CO_2.[19] Tropical forests may respond more readily to elevated CO_2 because their growth is not as limited by nutrients or temperatures as are other forests worldwide.[20]

Other reports draw opposite conclusions.[21] One report put out by the Hadley Centre, a British research agency, shows that the Amazon may be stressed by climate change in coming years, releasing its carbon at some future time.[22] Some negotiators and nongovernment groups cited this article as evidence that conserving carbon in tropical forests is an unsafe bet. After all, why protect forests if climate change may kill off large tracts in half a century? The possibility that the Amazon may be stressed in 50 or more years is quite real (although the IPCC has cautioned against placing too much confidence in forecasts past a few decades).[23] However, using this as an argument against providing incentives for sustainable forest management is logically suspect. Already, every year the Amazon shrinks by 4–5 million acres; plans for development of the Amazon could accelerate this trend.[24] Currently, forest degradation is a more accepted hypothesis for forest stress than is climate change. Consequently, the best strategy for safeguarding tropical forests is to finance and run conservation schemes in threatened forests. Protecting one area of forest can prevent degradation in surrounding areas, a process already implicated in substantial carbon emissions.[25] Forest conservation measures will help maintain a sink for preexisting atmospheric carbon, minimizing climate change. The earlier forests are destroyed, the sooner emissions enter the atmosphere, possibly leading to more rapid climate change. Additionally, any benefit from CO_2 fertilization—which models show growing for several decades in tropical forests—will be lost.

How Much Carbon Is Being Absorbed in Mature Tropical Forests?

Mature tropical forests may be sequestering through CO_2 fertilization approximately 750 million tons of carbon per year (Table 13.1), a significant brake on

TABLE 13.1. Estimates of Current Carbon Uptake in Mature Tropical Forests

		*Tropical Forest Sink (original estimate in bold)**		
Study	*Method*	*Gt C Worldwide*	*t-C/ha*	*Comment*
IPCC, 2000†	Literature review	1.58	**1.00**	From studies of mature forests; median of reported values.
Phillips et al., 1998‡	Tree measurements	**0.77**	0.49	Mean value; some regions stronger sinks than others.
Malhi & Grace, 2000§	Integrated analysis	**2.00**	1.26	Estimated global value for old-growth tropical forests and associated soils.
Average of studies		1.45	0.92	
Multiplied by 0.5 (to be conservative)		0.73	0.46	

*Values from these studies were recorded on a per hectare basis or as a global aggregate. To move between per hectare and worldwide values, the reported value (in bold) was either multiplied or divided by the estimated number of hectares remaining, approximated to be 1.58 billion hectares (adopted from: FAO, 1999: State of the World's Forests; Rome).
†This is the median of reported values for annual carbon uptake in mature tropical forests.
‡This is based on 600,000 individual tree measurements. This study was unique in having a global coverage of measurements for carbon uptake in mature tropical forests. Phillips, O. L., Y. Malhi, N. Higuchi, W. F. Laurance, P. Nuñez, R. Vásquez, S. G. Laurance, L. V. Ferriera, M. Stern, S. Brown, and J. Grace, 1998: "Changes in the carbon balance of tropical forest: Evidence from long-term plots," *Science* 282: 439–442.
§The value reported for carbon uptake in both soils and trees for mature tropical forests was used.

global climate change. The models chosen here do not represent all estimates of the sink in tropical forests. Some of the larger computer models that estimate carbon sources and sinks typically are parameterized by assumptions that may not be valid for tropical forests.[26] In particular, many models have a difficult time estimating the size of the tropical sink because of the range of uncertainty about the size of sources caused by land use change in the tropics.[27] Models used here are ones that have global coverage and are based on three separate analytical approaches. These studies have also been selected to remove the sequestration effect of recovering forests.

As a whole, these model estimations are not conservative in terms of the general consensus on the location of the terrestrial sink. Between 1989 and 1998, the terrestrial uptake of carbon was estimated to be 2.3 ± 1.3 Gt C per year.[28] Tropical forests are only one of many ecosystems suspected of acting as a sink, and current global carbon sink estimates are too high (hence the 50 percent adjustment used for the conservative estimate in Table 13.1).

Although there is still much uncertainty in this estimate, it seems plausible that tropical forests are actively sequestering substantial carbon out of the atmosphere. Using tree measurements, literature surveys, and integrated analysis, and then lowering the estimate by 50 percent, these studies suggest that approximately 750 million tons are being taken up annually in mature tropical forests worldwide. This translates (on the basis of an estimated 1.58×10^9 ha of natural tropical forests) to an average annual sink of 0.46 tons of carbon per hectare of tropical forest. Recent inversions of the atmospheric ^{13}C isotope record, though also subject to uncertainty, suggest a tropical sink of a size similar to the one reported here.[29] Some of the tropical sink observed with ^{13}C measurements results from regrowth and, as discussed earlier, should not be considered an added benefit for the atmosphere. Like other methods, isotopic calculations are complicated by the uncertainty in the size of the source of emissions from tropical forests. Whatever the actual size of the tropical forest sink, left standing, tropical forests serve as a temporary brake on global warming.

The longevity of terrestrial sinks probably is finite because ecosystems can't grow forever. Nevertheless, mature tropical forests assimilate a certain amount of GHG pollution, and any ecosystem that lowers GHG concentrations helps stabilize GHG levels.

Thus, in terms of minimizing atmospheric carbon concentrations, protecting tropical forests does two valuable things: It prevents a source of emissions and it maintains a sink for emissions already in the atmosphere.[30] In this regard, tropical forest conservation may be a better form of mitigation than equivalent reductions in fossil fuels. Compared with a ton of avoided fossil fuel emissions, every ton of carbon pollution avoided by tropical forest conservation results in emission reductions and some carbon drawn out of the atmosphere. Obviously, preventing fossil fuel emissions is critical. The point here is that reducing fossil fuel combustion will prevent emissions but will not draw other carbon out of the atmosphere.

Tropical Forest Climate-Related Ecosystem Services

GHG levels are only one factor controlling the earth's climate. Apart from their role as reservoirs, sinks, and sources of GHGs, tropical forests have numerous climate-stabilizing properties.[31,32] Heating and convection patterns generated at the tropics influence the global circulation of water, energy, and wind patterns. These patterns are influenced by solar radiation and other variables at the land surface. When forests are degraded and replaced by other land uses, some surface biophysical properties are disturbed. Any substantial change in a land surface will have some impact, outside the sphere of GHGs, on climate and weather patterns.[33]

Some of the climate-related ecosystem services that tropical forests provide are listed in Box 13.2.[34] This list is by no means exhaustive, and many of the parameters are intertwined. By convention, these parameters are listed in relation to grassland or agricultural land surfaces, the main land use types that replace tropical forests.

Equatorial circulation patterns and their subsequent driving of the planet's climate are dynamic, variable, and difficult to predict. This has limited the ability of climatologists to predict the precise impact of deforestation on climate patterns.[35] As a result, manipulating or protecting land surface conditions is not a legitimate mitigation option under the Kyoto Protocol. The protocol uses only the global warming potentials (GWPs) of the regulated GHG molecules listed

BOX 13.2

Ways in Which Tropical Forests Help Maintain a Historically Stable Climate

In terms of land surface parameters, tropical forests

- Influence the planetary boundary layer's height
- Alter sensible and latent heat fluxes
- Maintain a lower albedo (surface reflectivity)
- Influence heat, momentum, and water vapor exchange with the atmosphere
- Elevate soil moisture
- Provide heat storage
- Elevate surface humidity
- Reduce sunlight penetration to the surface
- Slow and elevate winds
- Moderate evapotranspiration (especially losses in dry seasons)
- Maintain vigorous convection and precipitation rates
- Inhibit anaerobic soil conditions

in its Annex A as its mitigation currency.[36] Land surface conditions and their influence on climate and weather are too challenging to build into a fungible climate change treaty. But as any climate modeler will note, our global weather and climate are functions of the gases in our atmosphere *and* the influence of the earth's surface properties on solar radiation.[37] Even if a climate change treaty does not explicitly include stability of land surface conditions in its goals, this is still an important aspect of climate change. Intact tropical forests stabilize biophysical processes that, though uncertain in terms of their net effect, maintain weather and climate patterns on which humans have come to rely.

The Insurance Value of Biodiversity

Tropical forests contain a wealth of genetic diversity that may ultimately lessen the impact of climate change on human and natural systems. Tropical forests contain a large number of the world's plant and animal species. Largely through deforestation, the planet is losing species at a rate 100–1,000 times the normal background rate of extinction.[38] Genetically distinct populations are at an even greater risk.[39] This lost genetic variability may substantially alter the severity that climate change has on humans and ecosystems.

On a warmer, more crowded planet (as might result from sea level rises and subsequent climate refugees), disease outbreaks may become more common.[40,41] New fungi and pests may stress major crops. In the 1970s, billions of dollars of rice crops were salvaged—and an unknown number of food emergencies averted—when a gene from a wild forest rice was found to be resistant to the virus.

Should climate change occur, especially at a rapid, nonlinear rate, we may want access to novel genetic combinations to address climate-related problems. Leaving the bioethical issues aside for the moment, new breeds of crops, drugs, and chemical applications could help counter some of the negative impacts of a warmer planet on health, agriculture, and the economy. Currently, approximately one in four commercial pharmaceuticals is partially derived from tropical plants.[42] Biological diversity housed in tropical forests serves as a pool of novel biochemical formulas (genes) that may help limit climate change's effect on humans and natural systems. If deforestation continues as is, these resources may not be there when we need them.

Finally, there is evidence that tropical deforestation may release and spread dangerous pathogens.[43] Thus, maintaining tropical forests preserves genes and species that may help combat climate change and may prevent the release of unwanted pathogens. Again, although fossil fuel mitigation is absolutely critical to addressing climate change, doing so will not safeguard enormous reservoirs of biological diversity. Thus, it may be helpful to look at fossil fuel emissions and tropical forest emissions not as an either–or proposition but as an and–plus situation.

Legal Context and Technical Challenges[44]

Because developing countries did not make legally binding commitments, their involvement in climate change mitigation is largely restricted to activities under the Clean Development Mechanism (CDM). Article 12 of the Kyoto Protocol creates the CDM to help developing countries (non–Annex I countries) achieve sustainable development while helping stabilize GHG concentrations in the atmosphere.[45] The CDM is also designed to help industrialized countries (Annex 1) achieve compliance with their quantified emission limitation and reduction commitments (QUELROs). For this purpose, Annex I parties may fund emission reduction projects in non–Annex I countries and get credit for Certified Emission Reductions (CERs) from these projects, which they can use to meet their QUELROs. In short, developed countries that work with willing developing nations to institute emission reduction projects will not have to reduce as many of their own emissions.

As written, Article 12 limits project eligibility to "activities resulting in [certified] emission reductions," omitting the key words "removals by sinks" used in Article 6. This omission may be legally important because selective wording may reflect a selective meaning. It may reflect a shared understanding of Kyoto negotiators to exclude carbon sequestration at least for the time being. However, this does not apply to forest conservation, which is an emission reduction. Article 12 does not discriminate for or against particular types of projects that would be eligible for mitigation (e.g., transportation, energy, forestry).

Deforestation for Reforestation?

One critique of the CDM is that it could encourage countries to cause deforestation and then claim credits for subsequent reforestation.[46] If poor rules were written this could indeed occur. A safeguard against this potential abuse is contained in the IPCC definition of reforestation.[47] It implies that reforested lands must have been used differently over a minimum period of 20 years. Therefore, deforestation must have occurred no later than 1988 (2008 minus 20 years) to be counted as reforestation during the first commitment period.

Another way to avoid the problem of "deforestation for reforestation" would be to enforce the CDM mandate for sustainable development (Article 12.2). This mandate is defined in Article 2, where forestry is said to be "protection and enhancement of sinks and reservoirs of GHGs, taking into account commitments under relevant environmental agreements; (and) promotion of sustainable forestry, afforestation and reforestation." Clear-cutting forests to get subsequent replanting credits would violate this definition.

A final approach would be to interpret the CDM such that it applies only to emission reductions and not include forestry sinks. As mentioned earlier, forest conservation is a form of emission reduction and is not a sink activity per se (although there is an additional sink benefit).[48]

Requirements for Certified Emission Reductions

Article 12 mandates a set of three requirements for the certification of emission reductions, the legal term for carbon offset credits that can be traded in the protocol:

Voluntary participation approved by each party involved,

Real, measurable, and long-term benefits related to the mitigation of climate change,

Reductions in emissions that are additional to any that would occur in the absence of the certified project activity.

The procedures to implement these requirements are not elaborated in the Kyoto Protocol; they are deferred to subsequent decisions. But the objectives for implementation are stated in Article 12.7 as "transparency, efficiency, and accountability through independent auditing and verification of project activities."

The following sections discuss how the certification requirements of the CDM and the principles for implementation could be applied to projects aimed at stopping tropical deforestation. Although forest conservation is currently ineligible for Kyoto-type crediting, this could be changed by subsequent decisions or in subsequent commitment periods.

Voluntary Participation

The first criterion concerns the noncompulsory nature of the CDM. No country will be forced to take part in the CDM. Moreover, each project is supposed to be consistent with the environmental and development priorities of both the host and the investor country. There is room for egregious interpretations of this aspect of the CDM, a point several indigenous groups raised during the negotiations. The protocol does require that any CDM project, forestry or otherwise, must be something that both the host and the investor want. Within a country, there are no explicit guarantees that local communities will not be trammeled, although the sustainable development and domestic support language throughout the protocol is relatively strong.

It is interesting to look at activities implemented jointly (AIJ) experience, which shows that different countries have different attitudes toward forestry activities (see Chapter 12). Individual countries are free to decide which projects to include in their overall portfolio of mitigation projects. Within Latin America, for example, the majority of countries wanted to have tropical forest conservation in the list of possible mitigation activities, although Brazil and a few other countries were opposed (see "Politics, Politics, Politics" section).

Real, Measurable, and Long-Term Environmental Benefits

CDM projects could be manipulated to reflect false, inflated, or temporary environmental benefits if analysis and verification are not carried out appropriately. Forest conservation projects are especially prone to elements that would negate real and measurable benefits for climate change, including baseline inflation, monitoring problems, the temporary nature of forest conservation, and leakage.

Baselines

For a project based in the CDM, the environmental benefits must be measured against a baseline, i.e. what would have occurred without the project. Governments or companies participating in CDM projects have an incentive to inflate the baseline, attributing more reductions to their projects than is accurate.[49] Baseline inflation is not unique to forestry projects and also can occur in energy-related projects.[50]

Historical deforestation trajectories can be used as baselines to avoid projects claiming that artificially high deforestation rates have been averted. Some have pointed out that 5- or 10-year baselines would more realistically capture the interannual variability of deforestation rates.[51] The Food and Agricultural Organization (FAO) has determined deforestation rates for the 1990s,[52] but there is some debate about the accuracy of these findings. At least two studies have criticized the FAO for underestimating deforestation rates.[53] Other studies have delineated hotspots of deforestation within individual countries.[54] Countries, nongovernment organizations, and scientists should be given access within the legal CDM framework to debate deforestation baselines and ensure that accurate ones are used. Huge expenditures on remote imaging over the past decades can be synthesized into reasonable baselines. Military images may be able to fill in data gaps.

MONITORING

It is also necessary to establish a monitoring program for CDM projects. Monitoring ensures that the claimed reductions in deforestation are actually achieved. Effective monitoring would couple remote sensing with on-the-ground measurements at a project site. For example, deforestation rates could be estimated from satellite imagery, and both air and ground monitoring could measure changes in deforestation rates from year to year.

The FAO Tropical Forest Assessment of 1990 used two phases of assessment to measure forest cover change.[55] Phase 1 of the project involved compiling a geographic information system of statistical information gathered in countries containing tropical forest. Phase 2 of the project used remote sensing to continuously assess forest cover throughout the tropics. Another project, the European Commission Tropical Ecosystem Environment Observation by Satellite (TREES) project, has compiled a map of tropical forest cover using 1-km (low) resolution satellite data.[56] These data were calibrated and verified using higher-resolution samples. Comparisons of TREES, FAO, and other imaging projects indicate differences in measured tropical forest extents but agree in many areas as well.[57]

Long-Term Environmental Benefits

The CDM states that transferable carbon offsets will be awarded only for projects that create long-term benefit. However, the CDM is not clear on the meaning of "long-term." It states only that emission reductions from CDM projects "shall be certified on the basis of real, measurable and long-term benefits to the mitigation of climate change."

From a scientific point of view, the long-term requirement of the CDM should cover the period of time it takes for carbon to decay in the atmosphere (e.g., 100–200 years in the case of carbon dioxide). Delaying emissions for any shorter period would produce only a minor benefit for climate change. However, forest conservation is never permanent. An area protected one year may be cut or burned the next year. This is a unique challenge that forest conservation faces in the legal and technical context of mitigation. If the long-term requirement of the CDM means permanence of preserved forests, this would be nearly impossible to guarantee at the outset of a project. Additionally, permanent conservation could be unacceptable to potential CDM host countries. It could be perceived by developing countries as de facto expropriation or an ultimate selling of lands. Several key developing countries, such as Brazil, have expressed this objection to CDM forestry.

Less than permanence could be sufficient if we approach the long-term issue from a practical point of view. "Temporary" mitigation activities will have a positive effect on climate change if emissions are reduced for a period of time while alternative abatement strategies become widespread. In this sense, the long-term requirement is defined by the expectation that structural changes in the world economy will arise in the future. These changes could be in the form of emerging carbon-free energy sources or in the valuation of forests outside their role in climate change mitigation. It may be possible to substantially reduce GHG emissions with widespread adoption of technologies such as solar power or hydrogen cells in the next 50 or 100 years. Alternatively, the value of tropical forests may rise in the future such that deforestation is no longer an attractive activity. Increases in value could be in the form of domestic values (i.e., watershed protection) or external values (i.e., biodiversity) that develop real markets. Forest conservation under the CDM could delay emissions until the incentive to deforest is lost or alternative technologies develop.

Practical experience with the long-term requirement was gained in the pilot phase of AIJs. The average duration of AIJ forestry projects is 39 years (median = 32.5 years).[58] In several aspects AIJs resemble the CDM; both mandate the requirement of "real, measurable and long term environmental benefits." Considering the precedence and experience of the pilot phase and economic projections of emerging technologies, it seems appropriate to apply a minimum period of 30–50 years to the CDM's long-term requirement.

Several approaches are available to enforce such a minimum period for CDM projects.[59] One way is to establish schemes of legal liability or trade sanctions. However, legal liability and sanctions are difficult to implement for environmental issues. Moreover, a system of international legal liability could create high transaction costs.

Forest-Secured Escrow Account

Another way to address the long-term requirement of the CDM is to design accounting and financial procedures for projects so a host country has an economic interest in lasting forest conservation. A specific way to do this is to place part of the money that a developing country will receive for protecting its forest into an escrow account during the lifetime of the project. An escrow account for forest conservation would work like this: The donor country sets up a savings account on behalf of the host country to secure interest in long-term forest conservation. Both the principal and the interest legally belong to the host country for agreeing to not destroy its forest. Interest from the escrow account is distributed to the host to provide interim conservation incentives. The principal in the account would be disbursed to the host country only when the forest conservation contract is fulfilled (e.g., after 30–50 years of verified forest conservation).

Conceptually, this is a different approach to financing forest conservation than traditional aid and donor programs. The forest-secured escrow account retains the decision making and the responsibility for the forest in the hands of the developing country. Some years into the contract, the host country might decide it is either unwilling or unable to maintain long-term forest protection. It can break the contract but will forfeit the escrow. Or the developing country may try to raise the price of the project. This should prevent investors from low-balling developing countries into accepting too little money for their forests. In addition to retaining decision making in the host country, such accounting systems allow multiple parties to jointly sponsor forest conservation. For instance, both an investor and a private environmental group could collaborate to offer a developing country a viable package.

Who should receive the payment for conserving a forest? The displaced farmer? The farmer who would have occupied the land? The forestry department? The state government? The federal government? This question of distribution—who gets the money for domestic forest conservation—is a difficult question. It is a real concern that has no easy answers.

To be effective, the payment for sustainable forest management should go to the party or parties that have the most influence over and reliance on the forest. The market forces behind the CDM may produce appropriate incentives for a domestic question of distribution to be resolved because for CERs to be awarded, host nations must prove verified reductions. This implies that local people who rely on the land must be either appropriately compensated or evicted. Although the latter is a possibility, there are mechanisms to control such abusive behavior. The CDM framework must facilitate equity and sustainable development discussions regarding these payments.

LEAKAGE[60]

Leakage is unintentional GHG fluxes resulting from mitigation projects or policies. Taking the forestry sector as an example, some argue that if one forest is protected, people causing the deforestation would simply switch to another forest, with no net GHG benefit occurring. This type of leakage can be thought of as *activity-shifting leakage* brought about by a specific project. An example of *market leakage* caused by a policy would be if the incentives for improved forest practices (e.g., less GHG emissions) apply exclusively to developed nations, excluding developing nations. This could result in logging companies exporting operations from developed to developing nations, displacing emissions and forest disturbance to poorer regions of the world. The original goal of applying pressure for developed nations to reduce emission from forestry could exacerbate

TABLE 13.2. Leakage Types

Causes	*Project* *Policy*
Effects	Positive Negative
Mechanisms	Activity shifting Market effects Life cycle effects Ecological
Scales	Local Regional Global
Sectors	Energy (fossil fuels) Land use (biomass)

*Market-driven leakage is mediated by a change in the price of goods; activity shifting occurs when capital moves to another location.

GHG emissions because some developing countries' forest industries are less efficient than those of developed countries.

There are other types of leakage as well (Table 13.2). Ecological leakage is a change in GHG fluxes mediated by ecological processes. For instance, protecting a tropical forest will maintain intact carbon within a project's area. However, this maintained forest might also stabilize land surface conditions, microclimate, pollinators, natural pest equilibriums, and other factors that maintain nearby forests.

In the recent climate change negotiations, many groups wary of, or opposed to, forest mitigation in the protocol used the leakage framework to suggest that tropical forest conservation would not produce real and lasting benefits for climate management.[61] The World Wide Fund for Nature (WWF) said,

> People in forest rich tropics often keep on clearing land for survival reasons as they lack fertile soil or just do not own land. They would simply switch to adjacent regions. However, such leakage of deforestation activities from one place to another, so far cannot be determined. Therefore, forest conservation efforts may be extremely useful for a certain region as biodiversity is concerned but may not be carbon-neutral if one considers the broader region or even national boundaries. Without addressing the underlying causes of land degradation, forest sink (sic) projects under the CDM are not likely to yield net GHG benefits.[62]

If a conservation project simply fences off a community from its forest, then leakage is a definite concern. However, this type of forest conservation approach is largely outdated. In practice, modern conservation projects rarely fence off forest from local people but rely on providing sustainable incomes and livelihoods not based on forest destruction. The contention of WWF that leakage cannot be measured or addressed stands in contrast to other analyses suggesting that although leakage is a concern, it is manageable and many policy tools can alleviate most leakage concerns.[63]

There is no evidence that leakage is a greater problem in forestry projects than in energy projects. In one study, energy sector leakage is calculated to be 4–15 percent[64] and is quoted elsewhere as 15–40 percent.[65] Thus, leakage appears to be a broad concern. It is complex, poorly understood, and poorly documented and is not unique to forestry.

Early climate mitigation projects have done a reasonable job of measuring leakage and taking it into account when estimating GHG benefits.[66] The Nature Conservancy and other private groups have been operating trial projects to see whether and how the challenges of carbon-financed forest conservation can work. And by its very nature, leakage is a second-order concern that should not derail mitigation efforts and certainly not just the forestry projects. It appears that attention focused on leakage in forestry projects is a mix of real concern, gut reactions (it is easy to visualize forest dwellers relocating to a new forest), and political arguments.[67]

Legal and Technical Summary

The legal and technical challenges to implementing forest conservation under the guise of climate mitigation will not be easy. As the world has seen in recent decades, efforts to stem tropical deforestation face substantial challenges. Many projects to help conserve forest have failed outright. There are risks that local communities could be manipulated and there are several ways for investors or host countries to "cheat" the system. There are also risks that even if a project succeeds within its borders, it will cause additional GHG fluxes outside the borders (leakage).

Given the technical challenges to forest conservation, allowing these projects to be credited could result in several project failures that may delay other emission reductions. An important question is whether tropical forest conservation should be used for some of the hard-fought emission reductions called for in the Kyoto Protocol. After all, the Kyoto Protocol requires only a limited number of emission reductions, and forest protection will detract from fossil fuel reductions. Good rules should catch most of the bad projects and not award spurious offset credits. Some bad projects may get through. However, although some people falsify their tax returns, we still have a self-reporting tax system that does a reasonable job collecting revenue for government. In the final analysis, the risk of abuses and misdeeds must be weighed against the potential for communities to be empowered to run good projects funded by the protocol.

On balance, it appears that with careful oversight, forest conservation projects can be shielded against most major abuses. The technical concerns for long-term forest protection are serious. Yet tropical deforestation is proceeding rapidly. Success in tackling this widespread cause of environmental damage and poverty will take time. If there are technical challenges, it will be useful to begin figuring out solutions as to how to effectively fit forest conservation into the climate mitigation rubric. Currently, there is no other viable international mechanism for launching widespread forest conservation.

The Economic Context

The Kyoto Protocol was put into place as a result of the UN FCCC signed at the 1992 UN Earth Summit (UNCED). Also signed at UNCED were two other pertinent international accords. The Convention on Biological Diversity (CBD) entered into force in December 1993 and has since been ratified by more than 130 countries. In Article 11 of the CBD, financial incentives were identified as a specific mechanism to help guide international efforts to conserve biodiversity. Also at Rio, the Declaration on Global Forests was drafted and has since been signed by several countries. By most accounts, nothing of substance has emerged from the Declaration on Global Forests, which has the ominous-sounding official title "Non-Legally Binding Authoritative Statement of Principles for a Global Consensus on the Management, Conservation and Sustainable Development of All Types of Forests."

The CBD and the Declaration on Global Forests have failed to secure the resources necessary to help stem biodiversity and habitat loss, especially in tropical forests. A UN report suggests that $30 billion in international aid is needed annually to conserve significant amounts of tropical forests.[68] In 1991, the world funneled only about $500 million toward forest and biodiversity conservation.[69] Without other significant financing alternatives for tropical forest conservation, some people looked to the Kyoto Protocol to transcend the uncertainties of valuing biodiversity by valuing forest habitats for their role in storing carbon.

A Cost-Effective Interim Strategy

Preventing or slowing deforestation is an economical way to mitigate climate change. Studies suggest that a blend of mitigation options from different sectors of the economy is likely to be optimal in the medium term.[70] Forestry could play an important role in this blend. Long-term projections show a sharp decline in mitigation cost in energy projects in the second half of the twenty-first century. In the long-term, then, the relative cost-effectiveness of forestry could disappear. Of course, by then, if we do nothing else to conserve tropical forests, they will be largely a matter of history.

Tropical forest conservation (and for that matter, reforestation of degraded lands and forest regeneration) is readily available and cost-effective. Estimates of the costs of carbon conservation and sequestration range from $2 to $25/t C.[71] Empirical results from the AIJ pilot phase support this estimate.[72] The cost per ton of carbon avoided through forest preservation and afforestation in the pilot phase has been around $10. From an investor's standpoint, the price of forest conservation looks pretty reasonable.

Opportunity Costs in Developing Countries

Will it be worth it for developing countries that want to conserve their forest to engage in carbon-financed forest protection? One way to assess this is to look at opportunity costs. Opportunity costs are costs one "pays" by forgoing alternatives. For forest conservation, this might mean the lost income one would have otherwise received through activities such as agriculture, logging, and ranching on these lands.[73]

Using Carbon to Finance Forest Conservation in Madagascar

One study on the costs and benefits of using carbon financing to protect Madagascar's last remaining large rain forest is informative.[74] This study compared a business-as-usual foreign aid project with several modeled logging scenarios that were based on real threats. The results were unequivocal, even when multiple discount rates and contingencies were assessed. In all scenarios, the local people did best under the existing foreign aid program of forest conservation compared to logging. However, the federal government lost large sums of money by forgoing logging concessions in favor of conservation. Madagascar gave up logging rights in favor of conservation and in doing so lost between $47–$334 million in taxes, fees, and other revenues. This is an enormous amount of money for a country in which many families subsist on less than a few dollars a day.

In this study, Madagascar's forgone revenue was compared with the global international benefit of avoided emissions. Multiple economic analyses suggest that carbon put into the atmosphere causes some economic damage.[75] Using a central estimate that every ton of carbon would cause $20 in damage worldwide,[76] the result was shocking. By paying only $5–$10 million to Madagascar, the international community was avoiding, albeit unintentionally, $72–$655 million in damages. Were carbon damages accounted for in the project analysis, the international donor community was paying only a fraction of what it should have been paying to conserve the rain forest and the carbon therein. Meanwhile, by forgoing large sums of money from timber sales, one of the poorest nations in the world bore 57–91 percent of the costs of keeping carbon out of the atmosphere.

If the world compensated Madagascar for the forgone logging revenue, it would have translated to $0.30–$3.94 per ton of carbon. Some groups opposed to forests conservation in the CDM used this as an argument against including forest conservation in the CDM.[77] It would have cost too little, they suggested, diluting the incentive for other types of emission reductions. But this conclusion obfuscates the key finding of the study. Under the existing aid packet, donors were paying only $0.07–$0.95 per ton of carbon. A few cents or dollars difference per ton of carbon, when multiplied by the vast amount of carbon protected in Madagascar's forest, would have generated tens of million of dollars of additional financing. Although this may not seem like a lot to some environmental groups, it would have been a more equitable way to protect a rain forest from the standpoint of the Madagascar people.

This study illustrates several points. First, current donor-driven forest conservation financing in Madagascar may provide some benefits but does not compare with more lucrative opportunities such as timber concessions. Second, the world community is paying only a fraction of the value of the benefit it receives by asking a developing country not to log when avoided damages of climate change are taken into account. Third, for a modest cost per ton of carbon—a few cents per gallon of gas, in fact—donors could match the cost of logging concessions. In doing so the donor community would have to pay more than they currently do to conserve tropical forests. But given the costs and benefits in light of carbon damages, this is a more just distribution of the cost of safeguarding a forest that is incredibly rich in biodiversity, full of vulnerable carbon, and a source of livelihood for thousands of rural Africans. Finally, the Kyoto Protocol's CDM is ideally situated to facilitate this financing by giving private or public investors a reason (inexpensive carbon offsets) to help stop deforestation.

Naturally, if such carbon financing were allowed, it would permit the investing company or country to not make equivalent emission reductions elsewhere. However, this is the premise on which the Kyoto Protocol was built: flexible mitigation that leads to sustainable development.

Realistic Estimate of Forest Conservation for Climate Mitigation

What is a realistic worldwide assessment of avoided emissions that could be realized under the CDM via tropical forest conservation? As noted previously, some groups raised the concern that including tropical forest protection in the Kyoto Protocol would swamp out other more important activities, such as fossil fuel emission reductions in developed countries. For example, Greenpeace suggested that 300 million tons of carbon via forest conservation could enter the "carbon market," which would lower the total cost of mitigation and prevent other emission reductions. Greenpeace remarked in a widely distributed paper that "the introduction of this amount (300 million tons) of very cheap projects in the CDM equation will drastically reduce the amount of money and technology transferred to developing countries, remove the incentive to develop less carbon intensive technologies in developing countries and allow industrialized countries to increase their emission close to business as usual."[78] A WWF briefing paper (which had several math errors but was circulated widely at negotiations) pointed out that forest conservation in the Amazon would allow industrialized countries to emit "in each of the five years of the commitment period more than 1.1 billion tons of carbon dioxide into the atmosphere."[79] Let's take a close look at the reality in many countries to see whether these are realistic concerns. The answer will help inform the question of whether forest conservation in the treaty would "drastically reduce the amount of money and technology" available for other important efforts.

To estimate a realistic amount of forest conservation possible,[80] the following formula is used:

> Rate of annual deforestation × Carbon emissions per hectare × Percentage of emission addressed × Percentage of attempted reductions that are successful in the long term = Potential carbon offset by forest conservation

Given a certain rate of deforestation in a country and associated carbon emissions, only a fraction of this deforestation can be stopped. There is limited institutional capacity, estimated in a prior study, for countries to arrest deforestation. Furthermore, of the measures taken to reduce deforestation and carbon emissions, only a portion will succeed in the long term. The history of development projects, and most entrepreneurial efforts in general, is one of high project failure rates. Historical data of project success rates, by country, from prior World Bank projects were used to approximate the likelihood of projects succeeding in the long term. Other estimates could be used and would yield different findings, so the results in Table 13.3 are a rough approximation. These represent the carbon credits from avoided deforestation that could realistically be claimed if there are reasonable verification and monitoring requirements.[81]

Given these assumptions and methods, approximately 40 million t C per year might be prevented from entering the atmosphere via forest conservation in the CDM. This is less than 0.5 percent of the estimated 8 billion tons of human-caused carbon emissions and about 5 percent of the approximate 800 million t C

TABLE 13.3. Estimated Plausible Annual Reductions of Carbon Emissions via Tropical Forest Conservation

Country	*Carbon Emissions from Deforestation (t C/yr)*	*Percentage of Emissions Addressed**	*Percentage of Attempted Reductions Successful in the Long Term†*	*Estimated Emission Reductions (t C)*	*Annual Host Country Income at $15/t C‡*
Angola	8,650,500	5%	10%	43,253	$648,788
Bolivia	66,815,000	10%	53%	3,529,785	$52,946,776
Brazil	347,575,293	10%	36%	12,651,741	$189,776,110
Cambodia	18,122,000	5%	10%	90,610	$1,359,150
Cameroon	13,996,500	10%	43%	603,110	$9,046,652
Central African Republic (CAR)	12,800,000	10%	10%	128,000	$1,920,000
Colombia	26,200,000	10%	27%	710,102	$10,651,528
Congo (DR)	127,280,000	5%	10%	636,400	$9,546,000
Ecuador	17,199,000	20%	43%	1,482,212	$22,233,182
Indonesia	142,004,000	10%	47%	6,710,958	$100,664,370
Madagascar	12,740,000	10%	10%	127,400	$1,911,000
Malaysia	46,200,000	10%	80%	3,697,752	$55,466,276
Mexico	42,333,333	5%	60%	1,269,264	$19,038,966
Myanmar	44,698,500	5%	43%	963,031	$14,445,460
Nicaragua	17,742,500	5%	10%	88,713	$1,330,688
Nigeria	2,964,500	10%	10%	29,645	$444,675
Papua New Guinea	14,896,000	5%	43%	320,935	$4,814,022
Paraguay	32,700,000	20%	10%	654,000	$9,810,000
Peru	20,832,000	5%	57%	591,052	$8,865,779
Philippines	22,932,000	10%	41%	942,472	$14,137,075
Tanzania	7,267,500	10%	38%	274,183	$4,112,744
Thailand	30,432,500	10%	72%	2,191,795	$32,876,932
Venezuela	50,300,000	10%	43%	2,167,430	$32,511,455
Vietnam	17,685,000	5%	43%	381,024	$5,715,359
Zambia	6,204,000	10%	10%	61,420	$921,294
TOTAL	1,150,540,126			40,346,287	$605,194,305

*This is an estimate of the percentage of deforestation that could be addressed under a policy similar to the CDM. In countries where there was no estimate or the estimate was less than 5%, a rate of 5% was used. These data were modified from M. Trexler and C. Haugen, 1995: *Keeping It Green* (Washington, DC: WRI & EPA).

†This estimates the percentage of emissions addressed that could be certified as long-term reductions. This estimate was derived by multiplying the percentage of World Bank projects in a country that were evaluated as successful by the percentage of projects in a country judged sustainable. Globally averaged, this may be a reasonable approximation of the likelihood that complex, multi-institutional forest conservation projects can produce real, measurable long-term emission reductions. However, for any specific country, these estimates should be regarded with skepticism. In calculations where a nation's estimate was less than 10% or not available, a rate of 10% was used. This material comes from World Bank, 1999: *1999 Annual Review of Development Effectiveness: Toward a Comprehensive Development Strategy* (Washington, DC: World Bank), Table 16.

‡This is equivalent to a price of $4/ton of CO_2. These represent undiscounted revenue streams.

of annual reduction commitments by Annex I countries under the protocol.[82] At a price of $15/t C these avoided emissions would generate an estimated $605 million annually for these countries, ranging from less than half a million dollars for Nigeria to almost $200 million for Brazil. Although the money is substantial for some countries, this level of mitigation will not swamp the market and prevent other important emission reduction activities. Furthermore, this estimate does not account for additional constraints to forest conservation projects initiated with carbon financing. Some countries may not find willing investors for forestry projects. Two of the three countries with the highest rates of deforestation face particular challenges (Indonesia is undergoing a difficult political transition, and the Democratic Republic of Congo has several countries' troops within its borders). The country with the highest rate of deforestation, Brazil, has repeatedly made clear that it is not going to embrace forest conservation projects within its borders.[83] Thus, the top three countries in terms of gross deforestation probably will not be amenable to significant forest protection projects within their borders.

Politics, Politics, Politics

At the COP6b negotiations, forest conservation was deliberately excluded from major funding via the Kyoto Protocol and the CDM. How did this come about? How were roughly 20 percent of worldwide GHG emissions excluded from a climate change treaty? The "forest conservation in the CDM" debate was one of the top remaining points of disagreement between key negotiating blocs when the talks were called off at COP6, before the withdrawal of the United States from the Kyoto process. The discussion reached the highest levels of government.

Three key interest groups opposed to forest conservation in the CDM largely shaped this outcome: Brazil, the EU and, ironically, some environmental groups such as WWF, Greenpeace, and parts of the Climate Action Network (CAN). Other interest groups, such as the United States and its negotiating allies (the Umbrella group), most Latin American countries, some African countries, and other environmental groups lobbied to have forest conservation included. Other groups and nations lined up on either side of the debate, but here I focus on Brazil, the EU, and the environmental groups opposed to forest conservation, whom ultimately prevailed.

Brazil

Brazil's negotiator at the talks, Dr. Luiz Gylvan Meira Filho, is a powerful speaker with a command of the technical aspects of climate change. He is also an influential negotiator and was the principal voice in backroom negotiations opposing forest conservation in the CDM. Brazil was adamantly opposed to forest conservation being part of the Kyoto Protocol and the CDM. There are several explanations for the Brazilian position. First, it seems Brazil had serious concerns about the technical aspects of forest conservation and emphasized the fact that forest protection can never be permanent.[84] Brazil's position may also be based on the question of national sovereignty: Projects that seek to maintain forests in the Amazon could be an encroachment on the right of Brazil to determine its use of the Amazon. Brazil currently has large plans for the Amazon, including Advance Brazil, a national development strategy that calls for up to $40 billion in new roads, hydroelectric power, railroads, and housing in the Amazon.[85] If completed, Advance Brazil would deforest large areas of the Amazon, resulting in billions of tons of carbon emissions. Clearly, efforts supported by the international community to conserve tropical forests would impede Advance Brazil. A final reason is that Brazil is already the largest emitter of carbon dioxide from deforestation in the world. Brazil may have wanted to set a precedent that tropical deforestation emissions would not be part of the global climate change regime.

Brazil was an influential member in the G77 negotiating bloc and was able to predispose the developing nation bloc to oppose (or at least not support efforts to include) forest conservation in the CDM. After the political agreement at COP6b was reached excluding forest conservation from the first commitment period, Brazil tried to remove any language that would have allowed the issue to be reconsidered for subsequent commitment periods. Although this effort was ultimately blocked, it suggests that Brazil is strongly opposed to forest conservation becoming a part of the climate change regime. As one scientist who followed Brazil's position stated, trying to convince the government otherwise is like "giving a blood transfusion to a cadaver."[86]

Brazil's official position does not necessarily reflect the entire Brazilian voice. One statement released during the negotiations and signed by 13 Brazilian institutions and other individuals commented,

> [We] understand that official Brazilian representatives in the negotiations have played an important role in overcoming impasses and convincing the principal historical source of emissions to accept their responsibility before the international community. Brazil was the author of the proposal that led to the incorporation of the CDM into the Protocol. But for the government to oppose including forest conservation projects in the CDM is not coherent with the gains it has achieved. . . . We expect that, should negotiations move for-

> ward on issues surrounding the implementation of the flexibility mechanisms, the Brazilian official position will allow the implementation in the CDM of projects in native forests, insofar as these comply with the principles of additionality, transparency, control of leakage, verifiability of results and such other rules and controls as may yet be determined.[87]

One of the institutions behind this letter was the National Council of Rubber Tappers of Brazil, an organization founded by Chico Mendes in 1985 with more than 270 local member organizations. Other indigenous and environmental groups within Brazil took various positions in favor of, or against, forest conservation programs in the Protocol.

Europe

The EU's position was that forestry activities, domestic or abroad, would not be a priority for meeting its Kyoto target.[88] The EU probably felt that the United States and other "umbrella" nations were trying to dilute the Kyoto Protocol with accounting loopholes. In a large sense, the EU held a stand it believed was principled, namely that most emission reductions in the protocol should occur as fossil fuel emission reductions in developed nations. The European nations had a few advantages in the protocol, such as free trading within the EU, serious cutbacks in coal use in the United Kingdom because of prior policies, and a collapsed East German economy leading to serious drops in emissions. Although it appears that there was some division on the issue within the EU, for the most part Europe united behind opposing forest conservation in the CDM.

Europe's position was buttressed by scientific information coming from the Hadley Centre for Climate Prediction and Research, the primary UK government climate science office. The Hadley Centre research on the issue of tropical forests used in negotiations was dominated by studies that concluded that forest mitigation was suspect for a variety of reasons (including albedo impacts in northern forests and the potential for die-offs of Amazonian forests by the end of this century). The EU government also listened carefully to what some environmental groups were saying on the issue.

Greenpeace, WWF, and CAN

The most vocal opposition to including forest conservation in the CDM came from an alliance of key environmental groups, largely headed by Greenpeace, the WWF, and members of the CAN. These groups were generally opposed to any measure that would reduce incentives for fossil fuel reductions.[89] For a long time, staff members of these groups did not publicly differentiate between forest activities that were sinks (plantations) and forest activities that were emission reductions (forest conservation). Whether this was a misunderstanding of carbon dynamics or a political strategy (plantations are widely seen as environmentally regressive in the negotiating atmosphere) remains uncertain. As of May 2002, the WWF climate change Web page did not acknowledge that tropical deforestation is even part of the cause of climate change. On their Web page under "Where Does CO_2 Come From" there is no mention of tropical deforestation.[90] And in the "Solutions" part of their Web page, conserving tropical forests is not mentioned.[91] This is certainly odd given that behind fossil fuel combustion, tropical deforestation is the leading cause of CO_2 emissions. It is especially odd given that WWF often tries to raise money for forest conservation programs in developing countries.

Greenpeace was a leading opponent to forest conservation in the CDM. It released dozens of statements, analyses, and reports that consistently showed how forest conservation could not work, and if it did, it would wipe out a majority of more important (in Greenpeace's view) emission reductions. In one report, Greenpeace asserted that forest projects in the CDM would have "no net benefits" for climate protection.[92] The rationale behind their arguments is difficult to ascertain, but it is clear that they believed the technical obstacles to long-term sustainable forestry in the protocol were insurmountable. This is not to say they were not acting in the interests of the global environment; they had a savvy political team that helped safeguard many important environmental components of the Kyoto Protocol. By and large, Greenpeace and other groups saw tropical forest conservation as another way for the wealthy countries that caused the bulk of the GHG problem to avoid concrete domestic steps or steps to reduce fossil fuel reliance. Greenpeace, more than any other environmental group, was able to pass on key information and positions to the European negotiators. For instance, at a critical early-morning moment on one of the last days of COP6, Greenpeace and WWF provided key analyses that helped persuade the EU to reject the American offer on domestic U.S. forestry proposals.

Other environmental groups (Environmental Defense, the Nature Conservancy, and the Union of Concerned Scientists) supported tropical forest measures during the negotiations. Some groups, such as the Sierra Club and the World Resources Institute, did not take particularly strong positions either way.

Although other players were obviously critical, these three constituencies were the driving force behind keeping tropical forest conservation out of the Bonn agreement on the Kyoto Protocol.

Conclusions

The inclusion of tropical forest conservation in a climate change framework is a matter of weighing opportunities against risks. The importance of tropical forests for climate change should be apparent. The continued destruction of these ecosystems is responsible for a significant portion of carbon emissions implicated in global warming. These systems also are likely repositories for carbon that is already in the atmosphere. These ecosystems, if protected, have the potential over many years to take up excess carbon dioxide in the atmosphere. Tropical forests also influence complicated atmosphere–biosphere exchanges that are vital to the long-term stability of current climatic patterns. Tropical forests are also enormous stores for biodiversity, which may be valuable in the future as we look for resources to combat diseases with expanded ranges and other symptoms of climate change. By almost any measure, tropical deforestation destabilizes climate and cultures; efforts to protect tropical forests will help stall climate change and its consequences.

The risks from flawed policies and implementation of forest conservation are real but seem manageable. Forestry projects in Costa Rica and elsewhere[93] use sophisticated monitoring and verification methods. This suggests that good protocols for forest projects can be carried out. By increasing the scale of projects and bundling domestic and international efforts, Costa Rica was able to reduce the transaction costs per unit of carbon mitigated, thereby allowing a higher overall level of control. But it is still too early to say whether the climate arena is the proper place to conduct tropical forest conservation. The few projects that have been undertaken have not been going long enough to make a judgment about their overall effectiveness.

Due largely to politics, sink forest activities (reforestation and afforestation) were allowed in the CDM by negotiators in recent rounds, whereas forest protection was kept out. It is not clear whether the political decisions at COP6bis are legally compatible with the language of the CDM; neither reforestation nor afforestation is a form of emission reduction. Yet for the near term, the Kyoto Protocol cannot be considered a complete GHG treaty; with the exception of some sinks, it is a fossil fuel treaty. Given that fossil fuels are the preeminent cause of global warming, this is not an unreasonable outcome.

However, given economic constraints and increasing marginal costs, a global treaty should address most major emission categories. It seems unwise to exclude almost a quarter of the world's emissions from a treaty *designed* to limit worldwide emissions. Nonetheless, that is where things stand for the Kyoto Protocol's first commitment period. Negotiators have decided that emissions from tropical deforestation will not be part of the coordinated climate change response. Given the overwhelming peripheral benefits of stopping tropical deforestation, inadequate funding for developing countries to maintain forests, and the affordable price tag of forest conservation, the results from the negotiations are that much more disheartening.

The arguments against including tropical forest conservation in the protocol rested primarily on technical matters or on the fact that including forest preservation would swamp the market (e.g., prevent fossil fuel reductions). These two arguments are logically opposed. The first is essentially saying we don't know enough about how to establish rules that will ensure good projects. The second argument says that widespread forest conservation will inundate the emission reduction market, shrinking incentives to clean up fossil fuels.

I agree more with the former argument. There is not nearly enough known about how to promote communities' self-interest in saving their forests. Some argue that this is a reason to exclude forest protection from a GHG treaty, but I believe that this is a compelling reason to focus attention on tropical forests.

The trajectory of worldwide GHG emissions may mean that much larger reductions will be needed than are called for in the Kyoto Protocol. Tropical deforestation will not be solved any time soon. In fact, we will lose a substantial amount of these irreplaceable and ecologically vital forests. We should begin learning now how to stop deforestation. We need to learn more about carbon dynamics of tropical forests. Educating personnel in successful techniques that can help arrest this irreversible process will take decades. While we bemoan the loss of biodiversity and the loss of tropical forests, the international community must ante up more money to enable a sustainable path.

Polls suggest that international public support is stronger for protecting rain forests than it is for stopping climate change.[94] Done properly, it does not need to be an either–or dilemma. Measures that secure threatened tropical forests also help fight climate change.

Notes

1. Food and Agricultural Organization (FAO), 2001: *State of the World's Forests* (Rome: FAO).
2. World Resources Institute, 1998: *Climate, Biodiversity and Forests: Issues and Opportunities Emerging from the Kyoto Protocol* (Washington, DC: WRI).
3. For a comprehensive review of the issue, see Intergovernmental Panel on Climate Change (IPCC), 2000: *Land Use, Land-Use Change, and Forestry* (Cambridge, England: Cambridge University Press).
4. Ibid.
5. I use the term *deforestation* to mean the conversion of a forest to something other than a forest (e.g., a grassland or a pasture). However, for many of the points I

make, *deforestation* could be broadly interpreted to include forest degradation. Outright deforestation in the tropics often is driven by agricultural pressure and land clearing, logging, firewood collection, and cattle grazing.

6. Although only carbon dioxide and carbon are considered explicitly, the general approach here (and maybe even findings) *may* hold up similarly for methane and nitrous oxides, both of which are also released and absorbed by tropical forests.
7. Fearnside, P. M., 1997: "Greenhouse gases from deforestation in Brazilian Amazonia: net committed emissions," *Climatic Change,* 35: 321–360.
8. IPCC, 2000.
9. Nepstad, D., A. Verissimo, A. Alencart, C. Nobres, E. Lima, P. Lefebvre, P. Schlesinger, C. Potter, P. Moutinho, E. Mendoze, M. Cochrane, and V. Brooks, 1999: "Large-scale impoverishment of Amazonian forests by logging and fire," *Nature* 398: 505–508.
10. Malhi, Y. and J. Grace, 2000: "Tropical forests and atmospheric carbon dioxide," *TREE,* 15 (8): 332–337.
11. Malhi, Y., D. D. Baldocchi, and P. G. Jarvis, 1999: "The carbon balance of tropical, temperate, and boreal forests," *Plant, Cell and Environment,* 22 (6): 715–740.
12. Polglase, P. J. and Y.-P. Wang, 1992: "Potential CO_2-enhanced carbon storage by the terrestrial biosphere," *Australian Journal of Botany,* 40: 641–656. See also note 15.
13. Foody, G. M., G. Palubinskas, R. L. Lucas, P. J. Curran, and H. Miroslav, 1996: "Identifying terrestrial carbon sinks: classification and successional stages in regenerating tropical forests from Landsat TM data," *Remote Sensing Environment,* 55 (3): 205–216.
14. Hashimoto, T., K. Kojima, T. Tange, and S. Sasaki, 2000: "Changes in carbon storage in fallow forests in the tropical lowlands of Borneo," *Forest Ecology and Management,* 126 (3): 331–337.
15. Brown, S., J. Sathaye, M. Cannell, and P. Kauppi, 1996: "Management of forests for mitigation of greenhouse gas emissions." In. R. T. Watson, M. C. Zinyowera, and R. H. Moss (eds.), *Climate Change 1995: Impacts, Adaptations and Mitigation of Climate Change: Scientific-Technical Analyses,* contribution of Working Group II to the Second Assessment Report of the Intergovernmental Panel on Climate Change (Cambridge, England: Cambridge University Press), pp. 773–798.
16. Saugier, B. and M. Mousseau, 1998: "The direct effect of CO_2 enrichment on the growth of trees and forests," in G. H. Kohlmaier, M. Weber, and R. A. Houghton (eds.), *Carbon Dioxide Mitigation in Forestry and Wood Industry* (Berlin: Springer-Verlag).
17. A fact that many fossil fuel companies have advertised as one of the many "benefits" of increased carbon emissions.
18. Wang, Y. P. and P. J. Polglase, 1995: "Carbon balance in the tundra, boreal forest and humid tropical forest during climate change: scaling up from leaf physiology and soil carbon dynamics," *Plant Cell and Environment,* 18 (10): 1226–1244.
19. Lloyd, J. and G. D. Farquhar, 1996: "The CO_2 dependence of photosynthesis, plant growth responses to elevated CO_2 concentrations and their interactions with soil nutrient status, I. General principles and forest ecosystems," *Functional Ecology,* 10: 4–32.

20. Lloyd, J., 1999: "The CO_2 dependence of photosynthesis, plant growth responses to elevated CO_2 concentrations and their interaction with soil nutrient status, II. Temperate and boreal forest productivity and the combined effects of increasing CO_2 concentrations and increased nitrogen deposition at a global scale," *Functional Ecology,* 4: 439–459.
21. Mooney, H. A., B. G. Drake, R. J. Luxmoore, W. C. Oechel, and L. F. Pitelka, 1991: "Predicting ecosystem responses to elevated CO_2 concentrations," *BioScience,* 41: 96–104.
22. Cox, P. M., R. A. Betts, C. D. Jones, S. A. Spall, and I. J. Totterdell, 2000: "Acceleration of global warning due to carbon-cycle feedbacks in a coupled climate model," *Nature* 408: 184–187.
23. IPCC, 2000.
24. Laurance, W. F., M. A. Cochrane, S. Bergen, P. M. Fearnside, P. Delamonica, C. Barber, S. D'Angelo, and T. Fernandes, 2001: "The future of the Brazilian Amazon," *Science,* 291: 998.
25. Laurance, W. F., S. G. Laurance, L. V. Ferreira, J. M. Rankin-De Merona, C. Gascon, and T. E. Lovejoy, 1997: "Biomass collapse in Amazonian forest fragments," *Science* 278 (5340): 1117–1118.
26. Malhi and Grace, 2000.
27. Aiguo, D. and I. Y. Fung, 1993: "Can climate variability contribute to the 'missing' CO_2 sink?" *Global Biogeochemical Cycles,* 7 (3): 599–609.
28. IPCC, 2000.
29. Townsend, A., G. P. Asner, P. P. Tans, and J. W. C. White, 2000: "Land use effects on atmospheric ^{13}C imply a sizeable terrestrial CO_2 sink in tropical latitudes," unpublished manuscript.
30. Alternatively stated, destroying tropical forests not only releases carbon into the atmosphere but also destroys an ecosystem that is absorbing some of the excess carbon already in the atmosphere.
31. Myers, N., 1997: "The world's forests and their ecosystem services," Chapter 12 in G. C. Daily (ed.), *Nature's Services: Societal Dependence on Natural Ecosystems* (Washington, DC: Island Press).
32. Myers, N., 1996: "Environmental services of biodiversity," *Proceedings of the National Academy of Sciences,* 93: 2764–2769.
33. Hastenrath, S., 1985: *Climate and Circulation in the Tropics* (Dordrecht, The Netherlands: D. Reidel).
34. Compiled from: O'Brien, K. L., 2000: "Upscaling tropical deforestation: Implications for climate change," *Climate Change* 44 (3): 311–329. Betts, R. A., 2000: "Offset of the potential carbon sink from boreal forestation by decreases in surface albedo," *Nature* 408 (6809): 187–190. Sellers, P. J., L. Bounoua, G. J. Collatz, D. A. Randall, D. A. Dazlich, S. O. Los, J. A. Berry, I. Fung, C. J. Tucker, C. B. Field, and T. G. Jensen, 1996: "Comparison of radiative and physiological effects of doubled atmospheric CO_2 on climate," *Science* 271 (5254): 1402–1406. Shukla, J., C. Nobre, and P. Sellers, 1990: "Amazon deforestation and climate change," *Science* 247: 1322–1325. Lean, J. and D. A. Warrilow, 1989: "Simulation of the regional climatic impact of Amazon deforestation," *Nature* 342: 411–413. Nepstad, D. C.,

C. R. De Carvalho, E. A. Davidson, P. H. Jipp, P. A. Lefebvre, G. H. Negreiros, E. D. Da Silva, T. A. Stone, S. E. Trumbore, and S. Vieira, 1994: "The role of deep roots in the hydrological and carbon cycles of Amazonian forests and pastures," *Nature* 372 (6507): 666–669. Xue, Y, 1997: "Biosphere feedback on regional climate in tropical north Africa," *Quarterly Journal of the Royal Meteorological Society* 123: 1483–1515. Kleidon, A. and M. Heimann, 1999: "Deep-rooted vegetation, Amazonian deforestation, and climate: Results from a modeling study." Kleidon, A., K. Fraedrich, and M. Heimann, 2000: "A green planet versus a desert world: Estimating the maximum effect of vegetation on the land surface climate," *Climate Change* 44: 471–493. Myers, N., 1996: "Environmental services of biodiversity," *Proceedings of the National Academy of Sciences* 93: 2764–2769.

35. Hastenrath, 1985.
36. IPCC, 1994: "Radiative forcing of climate change: The 1994 report of the Scientific Assessment Working Group of the IPCC," Meteorological Office Marketing Communications Graphics Studio.
37. Luxmoore, R. J. and D. D. Baldocchi, 1995: "Modeling interactions of carbon dioxide, forests, and climate," in G. M. Woodwell and F. T. MacKenzie (eds.), *Biotic Feedbacks in the Global Climatic System* (Oxford, England: Oxford University Press).
38. Pimm, S., G. J. Russell, J. L. Gittleman, and T. M. Brooks, 1995: "The future of biodiversity," *Science,* 269: 347–350.
39. Hughes, J. B., G. C. Daily, and Pr. R. Ehrlich, 1997: "Population diversity: Its extent and extinction," *Science,* 278 (5338): 689–692.
40. Epstein, P. R. (ed.), 1999: *Extreme Weather Events: The Health and Economic Consequences of the 1997/1998 El Niño and La Niña* (Boston: Center for Health and the Global Environment, Harvard Medical School).
41. Epstein, P. R. (ed.), 1998: *Marine Ecosystems: Emerging Diseases as Indicators of Change* (Boston: Center for Health and the Global Environment, Harvard Medical School).
42. Ballick, M., E. Elisabetsky, and S. Laird (ed.), 1996: Medical Resources of the Tropical Forest (New York: Colombia University Press).
43. Garrett, L., 1995: *The Coming Plague: Newly Emerging Disease in a World Out of Balance* (New York: Penguin Books).
44. This section benefited from some early work with Reimund Schwarze and Michael Mastrandrea.
45. The ultimate objective of the UN FCCC and the Kyoto Protocol is to stabilize GHG concentrations at a level that would prevent dangerous anthropogenic interference with the climate system.
46. German Advisory Council on Global Change (WGBU), 1998: "The accounting of biological sinks and sources under the Kyoto Protocol: a step forwards or backwards for global environmental protection?" WGBU, Bremerhaven, Germany, available online: http://www.wbgu.de/wbgu_sn1998_engl.html.
47. IPCC, 1996: "The revised 1996 IPCC guidelines for national greenhouse gas accounting," available online: www.iea.org/ipcc.htm.
48. Bionergy, or the planting of trees to make electricity that replaces fossil fuels, is also potentially a forest-based emission reduction.

49. Michaelowa, A., 1998: "Joint implementation—the baseline issue: economic and political aspects," *Global Environmental Change,* 8 (1): 81–92.
50. Chomitz, K. M., 1998: "Baselines for greenhouse gas reductions: problems, precedents, solutions," unpublished report for the Carbon Offsets Unit, World Bank.
51. Filho, L. G. M.: "Sinks and sinks," unpublished document.
52. FAO, 2001. See footnote 2.
53. Myers, N., 1992: "Future operational monitoring strategy of tropical forests: an alternate strategy," Proceedings from the World Forest Watch Conference, San Jose dos Campos, Brazil, May 27–29, 1992; Mathews, E., 2001: "Understanding the FRA: Forest briefing #1," (World Resources Institute, Washington DC), http://www.igc.org/wri/pdf/fra2000.pdf, last viewed October 2001.
54. Tropical Ecosystem Environment Observations by Satellites (TREES), 1998: "Identification of deforestation hot spot areas in the humid tropics," TREES publications series b, no. 14. Joint Research Centre, European Commission, Italy.
55. FAO, 1993: "Forest resources assessment 1990: tropical countries," FAO Forestry Paper 112, Rome.
56. Mayaux, P., F. Achard, and J. P. Malingreau, 1998: "Global tropical forest area measurements derived from coarse resolution satellite imagery: a comparison with other approaches," *Environmental Conservation,* 25 (1): 37–52.
57. Ibid.
58. Schwarze, R., 2000: "AIJ: Another look at the facts," *Ecological Economics* 32: 255–267.
59. Sedjo, R. G. M. and K. Fruit, 2001: "Renting carbon offsets: the question of permanence," *Resources for the Future,* available online: http://www.weathervane.rff.org/features/feature136.htm, last viewed October 2001.
60. Leakage is more fully described in a forthcoming paper: Schwarze, R., J. Niles, and J. Olander, 2002 (in print): "Understanding and managing leakage in forest-based greenhouse gas mitigation projects." Philosophical Transcations of the Royal Society, Series A, vol. 1797. I am grateful to my co-authors R. Schwarze and J. Olander and the Nature Conservancy for allowing me to use this work here.
61. Greenpeace International, 2000: "Should forests and other land use change activities be in the CDM?" November 2000, update for COP-6; Hare, B. and M. Meinshausen, 2000: "Stop cheating! Cheating the Kyoto Protocol: loopholes undermining environmental effectiveness," Greenpeace International.
62. WWF, 2000: "'Sinks' in the CDM? Implications and loopholes," WWF discussion paper.
63. Schwarze, R., J. Niles, and J. Olander, 2002.
64. The original information for this estimate comes from K. Chomitz, 1999: "Evaluating carbon offsets from forestry and energy projects: how do they compare?" World Bank Development Research Group, viewed October 2001 at http://econ.worldbank.org/docs/1111.pdf. For other calculations, see Schwarze, R., J. Niles, and J. Olander, 2002.
65. http://www.weathervane.rff.org/features/feature107.html.
66. For references, see Schwarze, R., J. Niles, and J. Olander, 2002.
67. The organizations that were adamant about a leakage concern in forestry were the

same ones opposed to forestry in the protocol for other reasons.

68. United Nations, 1999: "Matters left pending on the need for financial resources," Secretary General's Report, UN Commission on Sustainable Development, Intergovernmental Forum on Forests, Third Session, May 3–14, 1999, Document E/CN.17/IFF/1999/4.
69. Frumhoff, P. C., D. Goetze, and J. Hardner, 1998: *Linking Solutions to Climate Change and Biodiversity Loss Through the Kyoto Protocol's Clean Development Mechanism* (Cambridge, MA: Union of Concerned Scientists).
70. Mulongoy, K. J., J. Smith, P. Alirol, and A. Witthoft-Muehlmann, 1998: "Are joint implementation and the Clean Development Mechanism opportunities for forest sustainable management through carbon sequestration projects?" Paper prepared for the Policy Dialogue organized by the International Academy of the Environment and the Center for International Forestry Research in Geneva, 1998, www.iae.org/pd/forest-background.pdf.
71. Hourcade, J. C., K. Halsnaes, M. Jaccard, W. D. Montgomery, R. Richels, J. Robinson, P. R. Skukla, P. Sturm, and L. Schrattenholzer, 1996: "A Review of Mitigation Cost Studies," in J. P. Bruce, H. Lee, and E. F. Haites (eds.). *Climate Change 1995: Economic and Social Dimensions of Climate Change.* Contribution of Working Group III to the Second Assessment Report of the IPCC (Cambridge, England: Cambridge University Press). Smith, J., K. Mulonguy, R. Perrson, and J. Sayer, 1998: "Harnessing carbon markets for tropical forest conservation: Towards a more realistic assessment," International Academy of the Environment working paper, Geneva.
72. Schwarze, 1999.
73. Fearnside, P. M., 1997: "Environmental services as a strategy for sustainable development in rural Amazon," *Ecological Economics,* 20: 53–70.
74. Kremen, C., J. Niles, M. Dalton, G. Daily, P. Ehrlich, P. Guillery, and J. Fay, 2000: "Economic incentives of rain forest conservation across scales," *Science,* 288: 1828–1832.
75. Fankhauser, S., 1995: *Valuing Climate Change: The Economics of the Greenhouse* (London: Earthscan).
76. Roughgarden, T. and S. H. Schneider, 1999: "Climate change policy: Quantifying uncertainties for damages and optimal carbon taxes," *Energy Policy,* 27: 415.
77. Greenpeace International, 2000. Note that in this paper they actually confused the current cost that donor countries should be paying with the cost that donor countries are paying. Thus, even though the cost per ton of carbon still is pretty "cheap", it is higher than the numbers they used to show how "cheap" these offsets would be.
78. Greenpeace International, 2000.
79. Singer, S., 2000: "Sinks in the CDM? Implications and loopholes," WWF discussion paper.
80. Assuming unlimited financing.
81. Niles, J., 2000: "Preliminary estimate for 25 countries of the potential for forest conservation in the Clean Development Mechanism," unpublished report.
82. The Royal Institute for International Affairs, 2000: "Quantifying Kyoto: How will COP-6 decisions affect the market?" Proceedings from a workshop held August

30–31, 2000, Chatham House, London. Results are available online: http://www.riia.org/Research/eep/quantifying.html#conclusions (viewed November 2000).

83. Confidential notes of a private statement made by Jose Gonzalez Miguez, coordinator of global change research for the Ministry of Science and Technology. As both a negotiator with the Brazilian government and someone in a position needed to endorse CDM projects, Miguez was reported to have said that he "would not sign any letters for forest sector activities as long as he held the position."
84. Filho, L. G. M., "Sinks and sinks," unpublished document.
85. Laurance, W. F., M. A. Cochrane, S. Bergen, P. M. Fearnside, P. Delamônica, C., S. D'Angelo, and T. Fernandes, 2001: "The future of the Brazilian Amazon," *Science,* 291: 438–439.
86. Private papers, author.
87. Multiple authors, 2000: "Declaration of Brazilian Civil Society on the relations between forests and climate change and expectations for COP-6," unpublished document.
88. Commission of the European Communities, 1998: "Climate change—toward an EU post-Kyoto strategy," communication from the commission and the European parliament, Com. 98-353.
89. Not all groups in CAN were opposed to forest conservation in the CDM. As a minority voice within CAN, several groups wrote in a letter dated July 13, 2001, "The current negotiating text proposed by Dr. Pronk, which includes reforestation and afforestation projects in the CDM but textually excludes native forest protection, from our perspective represents a huge threat to the forest and the peoples of the forest. It is precisely this that the current Brazilian government wants, and exactly that with which we do not agree. . . . The most prudent and coherent position with respect to tropical forests in the CDM is against the inclusion of industrial forest plantations (the true 'sinks') and in favor of the inclusion of native tropical forest protection, the reduction of deforestation, and support for community projects that make the presence in the forest of indigenous and agro-extractive populations economically" viable.
90. http://www.panda.org/climate/co2_source.cfm, viewed May 2002.
91. http://www.panda.org/climate/solutions.cfm, viewed May 2002.
92. Greenpeace, 2001: "Should forests and other land use change activities be in the CDM?" Reissued for COP6b (Part 2), Bonn, Germany, July 16–27, 2001.
93. The Nature Conservancy is sponsoring numerous climate change forest projects.
94. Bloom, D. E., 1995: "International public opinion on the environment," *Science* 269: 354–358.

PART V

DEVELOPMENT AND EQUITY

CHAPTER 14

A Southern Perspective on Curbing Global Climate Change[1]

Anil Agarwal

Climate change may be the biggest North–South cooperation challenge the world has ever faced. While facilitating the affluence of industrialized countries, fossil fuel–based development has been largely responsible for causing climate change. Developing countries, on the other hand, are late entrants to western-style economic development; their populations remain economically poor, and their per capita emissions are far less than those of industrialized countries. Because developing countries did not create the environmental problem in the first place, industrialized countries should take the lead in remedial action. Climate change is not only a global environmental issue but also a North–South equity issue. Therefore, negotiations should work to check climate change and right global inequity.

The current approach to controlling climate change, as illustrated by the Kyoto Protocol, seems flawed at multiple levels. It may re-entrench the carbon-based energy infrastructure on a global level and perpetuate inequity between industrialized and developing countries. Industrialized countries, especially the United States, have pushed for flexibility mechanisms that allow them to get credit for national emission reductions without taking domestic action. Two such mechanisms, the Clean Development Mechanism (CDM) and emissions trading, fail to address southern equity concerns or promise to significantly reduce global greenhouse gas (GHG) emissions.

An alternative supported by developing countries—per capita emission entitlements—would be ecologically, economically, and socially sound. All nations would reduce their per capita GHG emissions substantially, but the burden would be shared equitably. The southern perspective has been long neglected by leading industrialized countries in climate change negotiations. It is high time that North and South worked together to create a climate change solution that all people and the environment can live with.

Equity in Climate Change Negotiations

Industrialized and developing countries must agree to share atmospheric space in an equitable manner. Although some nongovernment organizations (NGOs) go so far as to say that the climate treaty was not meant to deal with inequity in the world, equity must not be overlooked. Afraid that any debate on equity and entitlements would stop the United States from sending the treaty to Congress for ratification and end the protocol, most westerners—including the usually outspoken western environmental NGOs—have been largely mum on the issue. When it comes to dealing with a common resource such as the atmosphere, the concept of equity cannot remain in the background. It has to form the basis of any workable system. Inequity makes it very difficult for political leaders, especially in nations with an electoral democracy, to agree to a common action plan. It is fundamental to human nature that people cooperate only when there is a sense of fairness among them. Without equitably sharing, global solidarity will not be possible. Per capita emission entitlements are critical for equity in a climate change regime.

Equity is not only a moral issue but also a policy concern. According to the Stockholm Environment Institute (SEI), even if the North fails to curb emissions and relies on adapting to climate change, it must still face the geopolitical, demographic, economic, and human problems that will spill over from the South's likely inability to similarly adapt. Alternatively, if the countries of the North decide to avert climate change by forcing an inequitable burden on developing countries, they court similar problems.[2] In an increasingly globalized and interdependent world, industrialized countries cannot be insulated from the effects of climate change.

Historical and Future Responsibility for Climate Change

Industrialized countries owe their current prosperity to years of historical emissions, which have accumulated in the atmosphere since the start of the industrial revolution, and also to a high level of current emissions. Developing countries have only recently set out on the path of industrialization, and their per capita emissions are still low. The GHG emissions of one U.S. citizen were equal to those of 19 Indians, 30 Pakistanis, 17 Maldivians, 19 Sri Lankans, 107 Bangladeshis, 134 Bhutanese, or 269 Nepalis in 1996.[3]

With such high levels of GHG emissions, industrialized countries are holders of natural debt, borrowing from the assimilative capacity of the environment by releasing waste gases faster than they can be removed naturally. These countries therefore should not think of resources devoted to curbing climate change as a sudden extra cost being imposed on them but as the inevitable need to repay the ecological debt that has helped them achieve their present wealth.[4] Yet leaders of industrialized countries usually view emission reductions as an economic threat, not an ecological necessity.

Under these circumstances, any limit on carbon emissions amounts to a limit on economic growth, turning climate change mitigation into an intensely political issue. International negotiations under the UN Framework Convention on Climate Change (FCCC) aimed at limiting GHG emissions into the atmosphere have turned into a tug of war, with rich countries unwilling to compromise their lifestyles, and poor countries unwilling to accept a premature cap on their right to development.

Developing countries have demanded their space to grow while refusing to take on emission cuts at their current stage of development. The atmosphere is a common property resource to which every human being has an equal right. The people of industrialized countries have more than used up their share of the absorptive capacity of this atmosphere through their high emission levels in the past and in the present. To that extent, the global warming problem is their creation. So it is only right that they should take the initial responsibility of reducing emissions while allowing developing countries to achieve at least a basic level of development. Moreover, asking developing countries to reduce carbon emission levels now amounts to asking them to freeze their standards of living at their current stage of development. And this would amount to freezing inequality by ensuring that some countries will always be more developed than others in the world.

Developing countries will continue to grow, making huge energy investments in the next three to four decades. If these investments lock developing countries into a carbon energy economy like industrialized countries, it will be very difficult for them to get out of it. But if proper policies are put in place, developing countries can take a lead in creating a global market for zero-carbon energy technologies because they have two distinct advantages: They have more solar energy than most western countries, and they provide a huge niche market in several hundreds of thousands of their villages that are not yet touched by the power grid. Experts at SEI point out that because of the fossil fuel–based historic industrialization of the North, the South today finds itself facing a severely compromised climatic system if it follows the well-trodden path of the North. The South therefore has to bear the extra cost of taking a different path and has to get it right the first time.

This raises several critical issues. Energy production is based on long-lived capital, which, once built, commits a society to a lifetime's worth of emissions. A power plant built today will still be emitting 30 years from now, by which time global carbon emissions must be reduced by 25 percent from the business-as-usual scenario. The South is witnessing rapid economic growth, and its major energy investment decisions will significantly contribute to the majority of global emissions in the decades ahead. There is very little that can be done to change the fossil fuel–based path for the next 20 years. But if efforts to make renewables competitive by 2020 are not made now, then the world will stay committed to a carbon-based energy economy well into the next century. A slower rate of reduction today will mean either faster rates of reduction later or a higher risk of climate change, passing on a very heavy burden to future generations.[5]

The United States, European Union, and G77 at Climate Change Talks

As a result of these political complexities, negotiations under FCCC have turned into a game between unequal partners. G77, the negotiating bloc of developing countries, has often found itself politically outmaneuvered by alliances between the two main industrialized country groups: the United States and the European Union (EU). Although the EU and the United States often come to the negotiating table with divergent viewpoints, with the EU pressing for tighter commitments and the United States unwilling to give in, the two have almost formed a habit of resolving issues among themselves. The EU usually ends up giving in to the lax U.S. position—the recent Hague conference being an exception—and the two expect the developing world to accept their conclusions.

In past negotiations, the United States and EU have sorted out their differences and presented developing countries with a take-it-or-leave-it deal on climate change. To prevent this from happening in the future, there should be greater coordination of strategy between the EU, G77, and China. At the November 1998 COP-4 meeting in Buenos Aires, a positive development from the point of view of developing countries was a perceptible shift in the EU's position away from the United States. Since Kyoto, the United States has pushed for being allowed to meet its entire emission reduction commitment through flexible mechanisms. The EU, G77, and China have resisted such a policy. At COP-6 in the Hague, developing country representatives feared that "another climate meeting would end up serving the economic interests of the US more than the threat of global warming and climate change."[6] Yet EU ministers did not capitulate to U.S. demands, arguing that it was better to have no agreement than to be stuck with a bad one.

The G77 finds itself sidelined by the United States and EU in climate change negotiations. Southern governments participate as junior partners, worried about lectures and dictates from industrialized countries. After COP-6, the director of the Nigerian Conservation Foundation, Mutkar Aminu-Kanu, said, "We are beginning to think these conventions are no longer a negotiating process, that the West, in particular the US, calls the rest of the world to tell them what to do and if they won't do it the whole thing folds."[7] The West takes G77 consent for granted, without the group's participation in actual negotiations, and continues an extremely dangerous and undemocratic trend in international negotiations.[8]

Science Biased by the North

Added to these political complexities is the fact that tracking climate change, predicting the adverse affects with some degree of reliability, and pinpointing responsibility entails a degree of investment and scientific expertise that is available mostly to industrialized countries. This leaves developing countries, which have made little effort to expand their scientific capacity, dependent on northern scientists and institutions to tell them the extent and fallouts of global warming and to lead the negotiations in an intensely science-driven convention. Science has been used several times in the past to implicate developing countries, either by showing their future GHG contributions as increasing and counterproductive to industrialized country action or by making no distinction between the survival emissions of the South and the luxury emissions of the North. Also, there is an enormous disparity in North–South participation in the IPCC, with U.S. and European scientists making up most of all three IPCC Working Groups.

Moreover, the North-driven scientific process often places developing country concerns low on the priority list. For example, very little research has been conducted on the possible impacts of climate change on different countries and regions, leaving them unprepared to handle the adverse effects of climate change. Some scientists have even alleged that there seems to be a conspiracy of silence on this count because it may show that the most damage will take place in the developing countries. If this is true, there is a danger that the incentives for industrialized countries to take action against global warming will be low. A team of scientists sponsored by the UN have reported that on a vulnerability index, developing countries are, on average, twice as vulnerable as industrialized countries and small island developing countries are three times as vulnerable.[9]

To add to this political and scientific confusion, industrial groups with a vested interest continue to generate science disputing even the fact that global warming is a threat to the world. According to the U.S. Internal Revenue Service, business groups in the United States have spent millions since 1991 to persuade the public and policymakers that there is too much uncertainty about climate change to warrant action. They claim that the world should wait for more conclusive evidence before taking any preventive measures. Their bankrolling of skeptical scientists and visible ad campaigns has fueled inaction by industrialized countries, especially the United States.

The U.S. Stance: Obstacle to Effective International Action

From Rio to Kyoto to the present, the United States has hindered efforts to curb climate change. The list of U.S. demands includes developing country participation, low commitments, and the flexibility to meet their entire commitment through emission trading and the CDM. Whereas the first demand questions social justice and equity, the very basis on which any global negotiation should be built in a civilized world, the latter two threaten the ecological effectiveness of the treaty.

In 1992, the FCCC committed the West to no more than what one country, the United States, was willing to commit. Industrialized countries accepted the "common but differentiated responsibilities" principle, a very diluted version of the polluter-pays principle. Through the framework convention, industrialized countries got away with not having to account for their historical emissions.

Before Kyoto, the U.S. negotiating position demanded "meaningful participation of key developing countries." This served not only as a way to delay substantive action on climate change but also as a wedge in G77 unity. After getting host country Argentina and South Korea, members of the G77, to agree to "voluntary" commitments, the United States upheld them as examples of developing countries that wanted to see the Kyoto Protocol work. The definition of meaningful participation was left purposely obscure: Even if it eventually resulted only in developing countries agreeing to trade in emissions credits, it would give the United States and its allies a chance to meet their Kyoto Protocol commitments without domestic action.

The U.S. position at Kyoto, conditional on developing country participation, seeks to move the onus from the world's biggest polluter to countries that are likely to be major polluters in the future. The U.S. stance shifts NGO and media attention to developing countries, which are seen as holding up ratification by the United States. The terms for "meaningful participation" have been purposely left undefined but threatening, so that any offer from developing countries could be easily dismissed as "not meaningful enough."

The United States argues that southern emissions will surpass northern emissions in 2035, but this claim must be put into perspective. Statistically, this means that in 2035, 20 percent of the world's population living in the North will be emitting 50 percent of the carbon emissions and 80 percent of the world's population living in the South will be emitting only 50 percent of the carbon emissions.[10] In energy system changes, large developing countries such as India, China, and Brazil are not doing badly—in comparison with industrialized countries—with regard to reducing GHG emissions, according to a report published by the Worldwatch Institute in November 1997. All three have implemented meaningful policy reforms in the past decade, including politically difficult reductions in fossil fuel subsidies and improved efficiency in China.

Meanwhile, the United States is capitalizing on the fact that a protocol without their ratification is virtually meaningless because they are the world's largest emitters of carbon dioxide. The U.S. Senate, negotiators, and industry have capitalized on their ability to hold negotiations hostage to their demands. Before Kyoto, the Byrd–Hagel resolution sent a clear message to the rest of the world: "Give us something we do not like and we won't ratify. Let's see where that leaves you." This attitude had a significant effect in shaping the Kyoto Protocol.

U.S. Responsibility for Weakening the Kyoto Protocol

At the Kyoto negotiations in 1997, the United States came out the undisputed victor, having totally outwitted both the EU and G77 and China, the two major blocs opposing it. Kyoto was a "grand bargain" between a magnanimous U.S. commitment to reduce its emissions below its 1990 levels—something that the world media immediately hailed—and the acceptance of various trading mechanisms by other groups. Everything was contorted to fit this bargain. Brazil's proposal for a punitive Clean Development Fund miraculously turned into a market-based North–South tool for emission trading called the CDM. Emission trading between nations got into the protocol literally in the last hour of the conference, well after the official clock had been stopped. Russia and Ukraine, despite their extremely low emissions compared with 1990, calmly walked away with no commitments to reduce below their 1990 levels, making a huge amount of emission trading a reality at throwaway rates.

In current discussions, negotiators may be missing the forest for the trees in trying to appease the United States. Starting with the third conference of the parties (COP-3) in Kyoto, the FCCC process seems to have lost sight of its objectives. Mostly NGOs, but also governments, now seem to be working toward bringing the United States on board instead of looking for a sustainable solution to the climate change problem. The world's civil society, as represented in the climate negotiations, seems to be willing to give up on equity. Doing so will hinder the world's ability to transition to renewable energies and achieve the emission reductions needed to prevent major climate change.

To be specific, the G77 and China have consistently opposed the U.S. demand for including forest management and changes in sinks as a mitigation method in a climate change treaty. Sinks became the center of controversy in Kyoto when the United States, France, Australia, and New Zealand demanded that land use changes and forestry (LUCF) be included while calculating commitments by countries.[11] But in November 2000, the IPCC released a report stating "that there are too many complications associated with the use of LUCF to 'fix' carbon."[12] Developing countries generally believe that land use changes should not count toward emission reductions in CDM and in emission trading schemes. Counting LUCF would favor northern countries with large boreal forests, create a perverse incentive to deforest in tropical areas to receive credit for reforestation later, and allow industrialized countries to get credit for planting trees in developing countries under CDM. Most of all, allowing LUCF to count for emission reductions would not combat additional GHG emissions.

The Kyoto Protocol: A Weak and Flawed Solution

The Kyoto Protocol promises to be a weak agreement because of flexibility mechanisms, lack of a compliance mechanism, small mandated emission reductions, sink loopholes, and inequity within the accord. Nine prominent U.S. scientists and economists, including John Holdren, member of President Bill Clinton's Committee of Advisers on Science and Technology, note that the Kyoto Protocol assigns emission caps to the industrialized countries based on their 1990 emission levels. This "rewards historically high emitters and penalises low emitters . . . by basing future emission caps on past levels."[13] This agreement based on historical levels would allow high emitters to impose environmental damages on other countries, in violation of the polluter-pays principle. "This contravenes international environmental law," says this group of experts on climate and energy policy. They argue that the U.S. government's insistence on "meaningful participation" of developing countries will block the implementation of the Kyoto Protocol because the long-term equity concerns of the South have not been addressed. Southern countries cannot reasonably be expected to restrict their future emissions without being assured of a fair allocation scheme that will not impair their ability to develop.

Through flexibility mechanisms such as emission trading and the CDM, Kyoto would ignore equity issues, allow industrialized countries to avoid domestic emission reductions, and lock renewable energy out of the market. Most southern countries remain wary of flexibility mechanisms for a variety of reasons, including their impression that FCCC does not call on them to take the lead in GHG emission reduction, and that "meaningful participation" could be the first step on a slippery path toward voluntary commitments.

SEI experts argue that if northern countries rely heavily on flexibility mechanisms, they risk being unprepared for much deeper cuts ultimately needed to prevent climate change.[14] This is because any strategy that seeks to obtain least-cost carbon emission reduction options inevitably will focus on improving energy efficiency in the carbon energy sector. It will give the North least-cost options to meet emission reduction targets and allow them to continue on a carbon-intensive path.[15] Therefore, emission trading should be limited to projects that promote the zero-carbon energy system and should not be allowed for projects that promote the carbon energy system. Also, there should be a strict limit on the amount of credits that can be bought to count for domestic emission reductions. Such a cap on credits for emission trading would push industrialized countries toward domestic emission reductions.

In addition, the Kyoto Protocol lacks a compliance mechanism to make it enforceable. Because this is the first global agreement in which only the powerful industrialized nations have taken on commitments, it is not easy to conceive how poorer nations will be able to apply effective sanctions against the powerful nations if they do not meet their commitments.[16] In contrast to the Montreal Protocol and the Convention on International Trade in Endangered Sanctions, the Kyoto Protocol lacks a compliance mechanism based on trade sanctions. It is therefore unenforceable by "hard" law and subject to the voluntary participation of nations that ratify it.

Worse yet, the Kyoto Protocol by itself will do nothing to solve the climate change problem. As SEI experts put it,

> The direct GHG impact of the mandated reductions during the first budget period will amount to an almost negligible effect; they would reduce atmospheric carbon dioxide levels by only about one-third of one percent relative to where they would be in 2010 without a Kyoto Protocol.[17]

In fact, the protocol could even worsen the situation by locking renewable sources out of the energy market. As a global environmental agreement, it runs the risk of appeasing civil society, NGOs, and governments with a misplaced faith in a flawed accord.

Kyoto does not right North–South inequity but perpetuates it through three main inequities in the agreement. The first is that the protocol allows industrialized countries to bank emissions for future use. If an industrialized country reduces more than its target for 2010, then it can bank emissions even though it already has very high per capita emissions. But India, China, and Nepal, with extremely low per capita emissions today, cannot bank anything today for their future use. Second, a Dutch study points out that burden-sharing criteria that take into account historical emissions or a per capita approach favor developing countries, whereas the inclusion of all GHGs and land use–related emissions favors industrialized countries. The Kyoto Protocol does precisely the latter.[18] Third, if the Kyoto strategy is followed, then developing countries will soon have to undertake reductions at much lower baseline emissions than those industrialized countries had in 1990 or risk serious impacts of climate change that they will least be able to afford.

On a positive note, an SEI report concludes that the real importance of the Kyoto targets lies in their potential to motivate the North to determinedly direct resources toward developing and deploying technologies, infrastructure, and institutions that will build momentum toward long-term GHG mitigation options and progressively deeper GHG reductions. If Kyoto hastens a global transition to renewable energy, it will have served an important function for North and South.

Objections to the CDM

Of the three flexibility mechanisms, CDM promises to have the most impact on developing countries. Yet it is replete with flaws, making it particularly unpalatable to developing countries. Possibly the worst aspect of CDM is that it helps the North to buy up the cheap emission reduction options available today, leaving the South to pay a heavy price tomorrow. Economists predict that the carbon savings options that currently cost $10–$25 per ton of carbon could cost $200–$300 per ton in the long term.[19] When the South itself has reached high levels of energy efficiency and therefore its cost of curtailing emissions is high, the North will have no economic incentive to buy emission credits from it. And if global warming is still a threat—as it definitely will be because industrialized countries would have taken little action domestically—then the pressure will mount on developing countries to take expensive emission reductions themselves.

In other words, CDM encourages the current generations of developing countries to sell off their cheaper emission control options today, leaving future generations saddled with high-cost options tomorrow. It offers cash-strapped developing country governments an opportunity to discount the future, and nobody knows what would be the form of international cooperation at that time.

If developing countries participated in CDM, they would sell their cheap options for reducing emissions and not even get credit for it in the global balance sheet.[20] This buying and selling would take place without any property rights framework, essential for market-based systems. The South Asian Equity Group issued a statement warning that trading without property rights or entitlements would amount to a mortgaging of the future interests of the South. In addition, host countries of CDM projects cannot sell emission reduction credits. Instead, Annex I investors can carry out reduction projects in developing countries and sell the resulting credits at a higher price to countries in need of the credits. This represents yet another equity gap in the climate negotiations.

In terms of ecological effectiveness, CDM could ultimately prove to be a disaster. CDM will subsidize the very source of the problem, the carbon-based energy system, because all least-cost options are in the carbon-based system. By subsidizing carbon-based energy technologies, it will create further obstacles to the penetration of non–carbon-based energy technologies and could lock them out for several decades, thus ensuring that a high order of climate change becomes inevitable. Developing countries have expressed concern about whether CDM will end up promoting sustainable development or become yet another conduit for outdated technology. Developing countries stress that the host country should have the last word on what constitutes sustainable development, and CDM projects should spell out clearly their net contribution to development.

Developing countries have two main financial objections to CDM. As currently envisioned, a share of CDM projects will also be used to pay for the adaptation costs of developing countries. This provision amounts to taxing the poor to pay the affected poor. There is no such provision in the other mechanisms (JI and emission trading) meant for emission trading between industrialized countries. Also, developing countries demand "financial additionality" for CDM projects (i.e., that they use funds beyond official development assistance and direct investment flows to developing countries). Without additionality, industrialized countries could simply redirect funds currently earmarked for other southern development projects.

CDM: Bypassing Poorer Developing Countries

Another shortcoming of CDM concerns how it could bypass poorer developing nations. CDM under the present framework, without entitlements, is unlikely to benefit poorer nations among the G77 because industrialized countries are likely to give preference to projects in the more technologically rich countries among the G77, which will provide them with fast and cheap emission credits.[21] Within a purely market-driven framework, most CDM projects will go to larger and more industrially advanced developing countries such as India and China.

Africa has its own qualms with CDM as currently structured. African governments have argued that because Africa's carbon emissions are low and their energy consumption is only 2–3 percent of the global energy resources, there are few options for implementing CDM projects that reduce emissions from existing sources. So CDM should be designed to reward projects that promote socioeconomic development using clean technologies, and a concept of emission avoidance should be established. A project promoting infrastructure development in the energy sector would not only meet Africa's sustainable development needs but also avoid emissions. Africa can be meaningfully integrated into the Kyoto Protocol only if the principle of emission avoidance is incorporated.[22] African experts therefore express two key concerns about CDM: that a purely market-driven mechanism will bypass Africa and that even if it reaches Africa, it will not meet the region's priority concerns for sustainable development such as food and energy security of the poor majority. But a CDM that functions under an emission entitlement scheme will ensure that all poor countries can participate in it.

Economical, Ecological, and Equitable Action

To put a stop to this political, economic, and scientific game-playing—which currently seems to be concentrated on innovative and complicated ways to meet commitments without actually reducing carbon concentrations or to buy cheap options from developing countries—solutions that meet three criteria are needed. The first is their ecological effectiveness: whether they actually reduce the concentration of GHGs in the atmosphere. The second is their economic effectiveness: To be acceptable to both industrialized and developing countries, they must have the minimum possible impact on the global and national economies. And, finally, in the interests of fairness and global cooperation, the solutions must be socially just and equitable toward all countries. It is a challenge to all participating countries, and particularly to the world's civil society, to ensure that all measures agreed to under FCCC meet these three criteria.

This challenge is heightened because the world is divided into three key climate camps today. The first consists of nations that want to take serious action on global warming. For island states and European states with strong Green parties, the Kyoto Protocol must lead to ecologically effective action. The second camp consists of nations that believe that emission reduction will come at a high cost and are searching for lower-cost solutions. For the United States and other countries, Kyoto must lead to economically effective action. The third group is composed of poorer nations that depend on carbon emissions for their present and future development. Led by India, China, and other poor nations, they want the Kyoto Protocol to undertake equitable and socially just actions. The three objectives—of economic and ecological effectiveness and equity and global solidarity—can be put together to develop an action plan to keep climate change at tolerable levels.

One such way would be a per capita emission entitlement approach. An entitlement method might calculate the emissions absorbed annually by the global atmospheric sinks and distribute these emissions equally among all the people of the world, providing each person with an equal entitlement. Empirically, the EU's burden-sharing agreement shows how emissions can be equitably divided into entitlements. If the per capita emission entitlement were set at 0.38 tons of carbon per year, for example, industrialized countries would have to reduce their emissions sharply, and many developing countries would have room to grow. These entitlements could then be traded between countries. Those who consume more than their fair share of the world's environmental space would have to buy the extra space they want to use from those who do not consume their full share. In this way, the world will begin to value the unvalued commons.

The biggest advantage of tradable equitable emission entitlements is that they immediately engage developing countries and provide them with an incentive to keep emissions low. Trading of emission entitlements would immediately give them an incentive to move toward a low-emission developmental path so that the benefits from emission trading can stay with them for a long time. It would also provide an "enabling economic environment for technology transfer"[23] and serve as a strong disincentive against leakage because countries would be wary of allowing high-GHG economic activities to come into their countries. Entitlements ensure that North–South cooperation will remain open to southern countries as long as they are low emitters. They will not be entirely dependent on the least-cost options offered by the CDM. Thus, equal per capita emission entitlements offer the most just, effective, and meaningful way of getting developing countries to engage with the climate change problem.

What developing countries should not accept is a principle of emission trading built solely on the argument that they provide a lucrative opportunity today to reduce emissions cheaply. Emission trading cannot simply be carried out to achieve economic efficiency. It must be undertaken in an environment that also promotes ecological efficiency and global solidarity. The purpose of equity and an equal per capita entitlement principle is not to force industrialized countries to drastically curtail their economies. It is to create a framework for global cooperation so that the world can move as quickly as possible toward a world economy that can keep on growing by using renewable energy. A three-pronged combination of emission trading, equitable entitlements, and promotion of renewables thus constitutes a truly meaningful plan of action. Such an approach would help change consumption patterns and leapfrog into a technological world that is less carbon energy intensive.

In crafting an effective accord to curb climate change, northern and southern negotiators must find a way to significantly reduce emissions while not ignoring developing country concerns for equity or the global desire for economic development. Negotiations must give appropriate primacy to moral and ecological concerns instead of purely economic concerns. In relying on emission trading and the CDM, the Kyoto Protocol remains far from achieving this objective. A per capita entitlement approach would work far better at spurring a transition to renewable energy, addressing North–South inequities in fuel use, and ultimately curbing major climate change.

The Kyoto Compromise in Bonn and Marrakech

The meeting in Bonn in July 2001 to flesh out the Kyoto Protocol was predictably difficult. 180 countries finally reached an agreement on rules to implement the protocol, after almost six months of uncertainty on the issue. But we did not expect the world to give away so much to get so little.

George Bush, leader of the world's biggest economy and polluter, had already declared that the protocol was "fatally flawed in fundamental ways" and walked out of the multilateral discussions. The final permutation was that Japan, Canada, Australia, and Russia held the key to the agreement. These polluters played their cards well, prevaricating to the last moment to ensure that they got the deal they wanted.

First, these countries wanted major concessions on the use of vegetation to sequester carbon. They got it, to an amazing extent. Now every small area under trees can be calculated as a sink. Every scrubland is included because an area with 10–30 percent tree cover has been defined as a forest. And even areas with no trees temporarily, but which are expected to revert to being forests, can be included. Countries can also add up any management measures taken to improve productivity of forests, agricultural, and grazing lands as their contribution to cutting GHG emissions. For instance, if a new fertilizer use enhances carbon storage, then the impact it will have on the ability of the cropland to soak up carbon will be used to calculate the reduction in the country's emissions.

Under the final agreement Japan, for instance, can meet well over 50 percent of its reduction commitment by using better forests, grazing lands, and even better agricultural management practices. The same sink advantage is gained by all other polluters, which can either fix carbon in their own lands or buy their emission reduction targets by fixing carbon in developing country forests or agricultural or grazing lands. The enormous scientific uncertainties in measuring the effective reductions in emissions makes the Kyoto compromise a grand and shameless fudge account.

Second, given this extremely creative accounting, the polluters wanted an agreement in which the crooks, if caught, would not get penalized. The next big concession came on the issue of compliance. In the Kyoto Protocol, the world had to design an enforcement mechanism for the rich and powerful. The initial talk was for a punitive and legally binding compliance regime, which would put in place severe monetary penalties for not meeting the target. But the final agreement lacks teeth, with the enforcement branch politely called the facilitative branch. With an ineffective compliance regime, the Kyoto Protocol is now a voluntary agreement, not legally binding.

But why should we be surprised? The climate negotiations are not about the environment but the economy, and every nation is working overtime to protect its right to pollute. In this sham act, Japan has been the convenient ploy to get concessions. The EU (which makes much of its green commitment) has a history of caving in at the very last moment. In the same week when it was busy making euphoric proclamations about how it has saved the world by getting an agreement, the EU decided to postpone for another 10 years its program to remove subsidies on coal, the filthiest and most carbon-intensive fuel. Before the "historic" Kyoto agreement, the EU was going to phase out these subsidies starting July 2002. The EU has also decided to postpone its plan for domestic emission trading. Why? Because its own "green" companies complained that they would lose their competitive advantage.

After round 3 discussions this past November in Marrakech, the protocol still has no teeth in its realization, but marks the beginning of a new phase of action and implementation. By continuing to exploit their pivotal positions, Japan, Canada, Australia, and Russia managed to get more concessions from the EU. An agreement deciding upon the legally binding nature of enforcement mechanisms in the protocol, specifically if an industrialized country does not meet its GHG reduction commitments, was deferred to the first conference of parties after the protocol's implementation.

Still without US involvement and almost a year after the climate talks failed miserably at the Hague, Marrakech is a sign that countries are succeeding in resurrecting the protocol. Eligibility conditions using mechanisms, like emissions trading and project based investments, helping industrialized countries fulfill their production targets at a lower cost were hotly debated, with the four countries trying persistently to undermine the conditions. The final deal, however, upholds them. Countries will be allowed to bank credits generated from project-based investments in developing and industrialized countries, but by only up to 2.5 percent of the amount they are allowed to emit. Parties also decided that a developing country could unilaterally start a project and sell credits to industrialized countries.

The next grand compromise, we predict, will come when the world bows to the United States. Bush has made it clear that the most important part of his opposition comes from the fact that key developing countries such as China and India do not have binding commitments under the protocol.

At the next round of talks, which is predicted to happen at the end of October 2002, developing countries continue to be the next targets. The probability is that they will get a 10-year grace period to take on legally binding commitments.

G-77 countries are blissfully lost in the quagmire of discussions on funding and technology transfer. They fail to realize that without an effective climate convention they will lose a lot more than promises for a fistful of dollars. Emerging science tells us that climate change will result in greater climatic variation and extreme events such as floods, droughts, cyclones, and sea level rise, leaving poor people at the very margins of survival to become even more vulnerable. Therefore, it is in the interests of India and other developing countries to demand that the industrialized North take effective and measurable action to reduce its emissions.

Notes

1. Based on the introduction and first chapter in Agarwal, A., S. Narain, and A. Sharma, 1999: *Green Politics* (New Delhi: Centre for Science and Environment).
2. Kartha, S., K. Collin, D. Cornland, S. Bernow, and M. Lazarus (SEI), 1998: "'Meaningful participation' for the North and South," presented at Stockholm Environment Institute/Centre for Science and Environment Workshop Towards Equity and Sustainability in the Kyoto Protocol, Buenos Aires, November 8, mimeo.
3. Marland, G., et al., 1999: *National Carbon Dioxide Emissions from Fossil Fuel Burning, Cement Manufacture and Gas Flaring* (Oak Ridge, TN: Oak Ridge Laboratory).
4. Smith, K., et al., 1990: "Indices for a greenhouse gas control regime: Incorporating both efficiency and equity goals," mimeo, quoted in A. Agarwal and S. Nurain, 1992: *Towards a Green World: Should Global Environmental Management Be Built on Legal Conventions or Human Rights?* (New Delhi: Centre for Science and Environment).

5. Kartha et al., 1998, n.2.
6. Sharma, A., N. Singh, and S. Joshi, 2000: "Hold up!" in *Down to Earth,* 9 (15), December 31, available online: http://www.cseindia.org/html/dte/dte20001231/dte_analy.htm.
7. Ibid.
8. Ibid.
9. Anonymous, 1996: "Don't forget equity," in *Climate Change Bulletin,* UN FCCC Secretariat, 12: 3.
10. Basu, S., 1997: "India will not accept emission reduction targets," in *The Hindu,* November 22 (New Delhi: Kasturi and Sons).
11. Agarwal, A. and A. Sharma, 1997: "A farce or a face-off," in *Down to Earth,* 6 (15): 38, December 31 (New Delhi: Society for Environmental Communications).
12. Sharma et al., 2000, n. 6.
13. Baer, P., J. Harte, B. Aya, A. V. Herzog, J. Holdren, N. E. Hultman, D. M. Kammen, R. B. Norgaard, and L. Raymond, 2000: "Climate Change: Equity and Greenhouse Gas Responsibility," *Science* 289: 2287.
14. Kartha et al., 1998, n.2.
15. Agarwal, A., 1998: "Kyoto's ghost will return," in *Down to Earth,* 6 (16): 34, January 15 (New Delhi: Society for Environmental Communications).
16. Ibid.
17. Kartha et al., 1998, n.2.
18. Berk, M. and M. den Elzen, 1998: "The Brazilian proposal and other options for international burden sharing, workshop report: first answers towards Buenos Aires," Centre for Environmental Systems Research, University of Kassel, Germany, and the National Institute of Public Health and the Environment (RIVM), the Netherlands, presented at the Seventh International Workshop on Using Global Models to Support Climate Negotiations, September 21–22, mimeo.
19. Manne, A. and R. Richels, 1993: *Buying Greenhouse Insurance* (Boston: MIT Press).
20. Anonymous, 1998: "The (un) clean development mechanism," CSE Dossier Factsheet 3, November 5, p. 4 (New Delhi: Society for Environmental Communications).
21. Sharma, A., 1998: "No headway," in *Down to Earth,* 7 (14): 41, December 15 (New Delhi: Society for Environmental Communications).
22. Akumu, G. 1999: "Emissions avoidance and equity," in *ECO,* NGO Newsletter, Bonn, July 7.
23. Agarwal, A. and S. Narain 1998, "Sharing the Air," in *Down to Earth,* August 15, pp. 41–42 (New Delhi: Society for Environmental Communications), available online: http://www.cseindia.org/html/dte/dte.htm; Centre for Science and Environment, 1998: "Towards a non-carbon energy transition," Dossier Fact Sheet 3, New Delhi, November, pp. 1–4. See also Chakravarty, U., et al., 1997: "Extraction of multiple energy resources and global warming," University of Hawaii, mimeo; Chakravarty, U. 1997: "Clement climes," in *Down to Earth,* December 31, pp. 20–22 (New Delhi: Society for Environmental Communications), available online: http://www.cseindia.org/html/dte/dte.htm.

CHAPTER 15

Equity, Greenhouse Gas Emissions, and Global Common Resources

Paul Baer

The core of the climate change problem is as simple as it is daunting: We now know that it will be impossible for the whole world to obtain—or even to approach—the emissions levels of the industrialized countries, without gravely endangering our planetary life support systems. In the United States, emissions average over 5 tons of carbon per person per year;[1] even in more efficient European economies, average emissions exceed 2 tons of carbon yearly. Yet global annual emissions must fall by more than 50 percent—to a third of a ton per person or less—if atmospheric greenhouse gas (GHG) levels are to be stabilized in this century.

Meeting this target would be a big enough problem if it were just a matter of the industrialized nations reducing their emissions by a factor of 5 or 10, and bargaining among themselves about how to share the atmosphere. But this halving of total emissions must take place in a world where more than a billion people live on less than a dollar a day and 30 percent of children under 5 are malnourished.[2] This matters because it is still generally assumed that the solution to poverty is for the poor nations to "develop" along the same path that the rich nations have—for the South to become like the North. But this model depends on increasing energy use and, given current technology, increasing GHG emissions. Thus, if the developing nations follow the energy-technology path of the rich countries, the planet faces the risk of catastrophic climate change.

Since this risk puts a limit on allowable GHG emissions, one can think of that limit as defining the available "environmental space." And there is simply not enough environmental space for the South to develop the way the North has. Therefore, the particular environmental space at issue here—the atmosphere—must be brought under common governance; global rules for its use and allocation must be discussed, decided, and enforced. The UNFCCC and the Kyoto Protocol are steps in this direction.

The structure of the Kyoto Protocol, which establishes binding emissions caps on the developed countries, has made the distribution of those caps a central controversy both in the negotiations and in the ratification debate. Developing countries were explicitly exempted from caps because of their lower historical and current emissions, and because of their agreed need to devote their resources to sustainable development and poverty alleviation.[3] However, U.S. opponents of the Kyoto Protocol have vehemently argued that the protocol is not fair to the United States because developing countries have no caps and bear no costs.

The Clinton Administration did not submit the Kyoto Protocol for ratification and thus avoided this debate over fairness. On taking office, President Bush used the fairness argument as one reason for rejecting the protocol outright. Environmental groups in the U.S. made some effort to counter it, but their strategy seems largely to rely on ratification without the United States and on seeking domestic reductions outside the Kyoto framework. However, the unresolved debate over developing country commitments will continue to focus attention on the fair distribution of emissions rights.

If the Kyoto Protocol enters into force without the United States, caps for developing countries will be crucial to the negotiation of targets for the second commitment period (after 2012). If the protocol doesn't enter into force, the requirement that developing countries limit their emissions probably will be a key bargaining issue in negotiating an alternative agreement. And if the United States wants the developing countries to accept caps, it will have to propose an allocation formula that addresses the developing countries' fundamental concerns over equity. Everyone in the developing world cannot emit at the high rates of the North, but why should developing countries agree to restrictions that bind them to their current, much lower per capita rates or that restrict their economic growth? What is an equitable solution to this dilemma?

There are other important aspects of equity in the climate debate, such as the risks we impose on future generations (intergenerational equity) and liability for the harm that will be caused by climate change we are unable or unwilling to avoid.[4] However, because the question of equitable allocations among countries remains a major—and urgent—unresolved obstacle to an effective global treaty, I focus on the allocation issue in this chapter.

One can view the question of an equitable allocation of emissions rights as a political science problem or as an ethical problem. Much of what has been written about equity in the climate negotiations is relatively traditional political science; for example, authors attempt to analyze the rational (economic) interests of the parties and their relative power, and try to predict the outcome of the negotiations. In this framework, the analyst is a neutral observer, and equity (or the perception of inequity) is a possible variable to account for the negotiating outcome.[5] Sometimes authors will go so far as to suggest possible allocation formulas that they believe could be acceptable to all parties; however, the interests and preferences of countries are taken as given.

Alternatively, an ethical analysis looks at the justifications that competing parties offer for their negotiating positions and attempts to critically evaluate them.[6] In this framework, the analyst is a participant, and equity is something to be defined and argued for in order to influence the world. This is how I approach the problem in this chapter.

Framing the Problem: Burden Sharing Versus Resource Sharing in the Global Commons

The climate change problem can be posed as a question of burden sharing or as a question of resource sharing.[7] In the burden-sharing framework, the costs of protecting the atmosphere by reducing emissions to a safe level are a burden that must be shared globally. The costs come from the need to introduce lower-emitting technologies—presumed to be more expensive—and the requirement for reduced consumption. The issue of equitable allocations is usually framed this way;[8] U.S. opposition to the Kyoto Protocol is based on the argument that it imposes an unfair burden.

In this framework, it makes sense to say that the burden should be shared equally unless there are compelling reasons why it shouldn't be. If we accept a principle of equal sacrifice, and we believe that it is a greater sacrifice for a poor person to pay a dollar than it is for a rich person—in economic terms, the declining marginal utility of income—we might define a person's or country's fair share based on ability to pay. Like a progressive tax, this would mean that the wealthy pay a higher proportion of their income than the poor do, but the poor still pay something.

However, focusing on the burden of reductions obscures the question of who has been responsible for, and benefited from, the overuse of the atmosphere. Assessing responsibility requires us to focus on the atmospheric carbon sink as an economic resource, and to account for both its unequal appropriation in the past and its unequal use today. We need to ask who has used the resource, what benefits they have acquired from its use, and what losses will be suffered by those who cannot use as much as they otherwise would have. If the finite size of the available atmospheric space defines the total benefits that can come from its use, it is necessary to ask whether a person or country has received or will receive a fair share of the benefits. In this way we can meaningfully define overuse and underuse and define a party's obligation on this basis: Parties that have exceeded their share have obligations to parties that will therefore get less. To understand what would be a fair share, it will be necessary to look further at the nature of common resources.

The Nature of Common Resources

From one perspective, any system in which the use of a resource by one party causes harm to another can be viewed as a commons. Those harmed necessarily have a moral stake in the use or conservation of the resource, even if they don't have the ability to exploit it in kind and thus to cause a symmetric harm. However, it is when each party can cause harm to the others that we have a classic commons problem.

In a commons, individuals typically gain much more from their use of the resource than they suffer from the degradation their use causes; thus one can increase one's own well-being by overconsuming and harming the other users. Furthermore, restricting one's own use does not ensure protection against the harms caused by others' use of the resource. In these ways, a common resource establishes a moral community. To protect the resource and to protect themselves, the parties must grant each other the right to a fair share, and accept enforcement of a mutually agreed limit.

I argue that the fundamental principle of fairness in the governance of a commons is equality in decision-making and use, and in particular equality among people, not countries. This cannot be simply asserted or deduced, but rather that must be established through moral reasoning. By drawing on an extended analogy to a hypothetical common resource—in this case, a shared aquifer—and by rebutting common critiques, I will show how the principle of equal rights to common resources can be credibly justified.

Imagine two people—let's call them Nora and Sam—who share an island. Each of them has a well that pumps water from a shared aquifer. Nora discovers how to make a pump that pumps three times as fast, and is able to irrigate more farmland; soon she has a grain surplus, is feeding cattle, and is clearly healthier. Sam meanwhile is able to irrigate a much smaller plot and to feed only a few chickens, and is regularly falling ill. Eventually, however, he discovers how to make his own pump that is as powerful as hers.

Just before he installs his pump, they both find out that the level of the aquifer is starting to fall. Each is aware that the other is using the aquifer and at what rate. They get together and figure out how large the aquifer is and what its annual recharge rate must be. They know how much each of them has already pumped, and how long it will take to exhaust the remaining stock at the rate they will both soon be able to pump. They know also that if they are forced to immediately reduce to the recharge rate, it will seriously limit their food supplies. They are now forced to decide, whether individually or collectively, how fast to pump the water. Assuming that they decide that they can and must trust one another, what might we expect them to decide is a fair agreement, and why?

It seems likely that they would agree to share the aquifer equally unless one was willing to compensate the other. Nora does not have a good argument to make why Sam should continue to use less and she more; now that they each have a big pump, why shouldn't he be able to use his? Should he agree to remain permanently poorer? It would not make sense for Sam to agree to forever use a smaller share simply because he was using less at the time when the agreement was made.

On the contrary, Sam might point to the wealth Nora accumulated while she was living on an unsustainable share of the water and say that it is not fair for them now to use an exactly equal share. There was a fixed amount of water in the aquifer when they started, and it can only produce a finite amount of wealth before they are both required to learn to live off the sustainable flow; to divide only the remaining part of the aquifer equally would leave him perpetually poorer. Yes, Nora did not know that the level of water that she was pumping was unsustainable, but Sam did not agree to let her become wealthy at his expense. And indeed her wealth is at his expense; what she used, he cannot. He can make a good case that it is fair for him now to use more, or for her to compensate him for using less.

Some might recognize this situation as a version of the prisoner's dilemma and note that there is a noncooperative solution that is equally plausible. Nora or Sam might decide to pump as fast as possible, knowing that their use was unsustainable, but the other probably would do the same, leaving them both worse off. What matters here is that we see the situation of interdependence as necessarily creating a moral community: Each party can harm or be harmed by the other, and depends on the other's cooperation.[9] This, then, is the structure of a common resource: Even if we would like to get more than our fair share of the benefits, we know that it is not ethical for us to do so. Furthermore, absent any other compelling justifications, a fair share is a equal share.

What might constitute a justification for an unequal division of the aquifer? If it rains more on someone's part of the island, we might think it fair for him or her to accept a less than equal share of the aquifer. However, it is important to realize that such an argument for inequality in access to a particular resource is based on an appeal to equal opportunity more generally: No one should be better or worse off than anyone else simply because of which part of the island he or she happens to live on. In the real world, this principle is not given much weight since (for example) countries with large fossil fuel resources are not expected to give a free share to less fortunate countries. One might argue that fossil fuel reserves *should* be shared as common global resources, but the point here is that because in our hypothetical example the water lies underneath (and is equally accessible to) all parties, there is no way for one individual to physically exclude another from using it, and thus to charge for its use.

In this hypothetical example, I have placed the question of the allocation of the common resource into a very abstract context, as if it were the only resource in question and the only issue of negotiation between the two parties. It is in part through this abstraction that the principle of equality emerges so strongly; there is no possible gain to either party from accepting a less than equal share. However, the real world is much more complicated. For example, it could well be argued that, when there are a large number of unused common resources, each party would accept a principle of "first come, first served."[10] Allowing one party the right to claim a larger share of certain common resources, in exchange for allowing other parties a similar right to other resources, might be agreed to make everyone better off because it encourages innovation and investment in the development of those resources. This is a major justification for allowing homesteading or the establishment of mining claims or water rights.

However, this condition clearly does not hold in the case of the atmosphere. There have never been any negotiations between all the countries of the world, to say nothing of all the people, concerning general principles of allocation of global common resources. Not, that is, until today. The underusing countries have not agreed to allow the North's overconsumption.

An Ethical Analysis of Allocation Principles for Emissions Rights

This leads us back to the problem with which we started: the need for an international agreement to regulate GHG emissions and the controversy over what an acceptable allocation of rights would be. There is an extensive literature on this part of the equity debate, to which I cannot begin to do justice.[11] I focus on a relatively narrow but crucial aspect: whether the ethical arguments for various allocation principles are convincing. I address the fairness of various principles rather than the likelihood of their being accepted or the ease of implementing them, not because I believe that it is better to be morally righteous than to be practical but because what a government and its citizens believe is fair is one justification of country's negotiating positions. Especially because of the large role of the United States in the climate negotiations, this is not just an academic matter; the sense of fairness that eventually becomes dominant in the United States may have a significant influence on the future of the negotiations.

I assume in the following discussion that tradeable emissions permits are a plausible and desirable scheme for addressing climate change. There are reasonable arguments against tradeable permits, such as the practical difficulties of implementation and the possible negative impacts of market power (either buyers' or sellers'). Nonetheless, many analysts have concluded that the power of such a scheme to separate efficiency (making the most cost-effective reductions) from equity (determining who will pay for those reductions) makes it the best option for an international agreement.[12] Furthermore, the Kyoto Protocol itself explicitly includes mechanisms for emissions trading. Thus, in the remainder of this section I will assume that however emissions rights are allocated, they may subsequently be traded.

Why Emissions Rights Can't Be Equal by Country

I will begin by examining two principles that are seldom explicitly advocated but underlie many of the arguments against other principles (such as equal per capita rights). The first of these is the principle that every country should have a right to an equal share of the atmosphere. It simply isn't ethically plausible that the rights to use a common resource would be attributed in equal shares to every country. The benefits of the use of the resource fundamentally accrue to people; the allocation of emissions rights to countries is a pragmatic compromise. No one would argue that Fiji should have the same emissions rights as the United States.

I bring up this seemingly obvious point only because opponents of the Kyoto Protocol often argue that a reason for the United States to oppose Kyoto is that because the protocol restricts U.S. emissions but not China's, China will soon emit more than the United States. It is reasonable for the United States to be concerned that if China never accepts limits, U.S. emissions reductions will not prevent climate change. However, the Kyoto Protocol only addresses the period through 2012. One cannot claim that it's wrong if China some day emits more than the United States without a real argument about the basis for emissions rights, and the United States has a very weak argument. After all, China has more than four times the U.S. population; it must be acceptable for them to emit some amount more than we do.

Why Emissions Rights Can't Be Grandfathered

The second principle of resource allocation I will consider is grandfathering, the principle that a party's current level of use establishes a firm property right. Under this principle, if we need to establish a limit to use, a country is entitled to the same proportion of the limited resource that it had been using when use was unrestricted.

It is rare for anyone to make an ethical argument for pure grandfathering in the case of climate; freezing the relative emission rates of different countries at their current proportions can plainly be seen to be unfair to the low-emitting countries. Imagine being born in a poor country in the year 2050 and finding that you are allocated fewer permits than people in much wealthier countries simply because your country had been poor in 2005. You might very reasonably conclude that this was not a fair situation and might reconsider whether your nation should continue to abide by the agreement.

There is a weaker form of the argument for grandfathering that is more plausible, and that implicitly underlies a large number of proposed mixed or transitional allocation schemes: Because the high-emitting countries did not know that they were overusing a commons, it would be unfair to ask them to immediately restrict their use to a fair, sustainable share. However, this argument confuses two different points. The first is whether it would constitute an undue hardship on the high emitters to restrict their emissions sharply and rapidly (or to pay for their excess consumption). The second is whether the high emitters are entitled to the benefits of their current overconsumption. There are numerous precedents for allowing parties to stretch the repayment of their debts over time. But the legitimacy of the debt isn't determined by the harm that is caused by repaying it, and it is usually assumed to be up to the party who is owed to determine whether and how much to reschedule or reduce the debt. Thus this is at best an argument for temporary grandfathering as part of a transition.

Why Emissions Rights Can't Be Proportional To GDP

Others have argued that permits should be allocated at least in part proportionally to gross domestic product (GDP);[13] the greater a country's GDP, the greater its emission rights. This gives additional permits to the wealthier nations (attractive for getting them to buy in but not in itself an ethical argument), and it creates incentives to use one's allocation as efficiently (in the sense of reducing emissions per GDP) as possible. However, this principle has some unacceptable effects if carried to its conclusion; the wealthiest countries would always have the largest share of the permits, which, given their value, would have the effect of increasing inequality. Again, imagine a resident of a poor country some years from now, who may not burn as much coal as a resident of a rich country precisely because he is not as rich; it is hard to see how he or she would consider this to be fair.

Why Emissions Rights Should Be Per Capita

The central argument for equal per capita rights is that the atmosphere is a global commons, whose use and preservation are essential to human well being. Therefore, as I argued using the aquifer example, all people should hold both decision-making rights and use rights equally unless there is a compelling higher principle.

We might be able to determine what would count as a higher principle by considering the implications of the *reductio ad absurdum* of equal per capita rights.[14] No single *reduction* can capture all possible failures of an ethical principle, and there are several that might be interesting and relevant in this case. One possible *reduction* is that emissions permits are allocated immediately on a strict per capita basis and are not tradeable; this would clearly cause a harmful economic shock to the countries that had to make sharp reductions. This might well be judged unacceptable on utilitarian grounds if it caused more harm to those who were forced to reduce than it brought benefit to those who were not or if it actually harmed those it was meant to help due to global economic interdependence.

However, what we actually seem to care about here is outcomes, not principles of allocation; if an unequal allocation could be shown to permanently benefit those who receive lower allocations, few would argue that we should insist on strict equality. However, other than suggesting that developing countries might suffer if the North underwent economic contraction, no one has ever argued that poor countries actually would benefit from having lower emissions allocations than rich countries, especially not permanently.

Another possible *reduction* is that a global energy administration would actually issue a GHG emissions permit to every person on the planet and require them all to buy and sell them in a single enormous global market. This boggles the mind because of its impracticality, not its ethical failure. If the reason for an equal right is because each person is truly entitled to an equal share of the benefits, there are only practical reasons, not logical or ethical ones, for the permits to be issued to countries. Similarly, the idea that each person on the globe might vote on the total amount of emissions to be allowed seems absurd, but again for practical rather than ethical reasons.

Beyond Equal Annual Allocations: The Principle of Historical Accountability

Many analysts have also extended the principle of equal per capita rights to the principle of historical accountability. At the individual level, historical accountability could mean that each individual gets the same amount in their lifetime, regardless of when they are born; those who have already used more than their allowable share would have to purchase permits from those who haven't. Returning to the aquifer example, this is precisely the argument Sam has for why he should get a larger than equal share of the remaining water.

In practice, this would mean that a country's current allowed emissions are reduced if it has cumulatively overused the commons. There are many possible formulas for quantifying overuse and using it to modify current allocations;[15] however, the essential point is that countries are assumed to have benefited permanently (as by increased wealth and infrastructure) from that overuse and to have a debt to repay. There are some plausible ethical objections to historical accountability, such as the dubiousness of holding living persons responsible for the activities of their ancestors or the fact it hasn't been known for long that overuse was causing a problem.[16] Also, not everyone in wealthy countries has contributed equally to or benefited equally from their cumulative emissions. However, the correlation between the cumulative emissions of countries and their levels of overall wealth is clear, and the fact that wealth is unequally distributed within countries does not seem to justify ignoring the common benefits that have accrued.

In a hypothetical case, if we could identify the precise contribution that overuse of the commons had made to an individual's current wealth, it would be reasonable to consider that benefit to be an individual debt to those who will be unable to obtain a similar benefit. It seems reasonable that a country that has cumulatively but unequally overused the commons should be responsible for fairly distributing the debt among its citizens.

Practical Versus Ethical Objections to Equal Per Capita Rights

This leads to a more general consideration of the relationship between practical and ethical objections to equal per capita allocations. The three most common objections to equal per capita rights are that it provides an incentive to population growth, that poor people who would have a surplus of allocations would not benefit from their sale, and that the North (and the United States in particular) would never accept the financial burden. I will address each of these in turn and show that although they may indeed have some practical relevance, they do not in themselves constitute arguments that per capita allocations are not ethically justified.

Because governments would get permits to use or sell based on the size of their populations, opponents of equal per capita rights argue that it would give governments an incentive to increase, or at least not to limit, population growth.[17] However, this concerns the practical effect of a per capita allocation principle; it is not an ethical argument that people should not intrinsically have equal rights to the commons. This is not to say that practical arguments do not have ethical implications; again, if an equal per capita allocation led to substantial additional population growth that caused identifiable harm, we might therefore reject per capita rights because of the consequences. But because there are a variety of plausible solutions (one simple example being fixing the allocation to a base-year population), this argument does not carry much weight.

Another argument that has been made against equal per capita rights is that the resulting financial transfers would not aid the people they are supposed to help. Because the permits would be traded by governments, there is no guarantee that the poorest people who should be the owners of surplus permits would see much of the benefits from their sale. However, this again is a pragmatic, not ethical argument; the fact that there is not currently a channel for permit purchasers to pay the rightful owners of the resource does not mean they are not ethically obligated to do so. They certainly may not simply keep the money.

For better or worse, we generally accept national sovereignty as a basis for determining the internal allocation of resources; we do not judge the democratic nature of the Saudi royal family before we pay for the oil we import. Nor has anyone suggested that the United States did not deserve a large allocation because the benefits of emissions are unequally distributed domestically. Moreover, if we think that individuals should receive the benefits of the use (or sale) of their permits, we can help empower them to make that demand effectively by giving international recognition to the principle of equal rights.

Cost as an Objection to Per Capita Allocations

Finally, it is necessary to discuss what is usually given as the ultimate argument against equal per capita emission rights: that the industrialized countries, and the United States in particular, would never agree because of the high costs they would incur. In a tradeable permit system with a cap at today's global level, an immediate transition to an equal per capita allocation would result in trading of roughly 2 billion tons of carbon permits each year. Recent economic studies have estimated that for a cap based on the Kyoto framework, permits might trade in a global market at $20–$100 per ton,[18] but estimates of up to $200/ton or more have been made for more restrictive global caps. The United States alone exceeds its equal per capita share by more than a billion tons and thus could be required to purchase permits worth tens or even hundreds of billions of dollars.

Thomas Schelling, among others, has argued that because poor people in developing countries are the most likely to suffer from climate change, money spent by wealthy countries to prevent it is a form of foreign aid.[19] He also argues that it is unreasonable to expect the United States to pay so much more in this case than we currently do for other forms of foreign aid. However, if one accepts that there should be equal rights to global common resources, any costs associated with tradeable permits are payment for the use of resources, not foreign aid. The fact that the United States might not like or agree to such costs is not an ethical argument; I may not like the high price of oil, but that doesn't mean I can steal it.[20]

It is possible to argue that the harm that would come from paying a fair price for emissions rights is greater than the benefit that would come to those who sell the permits. There is some evidence that people feel that the loss of a given amount of income causes greater harm than the gain of the same amount of income causes benefit. One way to look at this is to consider that people have expectations built around their material lives and that even wealthy people suffer significantly when their expectations fail to be met.[21] However, in the case of global emissions trading, this is not a very plausible argument. The standard analysis of transfers between rich and poor, based on the declining marginal utility of income, is that the gain of $10 to a poor person means more than the loss of $10 to a rich person; the huge disparities of wealth between North and South suggest that the marginal utility of income in the South must be much higher.[22]

If rights to the global commons should be shared equally and paying for those rights would not cause more overall harm than good, there is little remaining justification for the North to refuse to agree to such payments. A country does not have the ethical right to opt out of the governance of a commons—to be a free rider—simply because it doesn't want to reduce its overconsumption. Because one country's use affects all the others, the moral community and moral obligations exist whether they are respected or not.

Conclusions

As I suggested at the beginning of this chapter, I consider myself to be not merely an analyst but also a participant in the process of defining equity in the climate change debate. Because the economic stakes are quite high, many parties are actively engaged in this process. It is my central claim that self-interest and ethical justification are not the same and that one can and must use reasoned argument to determine what is right. If the arguments for equal rights are justified, it follows that the U.S. government should change its negotiating position and agree to a treaty that establishes at least an eventual goal of equal per capita allocations.

With a commitment to equal per capita allocations, a global emissions cap covering developing and developed countries becomes possible, with enormous associated advantages. Such an agreement would create a large and (hopefully) efficient market for permits and thus bring down the cost of compliance worldwide. It would eliminate the need to establish baselines that dogs the Clean Development Mechanism and other project-based mitigation schemes. Per capita entitlements would eliminate the incentive for developing countries to delay reductions in emissions in order to increase their claim to atmospheric space. In all these ways, a transition to an agreement based on equal per capita rights would help us to stabilize atmospheric GHG concentration at lower levels and to limit the risks of dangerous climate change.

I do not presume to have addressed all the ethical questions concerning the equitable allocation of emissions rights. At the very least I hope I have made clear what a justification for a principle of equity must look like to be an ethical rather than practical (or selfish) argument. Finally, I hope that I have demonstrated that it is both possible and necessary for us to take part in the creation of new norms of international equity, and that the climate change debate offers us an opportunity to make an important contribution to a more just and sustainable world.

Acknowledgments

I would particularly like to thank Tom Athanasiou, Barbara Haya, Francesca Flynn, Richard Norgaard, and Paul Wilson for their thoughtful discussions and extensive comments, as well as three anonymous reviewers. All remaining philosophical and factual errors are my own.

Notes

1. In the form of carbon dioxide from fossil fuel burning and industrial activities; multiply by 44/12 to get the equivalent mass in CO_2.
2. World Bank, 2000: *World Development Report 2000/2001: Attacking Poverty* (Oxford: Oxford University Press).
3. This reasoning is embedded in the UN Framework Convention on Climate Change

itself, in the Berlin Mandate passed at the first Conference of Parties in 1995, and in the Kyoto Protocol.

4. For an excellent review of issues of intergenerational equity, particularly as it is connected with the theory of discounting in economics, see Arrow, K. J., W. R. Cline, K.-G. Mäler, M. Munasinghe, R. Squitieri, and J. E. Stiglitz, 1996: "Intertemporal equity, discounting, and economic efficiency," in J. P. Bruce, H. Lee, and E. F. Haites (eds.), *Climate Change 1995: Economic and Social Dimensions of Climate Change* (Cambridge, England: Cambridge University Press), 125–144. For a discussion of the question of liability for climate damages and other equity concerns, see Banuri, T., K. Göran-Mäler, M. Grubb, H. K. Jacobson, and F. Yamin, 1996: "Equity and social considerations," in the same book. See also Grubb, M. J., 1995: "Seeking fair weather: ethics and the international debate on climate change," *International Affairs,* 71: 463–496.
5. See Schelling, T. C., 1997: "The cost of combating global warming: facing the tradeoffs," *Foreign Affairs,* 76: 8–14; Rayner, S., E. L. Malone, and M. Thompson, 1999: "Equity issues and integrated assessment," in F. Tóth (ed.), *Fair Weather? Equity Concerns in Climate Change* (London: Earthscan Publications Ltd.), or Victor, D., 1999: "The Regulation of Greenhouse Gases: Does Fairness Matter?" in the same book.
6. Important articles using this framework include Ghosh, P., 1993: "Structuring the equity issue in climate change," in A. N. Achanta (ed.), *The Climate Change Agenda: An Indian Perspective* (New Delhi: Tata Energy Research Institute). Shue, H., 1993: "Subsistence emissions and luxury emissions," *Law and Policy,* 15 (1): 39–59; Bhaskar, V., 1995: "Distributive justice and the control of global warming," in V. Bhaskar and A. Glyn (eds.), *The North the South and the Environment: Ecological Constraints and the Global Economy* (New York: St. Martin's Press). Paterson, M., 1996: "International justice and global warming," in B. Holden (ed.), *The Ethical Dimensions of Global Change* (New York: St. Martin's Press).
7. I am indebted to Sunita Narain for clarifying this distinction.
8. More than a few articles have "burden sharing" in their titles: e.g., Grubb, M., J. Sebenius, A. Magalhaes, and S. Subak, 1992: "Sharing the burden," in I. M. Mintzer (ed.), *Confronting Climate Change: Risks, Implications and Responses* (Cambridge, England: Cambridge University Press); Parson, E. A. and R. J. Zeckhauser, 1995: "Equal measures or fair burdens: negotiating environmental treaties in an unequal world," in H. Lee (ed.), *Shaping National Responses to Climate Change* (Washington, DC: Island Press). Many more articles use it as their conceptual framework, and numerous workshops have been held on the subject.
9. The fact that it may be possible to cheat and that there is incentive to do so may lead to the need for an enforcement mechanism; it is precisely because the situation creates a moral community that each party may grant that it is fair for the other side to enforce a penalty on them if they are caught.
10. I thank Dick Norgaard for this insight.
11. Comprehensive bibliographies can be found in the IPCC's Second and Third Assessment Reports: Banuri et al., 1996; Banuri, T. and J. Weyant, 2001: "Setting the stage: climate change and sustainable development," Chapter 1 in B. Metz and

O. Davidson (eds.), *Climate Change, 2001: Mitigation* (Cambridge, England: Cambridge University Press). Important early work that I haven't cited elsewhere includes Krause, F., W. Bach, and J. Koomey, 1989: *Energy Policy in the Greenhouse,* Volume I (El Cerrito, CA: International Project for Sustainable Energy Paths); Rose, A., 1990: "Reducing conflict in global warming policy: the potential of equity as a unifying principle," *Energy Policy,* December 1990: 927–948; Agarwal, A. and S. Narain, 1991: *Global Warming in an Unequal World: A Case of Environmental Colonialism* (New Delhi: Centre for Science and Environment); Solomon, B. D. and D. R. Ahuja, 1991: "International reductions in greenhouse-gas emissions: An equitable and efficient approach," *Global Environmental Change,* 1 (5): 343–350; and Bertram, G., 1992: "Tradable emission permits and the control of greenhouse gases," *Journal of Development Studies,* 28 (3): 423–446. More recent discussions include Meyer, A., 1999: "The Kyoto Protocol and the emergence of 'contraction and convergence' as a framework for an international political solution to greenhouse gas emissions abatement," in O. Hohmayer and K. Rennings (eds.), *Man-Made Climate Change: Economic Aspects and Policy Options* (Heidelberg: Physica–Verlag); Baer, P., J. Harte, B. Haya, A. V. Herzog, J. Holdren, N. E. Hultman, D. M. Kammen, R. B. Norgaard, and L. Raymond, 2000: "Equity and greenhouse gas responsibility in climate policy," *Science,* 289: 2287; and an entire edited collection: Tóth, *Fair Weather?* 1999.

12. Some authors explicitly argue that tradable permits are more desirable than a carbon tax scheme, which in economic theory can also separate efficiency from equity. Epstein, J. M. and R. Gupta, 1990: *Controlling the Greenhouse Effect: Five Global Regimes Compared* (Washington, DC: The Brookings Institute). Many other authors assume the benefits of trading with little analysis.
13. For example, Sagar includes GDP in a hybrid proposal with population and historical emissions. Sagar, A., 2000: "Wealth, responsibility and equity: exploring an allocation framework for global GHG emissions," *Climatic Change,* 45: 511–527.
14. *Reductio ad absurdum* means roughly what it sounds like: "reduction to the point of absurdity." In ethical reasoning this means applying a moral principle in an extreme form to a hypothetical situation, typically to show that in cases related to the real example, there are higher principles that take precedence.
15. See Grübler, A. and Y. Fujii, 1991: "Intergenerational and spatial equity issues of carbon accounts," *Energy,* 16 (11/12): 1297–1416; Smith, K., 1991: "Allocating responsibility for global warming: the natural debt index," *Ambio,* 20 (2): 95–96; Hayes, P. and K. Smith (eds.), 1993: *The Global Greenhouse Regime: Who Pays* (London: Earthscan Publications Ltd.); Smith, K., 1996: "The natural debt: North and South," in T. W. Giambelluca and A. Henderson-Sellers (eds.), *Climate Change: Developing Southern Hemisphere Perspectives* (New York: John Wiley and Sons Ltd.).
16. For an articulation of the case against historical accountability, see Beckerman, W. and J. Pasek, 1995: "The equitable international allocation of tradable carbon permits," *Global Environmental Change,* 5: 405–413; for a response see Neumayer, E., 2000: "In defence of historical accountability for greenhouse gas emissions," *Ecological Economics,* 33 (2): 185–192.
17. In fact, this is raised by defenders of per capita rights more often than by opponents

because it is so easy to defeat; several solutions have been discussed in all the literature since Michael Grubb's seminal work. Grubb, M., 1989: *The Greenhouse Effect: Negotiating Targets* (London: Royal Institute of International Affairs).

18. Weyant, J. P. and J. H. Hill, 1999: "Introduction and overview," *The Energy Journal,* Special Kyoto Issue: vii–xiv.
19. Schelling, 1997.
20. David Victor has made a similar argument that a difficulty of adopting a tradable permit scheme is that wealthy countries will be reluctant to allocate assets worth hundreds of billions of dollars. Victor, D. G., 2000: "Controlling emissions of greenhouse gases," in D. Kennedy and J. A. Riggs (eds.), *U.S. Policy and the Global Environment: Memos to the President* (Washington, DC: The Aspen Institute). Others have pointed out that if the assets indeed have this value, the overconsuming countries are free-riding and increasing their unfair appropriation by delaying the transition to equal per capita rights. Parikh, J. and P. Parikh, 1998: "Free ride through delay: risk and accountability for climate change," *Environment and Development Economics,* 3: 384–389.
21. For an extensive discussion of this argument, see Wesley, E. and F. Peterson, 1999: "The ethics of burden sharing in the global greenhouse," *Journal of Agricultural and Environmental Ethics,* 11: 167–196.
22. In fact, if one uses a standard assumption about the declining marginal utility of income (i.e., that utility increases as the log of income), it can be shown that in a system of tradable permits, equal per capita rights are utility maximizing compared with grandfathering or any mixture of the two. Baer, P. and P. Templet, 2001: "GLEAM: a simple model for the analysis of equity in policies to regulate greenhouse gas emissions through tradable permits," in M. Munasinghe, O. Sunkel, and C. de Miguel (eds.), *The Sustainability of Long Term Growth* (Cheltanham, England: Edward Elgar Publishing Company).

PART VI

ENERGY CHOICES

CHAPTER 16

Renewable Energy Sources as a Response to Global Climate Concerns

John J. Berger

> Renewable energy sources are sufficiently abundant that they potentially could provide all of the world's energy needs foreseen over the next century.
>
> —Intergovernmental Panel on Climate Change[1]

This chapter presents an environmental and climatological rationale for the creation of a domestic and global economy based on renewable energy sources. Individual renewable energy sources and technologies are discussed to show that these technologies could cleanly, safely, cost-effectively, and indefinitely provide for our energy and transportation needs. The public policies described would help to transform the United States from a nation that is currently dependent on fossil fuels, to one that relies predominately on clean energy sources for electric power, transportation, and other services.

The United States gets 92 percent of its energy from fossil fuels and nuclear power. Less than 1 percent of primary U.S. and world energy supplies are provided by solar, wind, or geothermal sources (Table 16.1). The high level of fossil fuel dependency combined with current land use trends, such as urbanization and deforestation, are causing global environmental degradation and are widely believed to be responsible for destabilizing natural climate processes.

The environmental damage cited includes severe air and water pollution, destruction of certain ecosystems across large regions, pervasive losses of natural habitat, and the reduction of plant and animal biodiversity. Most of these

TABLE 16.1. Where the Primary Energy We Used Came from in 1997

	United States (%)	*World (%)*
Oil	38	34
Coal	24	24
Natural gas	25	20
Biomass fuels	3.8	13
Nuclear	7.7	6.4
Hydroelectric	1.3	2.3
Solar, wind, and geothermal	0.2	0.3

Source: President's Committee of Advisors on Science and Technology, Panel on International Cooperation in Energy Research, Development, Demonstration, and Deployment: "Powerful partnerships: the federal role in international cooperation on energy innovation," June 1999.

impacts are expected to continue or worsen dramatically in coming decades. Moreover, as ecosystems are degraded, so are the planet's air and water purification systems; pollination systems; natural flood control; pest control, soil creation, water storage capacity; and biological diversity. Yet it is biodiversity that is critical to nature's ability to withstand and recuperate from environmental stress and extreme conditions, such as those imposed by rapid climate change.

An extrapolation of current rising world population trends and increasing per capita energy use indicates that world energy use may well quadruple by 2100. Recent research by the Intergovernmental Panel on Climate Change (IPCC) concludes that the release of carbon dioxide and other heat-trapping gases from fossil fuel burning and other sources may cause mean global temperature to rise by as much as 10°F over the twenty-first century unless something is done to reduce the emission of heat-trapping gases. Numerous studies have also found that the elevated temperatures and air pollution likely to ensue from the expected fossil fuel combustion in coming decades will impair human health through physiological impacts on human respiratory, cardiovascular, and cerebrovascular systems.

However, if cleaner renewable energy technologies for heat, power, and mobility were broadly and intensively introduced, these grave threats could be reduced. This requires shifting the world from heavy dependence on fossil fuels to greater reliance on noncarbon energy sources, which generally are far kinder to the environment than fossil fuel technologies. It also requires using the most efficient energy supply and consumption technologies to restrain energy use. Not only does this curb pollution and environmental impacts, but it greatly reduces the total capital investment needed for building the clean energy economy's renewable generating capacity.

The costs and methods of reducing carbon emissions through 2010 are the subject of a major study by the U.S. Department of Energy's (DOE's) Interlaboratory Working Group on Energy Technologies. Its report, *Scenarios for a Clean Energy Future,* found that the nation could increase energy efficiency, reduce oil dependency, diminish air pollution, and restrain carbon emissions to levels approaching those of 1990 by 2010 at essentially no net cost to the economy through "smart public policies."[2] Whereas the coal industry would contract, and the railroads that depend on it would lose some revenue, national energy savings and the growth of the energy efficiency and renewable energy industries would more than compensate for those economic impacts. However, a more definitive "critical path" energy study does need to be done to estimate the costs and benefits of creating a renewable energy economy[3] and to map out the most cost-effective strategies and policies for doing so.

If the transition to a renewable energy economy were managed wisely, the United States might go from being the world's leading producer of greenhouse gases (GHGs) to being a world leader in commercializing clean energy systems. Modernization and reconstruction of the nation's energy system would induce hundreds of billions of dollars in new infrastructure investment and ancillary economic activity. In addition, as hundreds of billions in fossil fuel costs and related environmental and public health impacts were avoided every decade, the financial resources liberated could be used to help finance energy-efficient homes and businesses, clean industries, and zero-emission transportation systems.

Unfortunately, numerous economic, political, infrastructure, and regulatory problems must be overcome. These problems indirectly elevate the price of renewable energy relative to fossil fuel energy, slowing the adoption of renewables.

Distinguishing Renewable from Nonrenewable Energy Sources

Solar, hydro, geothermal, and wind technologies all operate without producing carbon emissions. They are called renewable or inexhaustible energy supplies because they are endlessly replenished by nature, like water flowing in a river.

By contrast, fossil fuels such as coal, oil, and natural gas were created over

TABLE 16.2. Renewable Energy Resources and the Services They Provide

Resource	*Energy Service*
Biomass	Electricity
	Gaseous fuels for heat, power, and transportation
	Liquid fuels, primarily for transportation
	Solid fuels for heat and power, or combined heat and power
Geothermal	Combined heat and power
	Electricity
	Heat
Hydropower	Electricity
	Mechanical power
	Electricity storage
Solar energy	Electricity
	Lighting
	Process heat
	Space conditioning (heating and cooling)
	Waste detoxification
	Water heating
Wind	Electricity
	Mechanical power

millions of years and exist in fixed quantities. We consume their stock, not their perennial flow, so the more we use, the less remains. Therefore, fossil fuels are neither renewable nor sustainable. High-grade economically exploitable deposits of these fuels will one day be exhausted. A sustainable economy therefore must rely primarily on renewable energy sources (Tables 16.2–16.6). Conventional nuclear power based on atomic fission is not renewable.[4]

Power from the Wind

Wind is an economical, pollution-free, inexhaustible domestic energy resource. As winds spin the blades of an aerodynamically sculpted hub mounted on a tall tower, an alternator behind the rotor generates clean electricity. Wind power is modular and hence scalable: Turbines can be used singly, in small clusters, or in large wind farms connected to the power grid. The largest turbines can produce up to 3 MW, but windmills can be made small enough to power an average house drawing 1 kilowatt. In addition, wind farms can be installed on cropland or ranches with minimal disruption of agricultural activities.

Wind turbines use no fuel and therefore produce no gaseous emissions, no

TABLE 16.3. Renewable Energy Resources, the Services they Provide, and the Technologies Employed for their Product

Resource	*Product, Use, or Energy Service*	*Technologies Employed to Use Resource*
BIOMASS	Alcohol fuels	Fermentation
		Simultaneous saccharification and fermentation of woody (and other cellulosic) material
	Other transportation fuels	Biodiesel
		Chemical synthesis of biodiesel and bio gasoline
	Electricity	Cofiring with coal
		Combustion turbine-generator
		Combined-cycle gas turbine
		Conventional gas turbine
		Integrated gasifier, combined-cycle gas turbine
		Integrated gasifier, combined-cycle gas turbine with cogenerated heat
	Combined heat and power	Conventional steam-turbine generator
		Steam turbine with cogeneration
	Fuel gas	Anaerobic digestion
		Biomass gasification
	Domestic heat	Woodstove, traditional and pellet
		Woodstove with emission control
	Industrial process heat	Direct combustion of farm and forest residue or factory waste in furnace or boiler
GEOTHERMAL	Combined heat and power	Piping of near-surface steam and hot water to provide local and district heat
		Using waste heat from geothermal power plants for district heating
	Electricity	Binary power plant
		Flash power plant
		Hot dry rock power plant (not yet commercial)
	Heat	Geothermal heat pump
HYDROPOWER	Electricity	Turbine generator in dam or run-of-river flume
	Electricity storage	Pumped storage
OCEAN POWER		
Marine currents	Electricity	Submerged turbine (not yet commercial)
Ocean thermal gradients	Electricity	Low–vapor-pressure turbine generator (not yet commercial)
Tides	Electricity	Tidal dam with turbine generator
Waves	Electricity	Float-activated hydraulic pump compressed-air turbine (not yet commercial)
		Oscillating water column compressed-air turbine (not yet commercial)

(*continues*)

TABLE 16.3. *Continued*

Resource	*Product, Use, or Energy Service*	*Technology Employed to Use Resource*
SOLAR	Electricity	Concentrator dish and engine
		Parabolic trough
		Photovoltaic modules
		Power tower
	Heat	Crop dryer
		Solar collectors for space heating and ventilation preheating
		Solar cooker
	Cooling	Refrigeration-cycle air conditioner
WIND	Electricity	Wind turbines in isolation, clusters, or wind farms, on land or offshore
	Mechanical power	Water pump

TABLE 16.4. U.S. Renewable Electric Generating Capacity, 1999 and 2000 Data.

Renewable Energy Resource	*Capacity (MW)*
Hydroelectric	80, 096*
Biomass	11,010*
Geothermal	2,669†
Wind	2,554‡
Solar	
Photovoltaic	49*
Solar thermal electric	325
TOTAL RENEWABLE GENERATING CAPACITY	96,703

†Year 2000 data from Lund, J. W., T. L. Boyd, A. Sifford, and R. G. Bloomquist. *Geothermal Energy Utilization in the United States–2000.* Although 2669 MW are installed, only 2020 MWe actually are in use. The difference is due to lack of sufficient steam at The Geysers, which is being replaced in part by water injection from a waste-water pipeline from Clear Lake and one under construction from Santa Rose, California.—J. W. Lund, personal communication, February 7, 2001.

*Year 1999 data from Energy Information Administration. U.S. Department of Energy. *Electric Power Annual 1999.*

‡Year 2002 data from "Wind Energy Growth Was Steady in 2000, Outlook for 2001 Is Bright." American Wind Energy Association, Washington, D.C.

TABLE 16.5. Theoretical Energy Output Possible from New and Existing U.S. Renewable Energy Sources in 2030 Assuming Installation of 20,000 MW of New Renewables per Annum

Energy Source	*Energy Service*	*Quantity*
Existing renewables (year-2000 estimate)	Electricity	ca. 100,000 MW
Biomass*	Electricity	100,000 MW
	Ethanol	More than sufficient to supply all cars and light trucks
	Methanol	More than sufficient to supply all cars and light trucks
Hydroelectricity (upgrades of existing dams)	Electricity	20,000 MW
Geothermal		
Hydrothermal	Electricity	25,000 MW
Hot dry rock	Electricity	5,000 MW
Solar thermal	Electricity	150,000 MW†
Photovoltaic	Electricity	100,000 MW‡
Wind	Electricity	200,000 MW§
TOTAL RENEWABLE ELECTRICAL GENERATION		700,000÷ MW
TOTAL LIQUID FUEL SUPPLY		Adequate for all cars and light trucks

Effects on Carbon Emissions

TOTAL NET CARBON EMISSIONS FROM VEHICULAR ENERGY SUPPLY	Near-zero
TOTAL CARBON EMISSION REDUCTION IN ELECTRICITY SECTOR RELATIVE TO 1998	›90%¶

Source: Office of Utility Technologies, Energy Efficiency and Renewable Energy and Electric Power Research Institute, December 1997: *Renewable Energy Technology Characterizations,* TR-109496 (Washington D.C.: U.S. Department of Energy, and Palo Alto, CA: EPRI).

*50 million acres of land in the United States, much of it now idle, could be used to grow biomass sufficient to produce 100,000 MW of electrical power per year—12% of U.S. 1998 generating capacity.
†This capacity is a somewhat arbitrary projection intermediate between PV and wind projections, reflecting the intermediate position of solar thermal electric costs relative to PV and wind.
‡This assumed that the 1999 installed base of 8 MW grows at 37% per year, 2.5 times the 1997 annual growth rate of U.S. solar cell shipments. U.S. production did increase by more than 40% annually in 2001 to 105 MW, and by 2001 global solar cell production was growing 37%, as calculated by P. D. Maycock, *Photovoltaic News,* V. 21 no. 2, February 2002.
§This assumes that installed capacity increases to 100,000 MW in 2020 (comparable to growth projections made for Europe and 20% above Energy Information Administration 2020 projections for the United States) and that from 2020 to 2030, growth equals 7%, which is less than a quarter of the current global growth rate of wind power capacity.
¶129,000 MW of year-2030 capacity is produced by natural gas. Additional capacity needed is provided by twentieth- and early-twenty-first-century fossil-fired units that remain in service.

TABLE 16.6. Energy Services and the Renewable Energy Technologies for Their Delivery

End-Use Energy Service	*Renewable Energy Technologies*
Cooling	Solar thermal-operated air conditioner
Electricity	Biomass power plant
	Fuel cell
	Geothermal power plant
	Hydroelectric power plant
	Integrated biomass gasifier and combined-cycle power plant
	Ocean wave, ocean thermal gradient, and marine current generators (not yet commercially available)
	Solar rooftop photovoltaic panels and power plant
	Solar thermal electric power plant
	Tidal power plant
	Wind turbine
Energy storage	Advanced batteries
	Compressed air
	Flywheel
	Pumped storage
	Superconducting magnetic energy storage
	Ultracapacitor
Heat	Air-source heat pump operating on renewably generated electricity
	Combustion of solid or gasified biomass
	Fuel cell operating on hydrogen or gasified biomass
	Geothermal energy (hot water or steam)
	Geothermal heat pump
	Hydrogen, direct combustion
	Electric resistance heater operating on renewably generated electricity
	Passive solar energy system
	Solar domestic hot water heater
	Solar space heater
Light	Building design features that capture solar light
Mobility	Conventional vehicle modified to consume alcohol from biomass or biodiesel
	Conventional vehicle modified to consume biogas
	Electric vehicle powered by renewably generated electricity
	Fuel cell powered by renewably generated hydrogen fuel or biomass-derived fuel*
	Hybrid vehicle using engine fueled by renewably derived liquid or gaseous fuel plus battery and electric motor

*Suitable for cars, buses, trucks, and possibly locomotives and ships.

particulates, no wastewater, and no solid waste. Moreover, they are largely independent of the volatile and at times soaring fuel prices that plague some owners of power plants generating electricity from natural gas.[5]

Wind Power Costs

How affordable is wind power? Can it provide a cost-effective alternative to fossil fuels? Currently, the average residential price of electricity in the United States is more than 8 cents/kwh. Peak-hour prices can be much higher—31 cents/kwh in early 2001 in the Pacific Gas and Electric Co. service area. By contrast, wind power can be generated for 4–6 cents/kwh, depending on the site quality (average wind speed), the project ownership structure (which affects the project's tax liabilities), the project's financial structure (proportion of financing provided by bondholders and stockholders, respectively), and terms of its financing.[6] When the current federal Renewable Energy Production Incentives or Production Tax Credits are taken into account, generation costs can easily fall from the 4- to 6-cent range to 3.5 cents/kwh or less.[7] Even allowing for normal profit and for transmission and distribution costs, wind power is an economically competitive energy source. Moreover, because it is fuel-free, and its cost is therefore highly predictable, users are insulated from potentially costly fuel price escalation (and the risk of future air quality–related environmental regulations).

By contrast, consider the recent price behavior of natural gas, the fuel for almost all new power plants under construction in the United States now and for the past few years. Natural gas, which averaged $1.94 per thousand cubic feet (mcf) at the wellhead as recently as 1998, reached the astonishing and unprecedented spot market price of more than $10/mcf for 4 days in December 2000 and, early in 2001, hit $13.22 on the spot market at Topock, California (on the border with Arizona). The average wellhead price in 2001 was $4.12/mcf, with California's natural gas prices more than twice national averages.[8]

The DOE in 2001 was estimating that the generation cost of combined-cycle gas generation would be about 4 cents/kwh in 2005, with gas selling at an average price of $4.20/mcf over the 20-year life of the plant.[9] Instead, if gas averaged a mere dollar higher (the year-2000 DOE estimate for 2001 wholesale prices), then the overall generation cost of natural gas power would jump to nearly 5 cents/kwh. For reference, wind power costs are expected to drop significantly by 2005 (perhaps another half cent per kilowatt-hour) and to continue declining through at least 2030 with improvements in wind power technology and large increases in wind turbine manufacturing volume. By 2005, wind power costs with federal incentives could be 3 cents/kwh or less in some cases. Given DOE's estimated natural gas power cost estimate of 4 cents/kwh in

2005, gas-fired power would then cost more than wind. However, natural gas experts expect the long-term cost of natural gas to fall again toward more normal levels once the 2000–2001 supply shortage is overcome through increased gas production and conservation. In the meantime, wind power, even at current wind power costs, is competitive with gas power. Wind power also offers greater cost predictability and invulnerability to future carbon caps or carbon taxes. Therefore, inasmuch as gas-fired power is cheaper than coal-fired, oil-fired, and nuclear power, wind is broadly competitive with power from *all* types of conventional fossil and nuclear fuels.

Wind Power Resources, Value, and Growth in Installed Capacity

Thanks to steeply falling wind power costs (from the early 1980s until today) and rapidly advancing technology, wind capacity is growing faster than any other energy technology in the world today: 22 percent a year during the 1990s and 40 percent for the past few years.[10] In the 1990s, wind power capacity tripled every 3 years. By the end of 1999, world wind capacity was 13,400 MW, and worldwide investment in wind power was roughly $11 billion. Further large expansions of wind capacity are planned over the next decade: Wind projects are under development in nearly 40 countries.

If appropriate policies are adopted, wind could produce 10 percent of the world's electricity by 2020 according to BTM Consult, an international wind energy consulting firm.[11] BTM calculated that 1.2 million MW of wind capacity could be installed in the next two decades. That would produce as much electricity as Europe now consumes—and more than all of Asia and Latin America consume combined—while creating 1.7 million new jobs and avoiding billions of tons of carbon dioxide emissions. Just with current trends and policies, Europe's wind capacity will rise to 40,000 MW by 2010 and to 100,000 MW by 2020, according to the European Wind Energy Association.

Prodigious untapped wind resources exist in central Asia, Europe, North America, and parts of Latin America. China's wind resources, for example, are sufficient to produce as much electricity as China consumes. The United States easily has sufficient wind resources to produce three times the nation's 1990 power consumption. Just the high-quality U.S. wind resources alone (Class 5–7) would be enough to site 3,500,000 MW of wind capacity, using only 1 percent of U.S. land, excluding sensitive and protected areas.[12]

As wind technology becomes more efficient and economical, lower grades of wind resources, which are far more ample, can be exploited. For example, Class 4 resources have several times the energy potential of the Class 5–7 resources.

The nation's huge wind resources represent an enormous energy bonanza. At the average systemwide residential electricity price of about 8.2 cents/kwh in the United States today, U.S. wind resources could produce gross revenues of more than $60 billion a year if 20 percent of U.S. electricity were wind generated.

Intermittency of Wind Power

Wind is created by differential solar heating of the Earth and its atmosphere. The resulting temperature differences involving the land, sea, and air masses and the interaction of air masses with the earth's irregular topography cause air pressure differences that, along with the earth's rotation on its axis, produce wind. Wind is intermittent yet quite predictable on a regional basis and at certain sites and could meet 20 percent or more of U.S. electricity needs without undermining the power grid's reliability. This percentage rises as other dispatchable generating technologies, such as dams and conventional power plants, are used to compensate for wind plant output fluctuations. Energy storage systems are also useful for mitigating the intermittence of renewable energy sources and are discussed later in this chapter.

With sufficient installed wind capacity, wind and other intermittent renewables could easily provide 30 percent of the capacity of an electrical grid without energy storage and a much greater proportion of the grid capacity if storage were provided. According to European researchers, "no absolute physical limit exists to the fraction of wind penetration on a large power system."[13] However, because wind power (and other intermittent renewables) have a lower capacity factor than fossil and nuclear power plants, wind capacity must be installed at twice or two and a half times the capacity of conventional generation to provide equal average annual energy output.[14] For example, for a wind power plant with a 30 percent capacity factor to equal the average generation of a conventional fossil-fired plant with a 60 percent capacity factor, twice as much wind power capacity would have to be installed. Whereas this would not alter the levelized per unit cost of the delivered wind power, it would require twice the initial capital investment for the wind generation.

Water Power

Before electricity and electric motors existed or were common, water power turned grindstones and often was harnessed by leather belts to sawmill blades. Since the 1930s, hydroelectricity has been produced in large dams, such as

Hoover Dam or Bonneville Dam. The contribution of hydroelectricity to our power supplies today belies the common notion that all renewables are new, experimental, and insignificant additions to the power grid.

Hydro facilities are the cheapest to operate of all conventional power plants, and they are dispatchable, meaning that they can be used to provide power on demand for any use, including peak or backup power. The disadvantages include high capital costs and the well-documented environmental and social costs of dam and reservoir building. Creation of a reservoir destroys preexisting ecosystems and may displace human settlements. Dams often block fish passage and decimate fish populations, especially of anadromous species. Dams can also destroy aquatic ecosystems through alterations in flow regimes, temperature, and turbidity.

The nation has about 80,000 MW of installed hydroelectric generating capacity—almost 10 percent of total U.S. generating capacity—and 19 percent of the world's electricity came from hydropower in 1997. Water power currently provides more than seven times the electrical capacity of biomass, the nation's next largest source of renewably generated electricity.

But hydro is a mature commercial technology, and most of the major large hydroelectric sites in the United States have already been developed. Those that remain typically are constrained from development for economic, regulatory, and environmental reasons. Little additional new large hydro construction is expected, and some large dams in Washington and Idaho are being considered for removal.[15]

Existing hydro capacity can be significantly expanded without new dam construction, however. Some 20,000 MW of undeveloped hydro capacity existed at U.S. dams in 1997. Upgrading dams by adding new turbines or rewinding old ones is inexpensive compared with constructing new hydro capacity and may also present opportunities for improving fish passage, downstream aquatic habitats, and water quality.

Solar Heat

When captured by suitable mechanical and electrical devices, the sun's energy can provide pollution-free hot water, space heating, and cooling services to homes, businesses, and industries. Not only can solar heat economically meet significant portions of the space-conditioning needs of buildings, typically 30–70 percent of residential heating or combined heating and hot water needs,[16] but it can also be concentrated to provide process heat for industry, water purification, detoxification of hazardous wastes, surface treatment of materials, and heat for cooking. It can also be used to preheat ventilation air, perform high-temperature water heating, and even furnish thermal power for air conditioning units.

Some 1.3 million U.S. buildings now use solar water heating. A quarter million commercial, industrial, and institutional buildings—including schools, military bases, offices, and prisons—use solar energy to heat or preheat space or water.[17]

Solar water heaters can economically provide 40–80 percent of an ordinary household's hot water demands, depending on the climate.[18]

Solar Cooling

Like solar heating equipment, active solar-powered cooling systems use solar energy gathered in various types of heat collectors, but instead of pumping heat into the house, they heat a pressurized refrigerant and cause it to evaporate to cool the surrounding air. Absorption evaporative cooling systems typically can provide 30–60 percent of a building's cooling requirements, and this technology is being actively pushed toward commercialization by the DOE.[19]

Solar Energy in Buildings

It has been well known for decades that when passive solar features, such as skylights or masses of heat-absorbing concrete, are coupled with sufficient insulation and advanced windows in a holistic building design, vast reductions in a building's energy need are possible. Energy experts at the Center for Buildings and Thermal Systems at the DOE's National Renewable Energy Laboratory state, "The nation [today] could reduce its energy use by 30–70 percent [in buildings] by simply incorporating advanced energy efficiency and renewable energy technologies into its buildings."[20] Unfortunately, because of historically low fossil fuel prices (for natural gas in particular), few homes today are outfitted with solar energy space heating or cooling systems.

Biomass Energy

Biomass literally means living matter, but the term is also used to refer to any organic material derived from plant or animal tissue. These resources are abundant and diverse, consisting of industrial and agricultural residues, including manure, trees, forestry wastes, municipal solid waste, and crops grown for energy production. Biomass is also an extremely versatile resource. It can be used to provide solid, gaseous, and liquid fuels, including ethanol and methanol fuels for powering vehicles. Thus it can be used for mobility or to produce heat or electricity in fuel cells and power plants.

Livestock manure and other organic waste can be decomposed (or "biodigested") by microorganisms in the absence of oxygen to form methane gas in a biogas plant. (Methane is the main ingredient in natural gas.) The biologically produced methane can then be burned for process heat or power generation. It can also be compressed and upgraded for distribution through natural gas pipelines or for conversion to electricity in a fuel cell. Capturing and using the methane is infinitely preferable to allowing the release of this powerful GHG during natural waste decomposition.

Global Importance of Biomass

Currently, biomass provides one-fourth of the world's total energy supply and is the primary energy source for three-quarters of the world's people, most of whom are in developing nations. This exceedingly important energy resource often is the only energy source in rural areas of the developing world or for its urban poor, and it is also a significant source in the developed world. Biomass will continue to have a major global energy supply role in the twenty-first century (and beyond), especially if given adequate R&D support and appropriate economic incentives for commercialization of advanced technology biomass conversion systems. Biomass gasification technology will make great improvements possible in the efficiency with which biomass is converted to energy in combined-cycle power plants, and gasified biomass can also be used in fuel cell power systems. Although biomass grown especially for energy production is not economically competitive with fossil fuels in the United States today, the IPCC found that "biomass has good prospects for competition with coal by 2020 in many circumstances, even if the price of biomass is somewhat higher than the price of coal."[21]

Energy Crops and Biofuels

Within two or three decades, biomass energy crops could be used to provide the United States with 100,000 MW of electrical generating capacity (12 percent of 1998 U.S. capacity). Raising the crops would take just 50 million acres. For comparison, the United States now plants 72 million acres in soybeans, and 128 million acres of cropland are projected to be idle in 2030. Today the United States has about 11,000 MW of biomass-fueled electric generating capacity. Almost all of the existing capacity is used to burn wood waste or agricultural residues.

In addition to solid and gaseous fuel for producing electricity, biomass can also be converted to carbon-neutral liquid fuels. Forestry wastes and biomass grown on uncultivated fields and idle farmland could produce 240 billion gallons of ethanol annually, which on an energy-equivalent basis would replace 160 billion gallons of gasoline. Because the United States uses only about 121 billion gallons of gasoline, biofuels could more than meet the fuel needs of cars and other light duty vehicles, according to the National Renewable Energy Laboratory. Moreover, fuel needs could be reduced to a third or a fourth of today's quantities by tougher fuel efficiency standards.

A study of the future price outlook for biomass ethanol under various production technology and world oil price assumptions found that by 2015, terminal prices for cellulosic ethanol might overlap with gasoline prices.[22] The study estimated that ethanol might range in price from $0.70 to $1.20 per gallon at the terminal (1998 dollars) while, under low and high world oil prices assumptions, gasoline at the terminal might range from $0.60 to $0.95 per gallon (1998 dollars). Under certain conditions of high world oil prices, rapid technological progress, and large volume production, ethanol could become cheaper than gasoline starting in 2010. The reference cases for ethanol and gasoline prices show a cost differential of only $0.20 per gallon in favor of gasoline from 2015 to 2020, which could easily be eliminated through tax exemptions and subsidies. Another forecast of bioethanol production costs using enzymatic processes found that in 2020 costs could range from $0.52 to $0.69 per gallon.[23]

Because it is more economical to produce ethanol from wastes than from grain crops, much research is devoted to producing ethanol from inexpensive cellulosic sources, including organic and municipal wastes. Hurdles to increased biomass energy production and use include competition for organic residues from other industries, such as wood products, and the low energy density of biomass relative to fossil fuels. Other significant drawbacks include particulate emissions and, for certain fuel cycles, the destruction of forests for fuel or the removal of excessive amounts of organic residues from fields or forests, to the detriment of soil fertility and integrity. Despite these risks, which are primarily a consequence of resource mismanagement, the use of organic wastes as biomass fuel has a promising future.

Hydrogen Fuel and Fuel Cells

Hydrogen is a clean, high-quality, convenient fuel that can be used to produce heat, electricity, or synthetic chemicals. Cars, buses, and trucks fueled with hydrogen not only have zero tailpipe emissions but can be almost pollution-free on a life-cycle basis.

Fuel cells are quiet, high-efficiency, modular devices. They can be flexibly fueled—by pure hydrogen, natural gas, or other hydrocarbon fuels—and they are already commercially available in large sizes for stationary power production. Only a few (mainly prototype) fuel cell vehicles are on the road today. (Daimler–Chrysler has developed one prototype, the NECAR 4, that reaches 90 mph and can go 280 miles on a single fueling. Ford has developed a prototype hydrogen-fueled sedan, the P2000HFC. Hydrogen fuel cells, which are costly now (an order of magnitude more expensive than automotive internal combustion engines), eventually will become competitive for use in homes for residential-scale, on-site power generation, space conditioning, and water heating.

Currently, the most economical way to produce hydrogen is by reacting natural gas with steam. However, that continues our dependence on fossil fuel. Hydrogen can also be produced at greater cost from ordinary water by electrolysis, the splitting of water into hydrogen and oxygen by electric current. (Both gases are collected separately at submerged electrodes.) To produce hydrogen economically by electrolysis with today's technology generally takes a very inexpensive source of power. Electrolysis can be powered with carbon-free renewable electric-generating technologies such as hydro, wind, solar thermal electric power, or solar cell (photovoltaic [PV]) power systems. Researchers are also trying to develop a photocatalytic process to produce hydrogen from water by sunlight.

Alternatively, hydrogen can be produced cleanly (without net carbon emissions) from gasified biomass or almost cleanly from gasified coal in plants equipped to separate and permanently store carbon dioxide. If hydrogen is produced from gasified biomass with carbon dioxide capture and storage, then the process on balance has a net negative impact on global atmospheric carbon.[24]

Capturing carbon in this manner is not yet commercial, but some researchers project that it would ultimately add little to overall gasification costs. Carbon capture would then allow coal to play a prominent role in the world's energy future during the twenty-first century, not as a raw fuel for direct combustion processes but as a source of clean-burning hydrogen-rich fuels. (Coal mining impacts would still remain, however, and carbon separation and storage might present as yet unstudied environmental risks.) As discussed elsewhere in this volume (see Chapter 17), geologic storage is unlikely to be permanent, and concentrated carbon dioxide is a toxic gas, so its escape can pose a safety risk.

Electric and Hybrid Vehicles

Like fuel cell vehicles, battery-powered electric vehicles (EVs) are extremely efficient, clean, quiet, smooth running, easy to maintain, and economical to operate. Tailpipe emissions are zero, and if the battery is charged with electricity from solar, wind, geothermal, hydro, or biomass resources, the entire fuel cycle will be emission-free (although emissions are likely in the vehicle manufacturing phase). Electric vehicles are also four to five times as efficient as gasoline vehicles when their efficiencies are compared by computing the efficiency of the gasoline vehicle from the nozzle of the fuel hose with that of the EV computed from the electric outlet at which it is charged. Thus, putting larger numbers of EVs on the road will catapult us toward a clean transportation system.

Unfortunately, early in 2001, the California Air Resources Board, under pressure from the auto industry, continued its policy of relaxing its zero-emission vehicle requirements for makers of automobiles sold in California, in effect reducing the number of electric vehicles that car manufacturers will be required to sell in the state. The commission's zero-emission mandate—first formulated in 1990—initially provided a strong impetus for the development of electric vehicles, but critics of the requirement successfully argued that advanced batteries could not be developed quickly and cheaply enough to meet consumers' needs and that because of high battery costs and limited vehicle range, consumers were not willing to buy electric vehicles in sufficient number to make their manufacture worthwhile. Supporters of electric vehicle technology maintained that demand would have been adequate had vehicles been produced in quantity and offered for sale at more reasonable prices.

The preceding objections about vehicle range and battery cost do not apply to hybrid electric vehicles. They have even longer ranges between refueling than conventional vehicles and, instead of a large, expensive battery pack, are equipped with fuel tanks and a small internal combustion engine plus an electric motor and battery. On-board regenerative braking systems are used to capture vehicle momentum during braking for conversion to electricity, which is stored in the vehicle battery for later use.

Hybrid electrics are cheaper to manufacture than pure EVs, and demand for them has been brisk. Because they are much more fuel efficient than conventional vehicles, they offer large potential fuel savings and the opportunity to avoid emissions. The Honda Insight, for example, gets more than 60 mpg in cities and more than 70 mpg on highways. Toyota began marketing its Prius hybrid in the United States in 2000, and demand has been brisk. Both hybrids and EVs are powerful weapons against smog, global warming, and acid rain. The pros and cons of electric, hybrid, and flywheel vehicles are discussed in depth elsewhere.[25]

Transforming Solar Heat into Electricity

Solar Troughs

Nine solar thermal electric trough plants were built in the 1980s by LUZ International Ltd. and its subsidiaries in the Mohave Desert of California. Solar thermal electric power plants concentrate sunlight on a heat receiver and then convey the heat to an engine that converts the heat to mechanical power and electricity. The three major types of solar thermal electric power plants are the parabolic trough concentrator, the dish and engine concentrator, and the solar power tower.

Whereas solar trough plants are already technologically proven and in commercial operation, dish and engine solar plants are on the verge of commercial use, and solar power towers (large central station facilities) are at least 5–10 years from limited commercial introduction.

The LUZ trough power plants are solar–fossil hybrids that not only turn the sun's heat into electricity but also burn natural gas for cloudy day or nighttime operation. With a combined electrical generating capacity of 354 MW, the plants send enough power over utility lines to meet the electrical needs of a small city. Three-quarters of the plants' yearly output is produced from sunlight; only a quarter is from natural gas. The output peaks in the afternoon, when customer demand in the Southern California Edison service area reaches a peak.

With a modest amount of low-cost heat storage, or by including biogas or even hydrogen fuel backup capability (instead of natural gas backup) for cloudy periods—solar thermal electric power plants can operate in a "load-following" mode 24-hours a day independently of short-term variations in sunlight.

This reliable utility-scale technology can produce almost unlimited amounts of 100 percent renewable electricity given the availability of vast areas of hot desert land throughout the southwestern United States and northwestern Mexico. Current life-cycle costs are 10–12 cents per kwh, twice the cost of coal power, but major cost reductions are possible through mass production, further R&D, and construction of larger plants.[26]

Solar Power Towers

Within a few years, another type of solar thermal electric power plant, the solar power tower, will enter the commercial power arena. This plant uses a field of specially designed solar-tracking mirrors called heliostats that concentrate sunlight from many angles simultaneously on a tower-mounted heat receptor. From there the energy goes to power an electric generator or to a coupled solar energy storage tank of molten salt for cloudy day and nighttime operation.

Power towers have higher solar concentration ratios and therefore higher temperatures and higher theoretical efficiencies than parabolic trough plants. Currently, they are also the only solar thermal technologies to offer significant thermal energy storage. In 1999, Solar Two, a 10-MW pilot plant operating at Barstow, California, for the first time successfully delivered grid-connected electricity around the clock for a week using a molten salt energy storage system. Sandia National Laboratory projects that power towers in Europe, Israel, and the United States should be able to produce power at under 10 cents/kwh in the near future. The DOE expects the costs of electricity from power towers to fall to 4 cents by 2030.

Dish Concentrator and Stirling Engines

The solar dish concentrator uses an external combustion heat engine mounted at the focal point of a dish-shaped mirror to heat a pressurized oscillating gas—usually hydrogen or helium—to more than 1,300°F. The contained gas expands and contracts to drive a piston. The engine may be a Stirling or a Brayton-cycle engine, the latter similar to a jet engine.

Solar dish and engine systems concentrate sunlight more than 2,000 times on the engine receiver, producing very high temperatures and efficiencies. At nearly 30 percent demonstrated efficiency, these systems are the most efficient of any solar electric power technology. They are also very reliable, quiet, easily installed, and dispatchable as hybrids, and they come in small sizes that can be scaled up or used in arrays. Solar dish and engine systems can be easily hybridized with natural gas or biogas for backup power production. Stirling engines when operationg on gas are expected to have low emissions because of steady operation at constant load, for which the engine is optimized. Plants like this could even be sited in suburban areas of the southwestern United States for grid support.

Costs of all the solar thermal electric technologies are above those of gas and coal but will fall substantially with mass production in the future (Table 16.7). Mass production and hybridization with gas-fired generating technology is likely to bring the costs of dish and engine systems down by an order of magnitude.

TABLE 16.7. Costs of Renewable Energy

		Levelized Cost of Energy (constant 1997 cents per kilowatt-hour)*				
Technology	*Configuration*	*1997*	*2000*	*2010*	*2020*	*2030*
Dispatchable technologies						
Biomass	Direct fired	8.7	7.5	7.0	5.8	5.8
	Gasification based	7.3	6.7	6.1	5.4	5.0
Geothermal	Hydrothermal flash	3.3	3.0	2.4	2.1	2.0
	Hydrothermal binary	3.9	3.6	2.9	2.7	2.5
	Hot dry rock	10.9	10.1	8.3	6.5	5.3
Solar thermal	Power tower	—	13.6	5.2	4.2	4.2
	Parabolic trough	17.3	11.8	7.6	7.2	6.8
	Dish engine hybrid	—	17.9†	6.1	5.5	5.2
Intermittent technologies						
Photovoltaics	Utility-scale flat plate thin film	51.7	29.0	8.1	6.2	5.0
	Concentrators	49.1	24.4	9.4	6.5	5.3
	Utility-owned residential (neighborhood)	37.0	29.7	17.0	10.2	6.2
Solar thermal	Dish engine (solar-only configuration)	134.3	26.8	7.2	6.4	5.9
Wind	Advanced horizontal axis turbines					
	Class 4 wind regime	6.4	4.3	3.1	2.9	2.8
	Class 6 wind regime	5.0	3.4	2.5	2.4	2.3

Source: Reprinted from Office of Utility Technologies, Energy Efficiency and Renewable Energy and Electric Power Research Institute, December 1997: *Renewable Energy Technology Characterizations,* TR-109496 (Washington, DC: U.S. Department of Energy, and Palo Alto, CA: EPRI), 7: 3.

*Costs are projected for a for-profit generating company (as opposed to a municipal power firm or other nonprofit.

†Cost is only for the solar portion of the year-2000 hybrid plant configuration.

Electricity from Solar Cells

PV devices turn light directly into electricity without fuel, fire, carbon dioxide, or pollution of any kind. They are silent, use no water, and have no moving parts to break down, and their main ingredient generally is recyclable silicon, the second most common element on Earth. Solar cells can contribute energy cleanly and indefinitely to a sustainable energy economy.

PV power is not only simple to use and produces few environmental impacts but is an extremely versatile technology. Solar modules and arrays provide commercial electricity for utility-scale power plants; they generate on-site power for grid-connected homes, offices, and schools; and they serve off-grid homes and remote power needs. They are cost-competitive for those remote uses against diesel generators and for myriad specialty applications, such as providing power for satellites, highway call boxes, traffic signs, street lights, signal buoys, and offshore oil drilling platforms. PV modules are also cost-competitive for peak period residential use. For example, whereas residential users in the Pacific Gas and Electric service area of California currently pay 11.7 cents/kwh, their peak power costs are 31 cents/kwh. Solar electric panels therefore are already competitive against peak rates in some areas of the United States (and in many foreign lands). In addition, solar electric panels also produce most of their power during peak hours, which are 12–6 P.M. in northern California, generating power when it is most valuable.

Solar electric panels with storage batteries can also be installed on rooftops or as an integral part of a building's roof, walls, or window glass. Used in this manner—providing power while "clothing" the building—PV panels help defray costs of structural materials such as roofing, windows, and wall cladding.

Grid-connected modules can provide for all or part of a building's electrical needs; the power grid serves as a backup power source and as a conduit through which to sell unneeded power. Solar panels with storage batteries can also be used as emergency power supplies to guard against systemwide power outages.

Ten million U.S. homes have ample (above-average) annual sunshine and suitable unshaded roofs to accommodate PV panels. These homes could produce 30,000 MW of solar electricity. Yet only about 100,000 American homes are equipped with PV panels.

To create a PV power plant, large numbers of modules are interconnected into arrays mounted on simple support structures, with or without tracking devices to keep the unit aimed at the sun for maximum power production. Because modules are factory built, solar plants can be constructed very rapidly compared with fossil-fuel plants. Once installed, solar power plants need minimal attention because the cells are long-lived and highly reliable.

Environmental Benefits of PV Systems

For every 4-kilowatt PV system installed on a residential rooftop in a prime sunny location, the PV system avoids the emission of 282,000 pounds of carbon dioxide, 1,500 pounds of sulfur oxides, and 900 pounds of nitrogen oxides during its useful life, compared with the average emissions produced nationally in power generation.

Present and Future Solar Cell Costs

Despite the fact that PV is the most expensive of the commercial renewable energy technologies when their wholesale bulk power generation costs at a power plant site are compared on a per kilowatt-hour basis, once PV is sited on a building, its cost must be compared to that of fully delivered retail power (from both other renewable and conventional sources), including all transmission, distribution, and utility overhead charges. On this basis, PV is competitive with the other generation sources. Moreover, PV module costs have fallen steeply from $1,000 per peak watt in the 1960s to only $3 per peak watt in 1999, all in current dollars (which understate the real cost reductions). Expected advances in solar technology and manufacturing are likely to reduce the production cost of PV to a tenth of today's levels and installed cost by two-thirds.

Thus, with continued R&D support, by the time most of today's power plants are due for replacement, it should be possible to generate power from PV systems for as little as 5 cents a kwh (1997 dollars).[27] Once the installed costs of solar electric systems are reduced to levels more competitive with coal, gas, and oil, PV growth will soar as thousands of megawatts of PV per year are sold, rather than the nearly 400 MW per year sold in 2001.[28] PV therefore is destined to become a very important energy source for the twenty-first century.

Geothermal Energy

Geothermal energy is produced mainly from the ancient heat in the Earth's core remaining from the planet's formation billions of years ago and, to a lesser extent, from radioactive decay heat (of elements such as uranium and thorium) within the Earth's crust. The friction of tectonic (crustal) plates sliding beneath each other at continental margins also supplies geothermal energy. In addition, the upper few feet of the Earth's crust are warmed by the sun's energy.

Because geothermal power plants burn no fuel, they make no significant contribution to global warming or acid rain. Binary-phase geothermal power plants emit no carbon dioxide, and in flash steam plants, carbon dioxide emissions resulting from the evaporation of dissolved carbon dioxide are less than a thousandth those of coal, oil, or natural gas plants.[29] Geothermal power therefore is a very reliable and almost carbon-free technology that produces continuous power on demand at reasonable costs (Table 16.7). However, constructing a geothermal power plant involves boring wells and building drilling pads, roads, and pipelines. Therefore, like any other potentially large industrial facility, a geothermal power installation is likely to be intrusive and incompatible with wilderness or other natural areas. Geothermal energy also is sustainable only if groundwater replaces hydrothermal fluids as quickly as they are withdrawn by a power plant. Electricity generation at some geothermal power plants has fallen because of overextraction, but smaller-scale use of geothermal energy for heat is invariably far below a reservoir's natural recharge capacity.

Geothermal energy can provide industrial process heat and direct space heating heat (and heat-driven chilling) for buildings as well as heat for electrical power generation. Some 95 percent of Iceland's buildings are geothermally warmed, as are buildings in at least 20 U.S. cities that use geothermal district heating systems. Their customers typically save 30–50 percent on their heating bills.

Geothermal energy is found as shallow permeable hydrothermal reservoirs of below-ground hot water and steam (or both) and as deeper hot dry impermeable rock, as well as geopressurized natural gas-laden brines and magma (hot molten rock). The exploitation of hydrothermal geothermal resources is a mature technology, but only a small portion of these resources are used today. Other geothermal resources will not be commercially viable until significant technical advances are made.

Known U.S. hydrothermal deposits could produce about 23,000 MW of electrical capacity; for comparison, the entire world had only about 8,000 MW of installed geothermal capacity in 1998. Inferred U.S. hydrothermal reserves could be used to generate 95,000–150,000 MW for 30 years, according to the U.S. Geological Survey. But this large resource is dwarfed by the size of the hot dry rock geothermal resource. The engineering feasibility of extracting energy from it has been demonstrated, but the rate and persistence of energy recovery from hot rock fracture zones must be demonstrated further before the future of hot rock geothermal power can be predicted. Additional deep drilling R&D is also needed if the technology is to become cost-competitive (Table 16.7).

The hot dry rock resource is believed to range from 3,000,000 quads (Q)[30] to 17,000,000 Q, which could supply all current U.S. electrical demand for up to half a million years. Ultimately, if further R&D is successful, geothermal energy could be among the nation's most valuable energy resources, offering a colossal energy payback in the form of pollution-free energy wherever advanced drilling technology can reach and profitably exploit the hot rock formations. Unfortunately, when federal energy research and development money was handed out in FY 1998, geothermal energy received only $29 million, or about 10 percent of the energy R&D budget, and the allocation for hot dry rock R&D was a small fraction of the U.S. geothermal budget.

Energy Efficiency as a Resource

Energy efficiency measures the amount of useful work derived from each unit of energy consumed. The greater the efficiency, the less energy needed for a given task. Energy efficiency is an energy resource because saved energy is available to do other useful work. Large efficiency improvements are possible on the supply side through improved energy conversion and transmission technologies and on the demand side through changes in residential, commercial, industrial, and transportation end-use technologies. Amory B. Lovins, founder and research director of the Rocky Mountain Institute, has calculated that the United States wastes $300 billion of the $505 billion spent annually on energy. He estimates that 75 percent of all electricity generated in the United States could be saved for less than the cost per kilowatt-hour of operating a coal power plant.

On the supply side, the efficiency of electric power generation can be doubled from the current global average of 30 percent. This can be accomplished by substituting combined-cycle gas power plants for less efficient conventional plants. Combined-cycle plants use the hot exhaust gases of natural gas combustion to operate a turbine and then use the waste heat to power a conventional steam cycle. However, because power plants typically have 30- to 40-year operating lifetimes, the replacement of existing power plants is likely to occur on a scale of decades.

Significant gains are possible on the demand side as well. Thirty-eight percent of U.S. energy is consumed in buildings and appliances, and industrial applications use another 36 percent and transportation 26 percent. A 1991 study[31] found that efficiency improvements could reduce residential energy demand by 40 percent at a cost less than the retail price of electricity. Larger commercial and industrial electricity customers, who can save the most money through energy efficiency improvements, have made significant recent progress in energy efficiency, but further opportunities still exist.

Motors, which consume 50 percent of total U.S. electricity and 60–70 percent of industrial electricity, can be retrofitted with adjustable-speed drives that reduce their electricity consumption by 20 percent.[32] Cogeneration—the simultaneous production of heat and electricity—and resource recycling can also contribute greatly to industrial energy efficiency.

Buildings present many opportunities to increase energy efficiency. Overall building energy use could be reduced 11 percent by 2010 and 22 percent by 2030 through innovations in space and water heating, space cooling, and lighting technology. Lighting, for example, consumes one quarter of the electricity in the United States.[33] At least 50 percent of this could be saved through technological changes, such as installing compact fluorescent lightbulbs in place of halogen or incandescent bulbs. With these improvements, carbon dioxide emissions in 2010 from fuels used directly in buildings would drop 27 percent from 1990 levels.

Future Energy Contributions from Renewable Resources

Under a business-as-usual energy forecast, renewable energy's share of total delivered energy worldwide is not projected to increase for at least the next two decades, according to the International Energy Agency's (IEA) 1998 *World Energy Outlook.* The EU also has projected that without a comprehensive renewable energy strategy and ambitious specific renewable energy installation targets, renewable energy will not make major additions to its share of total world energy supplies for the foreseeable future.[34] In the United States, renewable energy will not greatly increase its market share without a vigorous national effort over the next two decades, according to DOE projections.[35]

Only hydrothermal geothermal energy and wind are projected to be broadly competitive for bulk power generation under business-as-usual assumptions by 2030, according to both the DOE and the Electric Power Research Institute. Most other renewable electric technologies, they contend, will still cost two or more times as much as natural gas power. Some of that pessimism, however, is because conventional studies often (inappropriately) compare the fully delivered retail cost of power from technologies, such as PV, to the (cheaper) wholesale generation cost of central-station gas power. However, if the pessimistic forecasts prove true for certain renewable technologies, a panoply of renewable energy programs and incentives will be needed to insure those beneficial technologies are nonetheless adequately deployed.

Energy Policies to Promote Renewables

To accelerate the transition to a renewable energy economy, an energy strategy is needed that pulls renewables into the marketplace as it gradually reduces fossil fuel use and simultaneously restrains energy demand, all without curtailing the delivery of needed energy services.

Incentives that would bring renewables into the marketplace include expanded investments in renewable energy R&D; market stimulation through a minimum renewable energy requirement (known as a renewable portfolio standard [RPS]); energy production tax credits; investment tax credits; accelerated depreciation for certain energy-related investments; purchase aggregation, especially by government; commercial demonstration projects and programs; and education of the public and decision makers. These programs would be wholly or partially offset by a wire charge or carbon emission fees.

Fossil fuel use could be reduced by

- An outright prohibition on construction of new coal-burning facilities until carbon dioxide emission could be safely and economically eliminated, or vastly reduced, through carbon sequestration.
- More stringent vehicular fuel efficiency standards, which would dramatically reduce U.S. oil consumption.[36]
- Shrinking carbon emission caps for carbon-emitting power plants and other energy-using facilities, in conjunction with a system of tradable carbon emission permits (with penalties for noncompliance) to make fossil fuel energy more expensive, thereby constricting demand for fossil fuel power. As older fossil fueled capacity retired, spurred by accelerated tax depreciation allowances, new capacity replacements would then be primarily renewable.

A less politically feasible but simpler method than carbon caps for reducing fossil fuel use would be a direct per ton carbon emission fee.

To minimize the public financial support needed to induce the marketplace to provide the capital for a renewable generating system, public resources would be best spent on reducing the cost gap between renewables and conventional energy sources. When the prices of renewables fall relative to those of conventional fossil fuels, market demand for renewables will soar, prompting increased private investment in renewable generating capacity. Thus, public funds will go much further than if used to buy entire renewable generating systems outright.

Current U.S. Incentives for Renewables

Federal policies to encourage renewables currently include production tax credits, production incentives, demand aggregation, investment tax credits, accelerated depreciation, favorable financing (loans and bonding), and government-supported R&D, testing, and certification.

State policies include renewable portfolio standards; system benefit charges (SBCs); renewable electricity funds, net metering; income, corporate, sales, and property tax incentives; industry recruitment programs; disclosure of fuels and emissions; and grants and loans.

Renewable Portfolio Standards

Ten states have enacted various minimum renewable electricity purchase requirements known as RPSs. An RPS stipulates that a proportion of all new generation sold each year must be derived mainly from nonhydro renewables (an exception is sometimes made to include small hydro projects). Those who opt out of the system would have to pay a penalty or could purchase tradable renewable energy credits earned by generators of "excess" renewable energy. RPSs could add 5,000 MW or more of renewables to the U.S. generating system.[37]

The strengths of an RPS program are that it can provide predictable increments of least-cost renewable generating capacity over time and can be tailored to promote renewable resource diversity and achieve specific renewable capacity installation targets.[38] Costs of the RPS will be shared equitably throughout the power market with the help of tradable credits, and utility companies and independent power producers (IPPs) will have a strong incentive to facilitate the integration of the newly required renewable capacity into their power supply system: The better the integration, the easier it will be for the company to generate the renewable market share it must offer.

The weaknesses of an RPS program are that, unless carefully crafted, it will tend to promote only the least-cost renewable and at such a low profit margin that it will not stimulate the development of a domestic renewable industry infrastructure. As in Britain under the Non–Fossil Fuel Obligation (NFFO) bidding process, large multinational corporate players with extensive production experience in the least-cost technology may enter the domestic market, underbidding less experienced domestic bidders and precluding the development of experienced, low-cost, domestic IPPs.

The most successful RPS program in the United States to date appears to be a newly instituted program in Texas, where the state's electricity market is due to be deregulated in 2002. The Texas RPS places an obligation on electricity retailers to provide 0.5 percent of electricity from renewable sources this year, with the requirement increasing to 3 percent by 2009, at a cost to residential utility customers of less than 25 cents per month. The program provides an independent tradable renewable energy credit that gives the holder exclusive title to a quantity of renewable energy generated. The requirement already appears likely to be responsible for 1,000 MW of new renewable generation in Texas, mostly wind power but including some landfill gas and the renovation of some hydro facilities. Increasing utility interest in renewables and steep increases in natural gas prices are spurring on the program. Texas also uses an SBC. Its proceeds can be assigned or auctioned to support additional renewable generation.

Status of Federal RPS Legislation

Nationwide RPSs have been proposed for several years but failed to achieve congressional approval. The most ambitious national RPS plan, proposed by Senator James R. Jeffords (I-VT), would reduce CO_2 emissions, holding electricity sector carbon emissions in 2020 to year-2000 levels at a cost of only $18 per ton[39] and would exert much-needed downward pressure on natural gas prices by reducing demand for natural gas–fired electrical generation.

Jeffords' proposal is for gradual attainment of a 20 percent nonhydro RPS by 2020. (By contrast, the Clinton administration proposed a 5.5 percent RPS by 2010, phasing out the RPS by 2015.) Even under the Jeffords plan, the reduction in gas prices, a benefit enjoyed by all gas customers, would largely offset the increased electricity prices resulting from higher-cost renewable generation.

Jeffords' plan could be realized by increasing the RPS by as little as 1 percent per year. But to ensure that all types of renewables shared in the stimulus rather than merely the least costly technology type (wind), an SBC with proceeds earmarked for other renewable technologies competing in technology bands (similar types of generation vying against each other) would have to accompany the RPS.

Renewable Electricity Funds

Fourteen states have set up funds to support renewable energy, generally derived from small charges on utility bills, known as system benefit charges (SBCs). The funds are expected to provide more than $2 billion over the next 15 years through various mechanisms: production incentives in California, venture capital investments in Connecticut, and low-interest loans, grants to customers, subordinated debt, royalty financing, and equity financing. Importantly, SBCs can be used to support auctioned contracts (tender offers analogous to the British NFFO) in structured bidding rounds for potential suppliers of new renewable generating capacity.

Electricity Feed Laws

An electricity feed law sets a favorable purchase price for renewably generated power and establishes a utility system purchase obligation for that power. A meaningful feed-in tariff is unlikely to be easily accepted in the deregulating U.S. utility market because it exerts highly visible upward pressure on utility costs when utility systems are already under great pressure to reduce rates. In addition, the incidence of the costs is likely to fall unequally on utility companies according to how renewable resources in their service areas may be suitable for development and cost-effective. Utilities also will have an incentive to maintain barriers to the exploitation of these higher-cost resources rather than to facilitate their integration, as with renewables built under an RPS.

However, feed-in laws are stable, may be of long duration, and enable projects to deliver fixed, predictable revenue streams for 5 years or longer. Moreover, they are administratively simple mechanisms for supporting the development of a diversified renewable generating capacity (as occurred in California under the Public Utilities Regulatory Policy Act of 1978) and a domestic renewable industry infrastructure. To reflect the differing levels of commercial readiness among participating renewable technologies, different payment rates can be designated for different classes of renewables. Rates can be adjusted gradually over time in response to declining costs of target technologies and market responses to the feed-in program.

The weaknesses of feed-in laws are that they insulate renewable power producers from price competition and therefore fail to provide strong incentives to minimize costs. Another drawback is that the adoption of a feed-in law provides no guarantee as to how much new renewable capacity it will elicit. This makes the final cost of the feed-in law difficult to predict. The failure of a feed-in law to steadily reduce prices could be mitigated by lowering the feed-in tariff over time, but this understandably might arouse political opposition from feed-in law beneficiaries.

Net Metering

Net metering allows grid-connected utility customers who generate their own renewable power to send their surplus renewable generation back to their utility through their utility meter at a prearranged price. Thirty states have adopted net metering. However, most buyback rates in the United States are a fraction of retail rates and are far lower than EU buyback levels.

Disclosure of Fuels and Emissions

Fuels and environmental impacts of generation must be disclosed to purchasers by law in 15 states. This tends to benefit the more environmentally benign renewable generators.

Green Power

Consumers can voluntarily pay a premium for green power in 22 states. These programs have already resulted in 112 MW of new renewables; 107 MW more are planned.[40]

Specific Policy Proposals for Accelerating the Commercialization of Renewables

- Phase out all public subsidies to the fossil fuel and nuclear industries, freeing tens of billions of dollars of taxpayers' money annually. Redirect the savings to fund nonpolluting energy sources and to fulfill unmet social needs. A detailed study done by the Alliance to Save Energy and Douglas N. Koplow, *Federal Energy Subsidies: Energy, Environmental, and Fiscal Impacts,* found that in 1989 federal energy subsidies ranged from $21 to $36 billion a year, with almost 90 percent going to mature, conventional energy sources rather than to new emerging solar and wind technologies. An estimated $50 billion a year in military expenses in defense of U.S. international oil interests were not included.
- Give producers of electricity from nonpolluting energy sources generous energy production tax credits or incentive payments that reward them for every carbon-free or carbon-neutral kilowatt they produce. A federal production tax credit was created by the National Energy Policy Act of 1992 (EPAct) and was recently renewed, but it currently provides only 1.8 cents/kwh to wind and biomass energy producers for the first 10 years of plant operation. Public (nontaxable) power producers are eligible under EPAct for broader production incentives that also apply to solar and geothermal energy production, but only at the same low payment rates.
- Provide expanded investment tax credits to stimulate renewable energy investment for all renewable energy technologies whose performance has been certified by the National Renewable Energy Laboratory or other qualified body. EPAct provides only a 10 percent investment tax credit for equipment that uses solar energy to generate electricity or for solar heating and cooling, subject to various limitations and exclusions. The credit could easily be raised to 25 percent or more.
- Wherever consumers are able to choose their power providers because of retail deregulation, give consumers tax credits for choosing green power. Rewarding consumers financially for choosing environmentally desirable energy sources—rather than penalizing them by higher prices, as occurs today—will spur the market for green power.
- Establish a national wire charge on each kilowatt-hour of nonrenewably generated electricity and a national tariff on the transmission and transport of all nonrenewable fuel. A $0.01/kwh charge nationwide would yield a $30 billion windfall for investment in renewable energy; a $0.033 charge would provide $100 billion, neglecting the effects of price elasticity of demand. SBCs, a form of wire charge, have been adopted in several states. But the charges usually are

set at ridiculously low levels—often in the low tenths of a cent per kilowatt-hour or less—and some are scheduled to be abolished within 3–5 years.

- Establish an exacting national minimum renewable energy requirement or RPS. Anyone who sold electricity in the United States would have to provide a steadily increasing proportion of their power from nonpolluting energy sources or purchase surplus renewable energy credits earned by others. Starting with only a few percent of the total power offered, the standard would gradually rise until all major new additions to generating capacity would be from renewable sources. As existing generating capacity wore out—typically on a 20- to 40-year cycle—and was replaced by renewable capacity, the generation mix would be seamlessly transformed from conventional to renewable generation.
- Phase out fossil fuel plants that cannot meet new emission standards under the Clean Air Act. This simple step would increase demand for new generation that renewables could fill.
- Ban construction of new coal-fired plants, as Denmark has done. As a short-term expedient for reducing carbon emissions, convert existing coal-fired plants to natural gas wherever practicable. Substitution of natural gas for coal in power plants of comparable efficiency should reduce carbon emissions from these sources by roughly half. Fuel substitution and use of state-of-the-art combined-cycle natural gas plants will cut carbon emissions from the plant by two-thirds.
- Implement a tax on carbon emissions as Denmark, Finland, Netherlands, Norway, and Sweden have done. The carbon tax could be made more politically palatable than the Clinton administration's failed 1993 Btu tax if its proceeds were used to reduce payroll taxes. It could thus be revenue neutral but would shift taxes from labor to pollution.
- Establish a national renewable energy bank or trust fund to make long-term, low-interest revolving loans for renewable energy projects. Often the cost and availability of funding are critical to the financial viability of a renewable energy project, especially for capital-intensive technologies such as solar electric, geothermal, and wind, where most of the life-cycle costs are the upfront costs of the capital equipment; fuel is free and operation and maintenance costs are low.
- Expand and intensify renewable energy and energy efficiency research and development. In FY 2000, the U.S. Congress appropriated only $321 million for renewable energy—less than $1.19 per person.
- Shift federal government energy purchases from nonrenewable to renewable energy sources. The federal government purchases some $8 billion worth of energy per year. Why should not the federal government, which sets national

environmental standards for all, buy its energy from sources that meet the highest environmental standards? Why should our government greatly contribute to the air and water pollution it is trying to control by buying power from coal and natural gas or nuclear plants when it could insist on solar, geothermal, and wind electricity?

- Hold stranded costs hostage to stranded benefits. As portions of the utility industry are deregulated and restructured, the recovery by power companies of "stranded" costs (investments in large and costly nuclear or coal power plants that are no longer economically competitive) should not be permitted until the corresponding "stranded benefits"—utilities' waning support for renewable energy and energy efficiency research and development, low-income energy assistance (e.g., weatherization support), and purchases of renewably generated kilowatts—are increased or reinstated and guaranteed for the long term.
- Establish national net metering regulations that make it easy for households and other small generators of renewable energy to receive ample credits for selling back excess renewable power they generate to their local energy provider, paying only for the net power consumed in any billing cycle. Currently, 30 states have net metering laws, and several proposals for federal net metering legislation are being considered in Congress. The buyback rate should be at least 90 percent of the retail electricity price, as in Germany.
- Establish fair transmission and distribution rules for renewable energy. Take or pay transmission rules, known as capacity-based pricing, can be prejudicial to intermittent renewables. They force them to pay transmission charges whether they use transmission facilities at a particular time or not.
- Disclose energy sources and their emissions to utility customers with their utility bills so they can make informed decisions about their energy supplies. Many customers who are not aware of the environmental impacts of energy supply technologies would then switch to clean technologies.
- Require that all green power providers be properly certified to minimize the sale of nonrenewable energy under a phony green banner. The nonprofit Center for Resource Solutions (www.resource-solutions.org) has an independent accreditation program for green power plants.
- Modernize and upgrade turbines and generators of existing hydroelectric facilities while improving fish passage. This should make another 20,000 MW of renewable electrical power available to the economy at minimal environmental cost, except where removal of the dam itself is a better alternative.
- Use international carbon emission reduction trading mechanisms to cost-effectively reduce emissions. Inasmuch as investments in GHG emission con-

trol often are much cheaper in developing nations, international emission trading opportunities should be fully used to take advantage of as much early emission avoidance as possible. The world shares an atmosphere, and it does not matter to the Earth's heat balance whether the emissions are lowered in Bangkok, Kiev, or Shanghai. But nations should not use trading to avoid reducing their domestic emissions and enjoying the local benefits, such as cleaner air.

- Speed the transfer of energy efficiency and renewable energy technologies to developing nations. Just as we ought not export unsafe pesticides and pharmaceuticals banned for use in the United States, we should not export polluting fossil fuel technologies under the guise of helping other nations.

Some of the policies described already exist but need strengthening. Many states also offer renewable energy incentives (see www.solar.mck.ncsu.edu/dsire.htm for a comprehensive database). A few utilities also provide financial incentives, including grants, rebates, and equipment leases, to encourage customer use of renewable energy. But our current patchwork of renewable energy policies must be far better coordinated and strengthened.

Conclusions

Some representatives of the fossil fuel industry have proclaimed that curtailing fossil fuel combustion, even slightly, to reduce global warming would place an unbearable burden on the economy. In a 1997 study, however, five DOE national laboratories found that the United States could cut its carbon emissions to 1990 levels by 2010 with no net cost to the economy or even with a possible economic gain. This study was based on fuel cost savings and did not consider the additional economic benefits of reducing the negative externalities of fossil fuel combustion, including human health costs and air and water pollution. The study's results were confirmed in November 2000 by the Interlaboratory Working Group on Energy-Efficient and Clean Energy Technologies in their *Scenarios for a Clean Energy Future.*[41]

During the next 30 years, renewables will become even more affordable and economically competitive with fossil fuels. As initial capital and R&D costs are amortized, renewables will continue to provide electricity at low marginal cost. The cheapest reserves of fossil fuels have already been exploited, however, and these fuels will become increasingly expensive to use. This shift can be hastened by wise policy incentives and an increase in research and development funding for renewable energy so that renewable sources can meet the world's energy and environmental needs in the twenty-first century.

Overloading the atmosphere with carbon is not an inexorable process that humans are destined to continue. It is caused by conscious energy choices. Until now, those choices have been made according to narrow economic criteria : which fuels are most profitable for sellers and cheapest for buyers? But we need not project this short-sighted and irresponsible practice into the future. If we develop a national consensus that these criteria for selecting energy sources must be changed, we can make better energy choices. We now have the potent policy tools and superior technologies needed to eliminate excess carbon emissions and build a clean energy system.

Notes

1. Intergovernmental Panel on Climate Change (IPCC), 1996: *Climate Change 1995: Impacts, Adaptations and Mitigation of Climate Change: Scientific-Technical Analyses,* contribution of Working Group II to the Second Assessment Report of the IPCC (Cambridge, England: Cambridge University Press).
2. Interlaboratory Working Group on Energy Technologies, Office of Energy Efficiency and Renewable Energy, U.S. DOE, 2000: *Scenarios for a Clean Energy Future* (Oak Ridge, TN: Oak Ridge National Laboratory and Berkeley, CA: Lawrence Berkeley National Laboratory). ORNL/CON-476 and LBNL-44029, November.
3. Under varying economic, policy, and technology assumptions.
4. Nuclear fuel is manufactured primarily from high-grade uranium ores, which are limited in quantity and are therefore nonrenewable. To produce nuclear power indefinitely, uranium would eventually have to be replaced with reprocessed plutonium from highly radioactive spent reactor fuel. The plutonium economy would create serious safety, proliferation, and economic problems.

   In addition, the nuclear fuel cycle involves uranium mining, milling, fuel enrichment, and fuel fabrication, all of which use energy from fossil fuel (and generate radioactive waste). Contemporary nuclear fuel containing uranium-235 is manufactured with energy produced from burning coal.

   Nuclear power also presents many safety and environmental hazards ranging from routine radioactive emissions from nuclear power plants to catastrophic nuclear accident risks caused by power and coolant system malfunctions, earthquakes or terrorism, nuclear material transport, nuclear waste repository accidents, spent nuclear fuel pool ruptures and fuel melting; destruction of aquatic life in once-through cooling systems; and thermal pollution. Nuclear power plants are also costly to build compared with conventional power plants and have a history of enormous construction cost overruns. For all these reasons, almost no new nuclear power plants have been planned and constructed in the United States since the late 1970s. In addition, many nuclear plants today are so expensive to operate that they will be unable to compete in the nation's rapidly deregulating electric utility market and will have to be shut down or subsidized, as occurred when the California electric utility industry was restructured.
5. Naturally, plant owners with long-term natural gas supply contracts are insulated for a time from the sudden surge in prices.

6. The cost and availability of project financing is a critical determinant of overall project costs for capital-intensive renewable energy technologies.
7. The federal Renewable Energy Production Incentives and the federal Production Tax Credits are currently set at 1.8 cents/kwh.
8. Energy Information Administration (EIA) DOE: *Natural Gas Monthly,* December 2000; and EIA, DOE: *Short Term Energy Outlook,* January 2001. The spot market price of natural gas at the southern California town of Topock on the day this note was written was $13.22 per MBtu, which would imply an electricity cost of about 10 cents/kwh.
9. S. Sitzer, director, Coal and Electric Power Division, Energy Information Administration, U.S. DOE, personal communication, February 1 and 6, 2001.
10. Forty percent annually from 1993 to 1999, according to Tom Gray of the American Wind Energy Association.
11. Denmark already gets 10 percent of its electricity from the wind.
12. Class 5 wind speeds are 15.7–16.6 mph at 98.4 feet above the ground.
13. Office of Utility Technologies, Energy Efficiency and Renewable Energy and Electric Power Research Institute, 1997: "Renewable energy technology characterizations," TR-109496 (Washington, DC: U.S. DOE, and Palo Alto, CA: EPRI), December.
14. Capacity factor is the proportion of a power plant's maximum possible annual output that it actually delivers. For example, if a power plant with a 1,000-MW nameplate capacity operated 90 percent of the time night and day for a year at its maximum rated capacity of 1,000 MW, it would have a 90 percent capacity factor. Were the plant to shut down for a longer portion of the year or operate at less than maximum rated output, its capacity factor would be proportionately reduced.
15. Brower, M., 1992: *Cool Energy: Renewable Solutions to Environmental Problems* (Cambridge, MA: MIT Press).
16. U.S. DOE, Energy Efficiency and Renewable Energy Network, 1999: "Solar buildings program," www.eren.doe.gov/solarbuildings/space.html.
17. Solar Energy Industries Associations, n.d.: *Directory of the U.S. Solar Thermal Industry* (Washington, DC).
18. U.S. DOE, Energy Efficiency and Renewable Energy Network, 1999.
19. Ibid.
20. See "Buildings research," www.nrel.gov/lab/pao/building_energy.html, 1999.
21. Intergovernmental Panel on Climate Change, contribution of Working Group II, 1996: "Energy supply mitigation options," Chapter 19 in *Climate Change 1995: Impacts, Adaptations and Mitigation of Climate Change: Scientific-Technical Analyses* (Cambridge, MA: Cambridge University Press), p. 607.
22. DiPardo, J., 2000: "Outlook for biomass ethanol production and demand," (Washington, DC: Energy Information Administration, U.S. DOE), www.eia.doe.gov/oiaf/analysispaper/biomass-html.
23. Ruth, M. and K. Ibsen, 2000: "Bioethanol production using enzymatic processes: 20-year economic outlook," April 12, 2000, www.afdc.doe.gov/pdfs/4561.pdf.
24. The plants from which the hydrogen was extracted are replanted and thus are continuously removing carbon from the atmosphere; after each harvest, the carbon is

sequestered in permanent storage before the fuel is used, so the whole system acts like a pump extracting carbon from the atmosphere.

25. Berger, J. J., 1998: *Charging Ahead: The Business of Renewable Energy and What It Means for America* (Berkeley: University of California Press).
26. See Ibid. and Office of Utility Technologies, Energy Efficiency and Renewable Energy and Electric Power Research Institute, 1997: "Renewable energy technology characterizations," TR-109496, December, for more details.
27. Office of Utility Technologies, Energy Efficiency and Renewable Energy and Electric Power Research Institute, 1997: "Renewable energy technology characterizations," TR-109496, December.
28. Estimate by PV expert Maycock, P. D., 2000: "World cell/module production grows 38% to 277.90 MW," *Photovoltaic News,* January.
29. Berger, 1998, pp. 233–234.
30. A quad is 10^{15} Btu.
31. Koomey, J. The Potential for Electricity Efficiency Improvements in the United States Residential Sector. Lawrence Berkeley Laboratory Report LBL (30477): 48. As cited in *Climate Change, 1995: Impacts, Adaptations, and Mitigation of Climate Change, Scientific-Technical Analysis,* 1996.
32. Woodward, J., C. C. Place, and K. Arbeit, 2000: "Energy Resorces and the Environment," in W. G. Ernst (ed.), *Earth Systems: Processes and Issues* (Cambridge: Cambridge University Press), p. 398.
33. Ibid.
34. European Commission, 1997: "Energy for the future: renewable sources of energy, white paper for a community strategy and action plan," December.
35. Energy Information Administration, 1999: *International Energy Outlook 1999* (Washington, DC: Energy Information Administration, U.S. DOE).
36. In a national energy plan publicized in February 2001, the Natural Resources Defense Council, an environmental group, calculated that increasing fuel economy standards to 39 miles per gallon over the next 10 years for new cars, sport utility vehicles, vans, and light trucks would save the U.S. 50 billion barrels of oil by 2050.
37. Union of Concerned Scientists, 2000: "Clean power surge: ranking the states," http://www.ucsusa.org/energy/surge.html.
38. The discussion of renewable portfolio standards and electricity feed laws relies on Wiser, R., K. Porter, and S. Clemmer, 2000: "Emerging Markets for Renewable Energy: The Role of State Policies During Restructuring," *Electricity Journal,* January–February 2000; Rader, N. A. and R. H. Wiser, 1999: *Strategies for Supporting Wind Energy: A Review and Analysis of State Policy Options* (Washington, DC: National Wind Coordinating Committee).
39. Clemmer, Steven 1999: "A Powerful Opportunity: Making Renewable Electricity the Standard," January (Cambridge, MA: Union of Concerned Scientists).
40. Union of Concerned Scientists, 2000.
41. Interlaboratory Working Group on Energy Technologies, Office of Energy Efficiency and Renewable Energy, U.S. DOE, 2000.

CHAPTER 17

Fuel Cells, Carbon Sequestration, Infrastructure, and the Transition to a Hydrogen Economy

Michael B. Cummings

The most abundant element in the world may also be the best fuel for reducing the anthropogenic causes of global climate change. Hydrogen, the lightest element in the universe, has unique properties that make its use as a fuel attractive in the context of greenhouse gas (GHG) emission reductions. Although using hydrogen in combustion engines can produce energy with higher efficiencies and lower GHG emissions than natural gas (NG) or gasoline,[1] the idea of using hydrogen as the primary energy carrier in a hydrogen economy has long been contemplated for intermittent[2] renewable power sources and off-peak nuclear power.[3] The current potential for a significant shift to hydrogen has resulted in large part from advances in the enabling technology of fuel cells (FCs).[4]

Although environmentalists dream of a world where hydrogen is produced electrolytically[5] from renewable sources such as wind or solar photovoltaic (PV) electricity, and this hydrogen is then used to run FCs for electricity and heat, the cost of renewable resources currently hinders such renewable hydrogen production. A more likely scenario in the near term is the operation of FCs with hydrogen that was originally embedded in a fossil fuel. Methods exist for separating hydrogen from fossil fuels, using the hydrogen for energy consumption and then sequestering the separated carbon dioxide (CO_2) before it enters the atmosphere.[6] Such processes would prove useful only if a larger hydrogen infrastructure existed to aid in the production, transmission, and distribution of hydrogen.

The Fuel Cell

To be the impetus, or enabling technology, for a transition to a hydrogen economy, the advantages of FCs must be numerous enough to warrant attention, and fortunately they are. FCs are efficient and quiet and emit 170°F water as a pollutant.[7] They work by combining oxygen from the air and hydrogen provided in fuel form to make water. In this process, the FC uses the electron flow from the molecular bonding process, thus harnessing electricity.[8] Although FCs could be used for a wide range of applications including stationary power production and cogeneration (combined heat and power systems), and as batteries (e.g. cellular telephone power sources), current attention has surrounded their use in automotive applications.

Most of the focus for commercial FCs is on the proton exchange membrane (PEM) fuel cell.[9] PEMs, like most other FCs, convert energy at a higher efficiency than conventional conversion technologies, reduce or eliminate point source pollution, allow a diversification in fuel sources (provided that hydrogen can be produced from the primary energy source), and enable reductions in GHG emissions without substantial increases in the cost of the end product.[10] Other FC types under development include molten carbonate and solid oxide designs. These FCs probably will not be used in small-scale applications such as transportation because of their high operating temperatures (550–1,000°C), but they are well suited for cogeneration applications.[11]

Thus, by their nature, FCs can enable significant GHG emission reductions. Apart from electric power use, FCs could substantially reduce GHG emissions through transportation and heating applications, which account for roughly two-thirds of primary energy use.[12] Their scalability, or ability to be produced and operated at small sizes,[13] and flexibility enable their widespread use in an evolving economy, which demands mobility in fuels and conversion devices.[14]

The environmental benefits of hydrogen-fueled FCs are clear. They have even been acknowledged by vested interests in the existing petroleum and internal combustion engine paradigm. One of the world's largest oil companies, Royal Dutch Shell, created a hydrogen division in 1998, and has suggested through its scenario planning that 50 percent of the world energy supply could come from renewables by 2050.[15] Despite spending a lot of money in the past few decades on electric cars and recently on hybrid gasoline–electric vehicles, automotive companies have seen the clear advantages of FCs and have invested both money and institutional support in them.[16] By 1999, eight major automakers had announced plans to release commercialized FC vehicles around 2004 or 2005.[17]

Thus, the trend toward FC use is apparent in the high GHG-emitting transportation sector. However, many uncertainties still remain that separate the idealized hydrogen-using transportation sector from currently realistic options.

Why Pure Hydrogen?

Although hydrogen is needed to operate an FC, the source of this hydrogen does not necessarily have to be diatomic or liquid hydrogen. Many substances and fuels contain hydrogen that can be stripped off and used in an FC. These include gasoline, NG, and methanol.

Some argue that it is better to use some of these fuels as the hydrogen source because of the existing infrastructure. Most of the automotive industry favors using onboard reformers because the infrastructure for gasoline and methanol already exists.[18] One estimate by an official at automotive FC leader Daimler–Chrysler places the costs for retrofitting 30 percent of service stations in New York, Massachusetts, and California at $1.4 billion for hydrogen and $400 million for methanol.[19]

However, in such interim systems fuel reformers on the automobile would be needed to convert the hydrocarbons into hydrogen, and GHG (and other pollutant) emission reductions would be lower than those obtained from a pure hydrogen FC.[20] Reformed methanol FCs have small emission benefits,[21] and reformed gasoline FCs may offer little or no emission advantage over internal combustion engines.[22] Additionally, methanol reformers need a warmup period and can experience maintenance problems,[23] and methanol's toxicity is a potential concern.[24] The "delivered cost" of hydrogen to vehicles may be up to 50–100 percent more expensive than gasoline. However, because pure hydrogen FCs can be roughly 50 percent more efficient than FCs with reformers for gasoline, NG, or methanol, the total cost differential of the delivered energy service may not be significant, if it exists at all.[25] The increased efficiency of pure hydrogen FC vehicles results from lighter cars, increased FC performance from pure hydrogen versus reformed fuel, and the absence of a conversion energy penalty for fuel reforming.[26]

Iceland, the country most vigorously pursuing a transition to hydrogen, is considering whether to transition to an interim methanol system or to transition straight to hydrogen.[27] Participants in and observers of Iceland's transition worry about the potential for an inferior technology, methanol FCs, to "lock out" the superior technology of pure hydrogen FCs and thus delay the adoption of pure hydrogen FCs by decades.[28] It is hard to switch away from the current gasoline–internal combustion system today, and making two transitions in a short time period is highly unlikely.

Thus, although the interim methanol approach has received more attention and resources than a jump to pure hydrogen in both Iceland and the rest of the world, it is probably inefficient in the long term.[29] Because of the simplicity associated with their design, increased energy efficiency, lower production costs, quick refueling capacity, and fuel (hydrogen) production flexibility,[30] FC vehicles that store and use compressed hydrogen are preferable. Pursuing pure hydrogen FCs also avoids the multiple transitions inherent in an interim transition to methanol, which could prevent the waste of political and economic resources in the long run.

If pure hydrogen is chosen as the fuel of choice, the issues of hydrogen production and distribution still remain. When hydrogen is produced by truly renewable sources (i.e., electrolysis by wind or PV), it is too expensive under current market conditions.[31] In addition, hydrogen has a low energy density, making large-scale fuel storage difficult. Thus, for hydrogen to be commercially viable it must be manufactured cheaply, and a storage and distribution infrastructure must exist despite its low energy density. Fossil fuels can provide a transitional source of hydrogen, and it may be possible to sequester the CO_2 associated with reforming the hydrocarbon into hydrogen. Developing a hydrogen infrastructure using fossil fuels as the primary source of hydrogen could speed the transition to an eventual renewable hydrogen energy system.

Hydrogen Production and Carbon Dioxide Sequestration

In the context of this chapter, carbon dioxide sequestration, or carbon management (CM), involves capturing carbon in underground saline aquifers, depleted oil and gas fields, or coal beds or under the ocean floor.[32] This carbon dioxide may or may not have been generated from the reforming of hydrocarbons into usable hydrogen, but such systems probably would render the most net benefits at present. Although not ideal, CM may prove to be a necessary part of a diversified strategy for abating CO_2.[33]

Storage capacity for captured CO_2 appears to be sufficient within the context of Kyoto-scale emission reductions. Estimates for worldwide capacity are 150–500 gigatons of carbon (Gt C) in underground depleted oil and gas fields,[34] 100–300 Gt C in coal fields,[35] and 100–1,000 Gt C in deep saline aquifers.[36] Regardless of the estimates used, they are substantially larger than annual anthropogenic emissions of 6–8 Gt C.[37] Although there is apparent capacity, the long-term ability of these geologic structures to hold the carbon dioxide is uncertain. Current estimates predict that roughly 20 percent of the carbon will return to the atmosphere within 300 years,[38] so CM is unlikely to be permanent. Nonetheless CM could prevent some CO_2 from entering the atmosphere and provide a grace period while a more complete transition to renewable energy sources is made.

Carbon dioxide sequestration is justifiable only when it is cheaper than other emission abatement strategies, such as fuel switching, energy efficiency, or renewable energy. If the United States ratified the Kyoto Protocol and intended to meet its requirement, it would need to reduce its GHG emissions between 2008 and 2012 to 7 percent below its 1990 emission level. Part or all of this may be achievable through options that are cheaper than carbon sequestration. Emission reductions of 10–20 percent below 2010 baseline emissions probably would translate to marginal costs of abatement under $50/Gt C and probably could be accomplished through efficiency and fuel switching to NG.[39] This is cheaper than current postcombustion sequestration techniques available in electricity generation applications, which range from $70–$140/t C.[40]

However, reductions of 10–20 percent from 2010 baseline emission reductions may not be sufficient to meet Kyoto requirements.[41] Even if they are, Kyoto reductions are far from emission stabilization and definitely far from climate stabilization. CM by itself becomes attractive when reductions of more than 10 percent below 1990 emission levels are desired. Under these circumstances, when the marginal cost of alternative abatement strategies based on large-scale use of solar or nuclear power may be well over $100/Gt C, carbon sequestration by the mechanisms just mentioned could be a cheaper option than fuel switching to emission-free power sources.[42]

In light of the costs of carbon sequestration, it is not surprising that the first underground sequestration of CO_2 for the sole purpose of emission abatement occurred in a country with a high carbon tax. Norway's $170/t C (equivalent to 50 cents per gallon) tax induced the Statoil Company to capture CO_2 from its offshore Sleipner Vest gas field and inject it into an aquifer under the North Sea.[43] The project sequesters and injects roughly 300 kt C/yr. Indonesia is planning a similar project in the South China Sea that will sequester roughly 30 Mt C/yr, or roughly 0.5 percent of present global emissions from fossil fuel burning.[44]

Thus, under a certain level of carbon taxation, carbon sequestration could prove to be economical. However, CM appears to be most attractive when coupled with secondary uses. These include hydrocarbon reformation and hydrogen production, enhanced oil or NG recovery, and methanol production.

CO_2 is already used for enhanced oil recovery (EOR) throughout the world. In 1998 alone, 43 million tons of carbon dioxide were injected into the ground at more than 65 EOR projects.[45] It can also be used for enhanced NG recovery, and a new technology is being developed for use of CO_2 in recovering methane from coal beds.[46] Using existing technology but not necessarily existing infrastructure, NG can be extracted from an NG field or coal bed, reformed, and separated into hydrogen and CO_2 steam, and the CO_2 steam can be reinjected into the well or coal bed to enhance the recovery of remaining NG.[47]

When these methods are combined with the use of hydrogen for fuel, a triple dividend can be achieved.[48] The firm engaged in this activity can profit by selling the hydrogen, benefit from the enhanced recovery of the NG, and possibly gain a carbon sequestration credit under a trading scheme initiated by the Kyoto Protocol or similar agreement.[49] Besides EOR-type applications, captured CO_2 has other secondary uses, such as freeze-drying and carbonation.[50] However, some commercial uses such as carbonation do not offer the potential long-term sequestration benefits of geologic storage because the CO_2 will enter the atmosphere after consumption. Nonetheless, using fossil fuels to make hydrogen with carbon sequestration offers potential secondary benefits beyond GHG emission reductions, which increase its attractiveness in transitioning to a hydrogen economy.

CM is not without disadvantages or risks. We have not stored CO_2 in geologic structures often enough or for a long enough time to gain a complete understanding of its stability or its dangers. Possible risks include slow or rapid release of CO_2 from fields or aquifers and its subsequent effects on terrestrial or ocean acidity, organisms, and atmospheric conditions.[51] Because air that is 25 percent CO_2 is lethal,[52] leaks from sequestration spots could be catastrophic. However, experience gained from work in the oil and gas industries has shown us how to minimize the risk of such an event occurring.[53] From an emissions standpoint, an additional risk of carbon sequestration is long-term leakage. Because sequestration of CO_2 from fossil sources uses more energy than reducing fossil fuel consumption (or continued use of fossil fuel energy), long-term leakage of sequestered carbon poses the risk of emitting more CO_2 into the air than if nothing were done, which raises intergenerational equity issues.[54]

The potential for substantial CO_2 emission reductions through carbon management has significant impacts on the political feasibility of ratification of Kyoto or similar agreements. By reducing the threat to the politically mobilized fossil fuel industry and to countries with large reserves of fossil fuels, CM may enable progress in international and domestic emission reduction policy.[55] Because of their new capital requirements associated with CM and their lesser dependence on petroleum as a transportation fuel, developing countries are especially suited for using CM at low cost.[56] This alone could help end the stalemate in international policy negotiations. If emission abatement is made affordable for developing countries, these countries may consider participating in a global emissions reduction agreement. This would help neutralize political resistance to ratification of the Kyoto Protocol by developed countries such as the United States.

One policy implication of CM that warrants consideration is the difficulty of reversing a shift toward CM. Large-scale use of CM would entail large human and capital investments. Once these investments are made, a shift away from their use may be politically and technologically difficult.[57] It might be hard to convince the interested parties to transition to a more renewable system of hydrogen production once a CO_2 sequestration and hydrogen production system is in place. Because CM is only a temporary solution to carbon emissions, it can be dangerous to become heavily dependent on it for emission reductions and not focus enough on renewable energy or energy conservation.

Hydrogen Distribution

The lack of a significant infrastructure for hydrogen often is regarded as an obstacle to widespread FC use and transition to a hydrogen-based economy.[58] As discussed earlier, the original fuel for FCs can take many different forms as long as hydrogen is separable and usable for the FC. The development of this infrastructure must coevolve with the development and choice of FC technology use in automobiles. The form and nature of the infrastructure depend on the fuel that is chosen. If gasoline, methanol or NG reformers are incorporated into the design of FC cars, then the existing gasoline or NG infrastructures could be used to refuel the FC cars, or a methanol infrastructure (potentially a modified gasoline infrastructure) could develop. Conversely, if FCs are incorporated into automobiles without reformers, then an infrastructure that delivers pure gaseous or liquid hydrogen will be needed. There are obvious economic and political implications of this decision that affect many vested interests.

A pure hydrogen infrastructure is ideal. It would enable more efficient onboard FCs, thus reducing up-front costs and increasing fuel economy (estimated at 50 percent higher efficiency than methanol or gasoline FCs). These factors can combine to deliver life-cycle costs and costs per kilometer lower than or comparable to those of an FC vehicle running on gasoline or methanol.[59] Even if the costs for a hydrogen delivery system were equal to or slightly higher than a methanol or gasoline delivery system, the emission reduction benefits might outweigh the added capital costs. Because cars equipped with a methanol or gasoline reformer are not likely to be equipped with mechanisms to capture and sequester their emissions, they do not offer the same emission reduction potential of centrally distributed hydrogen. Ideally, in the future, hydrogen will be produced by renewable resources. Thus, a hydrogen delivery system of some sort will be needed to incorporate this renewable hydrogen into the energy system. An expansive pure hydrogen distribution network would allow the entry of renewable sources while still allowing hydrogen produced from fossil fuels to be used in the near future.

Centralized Versus Distributed Hydrogen Production

If FC automobiles are developed without methanol or gasoline reformers, the degree of centralization of hydrogen production still must be decided.[60] Under a CM scheme in which CO_2 is needed for enhanced NG or coal bed methane recovery, the hydrogen could be produced at the wellhead as part of a reforming and sequestering process. The hydrogen could then be piped through a network to a refueling station. Although it is more useful in terms of CO_2 sequestration, depending on distribution costs this approach may be more expensive than some of the alternatives.[61] Another option is to pipe NG from its recovery location to place of use and at a hydrogen fuel station reform the methane into hydrogen. Although this is probably cheaper because it uses an existing NG infrastructure, it would be difficult to perform carbon sequestration at the service station level.[62] Even if this was done, some level of carbon dioxide infrastructure would have to be built to pipe the CO_2 back to a burial site.

A hybrid system, or "city gate" infrastructure, would pipe NG in existing long-distance, high-pressure pipelines to a major hub or city, where hydrogen would be produced and distributed locally by a hydrogen distribution network.[63] If there is a sequestration site within the city's region (a few hundred kilometers), then CO_2 could be transmitted there for sequestration. If carbon sequestration is a goal, the city gate system may be the cheapest for both hydrogen production and carbon management with an uncentralized automobile transportation sector.[64] Another option that is especially applicable during a transition time is truck-delivered liquid hydrogen from NG reformation plants. The cost of liquefaction would be higher, although the capital costs of refueling infrastructure would be low and no pipelines would be needed (assuming there is hydrogen production within driving distance of the refueling stations).[65]

A system of centrally or regionally produced hydrogen probably is the cheapest way to deliver hydrogen to FCs in the long run when CO_2 sequestration is desired because collecting CO_2 from many distributed sources is cost-prohibitive.[66] Because CO_2 sequestration probably is necessary for cheap emission abatement and a cheaper transition to a hydrogen economy, central hydrogen production is more desirable than localized, distributed fuel reforming. The city gate option may provide a cheaper transitional or even long-term solution if

CO_2 burial or transmission options exist and renewable hydrogen production can enter the market and distribution system. Truck-delivered systems may be the cheapest under certain circumstances (limited volume and distribution), but because the goal of a hydrogen-based transportation system and economy is broad use of hydrogen as a fuel source, such systems will not provide a long-term solution. Central hydrogen production and distribution is also advantageous for developing countries because many of them do not have existing alternative infrastructures. They can start with a pure hydrogen infrastructure, leapfrogging to the most advanced distribution system.[67]

Cost of Distribution

Building a hydrogen distribution network is technically feasible, but it could be expensive, and this expense often is regarded as another impediment to hydrogen-powered vehicles.[68] Depending on geographic location and population density, the overall cost of a hydrogen transmission system could be one and a half to three times more expensive than the equivalent NG system, or $250,000–$1 million/mile.[69] However, because the first sector of the energy economy likely to be penetrated by hydrogen FCs is transportation, the piping and compressing infrastructure needed would be less than that for NG, and it could evolve with the demand for hydrogen in building power and heating applications.[70] Although the social benefits of building such a distribution network and using it to transform our energy system to a hydrogen-based one may well outweigh these initial costs, the cost factor may prove to be politically difficult to overcome in light of the perceived uncertainty surrounding global climate change and its impacts.

Technical Issues

Even if a central hydrogen production and distribution scheme is chosen, other hydrogen infrastructure issues remain. These include pipeline specifications, the potential for using existing pipelines, and hydrogen safety and storage.

The costs of developing a hydrogen infrastructure could be reduced significantly if hydrogen could be transported through the existing NG infrastructure and then separated out at the point of use. Unfortunately, some technical factors may limit the dual role of the NG infrastructure. Although some long-distance steel NG pipelines could be conditioned for hydrogen transport, low-pressure plastic pipelines used in local distribution are not suitable for hydrogen transport.[71] Additionally, some studies indicate that hydrogen transport over the NG infrastructure could foster hydrogen embrittlement (cracking) in some of the steel pipelines used for NG transport.[72] Thus, although technically feasible in some cases,[73] hydrogen transport through the existing NG infrastructure does not seem feasible on a widespread basis, so a separate hydrogen distribution network appears necessary.

Many people might be chary about the use, transport, and storage of hydrogen. Historical episodes such as the 1937 Hindenburg blimp explosion have contributed to a cultural fear of hydrogen. Ironically, a recent study found that hydrogen was not the cause of this explosion, and although hydrogen did burn in the explosion, the characteristics of burning hydrogen may have helped the survivors.[74] Nonetheless, like any other fuel, hydrogen does have characteristics that warrant attention from a safety perspective, particularly its lower flammability limit.[75]

Accordingly, studies have been undertaken to study hydrogen safety issues and to develop codes and standards for its use.[76] Preliminary studies concluded that hydrogen has some characteristics that are better than those of conventional fuels (such as its quick dispersal ability and nontoxicity) and some that are worse. On the whole, some experts believe that hydrogen is no more dangerous (and may be safer) than carrying a tank of gasoline in a car.[77] In terms of distribution, a limited amount of hydrogen is already safely transported long distances in pipelines and in trucks in gaseous and liquid forms for use in the chemical industry as well as other applications.[78] Risks associated with the use of hydrogen as a fuel, such as its lower flammability limit, are serious. However, widespread distribution and use of hydrogen does not appear to add unacceptable risks, especially when compared with risks associated with infrastructure for our present fuels.

Because of hydrogen's low energy density, storage is a major issue. Although there are many ways to store hydrogen, including liquid hydrogen and metal hydrides, compressed gaseous hydrogen seems to be the most likely form of use and storage in the foreseeable future.[79] Hydrides are very expensive, and liquefying hydrogen has an energy penalty of 30 percent, thus reducing overall efficiency and potentially increasing GHG emissions depending on the source or production method of hydrogen.[80]

Assuming that compressed gas storage is the medium of choice both on and off the vehicle or application site, three options are available for storing this compressed gas: aboveground storage, below-ground storage, and "pipeline packing."[81] For small-scale storage, aboveground pressure tanks are a viable option.[82] Assuming the absence of the technical limitations discussed earlier, another small-scale storage option is packing hydrogen into NG distribution pipelines.[83] However, if the use of hydrogen is to be widespread, large hydrogen storage capacity seems necessary. Although far from perfect, the only technically and economically viable option currently available is underground storage in depleted oil or gas fields, which has been done in commercial applications in England and France.[84] Barring any technological advancements, the best option for large-scale hydrogen storage appears to be underground storage. The existence of a comprehensive hydrogen distribution network obviously would facilitate the transport of hydrogen to and from these underground storage sites and from places of production and use. Nonetheless, advancements in hydrogen storage might hasten the transition to a hydrogen economy.[85]

Additional Sources of Hydrogen Production

Current projections of central hydrogen production focus on reforming NG, but if other sources became feasible, then producers would be able to sell hydrogen to users through an existing infrastructure. Although they are all at different stages of technical and economic feasibility, most alternative methods for producing hydrogen are more renewable than NG reformation.

One potential hydrogen source is biomass. Hydrogen could be produced from biomass through the production of synthetic gas ("syngas") from municipal solid waste (MSW).[86] Chemically processing this waste, two-thirds of it which could be renewable food or paper waste, could provide hydrogen at a negative cost because of the costs associated with solid waste disposal or storage.[87] Using 1995 NG prices, Professor Robert Socolow of Princeton's Center for Energy and Environmental Studies estimates that hydrogen production from MSW is comparable with that of NG reformation, assuming high disposal fees such as those in New York City.[88] With high global NG prices at present, MSW hydrogen production may even be more cost-effective than NG reformation today.

Hydrogen could also be produced through coal gasification or through more traditional methods of biomass energy production. One estimate claims that if all the cars in the United States had FCs, there would be enough hydrogen production capacity from the gasification of biomass from two-thirds of the current idle cropland in the United States.[89] Thus, gasification, particularly of biomass materials or wastes, could provide a viable source of potentially renewable hydrogen. There are associated climate and environmental issues that surround biomass fuels, such as agricultural degradation, GHG emissions from farming, and the degree of renewability of the practice, but in theory they provide a more renewable option than fossil fuel–derived hydrogen. However, because biomass hydrogen production is mostly independent of the fossil fuel industry, it might not garner the support of this politically active group.

From a climate perspective, the ideal way to produce hydrogen today is to generate electricity from a renewable, non–GHG-emitting source and use electrolysis to split water into oxygen and hydrogen. Through such a process, intermittent renewable energy sources such as solar, wind, and geothermal energy could become much more economically valuable and competitive with fossil fuels.[90]

One study estimates that global fuel needs in 2050 could be met through hydrogen from solar modules covering 0.5 percent of the total land area in the world.[91] Although this is a large land area, PVs can be integrated into buildings fairly effectively, thus reducing the need for solar farming on current open space. This study further estimates that current U.S. car fuel needs (assuming a fleet of 100 percent FC vehicles) could be powered by hydrogen produced from only 14 percent of developable U.S. wind power.[92] Geothermal electrolytic hydrogen is also a somewhat renewable option, although the total amount of geothermal capability is unknown.[93] Hydropower can also produce electrolytic hydrogen with some degree of renewability. Although far from commercial viability, biologically produced hydrogen (using algae) can also provide hydrogen in the future.[94]

Although renewable hydrogen production is technically possible, and the spatial capacity for such production probably exists, the cost of renewable electrolytic hydrogen is not competitive with that produced from reformed NG. Current cost estimates for solar or wind-derived hydrogen are two to three times that of NG-derived hydrogen.[95] Although the costs of wind and solar power should come down in the future with technical advances and mass production–induced cost reductions, currently they are much more expensive than fossil fuel–produced hydrogen in the absence of a carbon tax or similar instrument. Therefore, they are unlikely to play a large role in initial hydrogen production.

Inexpensive renewable electricity, as well as cheaper electrolyzers, seem to be prerequisites for the large-scale production of renewable hydrogen.[96] Estimates of the cost threshold for electricity to enable electrolytic hydrogen production range from 1 to 2 cents/kwh.[97] Thus, off-peak hydro is the most likely candidate for renewable electrolytic hydrogen production in the near future. This could be viable in developing countries with substantial amounts of off-peak power, such as Brazil,[98] thus helping these countries transition to a hydrogen economy at a lower cost. This estimate for the threshold price for electrolytic hydrogen production appears to be supported by some empirical evidence. Iceland, the only country to make a public commitment to a hydrogen economy, has the world's cheapest electricity, at 2 cents/kwh.[99]

The Near Future and the Transition

Although the technology for a hydrogen energy system exists today, time and economic factors indicate that a full transition to a hydrogen economy will take decades.[100] However, this does not preclude a rapid transition in certain sectors or geographic areas. Part of the needed hydrogen infrastructure already exists in terms of production and distribution in certain geographic areas and could be expanded quickly.[101] Although "merchant hydrogen" production and distribution is very limited, it could provide enough fuel for 2–3 million hydrogen FC cars in the United States.[102] Obviously, if demand for vehicle hydrogen increased substantially, a supply market would follow. However, demand for hydrogen is not likely to grow without existing infrastructure—thus providing another example of a "chicken and egg" bottleneck.

FC buses can take advantage of the centralized refueling that the current limited hydrogen infrastructure provides, and Ballard Power Systems demonstration buses are already being used in Chicago and British Columbia.[103] Worldwide, buses probably will be one of the first sectors to transition to hydrogen FCs. In fact, demonstration buses and total bus fleet conversion to hydrogen are the first two of the five stages laid out by Iceland hydrogen transformation scholars Bragi Árnason and Thornsteinn Sigfússon.[104] Early use of FC technology in buses is consistent with a possible hydrogen infrastructure scenario proposed by Princeton energy scholar Joan Ogden. Specifically, buses can use existing centralized refueling. Ogden proposes that once geographically concentrated demand evolves, hydrogen could be produced centrally, thus enabling carbon sequestration.[105] However, such a scenario may not evolve without policy action. Current projections estimate market-ready FC cars between 2004 and 2010, but that is only part of the picture. Without proper incentives, FC vehicles are not likely to take off on their own when refueling is not convenient for consumers. Thus, the demand necessary for market-driven infrastructure development might not emerge without government action.

Despite the seemingly large costs of infrastructure development, the distributional components of a hydrogen economy probably will not be the limiting factor in starting the transition. Rather, it will be the cost of FCs and FC vehicles. Ogden estimates that the hydrogen infrastructure capital costs in southern California[106] would be roughly $310–$620 per vehicle. Although this is not trivial, it is not unbearable. This is especially true in light of the alternatives, specifically methanol FC infrastructure ($550–$1,400/car) or gasoline FC infrastructure ($850–$1,200/car).[107] In places such as southern California, where there is a reasonable consensus that the current and projected levels of local air pollution from internal combustion engines are unacceptable, hydrogen infrastructure costs seem very tolerable. Although current PEM fuel cells are expensive because of their microscale production,[108] projections for high-volume production estimate capacity cost of $300/kilowatt, compared with $25–$50/kilowatt for internal combustion engines.[109] Thus, the FC costs are a few multiples above competitive internal combustion pricing in the absence of other incentives. Recent studies by major motor companies estimate that FC engines could drop in price to $50–$100/kilowatt with further improvements, which is still expensive compared with current automobile power sources.[110]

Conclusions and Policy Implications

A hydrogen infrastructure using FCs and carbon sequestration could have substantial benefits for GHG emission reductions. Such a system is not technically or economically impossible today and should be pursued earnestly. This infrastructure should strive to deliver pure hydrogen, not an interim hydrogen carrier such as methanol, to avoid wasted resources and unnecessary emissions. The use of carbon sequestration technologies cannot replace efforts to switch away from fossil fuels over the long term.[111]

Perhaps fossil fuel-derived hydrogen's greatest contribution will be its potential to make a transition to a hydrogen economy (and hopefully a renewable hydrogen economy) politically feasible. At this stage in the struggle to minimize the degree of future anthropogenic climate change, political opposition to GHG emission reduction appears to be the limiting factor. A hydrogen infrastructure that uses carbon sequestration would be less threatening to the fossil fuel industry, and they could profit from a transition to hydrogen. Because oil companies are the largest on-site producers and consumers of hydrogen, they would benefit in the short term from a transition to hydrogen because they could serve as the main suppliers.[112] Their experience with hydrogen production and handling, in addition to their knowledge of the transportation fuel sector, positions them well in the long-term hydrogen economy.[113] These factors offer some hope for the political feasibility of a transition to a hydrogen economy.

Regardless of what the first step toward a hydrogen economy entails, getting there suffers from a "chicken and egg" hurdle. Specifically, there is insufficient economic incentive from either the demand side (end-use technologies such as FC cars) or supply side (hydrogen gas production, transmission, and distribution) to assume the financial burden for developing this infrastructure.[114] Some argue that the development of the demand side would spur the rest.[115] In either case, the problem almost certainly requires government involvement to address this market failure.

Simple policy measures can induce substantial technological or behavioral changes. This was demonstrated by the carbon sequestration scheme that developed in response to Norway's carbon tax. By aligning financial incentives with emission reduction incentives, governments (especially the United States) can help speed the development of the triad of FCs, carbon sequestration, and a hydrogen infrastructure.

In addition to relying on policy- (or tax-) induced technical change, government can also play a role in developing hydrogen economy technologies through research and development. Because of the above-stated market failures as well as the "public good" nature of GHG emission reductions, public sector investments in energy R&D are justifiable.[116] Even if an increase in R&D spending is not possible, money from existing R&D programs focusing on fossil fuels should be directed toward research on technologies that offer help in dealing with climate change and other environmental problems caused by the use of fossil fuels.

Iceland, which has made a public commitment to fully convert to a hydrogen-based economy by 2030,[117] provides an example of government involvement in the transition to hydrogen. Partnerships between public agencies and private industry are paving the way to widespread hydrogen use in Iceland, including joint (but not equal) funding of projects.[118] In the view of Iceland's "Professor Hydrogen," Dr. Árnason, a public commitment is very powerful, if not necessary, in spurring business interest in partnerships that result in environmental benefits.[119]

The U.S. government can work with applicable industries to choose the best design for a hydrogen distribution network and then start work on implementing such a design. Although such a system would not be very practical for probably another decade because of the state of FC technology development and carbon sequestration technologies, rapid spread of these technologies would be greatly enhanced by an in-place hydrogen infrastructure. It took the Internet decades to be widely used, and its existence helped to spur a technological revolution. The U.S. government, and governments worldwide, together with industry have the potential to lay the foundation for the hydrogen revolution.

Acknowledgments

Special thanks go to Armin Rosencranz for his review and comments on multiple drafts of this chapter, to Stephen Schneider for insightful discussions in and out of class, and to Joan Ogden and Dan Kammen for their many helpful suggestions.

Notes

1. Socolow, R. (ed.), 1997: "Fuels decarbonization and carbon sequestration: Report of a workshop," Princeton University/Center for Energy and Environmental Studies Report No. 302, September, p. 32.
2. Renewable energy sources such as solar or wind power produce varying amounts of energy depending on the time of day or year.
3. Bockris, J., 1980: *Energy Options* (Sydney: Australia and New Zealand Book Company); Winter, C.-J. and J. Nitsch, 1988: *Hydrogen as an Energy Carrier* (New York: Springer-Verlag); Ogden, J. M. and R. H. Williams, 1989: *Solar Hydrogen: Moving Beyond Fossil Fuels* (Washington, DC: World Resources Institute); Ogden, J. and J. Nitsch, 1993: "Solar hydrogen," in T. Johannsson, H. Kelly, A. K. N. Reddy, and R. H. Williams (eds.), *Renewable Energy Sources of Electricity and Fuels* (Washington, DC: Island Press), pp. 925–1009. All reported in Ogden, J., 1999a: "Prospects for building a hydrogen energy infrastructure," *Annual Review of Energy and the Environment,* 24: 227–279.
4. Dunn, S., 2000: "The hydrogen experiment," *World Watch,* November/December: 14–25.
5. Electrolytic production of hydrogen entails the use of electricity to split water (H_2O) into the two elements that combine to form water, namely hydrogen and oxygen.
6. Socolow, 1997, p. 1.
7. Lovins, A. B. and C. Lotspeich, 1999: "Energy surprises for the 21st century," *Journal of International Affairs,* 53 (1).
8. Fuel Cells 2000: On-Line Fuel Cell Information Center: "What Is a Fuel Cell?" Available online: http://216.51.18.233/index_e.html, downloaded 10/30/00.
9. Williams, R. H., 1998a: "Hydrogen production from coal and coal bed methane, using byproduct CO_2 for enhanced methane recovery and sequestering the CO_2 in the coal bed," Princeton University Center for Energy and Environmental Studies Report No. 309, pp. 1–14.
10. Ibid., p. 1.
11. Williams, R. H., 1998b: "A technological strategy for making fossil fuels environment- and climate-friendly," *World Energy Conference (WEC) Journal,* July: 59–67.
12. Ogden, 1999a, p. 227.
13. Ibid., p. 230.
14. Herzog, H., B. Eliasson, and O. Kaarstad, 2000: "Capturing greenhouse gases," *Scientific American,* February: 72–79.
15. Dunn, 2000, p. 21.
16. "John Williams, then the leader of GM's internal team on global climate issues, has stated that this company has 'embraced fuel cells as the technology of choice' over the long term." As reported in Reed, D., 2000: "Bad news, good news: when it comes to climate, unexpected flip-flops cut both ways," *RMI Solutions Newsletter,* 16 (1, Spring): 1–22; quote on p. 20.
17. Ogden, 1999a, p. 229.
18. Dunn, 2000, p. 22.

19. Ibid., p. 23.
20. Ogden, 1999a, pp. 228, 229.
21. Lovins, A B. and B. D. Williams, 1999: "A strategy for the hydrogen transition," paper presented at the 10th Annual U.S. Hydrogen Meeting, National Hydrogen Association, Vienna, VA, April (Old Snowmass, CO: Rocky Mountain Institute), pp. 1–16.
22. Thomas, C. E., B. D. James, and F. D. Lomax, Jr., 1997: "Market penetration scenarios for fuel cell vehicles," *Proceedings of the 8th Annual U.S. Hydrogen Meeting,* Arlington, VA, National Hydrogen Association, March 11–13; Directed Technologies, Inc.; Thomas, C. E., I. F. Kuhn, Jr., B. D. James, F. D. Lomax, Jr., and G. N. Baum, 1998: "Affordable hydrogen supply pathways for fuel cell vehicles," *International Journal of Hydrogen Energy,* 23 (6, June): 507–516; Thomas, C. E., B. D. James, F. D. Lomax, Jr., and I. F. Kuhn, Jr., 1998: "Fuel options for the fuel cell vehicle: hydrogen, methanol or gasoline?" Fuel Cell Reformer Conference, Diamond Bar, CA, South Coast Air Quality Management District, November 20; Williams, B. D., T. C. Moore, and A. B. Lovins, 1997: "Speeding the transition: designing a fuel-cell hypercar," *Proceedings of the 8th Annual U.S. Hydrogen Meeting,* Alexandria, VA, National Hydrogen Association, March 11–13, Rocky Mountain Institute Publication #T97-9, www.rmi.org; Mark, J., 1997: "Environmental and infrastructure trade-offs of fuel choices for fuel cell vehicles," SAE-972693, Future Transportation Technology Conference, San Diego, August 6–8, Society of Automotive Engineers. As reported in Lovins and Williams, 1999, p. 2.
23. Dunn, 2000, p. 23.
24. Methanol is more toxic than gasoline. Jensen, M. W. and M. Ross, 2000: "The ultimate challenge: developing an infrastructure for fuel cell vehicles," *Environment,* 42 (7, September): 15.
25. Ogden, 1999a, pp. 262, 263.
26. Ibid., p. 263.
27. Dunn, 2000, p. 16.
28. Ibid., p. 24.
29. Ibid., p. 23.
30. Ogden, J., M. Steinburgler, and T. Kreutz, 1997: "Hydrogen as a fuel for fuel cell vehicles," *Proceedings of the 8th Annual Hydrogen Association Meeting,* Alexandria, VA, March 11–13, 1997. As reported in Ogden, J., 1999b: "Developing a refueling infrastructure for hydrogen vehicles: A southern California case study," *International Journal of Hydrogen Energy,* 24: 709–733.
31. Herzog, Eliasson, and Kaarstad, 2000, p. 78. For instance, Socolow, 1997, p. 18 reports electrolysis to be competitive when electricity costs between 1 and 2 cents/kwh. In 1995, the average price that U.S. utilities paid for PV and wind-generated electricity was 15.8 and 11.64 cents/kwh, respectively. As reported in Guey-Lee, L.: "Renewable electricity purchases: history and recent developments," available online: http://www.eia.doe.gov/cneaf/solar.renewables/rea_issues/html/renelec.html, downloaded 12/26/00.
32. Williams, 1998b, pp. 60–61; Herzog et al., 2000, p. 74; and Parson, E. A. and D.

W. Keith, 1998: "Fossil fuels without CO_2 emissions," *Science,* 282 (November 6): 1053–1054.

33. Parson and Keith, 1998, p. 1054.
34. Hendricks, C. A., 1994: "Carbon dioxide removal from coal-fired power plants," Ph.D. thesis, Department of Science, Technology, and Society, Utrecht University, Utrecht, The Netherlands. As reported in Williams, 1998b, p. 61. IPCC, 1996: Chapter 19, "Energy supply options," as reported in R. T. Watson, M. C. Zinyowera, and R. H. Moss (eds.), *Climate Change 1995—Impacts, Adaptations, and Mitigation of Climate Change: Scientific–Technical Analysis* (New York: Cambridge University Press).
35. Socolow, R. (ed.), 1997: "Fuels decarbonization and carbon sequestration," Princeton University CEES Report No. 302 (http://www.princeton.edu/~ceesdoe); Herzog, H., E. Drake, and E. Adams, 1997: *CO_2 Capture, Reuse, and Storage Technologies for Mitigating Global Climate Change* (Cambridge: Energy Laboratory, Massachusetts Institute of Technology); Williams, R. H., 1998: "Fuel cells, coal and China," paper presented at the 9th Annual U.S. Hydrogen Meeting, Washington, DC, March 4; Holloway, S., 1996: in "Third International Conference on Carbon Dioxide Removal," Cambridge MA, September, supplement to *Energy Conversion and Management,* 38 (1997): 193–198, www.ieagreen.org.uk/ghgt4.htm. All reported in Parson and Keith, 1998, p. 1053.
36. Ibid.
37. Parson and Keith, 1998, p. 1053.
38. Ibid.
39. Romm, J., M. Levine, M. Brown, and E. Peterson, 1998: "A Road Map for U.S. Carbon Reductions," *Science,* 279: 669; Edmonds, J. A., S. H. Kim, C. N. MacCracken, R. D. Sands, and M. A. Wise, 1997: "Return to 1990," Pacific Northwest National Laboratory Report No. 11819, Washington, DC, 1997. As reported in Parson and Keith, 1998, p. 1053.
40. Herzog, H., E. Drake, and E. Adams, 1997: *CO_2 Capture, Reuse, and Storage Technologies for Mitigating Global Climate Change* (Cambridge: Energy Laboratory, Massachusetts Institute of Technology); Watson, R. T., M. C. Zinyowera, and R. H. Moss (eds.), 1996: *Impacts, Adaptations, and Mitigation of Climate Change,* Report of IPCC Working Group 2 (Cambridge, England: Cambridge University Press). As reported in Parson and Keith, 1998, p. 1053.
41. Parson and Keith, 1998, p. 1053.
42. Edmonds et al., 1997; Houghton, J. T., L. G. Miera Filho, D. J. Griggs, and K. Maskell, 1997: *Stabilization of Atmospheric Greenhouse Gases: Physical, Biological and Socio-Economic Implications* (IPCC Technical Paper III, World Meteorological Organization); Hoffert, M. I., K. Caldeira, A. K. Jain, E. F. Haites, L. D. Harvey, S. D. Potter, M. E. Schlesinger, S. H. Schneider, R. G. Watts, T. M. Wigley, and D. J. Wuebbles, 1998: "Energy implications of future stabilization of atmospheric CO_2 content," *Nature,* 395: 881; National Academy of Sciences, 1991: *Policy Implications of Greenhouse Warming* (Washington, DC: National Academy Press); Gaskins, D. and J. Weyant, 1993: *American Economic Review,* 83: 318; Nordhaus, W., 1994: *Managing the Global Commons* (Cambridge: MIT Press), p. 61. All reported in Parson and Keith, 1998, p. 1053.

43. Parson and Keith, 1998, p. 1053; Williams, 1998b, p. 62; Socolow, 1997, p. 4.
44. Parson and Keith, 1998, p. 1053; Williams, 1998b, p. 62.
45. Herzog et al., 2000, p. 75.
46. Williams, 1998a, p. 2; Blok, K., R. H. Williams, R. E. Katofsky, and C. A. Hendriks, 1997: "Hydrogen production from natural gas, sequestration of recovered CO_2 in depleted gas wells and enhanced natural gas recovery," *Energy,* 22 (2–3): 161–168. As reported in Williams, 1998b, p. 61.
47. Williams, 1998b, p. 60.
48. Lovins and Williams, 1999, p. 9.
49. Ibid.
50. Herzog et al., 2000, p. 79.
51. Parson and Keith, 1998, p. 1054.
52. Socolow, 1997, p. 11.
53. Ibid.
54. Parson and Keith, 1998, p. 1054.
55. Ibid.; Herzog et al., 2000 p. 79.
56. Parson and Keith, 1998, p. 1054.
57. Ibid.
58. Lovins and Williams, 1999, p. 2.
59. Ogden, 1999a, p. 230; Ogden, 1999b, p. 724.
60. Ogden, 1999a, p. 255.
61. Ibid.; Socolow, 1997, p.6.
62. Socolow, 1997, p. 6.
63. Ibid.; Ogden, 1999, p. 249.
64. Socolow, 1997, p. 6.
65. Ogden, 1999a, pp. 258–259; Ogden, 1999, p. 718; Ogden, J., M. Steinburgler, and T. Kreutz, 1999: "A comparison of hydrogen, methanol and gasoline as fuels for fuel cell vehicles," *Journal of Power Sources,* 79: 143–168. As reported in Ogden, 1999a, pp. 258–259.
66. Ogden, J., 1999c: "Strategies for developing hydrogen energy systems with CO_2 sequestration," *Proceedings of the National Hydrogen Association Meeting,* 10th, Vienna, VA (Washington, DC: National Hydrogen Association). As reported in Ogden, 1999a, p. 269.
67. Ibid.
68. Ogden, 1999b, p. 721.
69. Socolow, 1997, p. 35; Directed Technologies, Inc., Air Products and Chemicals, BOC Gases, The Electrolyzer Corp., and Praxair, Inc., 1997: "Hydrogen infrastructure report," Ford Motor Co. U.S. DOE Contract No. DE-AC0294CE50389, Arlington, VA: Directed Technologies, Inc.; Ogden, 1999b; Ogden, J. M., E. Dennis, M. Steinbugler, and J. Strohbehn, 1995: "Hydrogen energy systems studies. Final report," U.S. DOE Contract No. xr-11265-2 (Princeton, NJ: Princeton University). As reported in Ogden, 1999a, pp. 249–251.
70. Ogden, 1999a, pp. 249–252.
71. Socolow, 1997, pp. 35, 36.
72. Blazek, C. F., R. T. Biederman, S. E. Foh, and W. Jasionowski, 1992: "Underground

storage and transmission of hydrogen," *Proceedings of the Annual U.S. Hydrogen Meeting, 3rd, Washington, DC,* (Washington, DC: Technology Transit Corporation), pp. 4–203–21; Jewett, R. P., R. J. Walter, W. T. Chandler, and R. P. Frohmberg, 1973: "Hydrogen environment embrittlement of metals," NASA CR-2163, prepared by Rocketdyne, Canoga Park, CA (Golden, CO: National Renewable Energy Laboratory); Hoover, W. R., S. L. Robinson, R. D. Stoltz, and J. R. Springarn, 1981: *Sandia National Laboratories Report No. SAND81-8006. Final Report, DOE Contract DEAC04-76DP00789* (Albuquerque, NM: Sandia National Laboratories); Holbrook, J. H., H. J. Cialone, M. E. Mayfield, and P. M. Scott, 1982: "The effect of hydrogen on low-cycle-fatigue life and subcritical crack growth in pipeline steels," Batelle Columbus Lab, BNL-35589, DE85006685 (Columbus, OH: Battelle Columbus Laboratory); Holbrook, J. H., E. W. Collings, H. J. Cialone, and E. J. Drauglis, 1986: "Hydrogen degradation of pipeline steels," U.S. DOE Report BNL-52049, DE87 005585, Washington, DC; Cialone, H., P. Scott, and J. Holbrook, 1984: "Hydrogen effects in conventional pipeline steels," in *Hydrogen Energy Progress,* V (Oxford, England: Pergamon), pp. 1855–1867. All reported in Ogden, 1999a, p. 249.

73. A study on the gas distribution system in Munich found "no technical barriers to using up to 100% hydrogen in existing low-pressure natural gas distribution systems." Buenger, U., W. Zittell, and T. Schmalschlager, 1994: "Hydrogen in the public gas grid: a feasibility study about its applicability and limitations for the admixture within a demonstration project for the city of Munich," *Proceedings of the World Hydrogen Energy Conference, 10th Hydrogen Energy Program, Cocoa Beach, FL* (Coral Gables, FL: International Association of Hydrogen Energy). As reported in Ogden, 1999a, p. 253.
74. Bain, A., 1997: "The Hindenburg disaster: a compelling theory of probable cause and effect," *Proceedings of the 8th Annual U.S. Hydrogen Meeting,* Alexandria, VA, National Hydrogen Association, March 11–13, pp. 125–128; Bain, A., personal communication, November 1, 1999. As reported in Lovins and Williams, 1999, p. 7.
75. Ogden, 1999a, p. 267.
76. The National Hydrogen Association with support from the U.S. DOE is developing hydrogen codes and standards. As reported in Socolow, 1997, p. 34.
77. Ringland, J. T., 1994: "Safety issues for hydrogen-powered vehicles," Sandia National Laboratories, March. As reported in Williams, 1998b, p. 60; Ford Motor Co., 1997: "Direct hydrogen fueled proton exchange membrane fuel cell system for transportation applications," Hydrogen Vehicle Safety Report Contract No. DE-AC02-94CE50389, Springfield, VA, National Technology Information Service, U.S. Department of Commerce. As reported in Ogden, 1999a, p. 268.
78. Approximately 5 percent of hydrogen produced in the United States is "transported to distant users via liquid-hydrogen truck, compressed-gas truck, or gas pipeline." Heydorn, B., 1994: "Byproduct hydrogen sources and markets," *Proceedings of the National Hydrogen Association Meeting, 5th* (Washington, DC: National Hydrogen Association); Kerr, M., 1993: "North American merchant hydrogen infrastructure," *Proceedings of the Annual U.S. Hydrogen Meeting, 4th* (Washington, DC: National Hydrogen Association), pp. 8–61. Both reported in Ogden, 1999a, p. 247.
79. Ogden, 1999a, p. 254.

80. Socolow, 1997, pp. 32–33.
81. Ogden, 1999a, pp. 241–243.
82. Ibid., pp. 242–243.
83. Ibid., p. 242.
84. Carpetis, C., 1981: "Estimation of storage costs for large hydrogen storage facilities," *International Journal of Hydrogen Energy,* 7 (2): 191–203; Taylor, J. B., J. E. A. Alderson, K. A. Kalayanam, A. B. Lyle, and L. A. Phillips, 1986: "Technical and economic assessment of methods for the storage of large quantities of hydrogen," *International Journal of Hydrogen Energy,* 11: 5; Blazek et al., 1992; Venter, R. D. and G. Pucher, 1997: "Modelling of stationary bulk hydrogen storage systems," *International Journal of Hydrogen Energy,* 22 (8): 791–798. As reported in Ogden 1999a, pp. 241–242.
85. Ogden, 1999a, p. 246.
86. Ibid., p. 233; Socolow, 1997, p. 19.
87. Socolow, 1997, p. 19.
88. Ibid.
89. Ogden and Nitsch, 1993. As reported in Ogden, 1999a, p. 266.
90. Lovins and Williams, 1999, pp. 9–10; Dunn, 2000, p. 17.
91. Ogden and Nitsch, 1993, as reported in Ogden, 1999a, p. 266.
92. Ibid.
93. Ogden, 1999b, p. 717.
94. University of California at Berkeley Press Release: "UC Berkeley and Colorado scientists find valuable new source of hydrogen fuel, produced by common algae," February 20, 2000, available online: http://www.berkeley.edu:80/news/media/releases/2000/01/01-27-2000b.html, downloaded November 5, 2000.
95. Ogden and Nitsch, 1993. As reported in Ogden, 1999a, p. 266.
96. "Hydrogen is made at large-scale via electrolysis of water at some dams, using low-cost off-peak power. Electrolysis is a competitive route to hydrogen only when electricity is inexpensive (roughly one to two cents per kilowatt hour), a situation that may occur with off-peak pricing. An expanded role for electrolysis will require lower-cost electrolyzers; one approach is to use the fuel cell in reverse to produce hydrogen at off-peak times, and then to produce electricity from that hydrogen at on-peak times," in Socolow, 1997, p. 18.
97. Ibid.; Ogden, 1999a, p. 240.
98. Ogden, 1999a, p. 240.
99. Dunn, 2000, p. 19.
100. Ogden, 1999a, p. 268.
101. Ogden, 1999b, p. 715.
102. Linney, R. E. and J. G. Hansel, 1996: "Safety considerations in the design of hydro-powered vehicles," *Proceedings of the World Hydrogen Energy Conference, 11th, Hydrogen Energy Program,* Stuttgart (Frankfurt: Schon Wetzel), pp. 2159–2168, as reported in Ogden, 1999a, pp. 246–247.
103. Ogden, 1999b, p. 712.
104. Dunn, 2000, p. 22.
105. Ogden, 1999a, p. 269.

106. Because of its historic air quality problems and public commitments to zero-emission vehicles, southern California may be an ideal location for a transition to a hydrogen FC transportation sector.
107. Ogden, 1999b, p. 728.
108. Williams, 1998b, p. 62.
109. Socolow, 1997, p. 35.
110. Ibid., p. 35.
111. Ibid., p. 3 of the Executive Summary.
112. Ogden, 1999a, p. 272.
113. Ibid.
114. Socolow, 1997, p. 36.
115. Ogden, 1999a, p. 272; Williams, 1998b, p. 59.
116. PCAST Energy R&D Panel, 1997: "Federal energy research and development for the challenges of the twenty-first century," report of the Energy Research and Development Panel of the President's Committee of Advisors on Science and Technology (PCAST), November. As reported in Williams, 1998b, p. 65.
117. Dunn, 2000, p. 20.
118. Ibid., pp. 20–22.
119. Ibid., p. 25.

CHAPTER 18

Energy R&D and Innovation: Challenges and Opportunities[1]

Robert M. Margolis and Daniel M. Kammen

The most promising way to reduce greenhouse gas (GHG) emissions to the atmosphere is to deploy new and cleaner technologies to deliver energy. Specifically, a transition is needed from fuels and technologies with a high carbon content to decarbonized fuels. Although there is wide national and regional variation, global decarbonization has progressed at about the rate of 1.3 percent per year.[2] This has been accomplished through transitions from coal and oil to gas, to renewable energy sources (to a far lesser extent), and through increased energy efficiency. One of the best ways to gauge the prospects for the deployment of these technologies is to measure the level of public and private investment in energy research. This chapter uses data on international trends in public sector energy R&D, U.S. (public and private) R&D investments and patents, and U.S. cross-sectoral R&D intensities to examine the relationship between R&D expenditures and innovation. The analysis presented here raises significant concerns about our commitment and capacity to develop renewable energy and low-carbon fossil fuel energy technologies. Advances in clean energy systems are critical to our ability to meet future energy supply and environmental needs, particularly in a GHG-constrained world.

A key finding of our analysis is that in the United States the total number of patents and total R&D funding have been highly correlated over the past two decades: Both roughly doubled between 1976 and 1996. Similarly, for the energy sector as whole, the total number of energy technology–related patents has exhibited a strong correlation with total energy technology R&D investments. However, unlike the upward trend seen in general R&D investments and patents, energy funding and patents issued have both declined precipitously since the early 1980s. A careful examination of fossil fuel and renewable energy sector patents and R&D investments by the U.S. Department of Energy (DOE) over the past two decades reveals some surprising results.

As one would expect, the total number of patents assigned to the DOE has decreased as budgets have declined; however, the total number of patents assigned to the DOE and those in which the DOE is a partner or has other financial interests has actually been increasing steadily during the past decade. This divergence is explained by the evolution of technology transfer–related laws and policies enacted by the U.S. Congress since 1980. A primary goal of these actions was to increase technology transfer from government-funded national labs to the private sector. The key point here is that both the level of R&D funding and government policies related to how R&D dollars are managed and spent are tremendously important. Although this supports the notion that it is possible to do more with less and that sound policies do matter, dramatic declines in the federal R&D investment portfolio and fluctuating or uncertain funding commitments fundamentally reduced our ability to nurture and implement promising technologies, programs, and partnerships.

Motivation

During the past decade, the end of the cold war and low fossil fuel prices have decreased the level of public and policymaker attention on energy planning. Yet during this same period, the domestic and global political and environmental challenges and the investments needed to develop clean energy technologies have increased significantly. This was illustrated by the controversy surrounding the U.S. decision to sign the Kyoto climate accord at the Fourth Conference of the Parties to the Framework Convention on Climate Change in Buenos Aries in November 1998. Given that energy extraction, transformation, and consumption are the primary source of GHGs, energy technology will clearly play a central, defining role in responding to the threat of climate change.[3]

A number of analysts have argued that because climate change depends more on cumulative GHG emissions over the next century than it does on the exact timing of those emissions, immediate or dramatic action is not needed to stabilize concentrations of atmospheric GHGs at an environmentally sustainable level.[4] However, we argue that the necessary steps to promote energy technology research and development and acquire the needed experience with new technologies through market penetration, pilot projects, and large-scale commercialization efforts will take significant amounts of time and resources. There is little time to waste. The record is not encouraging in this regard. In fact, in most

TABLE 18.1. Turnover Times for Selected Energy Supply and End-Use Technologies

Energy Technology	*Turnover Time (years)*
Industrial process equipment	3–20
Photovoltaic panel systems	3–20
Home appliances	5–15
Electric power plants	30–50
Residential and commercial building systems	50–100

Organisation for Economic Co-operation and Development (OECD) countries, government energy technology R&D budgets have been declining significantly in real terms since the early 1980s.[5]

Furthermore, even if policies are implemented to accelerate the rates of investment and hence innovation in R&D and in demonstration and commercialization (D&C), replacing existing technologies with advanced fossil or renewable energy technologies may take decades, given the long lifetimes of many energy technologies (Table 18.1). The transition to clean energy technologies is likely to be even slower in developing nations, where lack of funds and the tendency to be cautious about new and untested technologies (particularly where they must be imported) dampen the market for new innovations.

Although this provides a clear incentive to initiate climate change policies immediately, there have been a number of impediments. Key impediments have included calls for additional study (largely as a delaying tactic) and claims that the economic costs are too high and therefore voluntary targets are all that should be pursued.[6] Both arguments are incorrect.

The risks of global warming are sufficiently large that, at a minimum, a path of least-regrets (minimum cost) action or climate insurance should be initiated immediately.[7] In response to the potential risk of climate change and to the long atmospheric lifetime of CO_2 and several other GHGs, initiating emission reductions now will obviate draconian measures in the future. The second argument for inaction, based on economic costs, is also flawed. A number of studies conclude that significant reduction of anthropogenic GHG emissions could be achieved at costs that are comparable to current spending on environmental protection.[8] Current levels of economic growth suggest that such investments not only are possible but in many cases may result in unanticipated innovation and provide economic and political benefits.[9] Therefore, there are compelling reasons to initiate action now to reduce GHG emissions.

International and U.S. Energy R&D: Declining Trends

National funding levels for R&D vary significantly across industrialized nations. For example, as shown in Fig. 18.1, R&D as a percentage of GDP varies from roughly 1 to 3 percent for seven of the top R&D investing countries. As illustrated in the figure, countries have been able to change their levels in a short time. The variation between countries suggests that national policies can make a difference, and if R&D were recognized for the economic and scientific benefits it provides, policy leadership could have a major impact.

The United States consistently has had one of the highest ratios and has been a leader in terms of absolute spending levels. For example, in 1997 total U.S. government R&D expenditures roughly equaled the total government expenditures of the other six countries shown in Fig. 18.1. What is not apparent in the figure, however, is that compared with other major industrial countries the United States spends a disproportionate share of its R&D budget on defense-related R&D.

Defense-related R&D accounted for 55 percent of the total U.S. government R&D expenditures in 1997, as shown in Fig. 18.2. Although this represents a decline from the cold war period, it is nothing near the peace dividend expected.

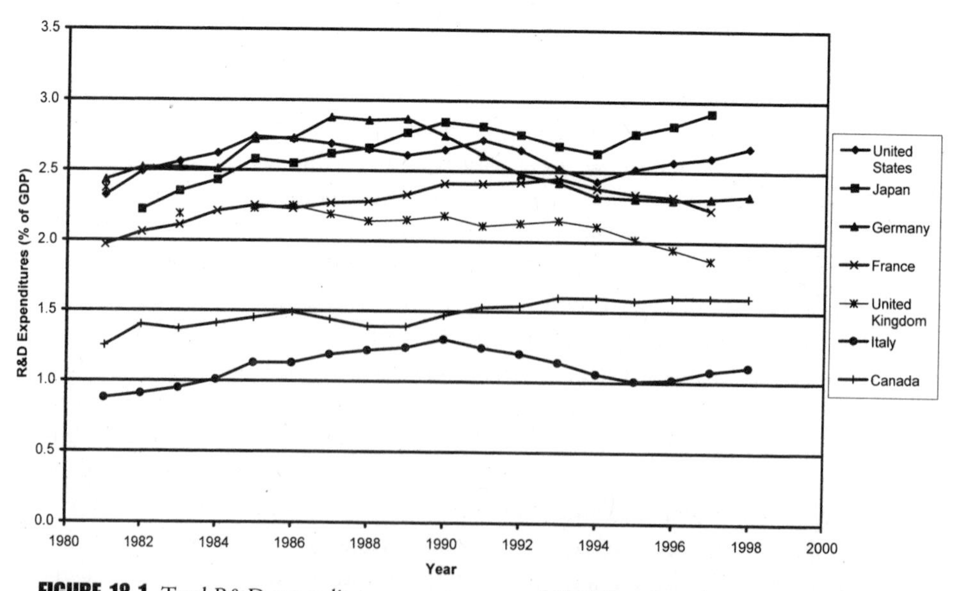

FIGURE 18.1. Total R&D expenditures as percentages of GDP for selected OECD countries. (From National Science Foundation, 2000: *Science and Engineering Indicators 2000* [Washington, DC: National Science Foundation], Appendix Table 2-63, available online www.nsf.gov/sbe/srs/seind00/.)

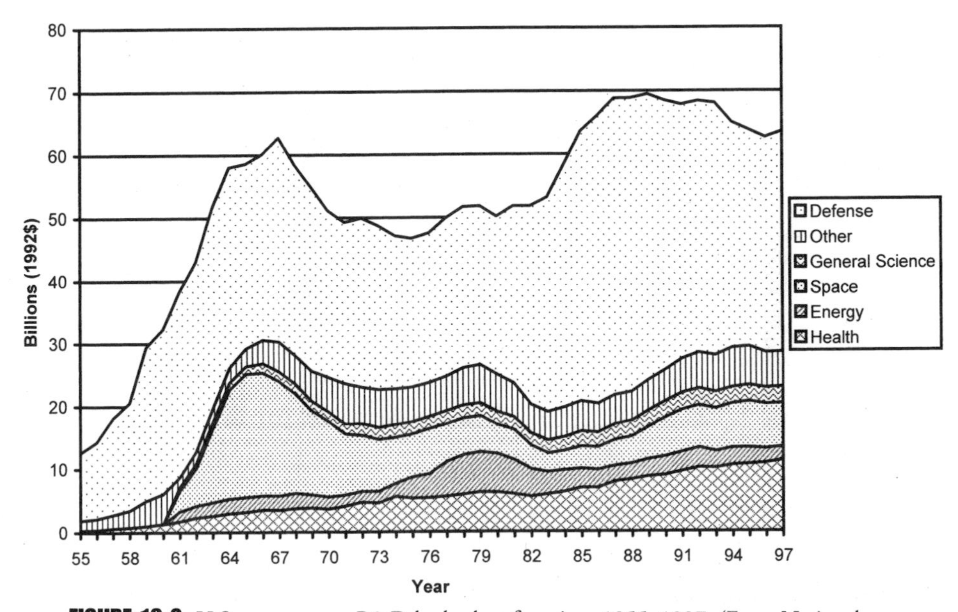

FIGURE 18.2. U.S. government R&D by budget function, 1955–1997. (From National Science Foundation, 1999: *Federal R&D Funding by Budget Function: Fiscal Years 1997–99* (Washington, DC: NSF, Division of Science Resources Studies), Table 25a, available at www.nsf.gov/sbe/srs/nsf99315/.)

In essence, in a dramatically changed world the U.S. R&D budget remains dominated by cold war spending priorities. The rising concern about global environmental issues and sustainable development has not translated into increased government R&D funds focused on addressing these issues. The picture becomes even more bleak when examining U.S. and international trends on energy R&D.

A recent survey of government-sponsored energy R&D in the 22 member countries of the International Energy Agency (IEA) clearly documents the dramatic real declines in energy R&D between 1980 and 1995.[10] In 1995, 98 percent of all IEA member country energy R&D was carried out by 10 countries. In rank order (highest to lowest public sector energy R&D budget) the countries were Japan, the United States, France, Germany, Italy, Canada, the Netherlands, Switzerland, Spain, and the United Kingdom. The government energy R&D budgets for these 10 countries in 1980 and 1995 are displayed in Fig. 18.3. As illustrated in the figure, the declines were particularly sharp in Germany, the United Kingdom, and the United States. Only two countries increased their energy R&D funding during this period: Japan and Switzerland. Overall, the changes represent at real decline of 39 percent in energy R&D funding.

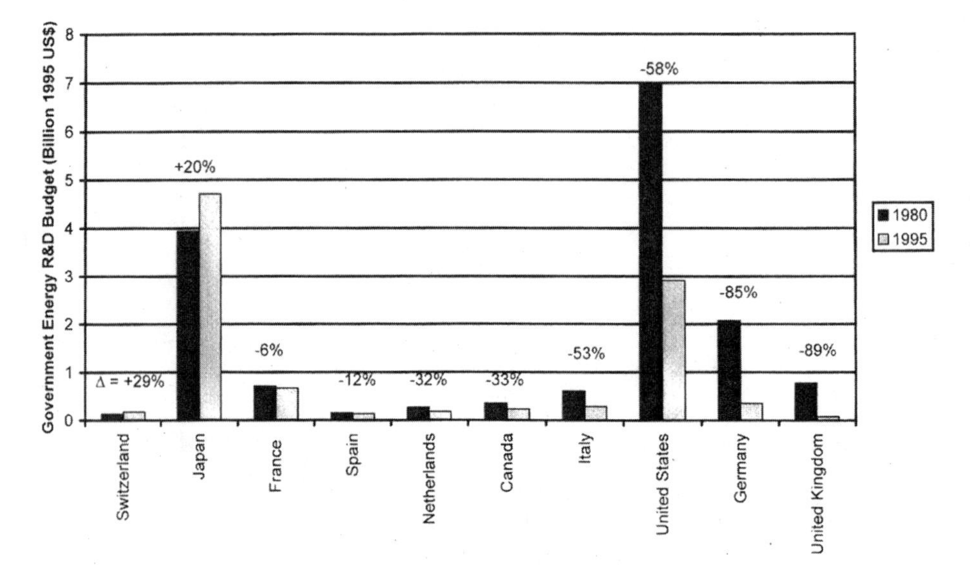

FIGURE 18.3. Government energy technology R&D budgets for selected IEA countries. *Note:* Data for France before 1990 are unavailable, so the figure displays 1990 and 1995 data for France. This is likely to understate the decline in R&D funding in France. (From International Energy Agency, 1997.)

Between 1980 and 1995, the overall trend across fuels has also been declining R&D budgets: Nuclear funding fell 40 percent, fossil funding declined 58 percent, and renewable funding fell 56 percent. In contrast, whereas energy conservation budgets were cut significantly between 1980 and 1990 (i.e., by 52 percent), they increased rapidly between 1990 and 1995. The post-1990 increases roughly returned total IEA energy conservation funding to its 1980 level (in real terms). However, there has been significant variation in the patterns among IEA countries.

Nuclear energy and energy conservation R&D provide good examples of this variation. In terms of energy conservation R&D, Japan, Spain, and Switzerland increased their budgets by 100 percent or more between 1980 and 1995, whereas France, Germany, and the United Kingdom cut their budgets by more than 80 percent. The variation between countries with respect to nuclear energy R&D was similarly diverse: The United States, Germany, Italy, and the United Kingdom cut their nuclear R&D budgets by at least 70 percent, whereas Japan and France increased their nuclear R&D budgets by 20 and 7 percent, respectively. Some countries have eliminated broad classes of energy R&D from their research portfolios, shifting their priorities toward a favored technology (i.e., nuclear energy in Japan and France), and other countries have cut energy R&D across the board (i.e., the pattern in Germany and the United Kingdom).

The decline was particularly pronounced in the United States, where the federal government's energy technology R&D budget decreased by 74 percent, from $5 billion to $1.3 billion, between 1980 and 1996.[11] In the United States, declining federal energy technology R&D budgets have been accompanied by declining private sector investments. In particular, the early phase of restructuring the electric utility industry has initiated an exodus from energy R&D and long-range strategic planning in the U.S. electricity sector. This abandonment of R&D is reflected in recent trends at investor-owned utilities (IOUs). For example, between 1994 and 1996 IOU investments in R&D decreased by 38 percent, from $650 to $403 million. During the same period the 10 largest IOU contributors to the Electric Power Research Institute (ERPI), the electric utility industry R&D consortium, cut their funding to EPRI by 47 percent, from $130 to $69 million.[12]

Perhaps even more telling, as shown in Fig. 18.4, during the same period the three major IOUs in California—the state leading the restructuring process—cut their total R&D funding by 61 percent and their funding to EPRI by 64 percent.[13] These utilities—Southern California Edison, Pacific Gas and Electric (PGE), and San Diego Gas & Electric—accounted for 79 percent of the total generating capacity and 87 percent of the total electricity generation in California in 1995.[14] During this period EPRI's funding from these three IOUs

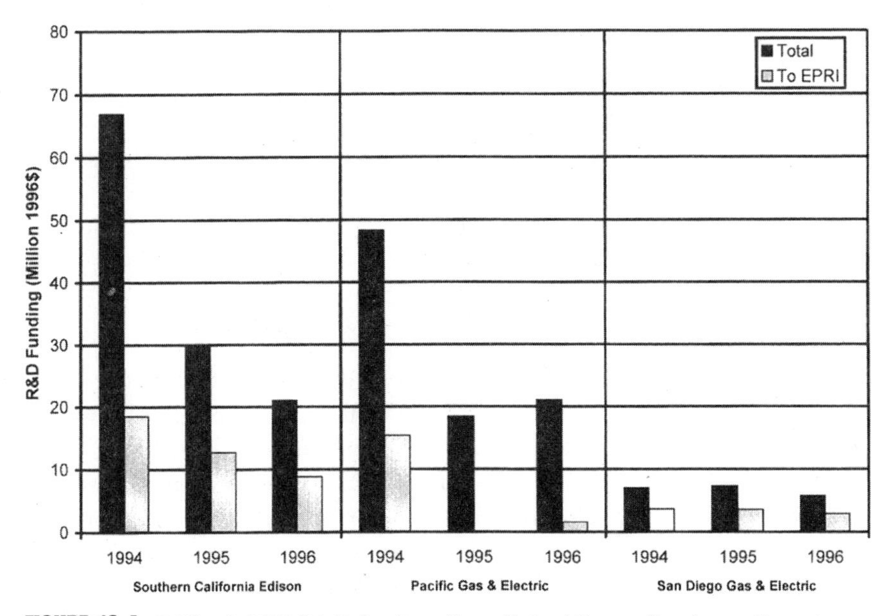

FIGURE 18.4. California IOU R&D funding. (From Federal Energy Regulatory Commission, U.S. Department of Energy, 1997: Form One Database 1994–1996.)

decreased more quickly than total utility investments in R&D. For example, PGE completely dropped out of EPRI in 1995 and then rejoined EPRI in 1996 at a significantly reduced level: Between 1994 and 1996 PGE's contribution to EPRI decreased by 90 percent. Meanwhile, during the same period PGE's total investments in R&D decreased by 56 percent.[15] The dramatic drop in PGE's funding to EPRI was especially significant because until this period PGE had been one of EPRI's most generous member utilities: In 1994 PGE was the third largest IOU contributor to EPRI.[16] The ongoing restructuring of the U.S. electricity industry and transition to a more competitive market are expected to lead to continuing declines in private sector investments in energy technology R&D.[17]

The drastic cuts in energy technology R&D funding among IEA member countries should sound an alarm: The wholesale dismantling of large portions of the industrial world's energy technology R&D infrastructure could seriously limit our ability to develop new technologies to meet emerging challenges linked to global climatic change.

R&D Investments, Innovation, and Patents

Three main approaches have been used to examine the relationship between investments in R&D and innovation: patent statistics, historical case studies, and econometric studies.[18] Each approach has strengths and weaknesses:

- Patent statistics are easily accessible and have a clear definition, yet the incentives to patent vary a great deal over time, space, and sector.
- Case studies provide a rich level of detailed information but lack the generalizability of larger, comparative data sets.[19]
- Econometric studies use production function models to examine the overall impacts of R&D on social output and productivity; however, they suffer from a range of problems associated with trying to infer causality from behavioral data based on correlation techniques.

In our analysis we choose to rely primarily on patent records because they provide a consistent metric over time and a sufficiently large data set for comparative quantitative analysis across economic and industrial sectors.

The rate of return on R&D in the U.S. economy has been estimated at 20–100 percent.[20] These estimates have been surprisingly consistent over time.[21] This high rate of return makes R&D one of the best areas for public and private sector investment. As illustrated in Table 18.2, estimates of the social rate of return on R&D investments are around 50 percent, and the private rates are around 20–30 percent. The clear message of Table 18.2 is that the spillovers

TABLE 18.2. Social and Private Returns to R&D Investments

Author (Year)	*Social Rate of Return (%)*	*Private Rate of Return (%)*
U.S. aggregate studies		
Bernstein-Nadiri (1988, 1989)	10–160	9–27
Bernstein-Nadiri (1991)	56	14–28
Griliches (1964)	35–40	—
Nadiri (1993)	50	20–30
Schere (1982, 1984)	64–147	29–43
Sveikauskas (1981)	50	10–23
Terleckyj (1974)	48–78	0–29
U.S. sectoral case studies		
Bredahl-Peterson (1976)	36–47	—
Evenson et al. (1979)	0–130	—
Huffman-Evenson (1993)	11–83	—
Mansfield (1977)	56	25
Schmitz-Seckler (1970)	37–46	—

Sources: Evenson et al., 1979; Griliches, 1995; and Nadiri, 1993.

from R&D are real and often large. In summary, economic studies have found that:

- The profitability of private R&D exceeds that of other investments (usually by a substantial margin).
- The social returns to private R&D are even larger.
- Both the social and private rates are significantly higher than the rate required for private sector investments in physical capital (typically around 10 percent).

Furthermore, because studies usually do not take into account all of the source of returns to R&D (such as improvements in quality of products to consumers and environmental benefits), they tend to underestimate the social returns to investments in private R&D.

How do we explain the fact that the rate of return for R&D investments is persistently high? R&D investments are inherently risky, so it might be expected that firms will require high rates of return from R&D investments. In addition, because it may be difficult for firms to communicate realistic expectations about an R&D project to potential investors, they may find it difficult to attract capital to R&D projects. Both of these phenomena are natural byproducts of the inherent uncertainty of R&D projects. As Cohen and Noll observe, "R&D risks are especially difficult to evaluate because research projects are necessarily designed to attack problems that have not been solved and for which there exists no directly relevant track record. Indeed, the more revolutionary the project's objective, the more difficult risk assessment will be."[22]

From an economic perspective, if risk and difficulty in communication are the dominant reasons for the unusually high rate of return for R&D investments, then the private sector should be viewed as effectively managing risks associated with R&D. However, a third factor often dominates the situation: Many of the benefits of R&D are difficult for private firms to appropriate and thus are realized by the broad public. This widely discussed form of market failure implies that the private sector is likely to underinvest in R&D and provides a rationale for a strong public role in encouraging R&D, either through government support for R&D activities or through policies aimed at creating incentives for the private sector to invest in R&D (i.e., through patent law or R&D tax incentives).

The rationale for a public role in the energy sector is particularly strong. The environmental, economic, and national security benefits to the public of investing in energy R&D are potentially very large.[23] In addition, as shown in Table 18.1, in the energy sector much of the existing capital stock has very long lifetimes. Therefore, it can take decades to commercialize new power systems. This makes the public role for R&D in the energy sector more critical.

Overall Pattern of Change

In Fig. 18.5 we present total U.S. patents granted and total U.S. funds invested in R&D between 1976 and 1996. Total U.S. patents include all patents granted in a given year.[24] Total U.S. investments in R&D include both public and private R&D.[25] As illustrated in the figure, during this period total U.S. investment in R&D increased from roughly $100 to $200 billion, and the total number of U.S. patents issued increased from roughly 70,000 to 110,000. Thus, between 1976 and 1996 both R&D investments and the number of patents issued in the United States roughly doubled.[26]

The fact that as R&D investments increased, patents increased proportionally over this period provides empirical support for the hypothesis that the United States has been underinvesting in R&D as a whole. If the United States had been investing in R&D at or near optimal levels at any time during the period, then further increases in R&D investments would be expected to result in diminishing returns. The absence of a saturation effect in the R&D investment relationship indicates that the United States has persistently underinvested in R&D.[27]

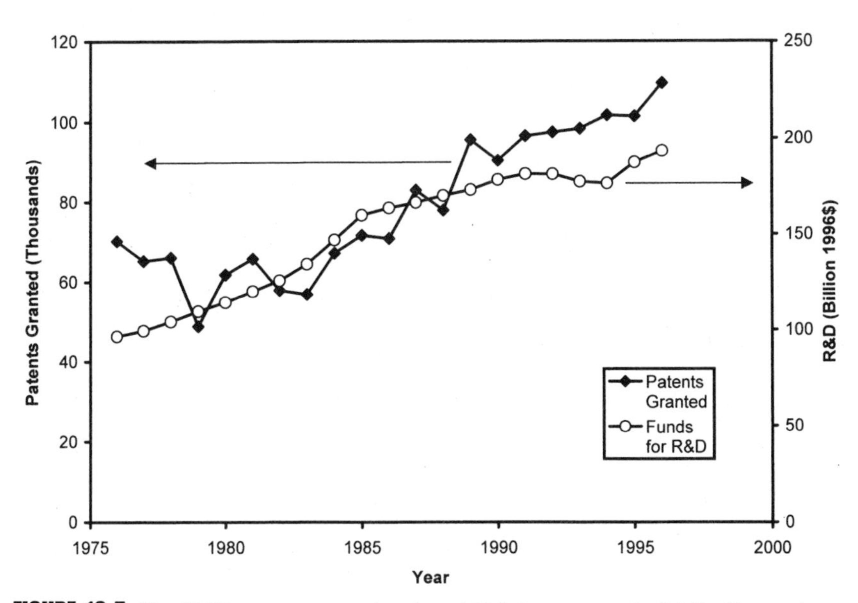

FIGURE 18.5. Total U.S. patents granted and total U.S. investments in R&D. (Patent data were drawn from Patent and Trademark Office, Patent Bibliographic Database; R&D data were drawn from National Science Foundation, 1998b.)

In Fig. 18.6 we present total U.S. energy-related patents and total (i.e., both public and private) U.S. investments in energy R&D between 1976 and 1996. Again we find that R&D investments and patents are highly correlated.[28] However, the trends in this figure are very different from the trends in Fig. 18.5. Between 1976 and 1996 U.S. energy R&D investments went though a dramatic boom–bust cycle, rising from $7.6 billion in 1976 to a high of $11.9 billion in 1979 and then decreasing through the 1980s and early 1990s to a low of $4.3 billion in 1996. Similarly, the number of patents related to energy technology experienced a boom–bust cycle, rising from 102 patents in 1976 to a high of 228 in 1981 and then declining to a low of 54 in 1994.

The divergence between the overall trends (Fig. 18.5) and energy sector trends (Fig. 18.6) during the 1976–1996 period is striking. Yet despite diverging trends they convey a similar message: For the U.S. economy as a whole and for the energy sector specifically, R&D investments and patents were highly correlated between 1976 and 1996. This again supports the hypothesis that the United States underinvests in energy related R&D. Furthermore, it illustrates that cuts in energy-related R&D have dramatic impacts on innovation in the energy sector.

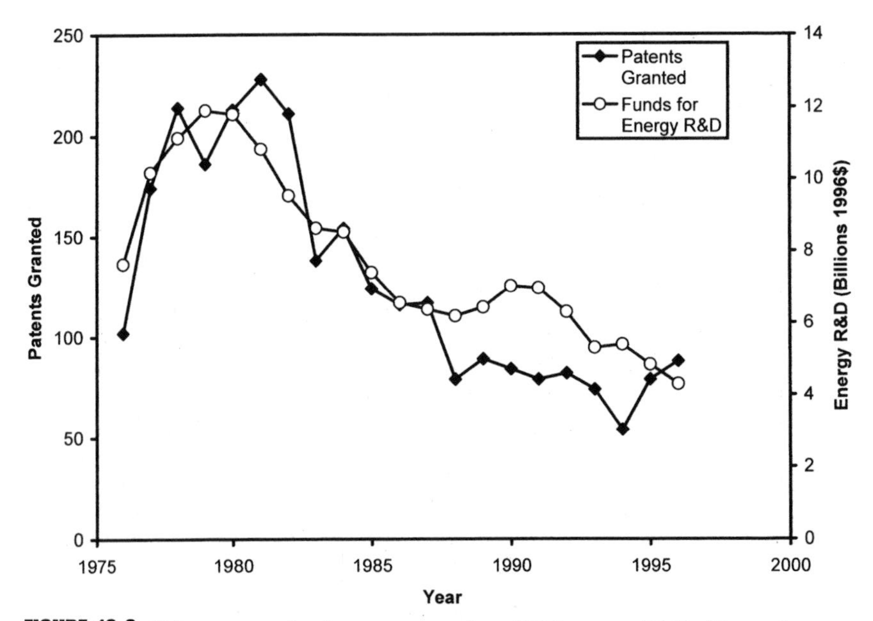

FIGURE 18.6. U.S. energy technology patents and total U.S. energy R&D. (Patent data were drawn from Patent and Trademark Office, Patent Bibliographic Database.) Total U.S. energy R&D includes both public and private R&D investments related to energy, defined as the sum of the DOE energy technology R&D (from National Science Foundation, 1998a; Meeks, R., 1997: *Special Data Compilation of NSF Historical Tables on Federal Energy R&D by Budget Function* [Washington, DC: NSF, Division of Science Resource Studies]), nonfederal industrial energy R&D (from National Science Foundation, 1998c: *Research and Development in Industry* [Washington, DC: NSF]; Wolfe, R., 1998: *Special Data Compilation of NSF Historical Tables on Industrial Energy R&D* [Washington, DC: NSF]), and EPRI R&D (from Electric Power Research Institute, various years: *Annual Report* [Palo Alto, CA: Electric Power Research Institute]).

Patterns of Change at the U.S. DOE

In Fig. 18.7 we compare DOE energy technology R&D with two measures of total DOE patents. The first measure, patents assigned to the DOE, roughly followed DOE energy technology funding between 1978 and 1996 (with a lag). As illustrated in the figure, patents assigned to the DOE increased between 1978 and 1985 and then decreased steadily through 1996.

The second measure, patents assigned or related to the DOE, is defined as all patents in the Patent and Trademark Office bibliographic database[29] that list "Department of Energy" in either the "patent assignee" or "government interest" field. The DOE typically is listed in the "government interest" field when it has funded research by an independent contractor resulting in a patent. Under these circumstances the patent usually is owned by the contractor and the DOE retains

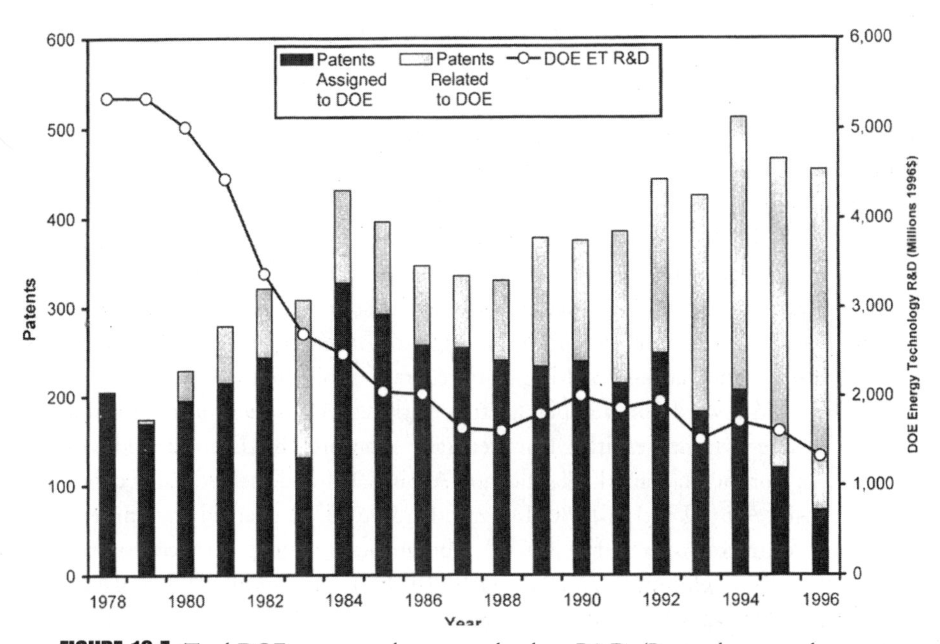

FIGURE 18.7. Total DOE patents and energy technology R&D. (Patent data were drawn from Patent and Trademark Office, Patent Bibliographic Database. R&D data were drawn from National Science Foundation, 1998a; Meeks, R., 1997: *Special Data Compilation of NSF Historical Tables on Federal Energy R&D by Budget Function* [Washington, DC: NSF, Division of Science Resource Studies].)

some rights to use or license the patent. As illustrated in the figure, the total number of patents assigned or related to the DOE increased throughout the period 1978–1996. The diverging trends between these two measures of total DOE patents can be explained by examining the increased efforts to encourage technology transfer from national laboratories and programs to the private sector.

Before 1980 the federal government largely retained the rights to patents resulting from federally sponsored R&D at national laboratories. The rationale for the government retaining the title to patents resulting from government sponsored R&D was that because federal funds were used to finance the work, they should be kept in the public sector, where they would be accessible to all interested parties. The government usually was willing to issue either an exclusive license or, more commonly, a nonexclusive license to companies on the patents it owned. During the late 1970s it was argued that this mode of operation was retarding the transfer of technology from federal laboratories to the private sector. Detractors of the existing system argued that without title (or at least an exclusive license) to an invention and the protection it conveys, a company would be unlikely to invest the additional time and money necessary for commercialization.

As summarized in Table 18.3, beginning in 1980 the U.S. Congress enacted a series of laws related to technology transfer that over time significantly modified the rules related to intellectual property resulting from R&D at national laboratories.[30] In 1980 Congress passed two important laws related to technology transfer: the Technology Innovation Act and the Patent and Trademark Amendments Act. The Technology Innovation Act made technology transfer a mission of all federal laboratories, and the Patent and Trademark Amendments Act relaxed existing restrictions on the transfer of rights to inventions resulting from government-sponsored R&D. Together these two acts created incentives and opportunities for national laboratories to loosen their control over the ownership of innovations resulting from federally sponsored R&D.

The trend toward a more open attitude with respect to the transfer of intellectual property rights resulting from federally sponsored R&D continued with the passage of the Trademark Clarification Act in 1984 and the Federal Technology Transfer Act (FTTA) in 1986. In particular, the FTTA enabled government-owned, contractor-operated (GOCO) laboratories to enter into cooperative research and development agreements (CRADAs) with nonfederal organizations.[31] Finally, in 1989 Congress passed the National Competitiveness Technology Transfer Act (NCTTA).

TABLE 18.3. Major Technology Transfer Initiatives by the U.S. Congress, 1980–1989

Year	*Legislation*	*Description*
1980	Technology Innovation Act (P.L. 96-480)	Made technology transfer a mission of all federal laboratories; also known as the Stevenson–Wydler Act.
1980	Patent and Trademark Amendments Act (P.L. 96-517)	Allowed universities and other performers of federally sponsored research to obtain title to inventions more easily; also known as the Bayh–Dole Act.
1984	Trademark Clarification Act (P.L. 98-620)	Granted broader authority to directors of government-owned, contractor-operated (GOCO) laboratories to engage in technology transfer activities; amended Bayh–Dole Act.
1986	Federal Technology Transfer Act (FTTA) (P.L. 99-502)	Allowed GOCO labs to enter into cooperative research and development agreements (CRADAs) with nonfederal organizations. However, under FTTA, GOCO labs could provide only material and personnel to projects, not direct funding to nonfederal organizations. Amended Stevenson–Wydler Act.
1989	National Competitiveness Technology Transfer Act (P.L. 101-189)	Extended authority to GOCOs to fully engage in cooperative research (i.e., sharing facilities, personnel, and funding for joint public–private projects). However, in practice very limited funding has been made available for CRADAs.

With the passage of the NCTTA, GOCO laboratories were allowed to fully engage in CRADAs (i.e., share personnel, equipment, or financing for R&D with private firms). The NCTTA also enabled GOCO laboratories to assign private firms the rights to intellectual property resulting from CRADAs. However, when entering a CRADA the federal government typically retains a nonexclusive license to any intellectual property resulting from the agreement.

As a result, between 1989 and 1995, the DOE signed more than 1,000 CRADAs.[32] It is not surprising that the progression from very tightly controlled to openly flexible ownership of intellectual property resulting from R&D at national laboratories parallels the increasing gap between patents assigned to the DOE and patents assigned or related to the DOE.

The divergence between patents assigned to the DOE and patents assigned or related to the DOE can also be seen in specific energy technology subsectors. In Figs. 18.8 and 18.9 we compare DOE energy technology R&D with patents (assigned to the DOE and assigned or related to the DOE) for the renewable and fossil energy technology subsectors. As illustrated in the figures, over time the number of patents assigned or related to the DOE for both of these energy

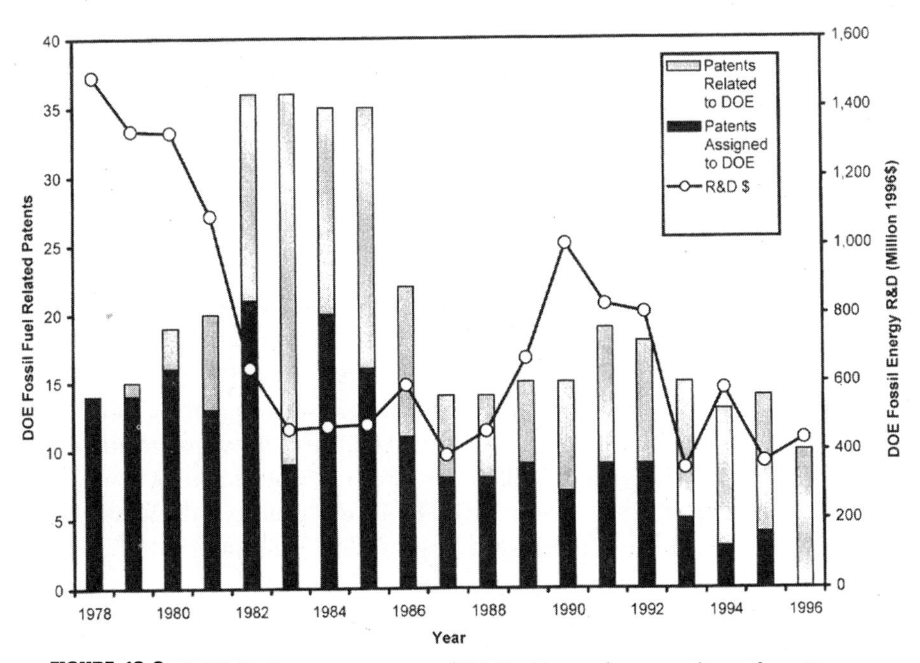

FIGURE 18.8. DOE fossil energy patents and R&D. (Patent data were drawn from Patent and Trademark Office, Patent Bibliographic Database. R&D data were drawn from National Science Foundation, 1998a; Meeks, R., 1997. *Special Data Compilation of NSF Historical Tables on Federal Energy R&D by Budget Function* [Washington, DC: NSF, Division of Science Resource Studies].)

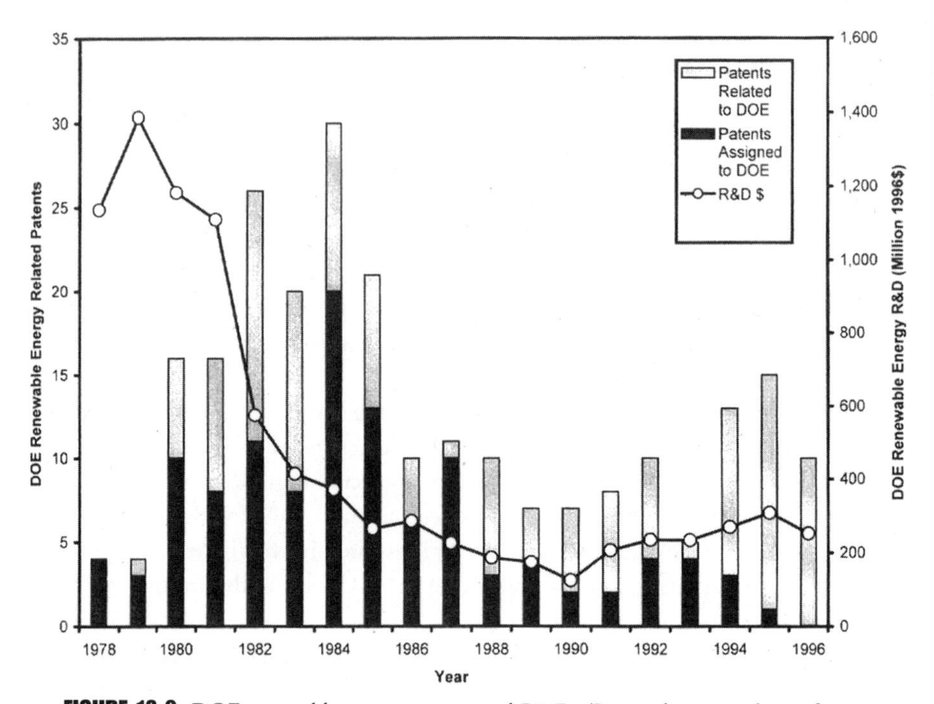

FIGURE 18.9. DOE renewable energy patents and R&D. (Patent data were drawn from Patent and Trademark Office, Patent Bibliographic Database. R&D data were drawn from National Science Foundation, 1998a; Meeks, R., 1997: *Special Data Compilation of NSF Historical Tables on Federal Energy R&D by Budget Function* [Washington, DC: NSF, Division of Science Resource Studies].)

technology subsectors has begun to diverge from the more traditional set of patents simply assigned to the DOE. In addition, both subsectors exhibit dramatic growth in the number of patents issued during the early 1980s and then a rapid decline during the late 1980s. The figures illustrate a time-delayed link between R&D investments and R&D output (in the form of patents) in the energy sector.[33] Furthermore, it illustrates the potential impact of dramatic boom–bust cycles on R&D productivity. This is consistent with the findings of the Yergin Commission report, which argued that historically volatility has worked against productivity in energy R&D investments.[34]

R&D Intensities Across Sectors

An alternative measure of the returns on investments is R&D intensity (defined as R&D as a percentage of net sales). R&D intensities for selected U.S. sectors in 1995 are shown in Fig. 18.10. Comparing R&D intensities across sectors reinforces our concern about the level of investment in energy technology R&D in the United States. As illustrated in Fig. 18.10, the energy sector's R&D intensity is extremely low in comparison to many other sectors. In fact, the drugs and medicine, professional and scientific equipment, and communication equipment sectors all exhibit R&D intensities that are more than an order of magnitude above the 0.5 percent of sales devoted to R&D in the energy sector. This low level of investment is particularly troubling given the high capital costs and

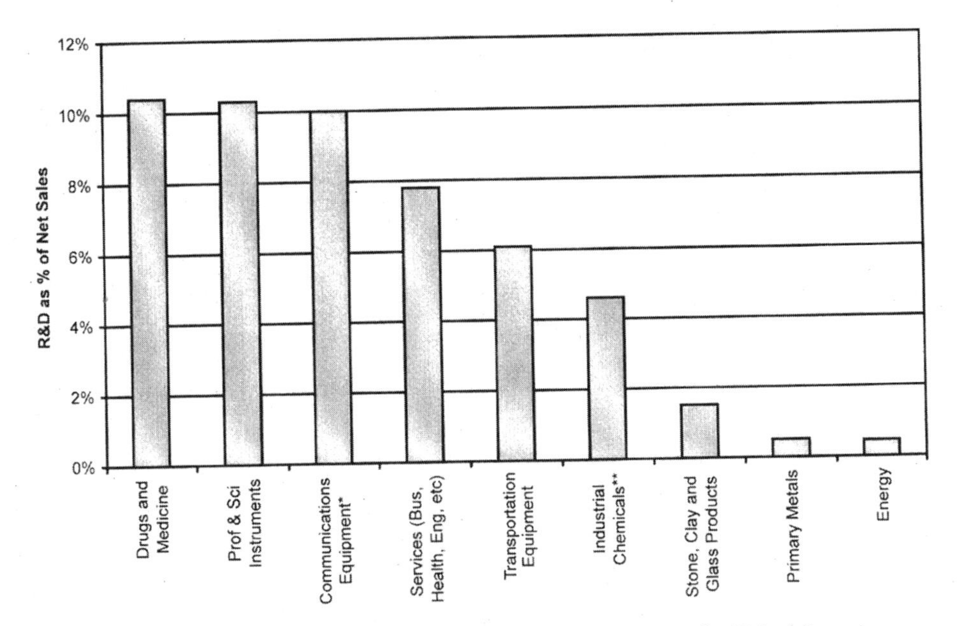

FIGURE 18.10. R&D as percentage of net sales for selected sectors in the United States in 1995. (Data for each industrial category in the figure, except energy, were drawn from National Science Foundation, 1998c: *Research and Development in Industry* (Washington, DC: NSF].) The data in the figure include both public and private funding for R&D. Energy R&D as a percentage of net sales was calculated from total (public and private) industrial energy R&D (from NSF, 1998c) and total energy expenditures in the United States (from Energy Information Administration, 1997. *State Energy Price and Expenditure Report 1995* [Washington, DC: EIA, U.S. Department of Energy]). The energy R&D data are gathered across industrial sectors (i.e., they are for industry as a whole).
Notes: The industrial sectors in the figure correspond to the following Standard Industrial Classification (SIC) codes: Drugs and Medicine (283), Professional & Scientific Instruments (38), Communications Equipment (366), Services (701, 72, 73, 75–81, 83, 84, 87, 89), Transportation Equipment (37), Industrial Chemicals (281–282, 286), and Stone, Clay, and Glass Products (32).
*The most recent year in which data are available for communication equipment is 1990.
**The most recent year in which data are available for industrial chemicals is 1992.

long planning horizons needed to bring new energy technologies to commercial application and the central role of energy in the environment–economy nexus.

Given that the energy sector is very capital intensive and produces a commodity that has small margins, one might expect it to a have a low R&D intensity. However, the differences between sectors in Fig. 18.10 are so striking that they force us to confront a critical question: In terms of encouraging technological change, is the energy sector more like a low-technology sector (i.e., the primary metals sector) or a high-technology sector (i.e., the communication equipment sector)? Because technology plays a such a critical role in finding, transforming, and exploiting energy, especially in an environmentally sound manner, we would expect the energy sector to be at least somewhere in the middle. The energy sector's extremely low R&D intensity clearly is another indicator of underinvestment in R&D in the energy sector.

Linking the Laboratory and the Market

One of the most dramatic—and in many ways obvious—findings in recent studies of the energy sector in both developed and developing nations is that excessive attention to basic R&D or applied work alone will not produce the best returns on the investment. Energy technologies are inherently applied, and programs and policies intended to decarbonize the economy by disseminating cleaner energy systems need to reflect that real-world endpoint. New strategies that foster market transformation, namely the accelerated introduction of new technologies, have been found to be effective in promoting both energy efficiency and renewable energy technologies.[35] A number of approaches are possible, of which we outline the most prominent in this section.[36]

Development Subsidies

Subsidies in the product development phase typically support the classic R&D phase, premarket design, or possibly diversification from a prototype to models tailored to particular market niches. These subsidies often are in the form of a direct grant or loan to a particular manufacturer, often on the basis of a promising engineering design. A benefit of this approach is that it can be simple to evaluate the proposal and to chart the impact of the subsidy in terms of product development. One drawback is that funding institutions may fall into the trap of picking winners before any feedback from end users is available.[37] A number of recent technology and environmental policy efforts have illustrated opportunities to move beyond this roadblock by promoting technologies in competitive programs in which subsidies are provided for combinations of technical and managerial innovations. Recent efforts to promote improved cookstoves in China[38] and national-level programs in sub-Saharan Africa[39] were based on provincial-level competitions to best meet the energy efficiency and economic needs of households.

Technology Sales Subsidies

Sales subsidies are also a traditional mechanism to support and develop the market for a new technology. In the classic formulation, end users receive a rebate from a third party (often the government) for the purchase of a technology. The benefits of this approach are that the subsidy can directly reduce up-front capital cost, which is often the critical obstacle for the dissemination of new technologies. Conversely, the drawback of this approach is that lump-sum subsidies may not provide an incentive for the performance of the technology, only the initial sales. In Nepal, however, subsidies for biogas digesters have been provided in stages over several years to guarantee that the systems perform well. These subsidies are incremental to provide the most support to the poorest and most remote households. Further, the biogas digester subsidy is provided to the installer, who also holds the loan to cover end-user purchases. The advantage of this arrangement is that the risk of a novel, often untested technology does not fall on the end user.

Market Support and Educational Subsidies

There has been a recent explosion of interest in subsidies that avoid direct financial subsidies while still supporting an emerging new technology or clean energy practice. One way to accomplish this is to subsidize the educational, training, or other knowledge-based aspects of the R&D-to-commercialization pipeline. For many technologies, particularly in developing nations, there is only a weak link between a promising new technology and the marketing skills and resources needed to achieve commercial success. Training programs, efforts to assist with market development, and other such "soft" subsidies can make a great deal of difference. The benefits often are far greater than direct hardware subsidies. An example of this approach is the development of improved cookstoves in Kenya, where marketing was subsidized but the cost of the stoves themselves reflected the actual production and market costs.

An important example program that integrates pieces from each of the three subsidy categories is that of the Greenfreeze refrigerator program.[40] The Greenfreeze program in Europe brought together scientists who had extensively researched the use of propane and butane as refrigerants with an East German company, DKK Scharfenstein. The meeting between the scientists and DKK Scharfenstein resulted in the birth of Greenfreeze technology for domestic refrigeration.

When DKK Scharfenstein (renamed Foron) announced its intention to mass produce Greenfreeze refrigerators, Greenpeace gathered tens of thousands of preorders for the yet-to-be produced product from environmentally conscious consumers in Germany. This overwhelming support from the public secured the capital investment needed for the new Greenfreeze product. The major European household appliance manufacturers, who had already invested in hydrofluorocarbon-134a refrigeration technology as the substitute for chlorofluorocarbons, were at first resistant to the hydrocarbon technology. However, once DKK Scharfenstein proceeded with its plans, the major manufacturers also began to convert to hydrocarbons. Within two years Greenfreeze became the dominant technology in Europe.

Many models of Greenfreeze refrigerators are now on sale in Germany, Austria, Denmark, France, Italy, Netherlands, Switzerland, and Britain. All major European companies—Bosch, Siemens, Electrolux, Liebherr, Miele, Quelle, Vestfrost, Whirlpool, Bauknecht, Foron, and AEG—are marketing Greenfreeze technology–based refrigerators. The German market has been fully converted to Greenfreeze technology, and in countries such as Germany and Denmark, more than 100 different Greenfreeze models are available for purchase.

These cases illustrate the need for a new approach to disseminating clean energy technologies: integrating energy R&D in the classic sense with efforts to actively understand and interact with the market. The danger lies in too large a public sector role directly in the market. However, numerous success stories exist, i.e., when public sector support has enhanced and encouraged active market development of clean energy solutions. This approach, called mundane science,[41] explicitly recognizes the need for a real and interactive connection between investments and needs rather than an excessively academic or theoretical decoupling of these actions.

Conclusions

The data on international trends in energy technology R&D funding, U.S. energy technology patents and R&D funding, and U.S. R&D intensities across selected sectors present a disturbing picture. First, energy technology funding levels have declined significantly over the past two decades throughout the industrialized world. The most dramatic reductions have taken place in the United States, Germany, and the United Kingdom. In the long run these cuts are likely to reduce the capacity of the energy sector to innovate. Second, our examination of energy technology R&D and patents in the United States reveals a significant correlation between R&D investments and patents. This finding is consistent with and extends previous work examining the relationship between R&D, patents, and innovation. Furthermore, the data support the assertions that investments in R&D provide significant and important returns and that the United States underinvests in energy technology R&D. Again we find that declining investments in energy technology R&D are likely to reduce our capacity to innovate. Finally, we observe that the R&D intensity of the U.S. energy sector is significantly below that of other technology-intensive sectors.

One surprise in the data is that over the past two decades, while DOE R&D investments and the number of patents assigned to the DOE declined, the total number of patents assigned or related to the DOE increased. Similarly, we find a divergence between the number of patents assigned to the DOE and the number of patents assigned or related to the DOE for the renewable and fossil energy technology subsectors. We trace this divergence to the range of technology transfer initiatives put in place between 1980 and 1989. Policies can make a difference, which is why proper R&D planning is so critical.

Although efforts to encourage technology transfer during the past two decades have been successful at increasing the total number of patents assigned or related to the DOE, this shift in policy cannot ameliorate the problems created by a declining federal energy R&D portfolio. This shift in policy has resulted in the transfer of ownership, from the public to the private sector, of intellectual property resulting from R&D at national laboratories. Although private ownership often is critical to commercialization, proprietary control of the majority of advances in basic energy research may be a disincentive for the further development, dissemination, and implementation of clean, efficient energy systems.

U.S. underinvestment in an area at the heart of the environmental–economic nexus is detrimental for both long-term U.S. energy security and global environmental sustainability. In particular, because the U.S. path is intimately tied to the evolution of global energy systems, this underinvestment in energy technologies is likely to reduce the options available in the future to the global community to address the environmental impacts of energy production and climate change. Ultimately, meeting emerging domestic and international challenges will entail increasing both U.S. and international energy technology R&D.

We conclude with two recommendations that will help to move us in a direction that addresses the challenges posed by global climate change, First, a broader view of the energy R&D process must be developed. This broader view would include focusing on pressing but often overlooked problems—what one might call the mundane research on sustainable energy technologies and the policies conducive to bringing these technologies into use.[42] This is particularly pressing given the fact that more than 2 billion people worldwide (roughly 35 percent of the world's population) rely primarily on wood, charcoal, and other traditional biomass fuels to meet their energy needs. Many more rely on kerosene lanterns and diesel generators. Meanwhile, a disproportionate share of energy technology R&D resources are focussed on advanced combustion systems, commercial fuels, and large centralized power facilities. Small-scale, decentralized energy systems can play a significant role in meeting the combined challenges of development and environmental conservation, yet there has been a general pattern of neglect of and underinvestment in such systems. There is now an important opportunity for even small investments in mundane energy technology R&D to produce large environmental and social returns.

Second, there is a need to broaden the definition of R&D to include the dissemination and sustained use of new energy technologies. Many analysts argue that governments should support only the development of new technologies, not their commercialization, but there are compelling reasons to look beyond the traditional role of government in R&D. In particular, there is a legitimate role for public funding of market transformation programs focused on clean energy technologies.[43] These programs should focus on technologies with steep experience curves, high probabilities of market penetration once the subsidies are removed, and price elasticities of demand of 1.0 or more.[44] Limiting market transformation programs to clean energy technologies enhances their performance by providing environmental benefits. The other three conditions ensure a strong indirect demand effect.

Acknowledgments

We thank S. DeCanio, S. Devotta, J. Holdren, R. May, A. Rosenfeld, and V. Ruttan for comments and advice. This work was supported in part by the Summit Foundation and the Class of 1934 Preceptorship at Princeton University, both awarded to D. M. K. D. M. Kammen also gratefully acknowledges support from the Energy Foundation.

Notes

1. Much of the analysis presented in this chapter draws on material in Margolis, R. M. and D. M. Kammen, 1999: "Evidence of under-investment in energy R&D in the United States and the impact of federal policy," *Energy Policy,* 27: 575–584; and Mar-

golis, R. M. and D. M. Kammen, 1999: "Underinvestment: the energy technology and R&D policy challenge," *Science,* 285: 690–692. Other important studies that discuss the critical role of energy in responding to the threat of climate change include Parson, E. A. and D. W. Keith, 1998: "Fossil fuels without CO_2 emissions," *Science,* 282: 1053–1054; and United Nations Development Programme, 1997: *Energy After Rio: Prospects and Challenges,* A. K. N. Reddy, R. H. Williams, and T. B. Johansson (eds.) (New York: United Nations Development Programme).

2. The rate of decarbonization varies greatly across nations. However, the estimated rate of decarbonization (grams of carbon per megajoule of energy) needed to offset economic expansion is roughly 3 percent/year. Nakicenovic, N., A. Grubler, A. Inaba, S. Messner, S. Nilsson, Y. Nishimura, H. H. Rogner, A. Schafer, L. Schrattenholtzer, M. Strubegger, J. Swisher, D. Victor, and D. Wilson, 1993: "Long-term strategies for mitigating global warming," *Energy,* 18 (5): 401–609.
3. Hoffert, M. I., K. Caldeira, A. K. Jain, E. F. Haites, L. D. D. Harvey, S. D. Potter, M. E. Schlesinger, S. H. Schneider, R. G. Watts, T. M. L. Wigley, and D. J. Wuebbles, 1998: "Energy implications of future stabilization of atmospheric CO_2 content," *Nature,* 395: 881–884; Kinzig, A. P. and D. M. Kammen, 1998: "National trajectories of carbon emissions: analysis of proposals to foster the transition to low-carbon economies," *Global Environmental Change,* 8 (3): 183–208.
4. See Flavin, C., 1997: "Banking against warming," *World Watch,* 10 (6): 25–35; Nordhaus, W. D., 1994: *Managing the Global Commons* (Cambridge, MA: MIT Press).
5. International Energy Agency, 1997: *IEA Energy Technology R&D Statistics, 1974–1995* (Paris: IEA, Organisation for Economic Co-operation and Development).
6. In the United States this has been a prominent political position. For example see, Sommers, L. H., 1997: *Comments made at the White House Conference on Climate Change,* October 6, http://www.whtehouse.gov/WH/EOP/OSTP/html/OSTP_Home.html.
7. Blinder, A. S., 1997: "Needed: planet insurance," *New York Times,* October 22.
8. Two widely cited studies that make this argument, by prominent U.S. economists, include Cline, W., 1992: *The Economics of Global Warming* (Washington, DC: Institute for International Economics); Nordhaus, W. D., 1994: *Managing the Global Commons* (Cambridge, MA: MIT Press).
9. This argument has been made persuasively in Benedick, R. E., 1991: *Ozone Diplomacy: New Directions in Safeguarding the Planet* (Cambridge, MA: Harvard University Press); Keohane, R. O. and M. A. Levy, 1996: *Institutions for Environmental Aid* (Cambridge, MA: MIT Press); and Khosla, A. and K. Chatterjee, 1997: "Is joint implementation a realistic option," *Environment,* 39 (9): 46–47.
10. International Energy Agency, 1997: *IEA Energy Technology R&D Statistics, 1974–1995* (Paris: International Energy Agency, Organisation for Economic Cooperation and Development).
11. Energy technology R&D data were drawn from National Science Foundation, 1998a: *Federal R&D Funding by Budget Function* (Washington, DC: National Science Foundation); National Science Foundation, 1983: *Federal R&D Funding for Energy: Fiscal Years 1971–1984* (Washington, DC: NSF). Here we define DOE

energy technology R&D as the sum of the following DOE R&D categories in the NSF reports: fossil energy, nuclear energy, magnetic fusion, solar and renewables, and energy conservation. This excludes categories such as basic energy sciences, biological and environmental research, and other miscellaneous research. Using this definition, energy technology R&D accounted for 55 percent of the DOE's total R&D budget in 1996. Energy technology R&D is a narrower category than energy R&D, as shown in Fig. 18.3, hence its lower values. Also, note that all dollar values cited throughout this chapter (unless otherwise indicated) have been converted from current to constant 1996 dollars using the gross domestic product chain-type price index (available online: www.bea.doc.gov).

12. Federal Energy Regulatory Commission, U.S. Department of Energy, Form One Database 1994–1996.
13. Ibid.
14. Energy Information Administration, 1997: *Electric Power Annual 1996,* Vol. 1 (Washington, DC: EIA, U.S. Department of Energy).
15. Federal Energy Regulatory Commission, Form One Database.
16. Ibid.
17. For good discussions about the unfolding impacts of restructuring on energy technology R&D investments, see Dooley, J. J., 1998: "Unintended consequences: energy R&D in a deregulated energy market," *Energy Policy,* 26 (7): 547–555; General Accounting Office, 1996: *Federal Research: Changes in Electricity-Related R&D Funding* (Washington, DC: U.S. General Accounting Office); Margolis, R. M., 1998: "Addressing the emerging crisis in energy technology R&D," paper presented at USAEE/IAEE 19th Annual North American Conference on Technology's Critical Role in Energy and Environmental Markets, October 18–21.
18. There are other possible indicators, such as number of publications, prototypes, software, licenses, and cooperative agreements.
19. Good examples of energy sector case studies are included in Cohen, L. R. and R. G. Noll, 1991: *The Technology Pork Barrel* (Washington, DC: The Brookings Institution); Lawrence Berkeley Laboratory, 1995: *From the Lab to the Marketplace: Making America's Buildings More Energy Efficient* (Berkeley, CA: Lawrence Berkeley Laboratory, U.S. Department of Energy).
20. For example see DeCanio, S. J., 1998: *The Economics of Climate Change* (San Francisco: Redefining Progress); Griliches, Z., 1995: "R&D and productivity: Econometric results and measurement issues," in *Handbook of the Economics of Innovation and Technological Change,* P. Stoneman (ed.) (Oxford, England: Blackwell); Jones, C. I. and J. C. Williams, 1998: "Measuring the social return to R&D," *Quarterly Journal of Economics,* 113 (4): 1119–1135; Nadiri, M. I., 1993: *Innovations and Technological Spillovers* (Cambridge, MA: National Bureau of Economic Research); Stokey, N. L., 1995: "R&D and economic growth," *Review of Economic Studies,* 62: 469–489.
21. For example, see Evenson, R. E., P. E. Waggoner, and V. W. Ruttan, 1979: "Economic benefits from research: an example from agriculture," *Science,* 205 (September 14): 1101–1107; Griliches, Z., 1987: "R&D and productivity: measurement issues and econometric results," *Science,* 237: 31–35; Mansfield, E., 1972: "Contri-

bution of R&D to economic growth in the United States," *Science,* 175: 477–486.
22. Cohen and Noll, 1991, p. 20.
23. President's Committee of Advisors on Science and Technology, 1997: *Federal Energy Research and Development for the Challenges of the Twenty-First Century* (Washington, DC: Energy Research and Development Panel, PCAST).
24. U. S. Patent and Trademark Office, Patent Bibliographic Database, available online: http://www.uspto.gov.
25. National Science Foundation, 1998b: *National Patterns of Research and Development Resources* (Washington, DC: NSF).
26. A linear regression with R&D as the independent variable and patents as the dependent variable yields an R^2 of 0.72 and a t statistic of 7.0 (significant at 1 percent level).
27. For additional indicators see Jones, C. I. and J. C. Williams, 1998: "Measuring the social return to R&D," *Quarterly Journal of Economics,* 113 (4): 1119–1135. Jones and Williams estimate the underinvestment in R&D to be by at least a factor of four.
28. A linear regression with energy R&D as the independent variable and energy-related patents as the dependent variable yields an R^2 of 0.84 and a t statistic of 10.0 (significant at 1 percent level).
29. U. S. Patent and Trademark Office, Patent Bibliographic Database, available online: http://www.uspto.gov.
30. For an interesting discussion of the international implications of changes in technology transfer–related legislation, see Mowery, D. C., 1998: "The changing structure of the US national innovation system: implications for international conflict and cooperation in R&D policy," *Research Policy,* 27: 639–654.
31. There are 10 main DOE GOCO laboratories: Lawrence Berkeley Laboratory, Los Alamos Laboratory, Oak Ridge National Laboratory, Argonne National Laboratory, Brookhaven National Laboratory, Sandia National Laboratory, Idaho National Engineering Laboratory, Lawrence Livermore National Laboratory, Pacific Northwest Laboratory, and the National Renewable Energy Laboratory. This group is often called "the national laboratories." The Galvin Commission report noted that approximately 30 percent of DOE's energy technology R&D budget was directed to national laboratories in 1994. See Secretary of Energy Advisory Board, 1995: *Energy R&D: Shaping Our Nation's Future in a Competitive World* (Washington, DC: Secretary of Energy Advisory Board, U.S. Department of Energy).
32. Mowery, D. C., 1998: "Collaborative R&D: How Effective Is It?" *Issues in Science and Technology,* 15 (1): 37–44.
33. One would expect there to be a 4- to 6-year delay between investments in R&D and resulting patents (i.e., a few years for R&D to produce results and then couple of years between application and the granting of a patent). A 2- to 3-year delay between patent application and award is empirically confirmed by data on DOE patents. For example, of all the patents applied for by DOE in 1990, 409 were granted to the DOE between 1990 and 1996. Most of these patents were granted within 1, 2, or 3 years of their application date (cumulatively 43 percent were granted within 1 year, 87 percent within 2 years, and 96 percent within 3 years).
34. Secretary of Energy Advisory Board, 1995: *Alternative Futures for the Department of*

Energy National Laboratories (Washington, DC: Secretary of Energy Advisory Board, U.S. Department of Energy).

35. See Duke, R. D. and D. M. Kammen, 1999: "The economics of energy market transformation initiatives," *The Energy Journal,* 20 (4): 15–64.
36. This section draws heavily on Chapter 17 "Case Studies" by Anderson, S., A. Mathur, S. Devotta, M. Iyer, and D. M. Kammen (Section and Coordinating Lead Authors) in the Intergovernmental Panel on Climate Change Working Groups II and III Report, 2000: *Methodological and Technological Issues in Technology Transfer* (New York: Cambridge University Press).
37. See Cohen and Noll, 1991 and Margolis and Kammen, 1999.
38. Barnes, D. F., K. Openshaw, K. R. Smith, and R. van der Plas, 1994: "What makes people cook with improved biomass stoves?" *World Bank Technical Paper No. 242* (Washington, DC: World Bank); Cabraal, A., M. Cosgrove-Davies, and L. Schaeffer, 1995: "Best Practices for Photovoltaic Household Electrification Programs," *World Bank Technical Paper Number 324*: Asia Technical Department Series (Washington, DC: World Bank).
39. Kammen, D. M., 2000: "Case study #1: Research, development, and commercialization of the Kenya Ceramic Jiko (KCJ)", in *Methodological and Technological Issues in Technology Transfer* (New York: Cambridge University Press), pp. 383–384.
40. Greenpeace, 1999, available online: http://www.greenpeace.org/~ozone/unep_ods/8greenfreeze.html.
41. For a more detailed discussion of mundane science and its role in sustainable resource management, see Kammen, D. M. and M. R. Dove, 1997: "The virtues of mundane science," *Environment,* 39 (6): 10–15, 38–41.
42. Ibid.
43. For a more detailed discussion of the rationale for clean energy technology market transformation programs, see Kammen, D. M., 1999: "Bringing power to the people: promoting appropriate energy technologies in the developing world," *Environment,* 41 (5): 10–15, 34–41.
44. The price elasticity of demand is defined as the percentage change in demand stemming from a 1.0 percent decrease in price. Values higher than 1.0 imply an increase in revenues as the price of a good declines.

CHAPTER 19

Business Capitalizing on Energy Transition Opportunities

Orie L. Loucks

Business has always adapted to change in the consumer's needs for services and products. Competing to meet these needs is what drives this adaptation, and doing so profitably is the central idea in entrepreneurship. Fossil fuel combustion is seen as damaging human health, natural resources, and climate systems and creating dependence on a foreign import. This represents a great business opportunity for those who can provide alternative energy products and services, especially from domestic U.S. sources.

Work with business faculty and students at Miami University, through the Center for Sustainable Systems Studies,[1] has taught us three lessons. First, businesses large and small are committing to various forms of sustainable development, locally to globally, and doing so profitably. Second, new environmental audit capabilities are emerging that help ensure that steps toward sustainability are measurable and reportable. Third, business' view of sustainability can and does consider the long-term outcome for communities (social concerns), for continuity of jobs, infrastructure, and services (economic interests), and for natural resources (environmental stability). A balance of all three lessons must be apparent when we consider the response by business and industry to the challenge facing society from our use of fossil fuels.

Fossil Fuel Emission Effects on Human Health and Ecosystems

Alternatives to fossil fuels are being considered because of the seriousness of environmental risks and economic hazards from continuing fossil fuel use. The

1990s have seen three separate policy assessments in the United States involving the effects of emissions from fossil fuel use: damages and risks from acid deposition (the National Acid Precipitation Assessment Program),[2] an evaluation of ground-level ozone and aerosol particulate health risks (carried out by the U.S. Environmental Protection Agency [EPA]),[3] and the assessment of climate change impacts, first by the Office of Technology Assessment[4] and recently by the Intergovernmental Panel on Climate Change.

Combustion of coal and oil is a well-known source of nitrogen and sulfur oxides. These gases are the precursors of fine particles (aerosols) in the lower atmosphere that reduce visibility, cause corrosion, and wash out in precipitation as acid rain. In addition, high positive correlations have been known for years between elevated levels of sulfate and nitrate aerosols in industrial areas and high levels of bronchial or pulmonary illness and associated human mortality rates. The particles typically occur as a salt, such as ammonium sulfate, that can become engorged with water, reducing visibility and often creating highly acidic fog. Other substances including metals and unused carbon particles are also present and now known to be carcinogenic. These constituents, in combination, contribute to irritation of the bronchial passages, especially in sensitive children, asthmatic people, and older adults, and can lead to prolonged illness and death.[5]

Another pollutant derived from reactions between fossil fuel combustion products is ozone (O_3), a highly reactive form of oxygen that affects human health and most plants. It forms from precursors such as nitrogen oxides and volatile organic compounds, both originating from automobile emissions and other sources. High ozone levels often are associated with the acidic aerosols just noted, but the effects of O_3 on humans include stimulation of rapid respiration rates, causing asthma and asthma-like symptoms, even at the concentrations permitted by the U.S. Clean Air Act and the recent ozone standard (120 ppb). The effects of ozone pollution are understood well in agricultural ecosystems, where the rapid respiration caused by O_3 (compared with O_2) leads to loss of the accumulating carbohydrate in leaves before it is transferred to seeds and fruit. Observed reductions in soybean yields, for example, are as much as 35 percent for some varieties and average 15 percent over most of the midwestern and eastern United States under existing ambient O_3 levels.[6] The effect of O_3 on forest species varies widely, from negligible to significant reductions in tree growth. Effects on even moderately sensitive species tend to be expressed most clearly as reduced root growth, thereby increasing the sensitivity of these species to commonly occurring droughts and insect and disease infestations, especially after drought.[7]

The effects from washout of the acidic aerosols in rain and snow are varied and costly in the long term. They include leaching of essential cations (the positively charged nutrients such as calcium and magnesium) out of the surface soil, loss of small soil and aquatic animals (essential for nutrient cycling) because the low pH alters the ionic balance in tissues of these sensitive species, aluminum toxicity in plants and animals at the new low pH (high acidity), and through nitrogen enrichment, lower availability of high-carbon (and low-nitrogen) compounds necessary for insect and disease defense in plants, creating conditions that slow growth and increase plant and animal mortality rates.[8]

My research provides an estimate of the annual damage (externality of fossil fuel use) from ground-level ozone and from acid aerosols, including their washout as acid rain. The result for acid aerosols ranges from $2 to $10 billion annually, excluding health effects, and up to $250 billion more annually when health effects are included.[9] An average estimate is $92 billion/yr. In addition, damage to human health, agricultural crops, and forest growth from ground-level ozone during the past 15 years has been estimated at $30–$70 billion annually. A best estimate of damage from ozone pollution is $47 billion/yr. The forestry and agriculture sectors account for about $40 billion/yr of this figure.

By comparison, consider estimates of the annual externality costs of climate change. A recent report by the UN Environment Programme estimates annual global costs in 2050 at U.S. $300 billion,[10] of which the U.S. share would be $75 billion. Other authors have estimated the U.S. damages at up to $100 billion/yr by 2050, in constant dollars. Expressing these results as present value (discounted at 2 percent) yields an average for climate damages today of about $20 billion annually.[11] In other words, the health-related damages from fossil fuel use and acid gas emissions should be seen as a much greater cost for the United States than the discounted present value of physical and health-related effects from climate change, at least over the next 10–20 years.

Let us consider how large these externalities are in relation to the value of all goods and services exchanged in the United States (gross domestic product [GDP]) in 1999. The direct costs total $159 billion annually for the United States. Other kinds of costs, including regulating emissions, maintaining energy security, and replacing damaged ecosystem services would bring the total cost to at least $200 billion annually for the United States, and potentially twice that.[12] Expressed in relation to the 1999 GDP, $7 trillion, the combined fossil fuel externalities make up 2–5 percent of U.S. GDP. Avoiding this economic burden could be a boon to the U.S. economy. Damage charges against fossil fuel industries to compensate the public for health and property damage (similar to those now being levied on tobacco products), set equal to the minimal estimate of damages ($200 billion), would be $144/ton of carbon, equivalent to 96 cents per gallon of gas. The human, economic, and regulatory burdens from fossil fuel emissions are compelling reasons to pursue low-polluting or nonpolluting energy sources and services in an open marketplace.

Business Responses: British Petroleum

The British Petroleum Company (BP) is one example of a large fossil fuel producer that is beginning to invest in alternative energy products. Among the first steps being taken is an initiative to understand and control the company's own greenhouse gas (GHG) emissions.[13] For example, a CO_2 protocol is being promoted to standardize procedures for measuring GHG emissions from the company. In conjunction with these steps, a CO_2 management team has been established to identify CO_2 reduction options, mostly through energy conservation. BP maintains a database of carbon reduction options with more than 180 projects, of which 50 were scheduled for 1999.[14] The chief executive officer, John Browne, announced in September 1998 a company GHG target of 10 percent below 1990 levels by 2010. BP also is working with other companies and a private environmental group, Environmental Defense of New York, to develop an emission trading system and has designed a pilot internal brokering system, with the oil-trading arm of BP serving as broker.

More importantly, BP also has become a major investor in solar photovoltaics (PVs) and is seeking to build a new energy marketplace in which it can offer a range of products and services.[15] A wholly owned subsidiary of the British Petroleum group, BP Solar, is now a leading manufacturer of PV panels, accounting for 10 percent of the global PV market in recent years. With sales of $80 million in 1997, BP's solar business is growing rapidly. The company is planning to increase annual turnover in its solar business to $1 billion by 2007.

The costs of PV in comparison to other alternative energy sources such as wind and biomass are shown in Table 19.1;[16] PV technology is still about five times as expensive as its fossil fuel competitors, decreasing at 40–50 percent per decade. Still, in niche markets, this energy at today's costs is competitive. With continuing cost cuts in 10 years, PV energy could be cheaper than fossil fuels, especially after the externality costs cited earlier are considered. BP's business projections assume cost reductions of 5 percent per year. Looking ahead, BP expects the growth for PV systems to remain at 15–20 percent per year for the foreseeable future.

Experience shows that public awareness and a nation's environmental policy also are important in guiding choices for alternative energy. Increased public awareness of the damages avoided with renewable energy can encourage PV use, particularly if it is coupled with time-of-day metering and other pricing policies for fossil fuels that present consumers with the real costs of their electricity and transportation fuels.

However, BP acknowledges that even with rapid technical advances and aggressive efforts to increase the implementation of renewables, solar PV will

TABLE 19.1. Cost of Delivered Energy, Current Capital Costs, and Anticipated Cost Reductions for Selected Renewables

Technology	*Cost, Delivered*	*Capital Current Cost (typical)*	*Trend in Capital Cost, Past 10 Years*	*Expected Trend in Capital Cost, Next 10 Years*
Wood and biomass crops (combustion or other conversion)	$0.05–$0.08/kwh	$2,500–$3,500/kw	–10 to –15% (electricity production) –5 to –10% (heat)	–30 to –50% –5%
Landfill gas from wastes	$0.04–$0.06/kwh	$630–$1,170/kw	–10 to –15%	Slight increase
Biofuels: ethanol	$0.24–$0.37/liter	$0.06–0.13/liter	–5 to –10%	–25 to –50%
Solar photovoltaics	$0.25–1.50/kwh	$8,000–$35,000/kw installed	–40%	–40 to –50%
Wind	$0.04–$0.10/kwh	$800–$3,500/kw installed	–30 to –50%	–20 to –35%

Source: Adapted from World Resources Institute, 1998.

remain only a small part of its energy business in the near term. In the longer run, BP expects power from solar PV and other forms of renewable energy to be an ever-increasing share of the energy services the company provides. The company expects that share to be in line with the energy mixes depicted in most projections of risks of climate change under low-GHG emission scenarios.

Business Responses: Shell Oil Group

Shell Oil Group has also announced initiatives for taking action on climate change in its own operations and helping customers reduce emissions. Shell Oil Group established a new operating division in 1997 known as Shell International Renewables. This and other Shell initiatives seek to "support market mechanisms that will help countries grow their economies in an energy efficient manner."[17]

Although Shell Oil's businesses are expected to grow by 3 percent a year overall, they have taken steps to ensure that the potential rise in their GHG emissions will be kept in check by improved efficiencies and reduced venting and flaring of waste gases. The company reports that it already has met the Kyoto Protocol target by reducing emissions in relation to its 1990 levels, and it plans to reduce emissions a further 5 percent by 2002. Shell expects to exceed the Kyoto target throughout the next decade by working in four ways. First, it plans to reduce GHG emissions by investing in energy efficiency and ending continuous disposal of unwanted gas during oil extraction. Second, it plans to help customers reduce GHG emissions through greater availability of low-carbon fuels and renewable energy choices. Third, Shell Oil Group has committed to investing in renewable energy, especially solar power, biomass energy, and wind.[18] Finally, Shell is increasing the availability of natural gas and liquid fuels. For example, Shell foresees that gas and renewables will meet almost 50 percent of the fuel needs for power generation in Organisation for Economic Co-operation and Development countries by 2020.[19] Shell expects this trend to continue as new fuels, such as hydrogen and renewable energy, get cheaper and easier to use.

In terms of specific new products, the Shell Group is developing a range of renewable technologies while also investigating opportunities presented by the fuel cell. For example, Shell is working with Daimler Benz to develop a fuel processor to produce hydrogen feedstock for cars powered by fuel cells.[20] At the same time, people living far from the national electricity grid in rural South Africa will soon be able to use solar-powered electric lights at a cost no higher than conventional lighting fuels. In a joint venture with the state electricity supplier, Eskom, Shell Renewables is to offer solar electricity to about 50,000 homes over 3 years in areas that do not expect to be connected to the grid. The system, developed by Shell Solar and a local company, Conlog, has three parts: a solar panel, a charge-controlled battery, and a security and metering unit. Families will use magnetic cards storing prepaid power units at a cost of about $8 a month. This is the largest commercial solar rural electrification project ever undertaken. It is expected to create jobs and opportunities for education and entertainment in remote regions of South Africa.

Shell Group also is a partner in a venture in Iceland to test the potential of hydrogen as a replacement for fossil fuels and to create the world's first hydrogen economy. Another related project is the development of bus services in several countries using fuel cells. Even with these large investments in emerging technologies, Shell's annual return on capital has remained competitive with those of U.S. energy companies that have not yet undertaken such investments.[21]

Business Response: Other Energy Technologies

Wind Turbines

Like solar PV modules, wind turbines are a clean, flexible energy source being developed mostly by smaller electricity-generating companies.[22] Wind technologies can be applied in standalone or grid-connected units. Wind-generated electricity has grown by more than 20 percent per year for the past 12 years and in 1997 exceeded 12 billion kwh.[23] Although this is less than 1 percent of electricity in industrialized regions, wind already is competitive with fossil fuel–based generation in many markets (Table 19.1), even without including the fossil fuel externalities noted earlier. In the United States, the wind industry has relied in the past on tax credits (of about 1.5 cents/kwh) and voluntary green pricing premiums, which some companies are willing to pay to reduce their CO_2 emissions.[24]

Much of the growth in wind-generated electricity has come through private sector investments, facilitated in many countries by government incentives. In 1994, nearly 50 percent of the world's installed capacity was located in California, and Denmark and Germany together accounted for another 33 percent.[25] The proportion in Europe has increased greatly since 1994. In the United States, many utilities are entering into wind-based electricity production on a small scale. As with solar energy, bilateral or multilateral development institutions have helped to finance wind projects in developing countries.

Wind energy is the fastest-growing energy source.[26] The goal of the Shell Group is to establish offshore wind farms in many countries. Their first large pilot project will be at the Hamburg refinery in Germany, where the electricity generated will be sold to specific markets. In the United Kingdom, Shell will install two of the largest offshore turbines in the world (2 megawatts each).[27]

The intermittent nature of the energy supplied by wind turbines is a deterrent to some commercial uses, a barrier that improved storage and distribution technologies would address. Equally important, power sector regulations often have restricted grid access for independent energy producers, discouraging increased deployment of wind, solar, and biomass technologies.[28] However, these constraints are being removed with the gradual shift toward electricity deregulation and associated requirements that all available sources be used.

As barriers are overcome, wind is poised to make a substantial contribution to the world energy supply. Many parts of the world have suitable wind regimes, with an estimated energy potential of 20 trillion kwh per year, nearly twice the total world energy consumption in 1995.[29] If energy production from wind and solar PV were to grow at 8 percent a year through the coming century, these two technologies alone could meet most of the energy needed from the new renewable sources projected for low-emission energy transition scenarios. Such large-scale deployment of wind technologies will involve important social and environmental trade-offs, including accepting some noise, local aesthetic intrusions, and potential hazards to birds, balanced against the land area, water pollution, and aesthetics lost to current open-pit coal mining and oil field operations. Moreover, wind is well suited for many remote areas, making it a promising technology for rural electrification in developing countries and remote regions of the United States and Canada.

Fuel Cells

The fuel cell is an electrochemical device that combines hydrogen (from a range of sources) with oxygen, releasing energy and water. It entails no direct combustion and has no moving parts. Although fuel cells of different types have been used extensively in space exploration, in the past they have been seen as too expensive for commercial use. Improvements over the last 10 years have yielded designs with the potential to power cars and buses and produce electricity.[30]

Fuel cells have many advantages over conventional engines for transportation and as stationary sources of electricity. They produce little noise, making them ideal for use in cars or in buildings such as libraries and hospitals. When they use hydrogen fuel directly, the only byproduct is water. When they use methane (CH_4) as a fuel, there is release of CO_2 but no acid gases.

For the present, most commercial fuel cells are likely to derive their hydrogen from natural gas. Although carbon dioxide is released, it is at a lower density in relation to the power produced than the CO_2 from conventional fossil fuel engines. However, natural gas as a source of hydrogen should be seen only as a bridge to a future hydrogen fuel era.[31]

At up to $3,500 per kilowatt in the past, fuel cells were once viewed as being of demonstration interest only. This changed dramatically in 1999 and 2000. Companies such as Alsthom in Europe and Ebara in Japan are reducing the cost of electricity from fuel cells by developing lower-cost designs and mass producing the components.[32]

The commercial prospect for fuel cells in transportation is being accelerated by leading automakers such as Daimler–Chrysler, Toyota, Ford, and Volkswagen.[33] Some of these companies have proclaimed fuel cells, rather then electric vehicles, to be the successor to the internal combustion engine. At least four auto companies plan to have cars powered by fuel cell engines on the market by 2004, some in 2002. Many of the most important developments have focused on the proton exchange membrane fuel cell, developed by Ballard Power Systems, a company in Vancouver, British Columbia. This company has a $500-million joint venture with Daimler–Chrysler[34] and joint ventures with Ford Motor Company and Ebara. Another company, Shell Germany, has supported the European Union's first hydrogen filling station, in Hamburg.[35] It will provide fuel for experimental fuel cell vehicles being used by the local authorities.

The Shell Group is participating in a Californian public–private partnership to examine whether fuel cell–powered vehicles can be a safe, practical, and efficient alternative to conventional vehicles.[36] The California project investigates the potential of fuel cell vehicles by driving and testing them under real-world conditions in California. It will also evaluate the viability of integrating alternative fuels, such as hydrogen and methanol, into the existing commercial fueling infrastructure and investigate the commercialization of fuel cell vehicles.

Stationary fuel cell electricity generators must be more durable than auto engines, and for now most models need a separate processor to derive hydrogen from natural gas. Still, unit costs are projected to decline from $2,000 per kilowatt to $100–$300 within a few years.[37] Many power companies in countries around the world are actively pursuing fuel cell electricity generation. The Alsthom company in France, Ebara of Japan, Siemens of Germany, and a few U.S. utilities, on an experimental basis, are all moving in this direction, some in partnership with firms such as Ballard.[38] Their current focus is the 100- to 300-kilowatt commercial market, where gas turbine engines already are in vogue. When the facilities are kept small, they can be networked to create distributed power systems rather than depending on very large central generating facilities linked together in a large, sometimes unstable grid system.[39]

Clean Future, Messy Transition

Companies such as BP, Shell, and the electric utilities considering wind and hydrogen see a complex, even messy transition in the years ahead, as explained in Fig. 19.1.[40]

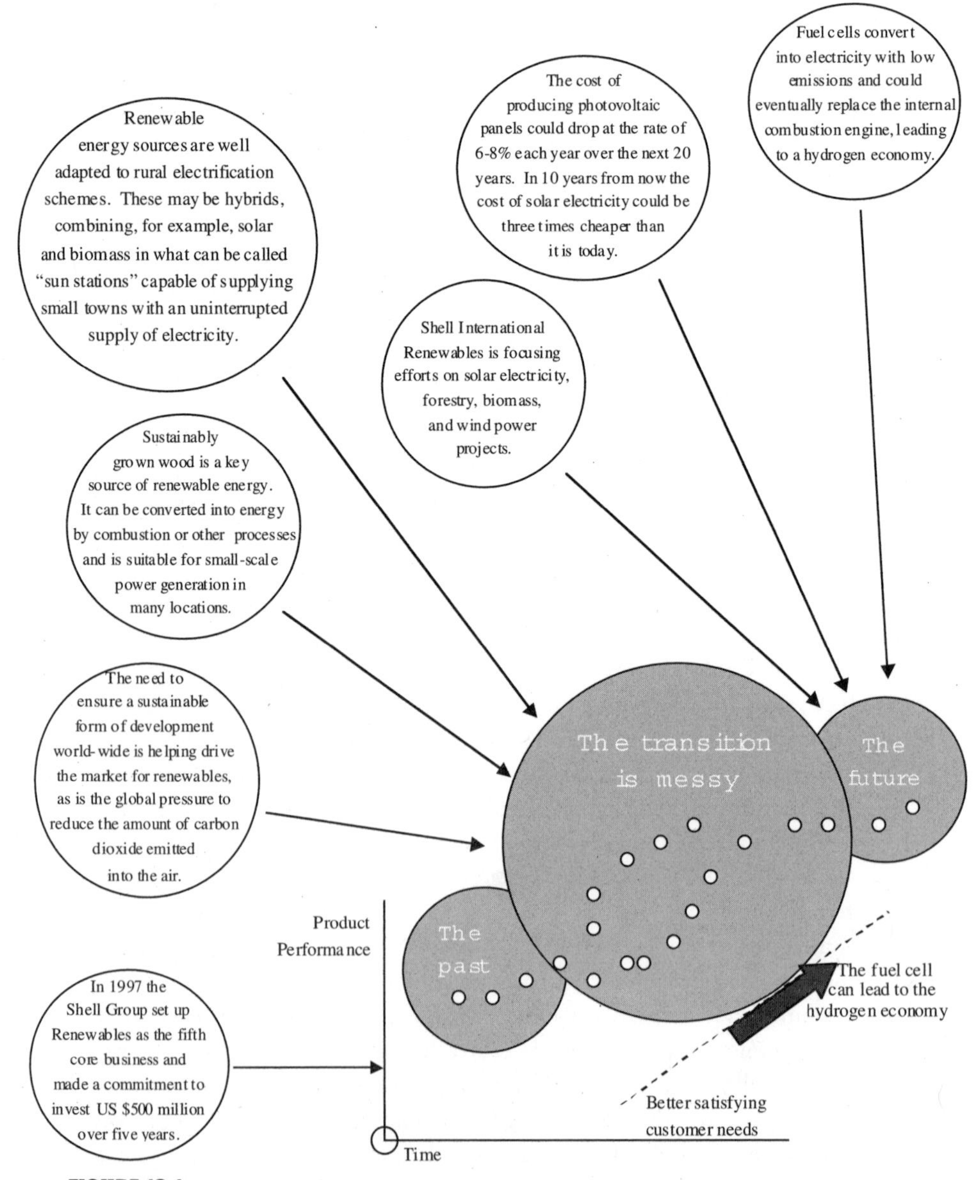

FIGURE 19.1. A representation by the Shell Group of how the transition in product performance may develop over the coming 20–40 years, a process that many see as a messy transition to a clean energy future. (Adapted from the Royal Dutch/Shell Group, 1999.)

Technical improvement in solar PVs, wind turbines, and fuel cells could create a hydrogen economy within the next 30 years,[41] despite the problems in storing and distributing hydrogen economically (which the Shell Oil Group is addressing).[42]

Environmentally, many positive aspects of the transition in energy forms will be seen in the next few decades because technical advances now being commercialized are leading to lower emissions of aerosol precursors and GHGs while increasing domestic employment.[43]

Commercially, however, the number of competing options make the transition from a carbon to a hydrogen economy look messy (Fig. 19.1) because it is far from certain which technologies will be viable in the marketplace.[44] The array of emerging options include the small-scale technologies powered by sunlight and wind, with local, high-reliability distribution systems.

Many conclusions are obvious from the information available now. One finding concerns the desire by businesses to recover a return on current investments being made in recent or new fossil fuel facilities before major new investments can be made for alternative fuels. This transition from one class of investment may take 30 years and is called "financial roll-over." It is recognized now as a major institutional barrier to a wind, solar or hydrogen economy. Some countries in Europe are addressing this issue. Interestingly, however, such barriers will not trouble many developing countries. Risk to existing investments presents very little problem where these investments have not yet been made and new infrastructure must be built for the first time. Developing countries can move directly to low-cost, nonpolluting energy economies while much of the developed world continues to bear the burden of nineteenth- and twentieth-century institutions and technology.

Considering the damage from intensive use of fossil fuels and the emerging availability of alternatives, one can be optimistic about the prospects for change. We may be about to see an exciting global competition between investors and businesses seeking to capitalize on the new energy technologies while established businesses defend their existing investments. We should each be asking ourselves which we think will be the likely winner.

Notes

1. Loucks, O. L., J. W. Bol, O. H. Ereckson, R. Gorman, P. Johnson, and T. Krehbiel, 1999: *Sustainability Perspectives for Resources and Business* (Boca Raton, FL: CRC/Lewis/St. Lucie Press).
2. National Acid Precipitation Assessment Program, 1991: *Acidic Deposition: State of Science and Technology,* P. M. Irving (ed.) (Washington, DC: U.S. Government Printing Office).

3. National Research Council, 1991: *Rethinking the Ozone Problem in Urban and Regional Air Pollution* (Washington, DC: National Academy Press); Office of Air Quality Planning and Standards, 1996: *Review of National Ambient Air Quality Standards for Particulate Matter: Policy Assessment of Scientific and Technical Information,* Report No. EPA/R-96-013 (Washington, DC: U.S. EPA).
4. Houghton, J. J., L. G. Meiro Filho, B. A. Callander, N. Harris, A. Kattenberg, and K. Masksell, 1996: *The Science of Climate Change,* contribution of Working Group I to the Second Assessment Report of the IPCC (Cambridge, England: Cambridge University Press); Office of Technology Assessment, 1991: *Changing by Degrees: Steps to Reduce Greenhouse Gases* (Washington, DC: Congress of the United States); Office of Technology Assessment, 1993: *Preparing for an Uncertain Climate* (Washington, DC: Congress of United States).
5. Office of Technology Assessment, 1984: *Acid Rain and Transported Air Pollutants: Implications for Public Policy,* OTA-O-204 (Washington, DC: U.S. Congress); Chestnut, L. G. and R. D. Rowe, 1989: "Economic measures of the impacts of air pollution on health and visibility," Chapter 7 in J. J. Mackenzie and M. T. El-Ashry (eds.), *Air Pollution's Toll on Forestry and Crops* (New Haven, CT: Yale University Press).
6. Office of Technology Assessment, 1984; MacKenzie, J. J. and M. T. El-Ashry, 1989: *Air Pollution's Toll on Forests and Crops* (New Haven, CT: Yale University Press).
7. Ratzlaff, W. A., D. A. Weinstein, J. A. Laurence, and B. Gollands, 1995: "Simulated root dynamics of a 160-year old sugar maple (*Acer saccharum marsh*) tree with and without ozone exposure using the TREGRO model," *Tree Physiology,* 16: 915–921.
8. Sharpe, W. E. and J. R. Drohan (eds.), 1998: *The Effects of Acidic Deposition On Pennsylvania Forests,* Proceedings of the 1998 Pennsylvania Acidic Deposition Conference (University Park: Pennsylvania State University); Loucks, O. L., 1998a: "The epidemiology of forest decline in eastern deciduous forests," *Northeastern Naturalist,* 5: 143–154; Aber, J., W. McDowell, K. Nadelhoffer, A. Magee, G. Berntson, M. Kamakea, S. McNultry, W. Currie, L. Rustad, and I. Fernandez, 1998: "Nitrogen saturation in temperate forest ecosystems: hypotheses revisited," *Bioscience,* 48: 921–934.
9. Loucks, O. L., 1998b: "Comprehensive estimate of combined costs of fossil fuel emissions," Congressional Seminar Presentation, U.S. Global Change Research Program, Washington, DC; Loucks, O. L., 2000: "Fuel cells: centerpiece of the hydrogen energy alternative," Proceedings, Climate Change Congress, Skies Above Foundation, Victoria, B.C.
10. UN Environment Program, 2001: "Impact of climate change to cost the world U.S.$300 billion a year," Feb. 3, 2001, press release on an article in *Our Planet,* available online: http://www.unep.org and http://www.ourplanet.com.
11. Cline, W. R., 1992: *The Economics of Global Warming* (Washington, DC: Institute for International Economics); Smith, J., 1966: "Standardized estimates of climate change damages for the United States," *Climate Change,* 32: 313–326.
12. Loucks, 1998b.

13. Minano, D. R., J. Lash, L. J. Fisher, and B. J. Bulkin, 1998: *Safe Climate Sound Business: An Action Agenda* (Washington, DC: World Resources Institute).
14. Ibid.
15. World Resources Institute, 1998: *Building a Safe Climate, Sound Business Future* (Washington, DC: WRI).
16. Ibid.
17. Royal Dutch/Shell Group, 1999: *The Shell Report. People and Profits. An Act of Commitment* (London: Royal Dutch Shell Group).
18. Royal Dutch/Shell Group, 2000: *The Shell Report 2000. How Do We Stand?, People, Planet and Profits* (London: Royal Dutch/Shell Group).
19. Ibid.
20. Royal Dutch/Shell Group, 1999.
21. Ibid.
22. WRI, 1998; Royal Dutch/Shell Group, 2000.
23. Royal Dutch/Shell Group, 2000.
24. Asmus, P., 2000: "Trends in the wind: lessons from Europe and the U.S. in the development of wind power," *Corporate Environmental Strategy,* 7: 51–61.
25. Dunn, S. and C. Flavin, 2000: "Sizing up micropower," Chapter 8 in L. R. Brown, C. Flavin, and H. French (eds.), *State of the World* (Washington, DC: World Watch Institute).
26. Dunn and Flavin, 2000.
27. Royal Dutch/Shell Group, 2000.
28. WRI, 1998.
29. WRI, 1998.
30. Ballard Power Systems, 2000: *Annual Report 1999: The Next Generation of Power* (Burnaby, BC: Ballard Power Systems Inc.).
31. Jensen, M. C. and M. Ross, 2000: "The ultimate challenge: developing infrastructure for fuel cell vehicles," *Environment,* 42 (7): 10–72.
32. Ballard Power Systems, 2000.
33. WRI, 1998.
34. Loucks, 1998b; Ballard Power Systems, 2000.
35. Royal/Dutch Group, 2000.
36. Ibid.
37. Dunn and Flavin, 2000.
38. Ballard Systems, 2000.
39. Dunn and Flavin, 2000.
40. Royal Dutch/Shell Group, 1999.
41. Jensen and Ross, 2000.
42. Royal Dutch/Shell Group, 1999, 2000.
43. Jensen and Ross, 2000.
44. Royal Dutch/Shell Group, 1999.

CHAPTER 20

Earth Systems: Engineering and Management†

Stephen H. Schneider

Imagine that we could let the world's economy continue to grow, bring the disadvantaged classes up from poverty, and not threaten the atmosphere or global ecosystems with unprecedented buildup of greenhouse gases (GHGs) and the climatic risks of such growth. Earth system engineering and management may just be such a panacea, some have suggested. But can we anticipate the costs or ever truly predict the consequences?

Few people would ask us to accept that a growing world economy based on greatly expanded per capita energy consumption would be free of environmental side effects. But many have claimed that the anticipated severalfold increase in GHGs—and associated sea level rises, intensified hurricanes, and drought and flood stresses—can be largely overcome by human ingenuity. Their optimistic vision depends greatly on what had been called geoengineering and has more recently been called earth system engineering, the deliberate manipulation of earth systems to manage the climatic consequences of human population and economic expansion.[1]

To others, the notion of geoengineering—injecting dust into the stratosphere, for example, to reflect some sunlight back to space and counteract global warming—is an irresponsible palliative. It evades the need for a real cure, such as curbing consumption by the rich and population growth by the poor and charging polluters for their use of the atmosphere as a free sewer.

In response, defenders of geoengineering retort that two-thirds of the world's people use a small fraction of the energy per capita of the rich. Cheap primary energy (mainly coal) is needed, they say, to build the economies of less developed countries and improve their well-being. The negative environmental side effects of this will have to be tolerated or sidestepped by geoengineering if we are to have both a materialistic, growth-oriented world and an undisturbed climate.

At times this debate takes on an ideological tenor. Claims that the imperative of development cannot be impeded by the prospect of global warming are greeted with the assertion that inadvertently damaging nature is bad enough, but deliberately attempting to manipulate the climate just to let our old habits prevail is a violation of stewardship and an ethical transgression against the natural world. These sets of opposing world views—anthropocentric expansion versus stewardship—are not new. They flared in the 1970s with Club of Rome debates over the limits to growth and matured with the publication of the Brundtland Commission's middle path, aiming to pursue sustainable development.[2] Today, they continue in arguments over whether nations must meet their emission reductions agreed in the Kyoto Protocol by domestic cuts—even if not cost-effective—or be permitted to buy their obligations elsewhere in the world at lower costs.

Let us return to the central question of what best characterizes earth system engineering. Is it a panacea for sustainable development, built with vision and ingenuity, or a palliative to avoid fundamental limits and maintain the privileged status quo for special interests? There is no easy answer to this question, but I believe that both sides have merit in parts of their arguments. I will try to sketch out some opportunities and pitfalls that might help to clarify the role of geoengineering and carbon management strategies in the climate policy debate.

Historical Perspective

In Homer's *Odyssey,* Ulysses is the frequent beneficiary (or victim) of deliberate weather modification schemes perpetuated by various gods and goddesses. In Shakespeare's *The Tempest,* Prospero, a mortal (albeit one with magical powers), conjures up a tempest to strand on his mystical island a passing ship's crew. In literature and myth, only gods and magicians can control the elements. But in the twentieth century, serious proposals for the deliberate modification of weather or climate came from engineers, futurists, or those concerned with counteracting the inadvertent anthropogenic modification of the earth's climate.

About 1960, Rusin and Flit[3] from the former Soviet Union published a long essay titled *Man Versus Climate* in which they suggested "improving" our planet by, for instance, diverting rivers from the Arctic to the Russian wheat fields or from the Mediterranean to irrigate areas in Asian USSR. One of their ambitious projects was to create a "Siberian Sea" with water taken from the Caspian Sea and Aral Sea areas. Of course, flowery rhetoric with images of blooming arid zones stands in stark contrast to the ecological disaster that surrounds the Aral Sea today, where environmental degradation resulted from much less radical geoengineering projects.[4]

FIGURE 20.1
Some geoengineering projects, such as this plan for the irrigation of the Sahara by creating a "second Nile" to refill Lake Chad, have become part of geoengineering folklore. (Reproduced from ref. 3)

Other such proposals have become part of geoengineering folklore and include damming the Gulf Stream, the Bering Straits, or the Nile, or creating a Mediterranean drain back into central Africa, where a "second Nile" would refill Lake Chad, turning it into the "Chad Sea" after the Straits of Gibraltar were dammed (Fig. 20.1). But the potential side effects if these projects misfire—which is not unlikely, given the complexity of the highly nonlinear climate system—are rarely discussed.

In the early 1970s, Russian climatologist Mikhail Budyko[5] suggested that it was "incumbent on us to develop a plan for climate modification that will maintain existing climatic conditions." What he endorsed was a stratospheric particle layer to reflect away enough sunlight to counteract global warming. But, wisely, he added that deliberate climate modification would be premature before the consequences could be calculated with confidence, a task for which the current simplified theories were inadequate.

William Kellogg and I looked at many such schemes in the 1970s and concluded then[6] that tampering blindly with the weather system would be the height of irresponsibility. Moreover, it would lead to disputes because any natural weather disaster occurring during deliberate climate modification experiments might well be blamed on the climate modifiers. We offered a modest proposal for "no-fault climate disaster insurance": If a large segment of the world thought that the benefits of a proposed climate modification scheme would outweigh the risks, they should be willing to compensate those who subsequently lost their favored climate.

Ironically, perhaps, the term *geoengineering* seems to have been applied first to a scheme that is no longer called by that name. It was informally coined by Cesare Marchetti,[7] who outlined a proposal for tackling the problem of CO_2 in the atmosphere by a kind of extended "fuel cycle" for fossil fuels. Under this proposal, CO_2 would be collected at certain transformation points such as the smokestacks of principal fossil fuel–burning industrial centers. It would be disposed of by injection into sinking thermohaline currents (say, the Mediterranean undercurrent entering the Atlantic at Gibraltar) that would carry and spread it into the deep ocean. Today, this kind of a plan is called industrial carbon sequestration, which is part of carbon management: controlling the amount of GHGs in the atmosphere. The term *geoengineering* has evolved to mean deliberate modifications to biogeochemical or energy flows in the climate system. This kind of tampering with natural processes, not surprisingly, inflames passionate debate.

Since Marchetti's article, perhaps the most ambitious attempt to justify and classify a range of geoengineering options was associated with a U.S. National Academy of Sciences (NAS) National Research Council panel on the policy implications of global warming.[8]

As a member of that panel, I can report that the very idea of including a chapter on geoengineering led to serious internal and external debates. Many participants (including me) were worried that even the thought that we could offset some aspects of inadvertent climate modification by deliberate modification schemes could be used as an excuse to continue polluting. Critics instead favored market incentives to reduce emissions or regulations for cleaner alternative technologies. But Robert Frosch countered as follows: What if a pattern of change currently thought unlikely but of high consequence actually started to unfold in the decades ahead? It would take decades to develop the technical and political tools to reverse the risks. We would simply have to practice geoengineering as the lesser evil.

Although skeptical about the viability of specific engineering proposals and the questionable symbolism of suggesting that we could sidestep real emission reductions, I nonetheless voted reluctantly with the majority of the NAS panelists, who agreed to allow a carefully worded chapter on the geoengineering options to remain in the report.

Extending Budyko's focus on the injection of aerosol particles (particles suspended in a gas) into the stratosphere, the geoengineering chapter of the report suggested that a 16-inch naval rifle fired vertically could propel a 1-ton shell consisting of dust particles up to an altitude of 20 kilometers. Given an aerosol lifetime in the stratosphere of 2 years, 10 megatons could be placed in the stratosphere 20 times during a 40-year period until 2030. Over this time the NAS authors estimated geoengineering costs to be about $5 per ton of carbon (as CO_2) mitigated. This cost is somewhat comparable to carbon taxes proposed by Nordhaus[9] for modest control of CO_2 emissions. But for a major mitigation of CO_2 emissions (for example, a 20 percent cut), Nordhaus's study suggests that the carbon taxes needed could be hundreds of dollars per ton carbon. (Conventional calculations of the costs of CO_2 mitigation through carbon taxes use economic models that are likely to overestimate the costs of mitigation because these models still ignore the effects of climate policies in inducing technological improvements.)[10]

But is it even possible to inject dust in the stratosphere, for example, in a manner that would perfectly offset a given injection of GHGs in the atmosphere? Even though the 30 percent increase in CO_2, 150 percent increase in methane, and addition of unnatural chemicals such as chlorofluorocarbons have spread fairly uniformly over every square meter of the earth since the industrial revolution, the patterns of heat trapped as a consequence are not uniform. The primary reason is the nonuniform distribution of other optically active constituents of the atmosphere, especially clouds.

Furthermore, humans add aerosols as well—not primarily the stratospheric kind, but mostly tropospheric sulphate aerosols resulting from the burning of coal and oil. These short-lived, lower-atmospheric aerosols are patchy in distribution and probably reflect sunlight back to space at the rate of up to 1 watt per square meter averaged over the Northern Hemisphere,[11] enough to offset perhaps one-fourth to one-half of the extra infrared heat associated with the enhanced greenhouse effect globally. And biomass burned also produces patchy distributions of aerosols, some of which actually warm the climate because they contain light-absorbing soot, as do some industrial aerosols, such as diesel engine exhaust.

Because of the patchy nature of the greenhouse effect itself, even if we could engineer our stratospheric aerosol injections to balance on a hemispheric (or global) basis the amount of hemispherically (or globally) averaged heat trapped by human-contributed GHGs, we would still be left with some regions heated to excess and others left cooler. I am not saying that such anomalies arising from aerosol geoengineering would necessarily be worse than, say, an unabated 5–7°C warming. But this is why the strong caveats in the NAS report are reiterated by all responsible people who have addressed the question.

As a postscript to this question, a climatic model study at the Lawrence Livermore National Laboratory[12] attempted to simulate whether the zonal patterns of stratospheric aerosol cooling could offset the more patchy patterns of GHG heating. They concluded optimistically that within the sampling precision of the model—which is still quite noisy—the aerosol scheme might not generate major regional climatic anomalies relative to those of unabated climatic change. Although not definitive, such studies are needed to give confidence in the effectiveness of any geoengineering scheme. And without high confidence in the outcome, any implementation would be controversial and could even lead to overt conflicts.

Caretakers for a Century?

No institutions currently have the authority to enforce responsible use of the global commons. There are some partially successful examples of nation-states willing to cede some national sovereignty to international authorities for the global good (for instance, the Montreal Protocol and its extensions to control ozone-depleting substances, the nuclear nonproliferation treaty, or the atmospheric nuclear test ban treaty). The Kyoto Protocol, even if ratified (currently a questionable prospect), would address only a small fraction of the needed emission cuts if CO_2 concentrations are to be stabilized below a doubling from preindustrial levels (much of the primary energy needed in 2050 will have to be mobilized with carbon emissions well below current standards, or huge efforts made to remove the excess)[13] (Fig. 20.2).

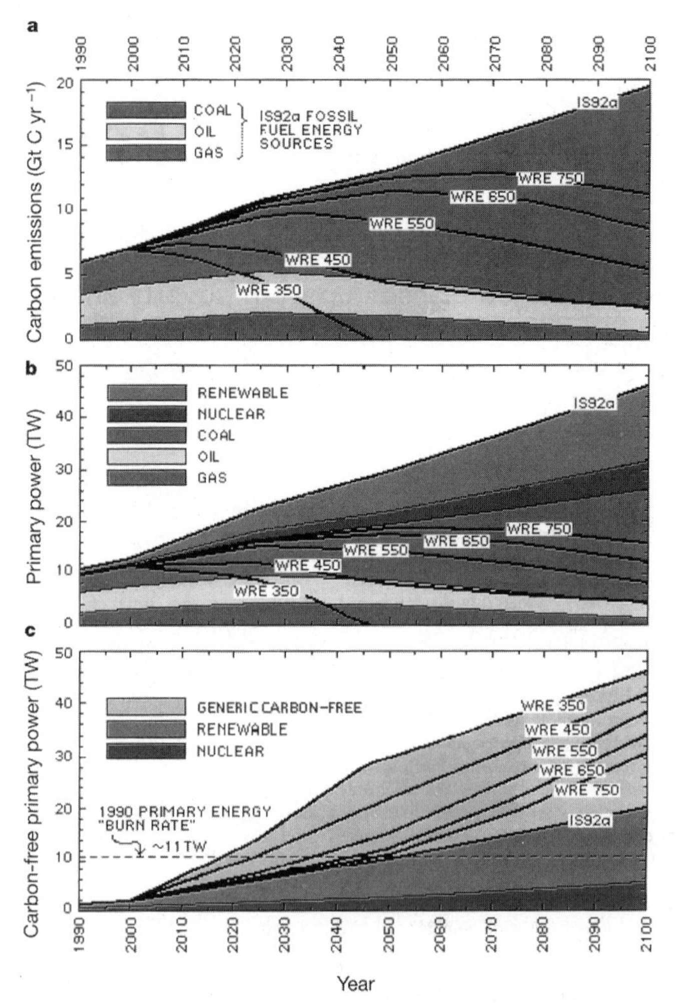

FIGURE 20.2. Fossil-fuel carbon emissions and primary power in the twenty-first century for various stabilization scenarios (IPCC scenario 1992a [dubbed "business as usual"], and stabilization of atmospheric CO_2 at 750, 650, 550, 450, and 350 p.p.m.v.). a, carbon emissions; b, primary power; and c, carbon-free primary power (see ref. 13 for further explanation). Carbon-free primary power is total primary power less fossil-fuel carbon power, calculated on the basis of net CO_2 emissions whether from sequestration or solar, nuclear or wind power. (Reproduced from ref. 13).

It would take a big increase in global-mindedness on the part of most nations to set up institutions to attempt to control climate and to compensate the losers should the interventions backfire (or even be perceived to have gone awry). Moreover, such an institution would need the resources and authority to make and monitor changes without interruption over a century or two—the time it will take the climate system to soak up the bulk of the GHGs we have injected. Thus, this is the time over which we would continuously need to inject measured amounts of dust in the stratosphere, iron in the oceans,[14,15] or sulfate aerosols into clouds to counteract the heat-trapping effects of long-lived constituents such as CO_2. So the most difficult obstacle in the path of geoengineering may be questionable governance rather than technical uncertainties.[16]

Varieties of Carbon Management

Two broad classes of carbon management can be distinguished. The first includes attempts to manipulate natural biogeochemical processes of carbon removal, or carbon sinks.[17] The second involves preventing carbon emissions into the atmosphere and instead disposing of carbon in stable reservoirs. David Keith (Box 20.1)

BOX 20.1. Geoengineering

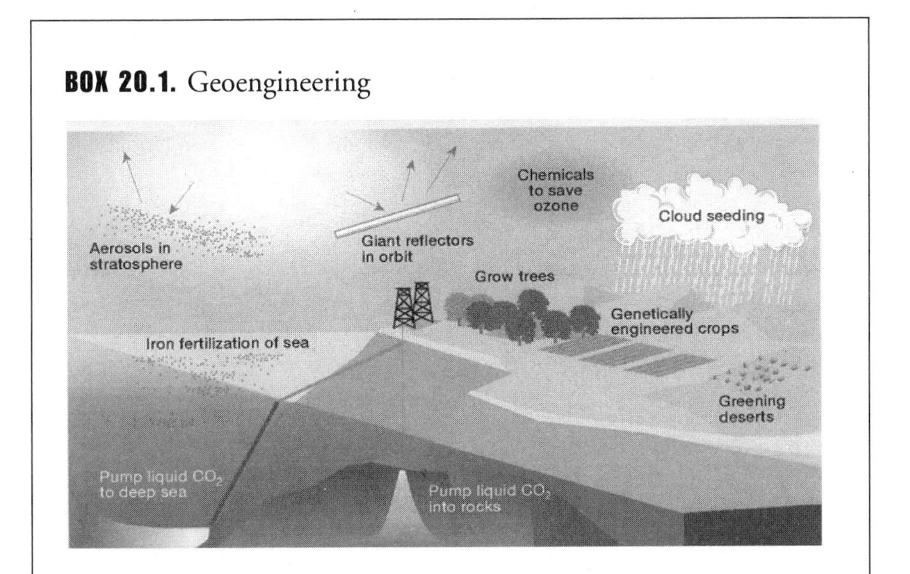

Geoengineering is planetary-scale environmental engineering, particularly engineering aimed at counteracting the undesired side effects of other human activities.[1] The term usually is applied to proposals for limiting the climatic impact of industrial CO_2 emissions by countervailing measures such as space-based solar

(continues)

shields. Scale and intent are both central to the common meaning of *geoengineering*, as the following examples demonstrate. The first is intent without scale: Ornamental gardening is the intentional manipulation of the environment to suit human desires, yet it is not geoengineering because neither the intended nor the realized effect is large-scale. The second is scale without intent: Anthropogenic CO_2 emissions will change global climate, yet they are not geoengineering because they are a side effect of the use of fossil fuels to provide energy services.

The distinction between geoengineering and more conventional responses to the CO_2 climate problem is fuzzy. *Geoengineering* has become a label for technologically overreaching proposals that are omitted from serious consideration in climate assessments. For example, few would object to applying the label to the first pair of examples that follow, but neither proposal rates serious consideration among climate policymakers. Conversely, the second pair receives serious consideration, but few would call them geoengineering.

Geoengineering Proposals

Enhancing Oceanic Sinks

Fertilizing the "biological pump" may enhance the flux of carbon into the oceans that maintains the disequilibrium in CO_2 concentration between the atmosphere and the deep ocean. Although use of nitrogen and phosphorus has been proposed, iron fertilization is the salient possibility because the ratio of iron addition to carbon fixation is very large (the Fe:C ratio is ~$1{:}10^4$, whereas for N:C it is ~1:6).

Iron fertilization experiments have produced marked increases in oceanic productivity,[2] and surveys have shown that biological productivity is iron-limited over substantial areas.[3] Although enhancement of surface productivity is possible, increasing the carbon flux into the deep ocean is highly uncertain; models suggest that even if iron fertilization were used at the largest possible scale, the carbon flux would not exceed ~1 Gt C/yr^{-1}. And problems abound because iron fertilization could produce anoxia in large regions of the deep ocean.

Shielding Some Sunlight

Warming caused by anthropogenic GHGs can be countered by deploying systems in the stratosphere or in space that scatter sunlight away from the planet. Stratospheric scatters are much cheaper but entail risks to stratospheric chemistry; space-based systems offer an expensive but clean alteration of the solar "constant."

Analysis has shown that it is possible to dramatically reduce the required mass and thus the cost of both scattering systems.[4] It had long been suggested that changes to the solar constant would compensate only poorly for the climatic effects of increased CO_2, even if mean surface temperature were accurately controlled. But a recent climate model experiment indicates that reduction of solar input can compensate for increased CO_2 with remarkable fidelity.[5]

Ambiguous Cases

Enhancing Terrestrial Sinks

Given the substantial human control over the terrestrial biosphere, the large natural carbon fluxes between the atmosphere and the terrestrial biosphere provide a powerful lever for manipulating atmospheric CO_2. A variety of methods have been proposed to exploit this leverage, including reforestation and sequestration in agricultural soils via no-till methods or the genetic modification of cultivars to enhance lignin content, thereby increasing the amount of CO_2 stored in such plants.[6]

Is it geoengineering? Enhancement of terrestrial sinks has been seen as green and low-tech, in sharp contrast with geoengineering. The idea has garnered wide support in industry and among environmental organizations. Yet, if implemented at the scale needed to capture a significant fraction of emissions, terrestrial sequestration would resemble planetary-scale environmental engineering and may well entail high-tech methods such as genetic modification of crops. The divergent treatment of terrestrial and oceanic sinks illustrates the inconsistencies that pervade discussion of planetary engineering.

Sequestering CO_2

We may use fossil energy without CO_2 emissions by first capturing the carbon content of fossil fuels while generating carbon-free energy products such as electricity and hydrogen and then sequestering the resulting CO_2 in geological formations or in the ocean.[7]

Is it geoengineering? The term *geoengineering* was coined in the 1970s to describe the injection of power-plant CO_2 into the deep ocean. Despite this etymology, it is unclear whether capture and sequestration is rightly classified as geoengineering. It is certainly an end-of-pipe technical fix, but injection into geological reservoirs resembles conventional pollution-mitigation technologies more closely than it resembles geoengineering because it limits emission of CO_2 to the biosphere rather than compensating for emissions after they occur. Put simply, if geological sequestration is an end-of-pipe solution, then biological sequestration is beyond-the-pipe.

Commentary

The post-war growth of the earth sciences has been fueled, in part, by a drive to quantify environmental insults to support arguments for their reduction. Paradoxically the knowledge gained is increasingly granting us leverage that we may use to deliberately engineer environmental processes on a planetary scale. The manipulation of solar flux using stratospheric scatterers is perhaps the best example of this leverage: We could reduce solar input by several percent—probably enough to initiate an ice age—at an annual cost of less than 0.01 percent of global economic output.[1,4] As remedies for the CO_2 climate problem, all proposed geoengineering schemes have serious flaws. Nevertheless, it is likely that this century will see serious debate about—and perhaps implementation of—deliberate planetary-scale engineering.

Notes

1. Keith, D. W., 2000: *Annual Reviews of Energy and the Environment,* 25: 245–284.
2. Boyd, P. W., et al., 2000: *Nature* 407, 695–702.
3. Behrenfeld, M. J. and Z. S. Kolber, 1999: *Science,* 283: 840–843.
4. Teller, E., L. Wood, and R. Hyde, 1997: Report no. UCRL-JC-128157 (Livermore, CA: Lawrence Livermore National Laboratory).
5. Govindasamy, B. and K. Caldeisa, 2000: *Geophysical Resource Letters,* 27: 2141–2144.
6. Rosenberg, N. J., R. C. Izaurralde, and E. L. Malone, 1998. *Carbon Sequestration in Soils: Science, Monitoring, and Beyond* (Columbus, OH: Battelle).
7. Parson, E. A. and D. W. Keith, 1998: *Science,* 282: 1053–1054.

suggests that the dividing line between geoengineering and mitigation is when a technology acts by counterbalancing an anthropogenic forcing rather than reducing it.

Carbon management by manipulating biogeochemical cycles overlaps with geoengineering. Ideas include iron fertilization of the oceans to enhance uptake of carbon by the resulting blooms of phytoplankton, planting vast forests of fast-growing trees to sequester carbon,[18] or altering agricultural practices to increase carbon storage in soils.[19]

The prevention of carbon emissions that otherwise would have been injected directly into the atmosphere is not geoengineering. Briefly, it includes preserving primary forests that otherwise might have been cut down (which also helps to preserve biodiversity; Fig. 20.3); processing fuels such as coal or methane to increase the hydrogen content and remove carbon, then injecting the carbon into storage reservoirs; using less carbon-intensive energy supply systems; and improving energy efficiency. The last two of these, of course, are what has come to be called "mitigation," usually favored by environmentalists. (The climate policy debate typically argues the costs of mitigation versus adaptation, although geoengineering has been mentioned as a third category from the outset.)[20]

One idea is to build on the chemical industry's existing experience of industrial-scale carbon removal and sequestration. Nitrogen fertilizer, for example, is

FIGURE 20.3. Keeping carbon in forests provides a "double dividend', as primary tropical forests contain good stores of CO_2 and also high biodiversity. But any carbon management scheme must take into account compensation for local people who lose their opportunity to convert the forest into economic product. In addition, monitoring is required to ensure that the carbon stays sequestered and that carbon "credits" are paid out to the donor to the project over time.

manufactured when carbon fuels such as natural gas or gasified coal are converted to secondary energy carriers such as hydrogen, although this is not done for the purpose of using the clean-burning hydrogen as a fuel but rather for chemical processing. And as carbon-intensive fuels such as coal are progressively converted to more hydrogen-based fuels such as methane, carbon dioxide is a byproduct that should be sequestered in a stable reservoir. The oil industry also has long experience with CO_2 sequestration through advanced oil recovery schemes. With one exception, these are not aimed at reducing atmospheric emissions of CO_2. Nonetheless, this experience can be built upon to develop carbon management for climate purposes.

The feasibility of CO_2 sequestration below ground has already been explored at small scales. The Sleipner West offshore platform in the North Sea, operated by the Norwegian company Statoil, is an interesting experiment in which about 1 million tons of CO_2 annually is stripped out of the natural gas mixture brought out of the earth. The CO_2 is reinjected into an aquifer about 1,000 meters below the ocean surface. As the CO_2 spreads along this geological formation, eventually—perhaps over hundreds of years—it may leak out, but this slow reinjection back into the climate system will avoid the acute build up of CO_2 that would have occurred under normal circumstances. Most interesting, perhaps, is why this first-of-a-kind plant was built: Norway had instituted a tax on carbon emissions of around \$50 per ton, and it seems that CO_2 removal and sequestration might be cheaper than the tax.

This, of course, is the crux of the climate policy debate: How can we create incentives to put a price on carbon or other heat-trapping gases? Debate rages about whether to provide incentives directly by a carbon tax, indirectly via targets and timetables (as in the Kyoto Protocol), or via subsidies to enterprises willing to develop carbon management schemes. But without such incentives, the extent to which technological options will be explored is questionable.

Carbon management thinkers have also suggested that industrial efforts should not be limited to centralized sources such as power plants or oil platforms but must consider distributed applications such as transportation systems. Perhaps we will see the development of a few centralized plants to produce hydrogen fuel for zero-emission vehicles. To be cost-effective, such plants must be in areas with abundant resources of fossil fuels and adequate storage reservoirs for the waste carbon. However, some have questioned whether sequestered carbon will remain buried and thus whether carbon credits should be given unless it is proved that the storage is lasting. To eliminate endless debate I propose an inexpensive fix: to add into the injected CO_2 an inert chemical tracer unique to each sequestration site. Thus, the nondetection of this tracer over time would serve to certify that carbon credit is deserved for such sequestration projects.

"Strong" or "Weak" Engineering

In 1992 at the Rio Earth Summit, the UN Framework Convention on Climate Change committed the nations of the world to avoid "dangerous anthropogenic interference with the climate system." At that time the thinking was primarily about inadvertent modification. But now it seems that world leaders may have to extend their value judgment about what is harmful by including deliberate interference with the climate system.

Naturally, there is a philosophical debate about whether there is anything ethically wrong with such tinkering. On one hand, it is argued, we have progressed from hunting and gathering by increasingly large-scale manipulations of the natural environment. In fact, some environmental writers have despaired that an altered climate is already "The End of Nature."[21] It has been argued[22] that earth system engineering and management is the approach "to rationally engineer and manage [the Earth] to provide the requisite functionality" and that this is not a logical transgression of naturalness because the earth is already an artifact of our manipulations. Of course, anything other than a preservation of current structure and function demands a definition of *improvement,* what constitutes an "improvement" over the natural, and this judgment will be very different across diverse cultures and over generations.

Moreover, there are still areas in the polar regions and deep tropical rain forest where there is, as Professor David Keith has observed, "essentially no visible human imprint; where the majority of species have evolved in situ . . . and where biochemical perturbations are small." Such landscapes are not "artificial" simply because a slight global climatic change has already occurred. We should avoid disrupting them further rather than using "light perturbation" as an excuse to turn over the future of all nature to the value judgments of the planetary managers.[23]

Given our growing inadvertent impact on the planet, adaptation alone is likely to prove inadequate. But I would prefer to reduce slowly our economic dependence on carbon-emitting fuels rather than to try to counter the potential side effects with centuries of non-stop injecting sulfuric acid into the atmosphere or iron into the oceans. Laying stress instead on carbon management, with little manipulation of biogeochemical or energy fluxes in nature, is a much less risky prospect, despite remaining uncertainties about the longevity of deep-earth or deep-ocean carbon storage and the possible ecological consequences of localized injections of vast quantities of CO_2 in the oceans or the potential damper on global economic development. If preliminary studies prove reasonable, then the cost penalties for closing the industrial cycles by reinserting waste CO_2 back in the earth might be only a few tenths of a percent of current energy system costs—something akin to $50–$100 per ton carbon. Such costs would have only a trivial impact on economic growth in the twenty-first century. The actual costs of various forms of carbon management will be crucial in determining how much climate change we and the unmanaged environment will have to adapt to in the decades ahead. But until national governments cooperate and provide incentives to both producers and users of climate-altering products, the potential for any carbon management enterprise will be limited and the likelihood of dangerous climatic changes increased.

To me, any stronger form of earth system engineering and management is a revision of Rusin and Flit's fantasy of 40 years ago to transform the earth system to achieve "improvements in climate." Those wanting to usurp the powers of ancient gods and conjurers should recall the ancient Greeks' warnings about human hubris embodied in the story of Prometheus.

Acknowledgments

†Adapted (with permission) from Schneider, S. H., 2001. Earth systems engineering and management, *Nature,* 409: 417–421.

Notes

1. Socolow, R. (ed.), 1997: *Fuels Decarbonization and Carbon Sequestration,* PU/CEES Rep. No. 302, September (Princeton, NJ: Centre for Engineering and Environmental Studies, Princeton University), available online: http://www.princeton.edu/~ceesdoe/.
2. World Commission on Environment and Development, 1987: *Our Common Future* (Oxford, England: Oxford University Press).
3. Rusin, N. and L. Flit, 1960: *Man Versus Climate.* Translated from Russian by Dorian Rottenberg (Moscow: Peace Publishers).
4. Glazovsky, N. F., 1990: *The Aral Crisis: The Origin and Possible Way Out* (Moscow: Naulca).
5. Budyko, M. I., 1977: "Climate Changes," *American Geophysical Union,* Washington DC. English translation of 1974 Russian volume, 244 pp.
6. Kellogg, W. W. and S. H. Schneider, 1974: "Climate stabilization: For better or Worse?" *Science,* 186: 1163–1172.
7. Marchetti, C., 1977: "On geoengineering and the CO_2 problem," *Climate Change,* 1 (1): 59–68.
8. Panel on Policy Implications of Greenhouse Warming, Committee on Science, Engineering, and Public Policy, National Academy of Sciences, 1992: *Policy Implications of Greenhouse Warming: Mitigation, Adaptation, and the Science Base* (Washington, DC: National Academy Press), pp. 433–464.

9. Nordhaus, W. D., 1992: "An optimal transition path for controlling greenhouse gases," *Science,* 258: 1315–1319.
10. For a critique, see Grubb, M., M. H. Duong, and T. Chapuis, 1994: in N. Nakícenovic, W. D. Nordhaus, R. Richels, and F. L. Tóth (eds.), *Integrative Assessment of Mitigation, Impacts and Adaptation to Climate Change,* CP-94-9 (Laxenburg, Austria: International Institute of Applied Systems Analysis), pp. 513–534.
11. Intergovernmental Panel on Climate Change (IPCC), 1996: *Climate Change 1995: The Science of Climate Change,* contribution of Working Group I to the Second Assessment Report of the IPCC, Houghton, J. T., L. G. Meira Filho, B. A. Callander, N. Harris, A. Kattenberg, and K. Maskell (eds.) (Cambridge, England: Cambridge University Press), 572 pp.
12. Govindasamy, B. and K. Caldeira, 2000: "Geoengineering Earth's radiation balance to mitigate CO_2-induced climate change," *Geophysical Research Letters,* 27: 2141–2144.
13. Hoffert, M. I., K. Caldeira, A. K. Jain, E. F. Haites, L. D. D. Harvey, S. D. Potter, M. E. Schlesinger, S. H. Schneider, R. G. Watts, T. M. L. Wigley, and D. J. Wuebbles, 1998: "Energy implications of future stabilization of atmospheric CO_2 content," *Nature,* 395: 881–884.
14. Watson, A. J., C. S. Law, K. A. Van Scoy, F. J. Millero, W. Yao, G. E. Friedrich, M. I. Liddicoat, R. H. Wanninkhof, R. T. Barber, and K. H. Coale, 1994: "Minimal Effect of Iron Fertilization of Sea-surface Carbon Dioxide Concentrations," *Nature,* 371: 143–145.
15. Monastersky, R., 1995: "Iron verses the Greenhouse: Oceanographers cautiously explore a global warming therapy," *Science News,* 148: 220–222.
16. Schneider, S. H., 1996: "Geoengineering: Could- or Should- We Do It?" *Climate Change,* 33: 291–302.
17. Intergovernmental Panel on Climate Change, 2000: *Land Use, Land-Use Change, and Forestry,* A Special Report of IPCC, Watson, R. T., I. R. Noble, B. Bolin, N. H. Ravindranath, D. J. Verardo, and D. J. Dokken (eds.) (Cambridge, England: Cambridge University Press), 377 pp.
18. Johansson, T. B., H. Kelly, A. K. N. Reddy, and R. H. Williams (eds.), 1993: *Renewable Energy: Sources for Fuels and Electricity* (Washington, DC: Island Press).
19. Rosenberg, N. J. and R. C. Izaurralde, 2001: "Storing Carbon in Agricultural Soils to Help Head-Off a Global Warming, Guest Editorial," *Climatic Change,* 51: 1–40.
20. Chen, R. S., E. M. Boulding, and S. H. Schneider (eds.), 1983: *Social Science Research and Climate Change: An Interdisciplinary Appraisal* (Dordrecht: D. Reidel Publishing Company), 255 pp.
21. McKibben, W., 1989: *The End of Nature* (New York: Random House), p. 226.
22. Allenby, B., 1999: "Earth Systems Engineering: The Role of Industrial Ecology in an Engineered World," *Journal of Industrial Ecology,* 2: 73–93.
23. Keith, D. W., 2000: "The Earth is Not Yet an Artifact," *IEEE Technology and Society Magazine* 19: 25–28.

APPENDIX A

Climate Negotiation History

Leonie Haimson

In 1992, at the Earth Summit in Rio de Janeiro, the UN Framework Convention on Climate Change (FCCC) was adopted. The treaty called for the nations of the world to prevent "dangerous anthropogenic interference" with the climate by stabilizing the concentration of greenhouse gases (GHGs) in the atmosphere. The agreement called for industrialized countries to take the first step by voluntarily reducing their emissions to 1990 levels by the year 2000 because they bore most of the responsibility for the problem. However, these voluntary measures were soon recognized to be ineffective because many nations, including the United States, continued to emit more GHGs than before.

The FCCC also instituted a process that required governments to regularly review the science of global warming and to assess the implementation of efforts to control climate change. Seven formal negotiating sessions, or Conferences of the Parties (COPs) have followed the Earth Summit. At COP-1 in Berlin, Germany, in March 1995, participating nations issued the Berlin Mandate, which acknowledged that the voluntary approach had failed, and agreed that the commitments by industrialized countries to reduce their emissions would have to be strengthened, including "quantified reductions" within specified time frames.

At COP-2, which convened in Geneva, Switzerland, in July 1996, the United States announced that it would support legally binding targets and timetables in the near future to reduce the accumulation of GHGs and challenged other industrialized nations to do the same. More than 100 countries signed onto the Geneva Declaration in support of such targets, involving significant reductions in GHG emissions.

In an interim negotiating meeting in Bonn, Germany, in March 1997, the

Europeans took the lead by proposing that all industrialized nations be required to reduce their GHG emissions by 15 percent from 1990 levels by the year 2010. The U.S. government proposed a system of multiyear emission budgets, with international trading in emission allowances that would significantly reduce the costs of reductions. And Bert Bolin, scientist and then-chair of the Intergovernmental Panel on Climate Change, made a forceful presentation showing that reductions undertaken solely by industrialized countries would not be sufficient to limit warming to environmentally sustainable levels; developing countries eventually would need to curb their increasing rates of emissions as well.

Although it was generally understood that developing nations would follow once the industrialized countries submitted to mandatory targets and began to cut their emissions, the need to involve them in a more immediate manner took on new political urgency in the United States when the Byrd–Hagel Resolution passed the Senate by a 95–0 vote in June 1997. This resolution proclaimed that the Clinton administration should not sign any agreement that would mandate reductions for industrial nations unless it included similar commitments from the developing countries because otherwise the competitiveness of the U.S. economy would be harmed. Nevertheless, the environmental community and other world leaders placed countervailing pressure on the White House to follow through on their previously stated positions and accede to an agreement that would contain binding reductions on industrialized nations, whether or not they could get the developing countries to immediately adopt targets as well.

COP-3 met in Kyoto, Japan, in December 1997. U.S. Vice President Al Gore, one of the world's first political leaders to recognize and speak out on the issue of climate change, made a dramatic appearance during the second week of the conference. He asked the delegations to do their best to come up with a workable agreement involving realistic and binding targets and announced that he had instructed the U.S. negotiators "to show increased flexibility." At the very end of the conference, many hours after it was originally supposed to adjourn, more than 150 nations adopted the Kyoto Protocol. This unprecedented agreement committed the industrialized nations to make legally binding reductions in their emissions of six GHGs: carbon dioxide, methane, nitrous oxide, hydrofluorocarbons (HFCs), perfluorocarbons (PFCs), and sulfur hexafluoride (SF_6). The mandatory targets varied from country to country, but together they would average cuts of about 5 percent below 1990 levels in emissions from industrial nations by the period 2008–2012.

The United States agreed to cuts of 7 percent, Japan to cuts of 6 percent, and the nations of the EU to joint reductions of 8 percent. Key to the U.S. assent to such an ambitious target was the agreement that a system of emission trading among industrialized countries would be established by which they could buy and sell allowances to emit GHGs.

The Kyoto Protocol also authorized the Clean Development Mechanism (CDM), in which industrialized countries could meet part of their obligations for reducing emissions by investing in projects that reduced carbon emissions in developing countries. As to the further participation of the developing countries, the United States pushed for a provision that would have allowed these countries to voluntarily opt in to binding commitments, but Saudi Arabia, China, and India blocked this measure.

In November 1998, COP-4 convened in Buenos Aires, Argentina. The goal was modest: to agree upon a plan to negotiate rules for key elements of the Kyoto Protocol, including emission trading, compliance, carbon sinks, and the CDM. By the end of the meeting, negotiators managed to do little but to set a new deadline of November 2000 for deciding on these rules.

In addition, Argentina and Kazakhstan announced their intention to adopt voluntary commitments to limit their emissions, seeing potential environmental and financial advantages in opting into an international emission trading system whereby their economies might benefit from increased foreign investment in clean energy.

COP-5 met from October 25 to November 5, 1999, in Bonn. Although there were few actual headlines, many of the participants, from government officials to members of environmental groups, felt a renewed momentum in the air. The biggest development occurred right away, when German Chancellor Gerhard Schroeder opened the conference with a dramatic challenge to the developed nations to ratify the Kyoto Protocol and bring the treaty into force by 2002, or "Rio + 10," the tenth anniversary of the Earth Summit.

The EU responded that it was "ready and willing" to meet this deadline, as did Japan, whereas Canada and the United States called for the treaty to enter into force "at the earliest possible date." The Clinton administration, already hesitant to set a firm date for ratification, was especially cautious because of the Senate's rejection of the Comprehensive Test Ban Treaty just weeks before.

Participants also noted an apparent easing of the three-way gridlock between the United States and its allies, the EU, and the developing countries. The United States ceased demanding that COP-6 be put off until after the November 2000 presidential elections. The EU began to relax about limiting the extent to which nations could use emission trading to meet their Kyoto commitments. Even some developing countries began to hint that they might more seriously consider agreeing to limitations on their GHG emissions in the future.

COP-6, held in The Hague in November 2000, was a great deal more contentious and ended in deadlock. At this meeting, a host of difficult issues were angrily debated, including how much trading in emission allowances would be permitted, how much carbon sinks such as forests could be counted toward emission reductions, and how compliance would be enforced. In particular, the United States fought hard to gain the maximum allowance on both sinks and trading so that it would have a better chance to comply with the Kyoto targets, because otherwise it would need to cut domestic emissions 35 percent from anticipated levels by 2010—a difficult if not impossible goal.

The U.S. position on sinks was supported by Japan, Canada, and Australia—and even a few environmental groups, which argued that giving some limited credit to sinks was a cost-effective way to help control global warming and encourage forest protection at the same time. On the other hand, European nations—which are less forested—countered that credit toward the Kyoto quota should be limited to direct emission reductions.

According to a detailed account of the negotiations (http://chicagotribune.com/news/nationworld/article/0,2669,SAV-0011260410,FF.html, Ray Moseley, Nov. 26, 2000), the United States lowered its initial call for a 310-million-ton annual carbon sink credit—most of which represented forest growth that would have occurred anyway—to 125 million tons, then to 78 million tons. Finally, in the wee hours after the last night of the conference, the United States agreed to a European counteroffer of 55 million tons, less than one-fifth of its original demand. This amount also represented less than 1 percent of the 6 billion tons of carbon dioxide released by humans each year and less than 10 percent of the total reductions required of the United States by 2010. According to some sources, the proposal had its source in a telephone discussion between President Clinton and British Prime Minister Tony Blair.

But when the European negotiators in the room, including British Deputy Prime Minister John Prescott, French Environment Minister Dominique Voynet, and German Environment Minister Jurgen Trittin, brought the offer to a larger group of EU delegates for ratification, there was bitter resistance, especially from Denmark. The resistance caused both Germany and France to cave in. When Frank Loy, undersecretary of state and chief U.S. negotiator, responded with an even lower offer of 40 million tons on the last morning, he was told that it was too late—most of the negotiators had gone home. Voynet said an agreement might very well have been reached had the negotiations continued for just another half-day. Bill Hare from Greenpeace International commented, "We're better off with no deal than a bad deal." From an entirely different perspective, the Global Climate Coalition (GCC), representing most of the U.S. oil and fossil fuel industry, agreed: "No deal is better than a bad deal."

However, many others were devastated that no compromise had been reached, particularly because George Bush, already on record as opposing the

Kyoto Protocol, seemed likely to become the next U.S. president. Not surprisingly, the collapse of the negotiations led to a torrent of recriminations among the participants. Prescott was furious with the other Europeans in the room who had initially agreed to the deal only to subsequently recant, saying he felt "gutted." He fiercely criticized Voynet, who he believed bore special responsibility in her role as head of the EU delegation.

In March 2001, President George W. Bush announced his decision to pull out of the Kyoto Protocol, saying it was "fatally flawed" and was not in the U.S. interest. This decision, by the leader of the world's largest emitter of GHGs, led to heated reaction at home and furious condemnation in the rest of the world. The European leaders, in particular, forcefully attempted to convince him to change his mind, but Bush continued to argue that complying with the agreement would be damaging to the U.S. economy and would have little chance of being ratified by the Senate. The success of the next negotiating session to finalize the rules of the Kyoto Protocol, put off until July at the request of the Bush administration, seemed even more in doubt because the leaders of both Canada and Japan began to signal that they might follow the United States and reject the Kyoto agreement as well.

In July 2001, the nations of the world met again in Bonn for a conference known as COP-6bis because COP-6 in The Hague had ended without agreement. Against all predictions and in a triumph of European diplomacy, negotiators from 178 countries agreed on the rules of the Kyoto Protocol, despite the continued intransigence of the United States. The nations of the world collectively decided that it was indeed time to act on the most critical environmental problem of the twenty-first century. As New Zealand delegate Peter Hodgson said, "We have delivered probably the most comprehensive and difficult agreement in human history" (Reuters, July 23, 2001).

The Europeans showed unprecedented flexibility at the meeting, offering industrial nations unlimited emission trading, giving special deals to wavering countries to count their growing forests as carbon sinks, and softening penalties for noncompliance—all primarily to enlist the support of Japan, whose participation would be necessary for the treaty to come into force. The Kyoto Protocol would become legally binding only when ratified by 55 countries, including those responsible for more than 55 percent of the carbon dioxide emitted by industrial countries in 1990. As of August 2001, only 37 nations, mostly developing countries, including Mexico, along with a scattering of former Soviet states and Romania, had ratified.

Yet after COP-6bis, many more countries prepared to ratify the accord. (The current list is available at http://www.unfccc.int/resource/kpstats.pdf.) Although ratification by Japan and especially Russia, was still far from certain, the odds of this occurring became much more likely after the agreement at Bonn was struck. With Europe, Japan, and Russia ratifying, the 55 percent requirement would also be satisfied. Although some environmental organizations grumbled that the emission targets had been unacceptably watered down in Bonn, nearly everyone agreed that these provisions were far preferable to having the treaty collapse, thereby leaving the world without a substantive strategy to control climate change. By May 31, 2002, more than 55 nations had ratified, including all the member countries of the European Union. With Japan and Russia likely to follow, the nations of the world (save the United States) appeared back on track toward bringing the Kyoto Protocol into force by 2002, the tenth anniversary of the Rio Earth Summit.

At the G-8 meeting that convened during COP-6bis, Bush again held his ground against the combined pressure of the other world leaders, arguing that it would be futile for him to push for U.S. ratification of Kyoto because the U.S. Senate was on record opposing the treaty by means of the Byrd–Hagel Resolution. But the political situation in the United States has evolved since 1997, and shortly after the Bonn meeting the U.S. Senate Foreign Relations Committee voted 19–0, resolving that the Byrd–Hagel Resolution should not "cause the U.S. to abandon its shared responsibility to help find a solution to the global climate change dilemma" and that because "American businesses need to know how governments worldwide will respond to the threat of global warming" the United States should take "responsible action to ensure significant and meaningful reductions in emissions of greenhouse gases from all sectors." The resolution also stated that the United States should participate in all upcoming international climate negotiations.

The Foreign Relations Committee resolution went on to note that the United States had already ratified the 1992 UN FCCC agreement, which required that industrialized countries develop plans to reduce their GHG emissions to 1990 levels and clearly stated that they should take the lead in combating climate change. Senator Chuck Hagel (R-NE) was on the committee, which voted unanimously for the new resolution.

The other sponsor of the 1997 Byrd–Hagel amendment, Senator Robert Byrd of West Virginia, was even more vocal in his opposition to Bush's rejection of Kyoto and came out in support of binding limits on GHG emissions because, as he argued, the voluntary approach had not worked. Byrd also cosponsored legislation, along with Senator Ted Stevens (R-AK), to jump-start U.S. efforts to forestall climate change and to require the White House to develop a long-term national plan to stabilize GHG levels. Byrd reiterated that Bush was mistakenly using the 1997 resolution as an excuse to do nothing: "I do not believe that this resolution should be used as an excuse for the United States to abandon its shared responsibility to help find a solution to the global climate change dilemma. . . . If we are to have any hope of solving one of the world's greatest challenges, we must begin now" (available online: http://www.planetark.org/dailynewsstory.cfm?newsid=11645, July 19, 2001).

Web Links to More Negotiation Information

- To keep current on the current state of the negotiations, check out my column in Grist Magazine at http://www.gristmagazine.com/grist/heatbeat.
- The official UN negotiating documents are on file at the Web site of the Climate Change Secretariat (http://www.unfccc.de/), as are a timetable of future negotiations and international meetings and a country-by-country account of what the signatories to the climate agreement have already done to combat global warming.
- The UN Environment Programme (http://www.unep.ch/iuc/) has extensive information about climate change and the climate treaty.
- To follow the details of past negotiations, you can look at back issues of the *Earth Negotiations Bulletin* (http://www.iisd.ca/climate/), an independent reporting service that provides coverage of official UN environmental negotiations, published by the International Institute for Sustainable Development.
- For an glimpse at what is happening while the sessions are taking place, check out the irreverent *ECO* (http://www.climatenetwork.org/eco), the newsletter of the international Climate Action Network, published daily during major meetings of the climate negotiations.

APPENDIX B

"Hot Air" and "Hot Air" Policies

Reimund Schwarze

What Is Hot Air?

The Kyoto Protocol sets quantified emission limitation or reduction objectives for 38 countries and the EU. Each country's specific reduction varies. As outlined in Annex B of the Kyoto Protocol, most countries are assigned a reduction of 8 percent from their 1990 baseline; Japan and Canada must achieve a 6 percent reduction. Certain countries were given emission targets, which will allow these nations to maintain or emit more than their 1990 emission levels. For example, Australia can increase its emissions by 8 percent and Iceland by 10 percent.

Emission reductions are required during the first commitment period (2008–2012), but the reductions are expressed relative to a 1990 baseline. Whereas most countries have increased GHG emissions since 1990, countries such as the Russian Federation and the Ukraine, which suffered economic decline after the collapse of their centralized governments, are dramatically below their 1990 emission levels and probably will remain below them for the near future. The Kyoto Protocol requires Russia and the Ukraine to maintain 1990 levels of GHG emissions. Therefore, these countries will have surplus assigned amounts that they will be allowed to sell in a tradable permit market.

This free surplus is what opponents call "hot air" because it will be treated as emission reductions in an emission market although no true abatement has occurred. Critics assert that emission reductions that are caused by factors outside the realm of environmental policy are not purposeful emission reductions. Consequently, they cannot be credited to the host country and must be banned from trading. Proponents of emission trading argue that "hot air" does not affect the overall Kyoto target. Even if the full surplus of emission rights from Russia and the Ukraine were traded to the West, GHG emissions from Annex B countries would still be on average 5.2 percent below 1990 levels.

Both views reflect different understandings of the nature of emission trading. Whereas the proponents of emission trading think of it as a cap-and-trade regime, in which only the overall emission target counts, critics perceive it as a regime of emission reduction credits in which tradable permits result from a decrease of emissions compared with business as usual.

How Much Hot Air Exists in the First Commitment Period?

The actual amount of "hot air" in the first commitment period is difficult to predict because it depends on several factors: the pace and timing of economic recovery in economies in transition (EITs), the availability of fossil and non–fossil fuels, and the stringency of environmental policies. The most comprehensive recent estimate of "hot air" is based on scenarios developed at the International Institute for Applied Systems Analysis (IIASA).[1] The main results of this study are shown in Fig. B.1.

Figure B.1 depicts the emission surpluses or deficits of different countries or groups of countries under different future scenarios. An emission surplus is defined in this study as a positive difference between the projected business-as-

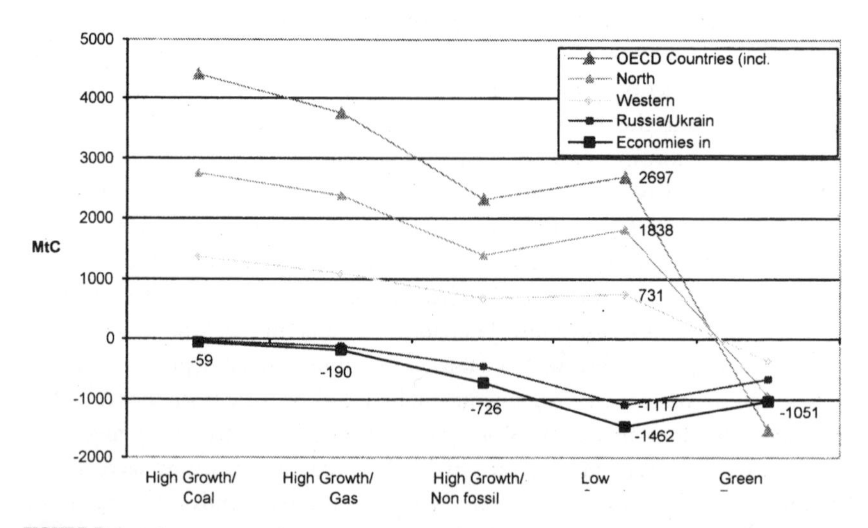

FIGURE B.1. Schematic representation of various-engineering proposals (courtesy B. Matthews).

usual emissions (BAU) in the 5-year period 2008–2012 and the Kyoto target for that period. An emission deficit ("hot air") is defined as a negative difference between BAU emissions and the Kyoto target.

From the IIASA study we display five possible future scenarios: three high-growth scenarios and two low-growth scenarios. The high-growth scenarios reflect different assumptions about dominant fuels with rapid growth. The first sub-scenario predicts coal as the dominant fuel source because of scarcity of oil and gas. The second sub-scenario assumes that oil and gas remain important fuel sources. In the third sub-scenario, improvements in non–fossil fuel technologies (renewables and nuclear) lead to economically competitive alternatives and long-term elimination of fossil fuels. The low-growth scenarios reflect different economic and environmental policy futures. The first sub-scenario captures the effect of a sluggish world economy and greater-than-expected difficulties in the economic recovery in Russia and the Ukraine. The "green future" sub-scenario, on the other hand, predicts the effects of a broad, unprecedented national and international effort to protect the environment (e.g., international carbon tax with revenue recycling to developing countries) combined with technological improvements in energy use and renewables.

Predicting the size of the "hot air" emission deficit is driven largely by assumptions of economic growth in EITs. Economic recovery fueled by coal or natural gas would shrink "hot air" in EITs to a negligible 59 Mt C or 190 Mt C, respectively. However, even with economic recovery there could be an emission deficit in EITs of 726 Mt C if these countries choose the non–fossil fuel or nuclear option. This result sheds some light on the difficult nature of "hot air" because, in the latter case, the emission deficit stems from domestic technology choice, not from an overgenerous allocation of assigned amounts.

The IIASA study suggests that the low-growth scenario is the most likely outcome. According to this prediction, the authors find that "hot air" in the first commitment period could be as high as 1,117 Mt C in Russia and the Ukraine and 1,462 Mt C in EITs as a whole. This amount would not fully cover the need to reduce emissions in Organisation for Economic Co-operation and Development (OECD) countries, which are estimated at 2,700 Mt C. However, depending on the no-regret potential and the amount of trade among OECD countries, it could make up 50 percent of the future market for assigned amounts. This share will be even higher (near 100 percent) if the United States remains steadfast in its refusal to ratify the Kyoto Protocol.

Policies to Address Hot Air

Allocating too many emission rights to Russia and the Ukraine created the "hot air" problem. A straightforward correction to this problem would be to renegotiate this initial allocation so that fewer emission rights are assigned. One way would be to request that Russia and the Ukraine agree to reduction targets instead of the present stabilization targets. Another way would be to choose a base year other than 1990 (e.g. 1995) in which most of the emission reductions from the economic collapse would be discounted. However, both ways seem politically infeasible. Renegotiating the initial allocations of Russia and the Ukraine probably would initiate a process of renegotiating all Kyoto targets, putting the greatest achievement of Kyoto—the move toward quantitative targets—at risk. Short of renegotiating Kyoto targets, there are three ways to respond to the problem of "hot air": restrict the sale, restrict demand, or collectively buy out "hot air" emissions. Each solution has severe shortcomings.

Sales Cap

Restricting the sale of "hot air" would have only a temporary effect if banking of emissions for the future were allowed. With banking, assigned amounts that cannot be sold in the first commitment period will be saved for sale in future periods. A simple supply cap therefore would only postpone the problem of "hot air" into the future. Because banking is legal under Article 3 of the Kyoto Protocol and economically important to prevent market disturbances, a supply cap would not be a viable solution. Also, the present wording of Kyoto Protocol would not cover it. "Hot air" is a legitimate entitlement of Russia and the Ukraine (if the protocol is ratified), and there is nothing in the protocol that would restrict the sale (as opposed to the purchase) of assigned amounts.

A fact that is often overlooked is that a supply cap would also impose a hidden cost. Because it is practically impossible to distinguish "hot air" from true emission reductions, it would suppress all sales of permits, not just "hot air" sales. Incentives for true low-cost emission reductions would be lost.

Demand Cap

Restricting the demand would be lawful but ineffective. A demand cap is legally supported by a provision of Article 17, which states that the use of this flexibility mechanism shall be "supplemental to domestic actions." This clause could imply that only a certain percentage of the reduction commitments can be achieved through purchases of assigned amounts. Such a concrete ceiling would apply to all flexibility mechanisms and to all potential sellers, not just Russia and the Ukraine. Therefore, it will have no specific effect on the sale of "hot air" and may even accelerate the problem. Because "hot air" will be the cheapest supply of emission rights (because it comes at zero cost to the supplier), a demand cap would affect primarily the true emission reductions from developing countries, which carry a positive price.

Buyout of "Hot Air"

A buyout of "hot air" could be done in different ways. One way would be an open market operation in which a group of concerned parties (e.g., the OECD) or a designated UN authority would buy assigned amounts from Russia and the Ukraine solely for the purpose of retiring them. Alternatively, it could be done through purchases of "hot air," which are paid in kind with emission reduction equipment or by revenues earmarked for purchases of emission reduction investments. In other words, Russia and Ukraine would get emission reduction equipment in exchange for "hot air" credits, which would be consequently subtracted from their emission budgets. Industrialized countries could also pay for "hot air" with nonquantifiable projects such as environmental education, emission monitoring devices, or research.

COP-6 Developments

The question of a concrete ceiling on the use of Kyoto mechanisms has been one of sharpest debates between the EU and the Umbrella Group countries at COP-6 in The Hague. The debate stems from language in the protocol stating that domestic actions should be the main means for reaching the reduction commitments, and emission trading should be "supplemental" (Article 17). In the view of the Umbrella Group countries, this language implies only that countries should not rely solely on use of the mechanisms (a qualitative cap). However, the EU insisted during the COP-6 negotiations that this language implies a quantitative cap on acquisitions and sales of emission rights. The negotiations in The Hague consequently failed to produce results. At the follow-up meeting in Bonn the EU assumed a weaker position. Facing the threat of the protocol's demise, the EU agreed to language that reflected only a preference for domestic action (FCCC/CP/2001/L.7, p.7, par. 5).

Some observers have asserted that this change of attitude applies also to the implications of a concrete ceiling for EU member states. The Energy Research Center of the Netherlands has calculated the cost consequences for EU member states that result from different types of ceilings for Kyoto mechanisms.[2] The study illustrates that ceilings will lead to a decrease in purchases of emission reductions and consequently a drop in the market price of emission reductions. The lower market price will result in lower profits for countries that are net sellers of emission reductions, such as Germany, France, Spain, and Portugal.

Countries that are net purchasers of emission reductions could profit from these lower prices if this ceiling is not binding for them. Therefore, the cost consequences for net purchasers differ between countries. For Austria and Denmark, all ceilings considered will result in higher costs of meeting Kyoto commitments. In case of Italy, the Netherlands, Finland, Sweden, Belgium, and the United Kingdom, some ceiling options considered lead to higher costs and others lead to lower costs. As these differing implications for its own member states have become clear to the EU, it has been wavering internally on the issue of concrete ceilings.

Notes

1. Victor, D. G., N. Nakicenovic, and N. Victor, 2001: "The Kyoto Protocol Emission Allocations: Windfall Surpluses for Russia and Ukraine," Climate Change, 49 (3): 263–277.
2. Ybema, J. R., T. Kram, and S. N. M. Van Rooijen, 1999: "Consequences of ceilings on the use of KMs: A tentative analysis of cost effects for EU member states," ECN-C-99-003, available online: http://www.ecn.nl/unit_bs/Kyoto/mechanism/ceilings.html.

Glossary

Albedo: The percentage of sunlight reflected from an entity, such as snow.

Anthropogenic: Of human origin.

Atmospheric aerosols: Minute particles suspended in the atmosphere.

Bonn Agreement (COP-6b): Agreement signed by 178 nations in Bonn, Germany, in July 2001 to implement the Kyoto Protocol. The United States was the lone nonsignatory.

Carbon dioxide (CO_2): An important natural trace gas that affects climate and life. Also the most significant anthropogenic greenhouse gas, produced predominantly from burning fossil fuels.

Clean Development Mechanism: A provision, defined in Article 12 of the Kyoto Protocol, that allows Annex I (industrialized) countries to earn credits to meet domestic emission reduction commitments by investing in projects in non–Annex I countries that reduce emissions below what would have occurred in absence of the project.

Climate surprises: Rapid, nonlinear responses of the climatic system to anthropogenic forcing, such as collapse of thermohaline circulation in the North Atlantic Ocean or rapid deglaciation of polar ice sheets.

Common but differentiated responsibility: A norm of international environmental law embraced by signers of the UN Framework Convention on Climate Change in 1992. It means that all nations share responsibility for protecting the global atmosphere and abating greenhouse gas production but that rich nations, having produced most of the emissions to date, would bear the lion's share of the abatement costs, at least at the outset.

Conservation tillage: Tillage and planting systems, such as no-till or ridge till, that leave 30% or more carbon-containing crop residue remaining on the soil surface after planting. In contrast, conventional tillage typically leaves less than 15% residue cover.

Contraceptive prevalence: The percentage of women of reproductive age (15–44) (married and in consensual unions) using either any method of contraception or modern methods, which include male and female sterilization, IUDs, the pill, injectables, hormonal implants, condoms, and female barrier methods.

Energy balance: The condition in which a system gains and loses energy at the same rate, resulting in a constant average temperature over long periods.

Enteric fermentation: Transformation of plant material in the anaerobic digestive systems of ruminant animals, such as cattle, sheep, buffalo, and goats, that produces methane.

Equilibrium simulation: A model calculation that gives the final equilibrium state that results eventually from the change in some condition (typically a doubling of atmospheric carbon dioxide).

Feedback: A process in which an initial occurrence results in an additional occurrence that may amplify (positive feedback) or diminish (negative feedback) the initial amount of change.

Framework Convention on Climate Change (FCCC), 1992: The major international agreement committing parties to abate anthropogenic greenhouses gases. The FCCC, a UN-sponsored activity often labeled UN FCCC, articulates the principle that the parties should avoid "dangerous anthropogenic interference with the climate system" and the principle of common but differentiated responsibility. It leaves implementation to be worked out at Conferences of Parties.

Frequentist or objective probability: The likelihood of an event occurring or a process operating based on a large set of replicable experiments from which frequency charts can be constructed (e.g., the probability of getting a head from a flipped unbiased coin can be objectively determined by a long set of flipping trials). This idealization of "objective" science is rarely fully applicable to complex systems in which uncertainties accompany even the structural description of the system; the climate system is a case in point (see *Subjective or Bayesian probability*).

General circulation model (GCM): A complex three-dimensional and time-dependent computer model used for detailed climate modeling.

Global warming potential: Numerical index based on integrated radiative forcing over time that reflects the ability of a greenhouse gas to trap heat in the atmosphere and the atmospheric lifetime of the gas. It is usually referenced to CO_2, which has a global warming potential of unity.

Greenhouse effect: The absorption of outgoing infrared radiation by clouds and greenhouse gases in the atmosphere, resulting in downward reradiation and warming of a planet's surface.

Greenhouse gas: An atmospheric gas that absorbs and emits infrared radiation, giving rise to the greenhouse effect.

Infrared radiation: An invisible form of electromagnetic radiation with wavelengths longer than that of red light, emitted by objects with Earth-like temperatures.

Intergovernmental Panel on Climate Change (IPCC): An international association of thousands of climate scientists, impacts specialists, and policy analysts, established by the World Meteorological Organization and the UN Environment Programme, which has produced three major reports on climate change to governments, which must approve the reports. It is often called the most credible climate consensus assessment body.

Kyoto Protocol (COP-3), December 1997: An agreement signed by most nations, including the United States. Parties (primarily developed countries) agreed to an average rollback of 5.2% in greenhouse gas emissions by 2008–2012, compared with the base year of 1990. It envisioned several new mechanisms, including the Clean Development Mechanism and a global system of tradable emission permits.

Methane (CH_4): The second most important of the anthropogenic greenhouse gases, arising primarily from anaerobic decomposition of organic matter; primary constituent of natural gas. It has a global warming potential more than 10 times that of CO_2.

Natural greenhouse effect: The effect of naturally occurring greenhouse gases (primarily clouds, water vapor, and carbon dioxide) that keep Earth some 33°C warmer than it would otherwise be.

Parts per million (ppm): A unit of concentration, expressing the number of volume units of a given substance in 1 million volume units of the ambient environment.

Radiative forcing: A measure of the effect of greenhouse gases and other radiation modifying substances or surfaces (e.g., atmospheric aerosols, land use changes) that gives them the capacity to alter both absorbed solar and outgoing infrared radiation, measured in watts per square meter.

Risk: A situation in which system behavior is well known and the chances of different outcomes can be quantified by probability distributions. Risk is typically defined as the condition in which the event, process, or outcome, and the probability that each will occur, is known. Specifically, risk equals probability times consequences.

Risk neutrality: The case in which the expected utility is equal to the utility of the expected value. *Risk aversion* is the case in which the expected utility is lower than the utility of the expected value.

Risk acceptance or risk tolerance is the case in which the expected utility is greater than the utility of the expected value.

Subjective or Bayesian probability: The degree of belief that an event would occur or a process would operate based on the subjective judgment of an assessor using available information, including objective probabilities of the behaviors of the subcomponents of a complex system. Complex systems usually cannot be fully tested with replicable experiments, so the likelihood of any outcome of necessity involves a degree of belief, preferably informed by as much objective information as can be made available so that such a subjective judgment can become an *expert opinion.* All probabilities of future events and those of the behaviors of complex systems are to some degree subjective rather than a fully objective likelihood.

Sulfate aerosols: Sulfur-based particles produced in industrial regions primarily by the burning of high-sulfur coal and oil (and by some natural processes worldwide) and exerting a local cooling effect because of their high reflectivity for sunlight.

Surprise: A condition in which the event, process, or outcome is not known or expected. *Imaginable surprise* is an event or process that departs from the expectations of some definable community and thus acknowledges that many events often are anticipated by at least some observers. It may be possible to identify *imaginable conditions for surprise* (e.g., rapid forcing of nonlinear systems) that might induce surprises even though the actual surprise events are unknown.

Tradable emissions or *tradable emission permits:* A global system of carbon emission permits that can be traded between nations. Nations needing more permits than they are allotted would buy unneeded permits from less developed nations with no targets or nations with carbon emissions below their targets.

Transient or dynamic responses: The evolutionary paths of environmental and socioeconomic systems in reaction to some forcing event. This is in contrast to the *equilibrium response,* which describes how these systems react to forcings after the system reaches equilibrium. The actual socionatural system is undergoing a transient evolution.

Transient simulation: A model calculation that follows changing conditions over time.

Uncertainty: The condition in which an event, process, or outcome is known (factually or hypothetically), but the probabilities that it will occur are not known but could be characterized by subjective estimates from expert opinions.

About the Contributors

ANIL K. AGARWAL was a mechanical engineer by training and spent most of his life working as a journalist and environmental activist. He became a scientific writer for the *Hindustan Times* in 1973, and in 1980 created the Center for Science and Environment (CSE), which is based in New Delhi. His first "State of the Environment in India" was published in 1982 and attracted wide attention. He became known internationally in 1991 when he intervened in the debate on climatic change by stressing that the problem of greenhouse effects was a worldwide one as well as a worldwide responsibility. He was a lead author of Working Group III of the Integrated Panel on Climate Change. He also founded the semi-monthly periodical *Down to Earth* in 1992. Agarwal died of cancer on January 2, 2002 at the age of 54.

KAI S. ANDERSON is legislative director for Senator Harry Reid (D-NV), Assistant Democratic Leader of the United States Senate. From 1998–1999 he worked for Senator Joseph I. Lieberman (D-CT) as a Geological Society of America/U.S. Geological Survey Congressional Science Fellow specializing in energy and environmental policy. He holds a bachelor of science degree in geology (1993) and a Ph.D in geological and environmental sciences (1998) from Stanford University.

PAUL BAER is a Ph.D candidate in the Energy and Resources Group at the University of California, Berkeley. His research in the area of ecological economics focuses on both scientific modeling and the equity of climate policy alternatives. He is also a co-founder of the group EcoEquity.

ALISON BAILIE is an associate scientist at Tellus Institute, a nonprofit research organization. Her expertise is in the area of energy policy modeling and analysis on both the supply and demand side, with a focus on policies and measures to reduce greenhouse gas and air pollutant emissions.

JOHN J. BERGER is a Ph.D and an independent energy and natural resources consultant who specializes in renewable energy. He is the author and editor of seven books and has consulted for the National Research Council (NRC) of the National Academy of Sciences, Fortune 500 companies, nonprofit groups, and governmental organizations, including the U.S. Congress. He has been a visiting associate professor of environmental policy at the Graduate School of Public Affairs of the University of Maryland, and an adjunct professor of environmental science at the University of San Francisco.

STEPHEN BERNOW is vice president and director of the Energy Group at Tellus Institute, a nonprofit research organization, which he co-founded in 1976. His work in North America focuses on integrated energy and environmental policy and planning within and across all sectors, as well as on policies and measures to reduce greenhouse gas emissions while decreasing air pollution and stimulating technological progress.

THOMAS G. BURNS is an independent energy and environmental consultant who focuses on strategic planning, energy economics, and global environmental issues. He retired from the Chevron Corporation in 2000 after a 37-year career that spanned all functions of the international petroleum industry.

MICHAEL CUMMINGS received a bachelor of science degree in earth systems, a bachelor of arts degree in economics, and a master of science degree in earth systems from Stanford University. During the 2002–2003 academic year he will complete an IIE Fulbright Scholarship in Iceland researching issues related to the use of hydrogen energy.

BILL DOUGHERTY is a senior scientist at Tellus Institute, a nonprofit research organization. He is active in local and national policy modeling in the electric and transport sectors for reducing greenhouse gas emissions, as well as in the assessment of environmental co-benefits in the U.S. and internationally.

LAWRENCE H. GOULDER is an associate professor of economics and a senior fellow in the Institute for International Studies at Stanford University. His research focuses on U.S. and international environmental policy. In collaborative work with biologists and climatologists, Dr. Goulder has examined methods for achieving sustainable economic growth and well-being, the philosophical and empirical bases for valuing ecosystem services, and the potential to stimulate cost-reducing technological innovation through environmental policies.

LEONIE HAIMSON writes a regular column on the science and politics of climate change for the online environmental journal *Grist Magazine*. She is also the coauthor of *The Way Things Really Are: Debunking Rush Limbaugh on the Environment*, and the editor of *Common Questions on Climate Change*, published by the U.N. Environment Programme.

DANIEL M. KAMMEN is professor of energy and society in the Energy and Resources Group, professor of public policy in the Goldman School of Public Policy, as well as director of the Renewable and Appropriate Energy Laboratory (RAEL) at the University of California, Berkeley. His research interests are focused on renewable energy, international development, and the science and policy issues surrounding sustainable development, risk analysis, and environmental change.

SIVAN KARTHA is a senior scientist at Tellus Institute, a nonprofit research organization, and Stockholm Environment Institute, Boston Centre, whose research pertains to assessments of renewable energy technologies and policy analyses relating to climate change and human development.

KRISTIN KUNTZ-DURISETI received a Ph.D in political science and a master's in applied economics from the University of Michigan. Her current research projects include modifying integrated assessment models (IAMs) of global climate change to account for the precautionary saving motive, incorporating climate system discontinuities into IAMs, and examining the assumptions for modeling adaptation in current IAMs.

MICHAEL LAZARUS is a senior scientist at Tellus Institute, a nonprofit research organization. He has nearly 20 years of professional experience in the analysis of climate change mitigation options, sustainable energy strategies, energy modeling, and the environmental impacts of energy projects in the U.S. and internationally.

ORIE L. LOUCKS is an Ohio eminent scholar of applied ecosystems studies and professor of zoology at Miami University. He also taught for many years at the University of Wisconsin. Loucks has collaborated with scholars in many fields beyond the dynamics of lakes and forests, including policy-related work on biodiversity conservation, air pollution effects, and the sustainability of ecosystems and commerce.

ROBERT M. MARGOLIS is on the research faculty at Carnegie Mellon University, where he is the executive director of the Center for the Study and Improvement of Regulation. He is broadly interested in the role of scientific and technical information in the policy-making process, particularly as it relates to environmental sustainability issues. His main research interests include: energy technology and policy; research, development, and demonstration policy; and energy–economic environmental modeling.

FREDERICK A. B. MEYERSON is a Ph.D, ecologist, demographer, and former attorney. His work focuses on the interactions between human population and the environment, particularly climate change and the loss of biological diversity and habitat. He is an American Association for the Advancement of Science (AAAS) fellow at the National Science Foundation (2001–2002) and assistant professor (adjunct) of ecology and evolutionary biology at Brown University.

BRIAN M. NADREAU received a bachelor's degree in economics from Stanford University in 2001 and is currently pursuing a master's degree in management science and engineering at Stanford. His academic interests include environmental economics and policy design, the use of mathematical modeling to improve business performance, and information technology.

JOHN O. NILES studies climate change and tropical forests. He has advised governments and organizations on carbon cycling, economics, and biodiversity. Niles has degrees from the University of Vermont and Stanford University and is a doctoral student at the University of California, Berkeley.

HOLLY L. PEARSON received a doctorate from the Department of Biological Sciences at Stanford University in 1999. She did a one-year post-doctoral Lokey Research Fellowship with Environmental Defense, where she worked with the climate change group. In 2002 she earned a jurist doctorate degree at New York University School of Law.

ARMIN ROSENCRANZ is a lawyer and political scientist, as well as a consulting professor in the Program in Human Biology at Stanford University. In 1987, he founded the nonprofit organization Pacific Environment, which he led until 1996. He has had two Fulbright awards to India, and is co-author of *Environmental Law and Policy in India* (2001). He teaches several global and U.S. environmental policy courses at Stanford. This book arose from the course

"Controlling Climate Change in the 21st Century," which he has co-taught with Stephen Schneider since 1999.

STEPHEN H. SCHNEIDER is a professor of environmental biology in the Department of Biological Sciences at Stanford University, as well as a professor by courtesy in the Department of Civil and Environmental Engineering at Stanford. He is a lead author for Working Group II, Intergovernmental Panel on Climate Change, Third Scientific Assessment, and is co-author of the IPCC guidance paper on uncertainties. He is also the founder and editor of the journal *Climatic Change*. Schneider was a MacArthur Fellow in 1992, and was elected to the National Academy of Sciences in April 2002.

REIMUND SCHWARZE is assistant professor for environmental economics at the University of Technology in Berlin, Germany, and was a research fellow at the Center for Environmental Science and Policy at Stanford University in 1999. He has served as environmental expert in German government hearings on strategies for sustainable development, and acted as a senior lecturer under the German–Turkish and the German–Vietnamese Environmental Cooperation Program. He received his Ph.D in economics from the University of Technology in Berlin (1994).

ELEANOR (BONNIE) TURMAN received a doctorate in mathematics from the University of Florida in 1980. She has worked in the telecommunications industry as a software engineer for the last 20 years. She is also pursuing a master of liberal arts at Stanford University.

JONATHAN BAERT WIENER is a professor of law and of environmental policy, as well as faculty director of the Center for Environmental Solutions at Duke University. From 1989–1993 he served at the U.S. Department of Justice, the White House Office of Science and Technology Policy (OSTP), and the White House Council of Economic Advisers (CEA) in both the first Clinton and Bush administrations, where he helped formulate U.S. climate policy and negotiate the international climate treaties.

RICHARD WOLFSON is a Benjamin F. Wissler professor of physics and professor of environmental studies at Middlebury College. He received his bachelor of arts degree in philosophy and physics from Swarthmore College, a master of science degree in environmental studies from the University of Michigan's School of Natural Resources, and Ph.D in physics from Dartmouth. Wolfson is the author of three books, more than 60 scientific publications, and several video courses for nonscientists.

Index